Undergraduate Texts in Mathematics

Undergraduate Texts in Mathematics

Abbott: Understanding Analysis.
Anglin: Mathematics: A Concise History and Philosophy.
Readings in Mathematics.
Anglin/Lambek: The Heritage of Thales.
Readings in Mathematics.
Apostol: Introduction to Analytic Number Theory. Second edition.
Armstrong: Basic Topology.
Armstrong: Groups and Symmetry.
Axler: Linear Algebra Done Right. Second edition.
Beardon: Limits: A New Approach to Real Analysis.
Bak/Newman: Complex Analysis. Second edition.
Banchoff/Wermer: Linear Algebra Through Geometry. Second edition.
Berberian: A First Course in Real Analysis.
Bix: Conics and Cubics: A Concrete Introduction to Algebraic Curves.
Brémaud: An Introduction to Probabilistic Modeling.
Bressoud: Factorization and Primality Testing.
Bressoud: Second Year Calculus.
Readings in Mathematics.
Brickman: Mathematical Introduction to Linear Programming and Game Theory.
Browder: Mathematical Analysis: An Introduction.
Buchmann: Introduction to Cryptography.
Buskes/van Rooij: Topological Spaces: From Distance to Neighborhood.
Callahan: The Geometry of Spacetime: An Introduction to Special and General Relavitity.
Carter/van Brunt: The Lebesgue–Stieltjes Integral: A Practical Introduction.
Cederberg: A Course in Modern Geometries. Second edition.
Chambert-Loir: A Field Guide to Algebra
Childs: A Concrete Introduction to Higher Algebra. Second edition.
Chung/AitSahlia: Elementary Probability Theory: With Stochastic Processes and an Introduction to Mathematical Finance. Fourth edition.
Cox/Little/O'Shea: Ideals, Varieties, and Algorithms. Second edition.
Croom: Basic Concepts of Algebraic Topology.
Curtis: Linear Algebra: An Introductory Approach. Fourth edition.
Daepp/Gorkin: Reading, Writing, and Proving: A Closer Look at Mathematics.
Devlin: The Joy of Sets: Fundamentals of Contemporary Set Theory. Second edition.
Dixmier: General Topology.
Driver: Why Math?
Ebbinghaus/Flum/Thomas: Mathematical Logic. Second edition.
Edgar: Measure, Topology, and Fractal Geometry.
Elaydi: An Introduction to Difference Equations. Third edition.
Erdős/Surányi: Topics in the Theory of Numbers.
Estep: Practical Analysis in One Variable.
Exner: An Accompaniment to Higher Mathematics.
Exner: Inside Calculus.
Fine/Rosenberger: The Fundamental Theory of Algebra.
Fischer: Intermediate Real Analysis.
Flanigan/Kazdan: Calculus Two: Linear and Nonlinear Functions. Second edition.
Fleming: Functions of Several Variables. Second edition.
Foulds: Combinatorial Optimization for Undergraduates.
Foulds: Optimization Techniques: An Introduction.
Franklin: Methods of Mathematical Economics.

(continued after index)

Saber Elaydi

An Introduction to Difference Equations

Third Edition

Springer

Saber Elaydi
Department of Mathematics
Trinity University
San Antonio, Texas 78212
USA

Mathematics Subject Clarification (2000): 12031

Library of Congress Cataloging-in-Publication Data
Elaydi, Saber, 1943 -
An Introduction to difference equation / Saber Elaydi.-3rd ed.
p. ; cm. — (Undergraduate texts in mathematics)
Includes bibliographical references and index.
ISBN - 0-387- 23059 - 9
1. Difference equation. I. Title. II. Series.
QA431.E431.E43 2005
515'.625 – dc22 2004058916

Elaydi: An Introduction to Difference Equations

First Indian Reprint 2008

ISBN 978-81-8128-904-9

This edition is manufactured in India for sale only in India, Pakistan, Bangladesh, Nepal and Sri Lanka and any other country as authorized by the publisher.

This edition is published by Springer (India) Private Limited,
A part of Springer Science+Business Media, Registered Office: 906-907,
Akash Deep Building, Barakhamba Road, New Delhi – 110 001, India.

Printed in India by Rashtriya Printers, Delhi.

Preface to the Third Edition

In contemplating the third edition, I have had multiple objectives to achieve. The first and foremost important objective is to maintain the accessibility and readability of the book to a broad readership with varying mathematical backgrounds and sophistication. More proofs, more graphs, more explanations, and more applications are provided in this edition.

The second objective is to update the contents of the book so that the reader stays abreast of new developments in this vital area of mathematics. Recent results on local and global stability of one-dimensional maps are included in Chapters 1, 4, and Appendices A and C. An extension of the Hartman–Grobman Theorem to noninvertible maps is stated in Appendix D. A whole new section on various notions of the asymptoticity of solutions and a recent extension of Perron's Second Theorem are added to Chapter 8. In Appendix E a detailed proof of the Levin–May Theorem is presented. In Chapters 4 and 5, the reader will find the latest results on the larval–pupal–adult flour beetle model.

The third and final objective is to better serve the broad readership of this book by including most, but certainly not all, of the research areas in difference equations. As more work is being published in the *Journal of Difference Equations and Applications* and elsewhere, it became apparent that a whole chapter needed to be dedicated to this enterprise. With the prodding and encouragement of Gerry Ladas, the new Chapter 5 was born. Major revisions of this chapter were made by Fozi Dannan, who diligently and painstakingly rewrote part of the material and caught several errors and typos. His impact on this edition, particularly in Chapters 1, 4, and Chapter 8 is immeasurable and I am greatly indebted to him. My thanks go to Shandelle Henson, who wrote a thorough review of the book and suggested the inclusion of an extension of the Hartman–Groman Theorem, and to Julio Lopez and his student Alex Sepulveda for their comments and discussions about the second edition.

I am grateful to all the participants of the AbiTuMath Program and to its coordinator Andreas Ruffing for using the second edition as the main reference in their activities and for their valuable comments and discussions. Special thanks go to Sebastian Pancratz of AbiTuMath whose suggestions improved parts of Chapters 1 and 2. I benefited from comments and discussions with Raghib Abu-Saris, Bernd Aulbach, Martin Bohner, Luis Carvahlo, Jim Cushing, Malgorzata Guzowska, Sophia Jang, Klara Janglajew, Nader Kouhestani, Ulrich Krause, Ronald Mickens, Robert Sacker, Hassan Sedaghat, and Abdul-Aziz Yakubu. It is a pleasure to thank Ina Lindemann, the editor at Springer-Verlag for her advice and support during the writing of this edition. Finally, I would like to express my deep appreciation to Denise Wilson who spent many weekends typing various drafts of the manuscript. Not only did she correct many glitches, typos, and awkward sentences, but she even caught some mathematical errors.

I hope you enjoy this edition and if you have any comments or questions, please do not hesitate to contact me at selaydi@trinity.edu.

San Antonio, Texas — Saber N. Elaydi
April 2004

Suggestions for instructors using this book.

The book may be used for two one-semester courses. A first course may include one of the following options but should include the bulk of the first four chapters:

1. If one is mainly interested in stability theory, then the choice would be Chapters 1–5.
2. One may choose Chapters 1–4, and Chapter 8 if the interest is to get to asymptotic theory.
3. Those interested in oscillation theory may choose Chapters 1, 2, 3, 5, and 7.
4. A course emphasizing control theory may include Chapters 1–3, 6, and 10.

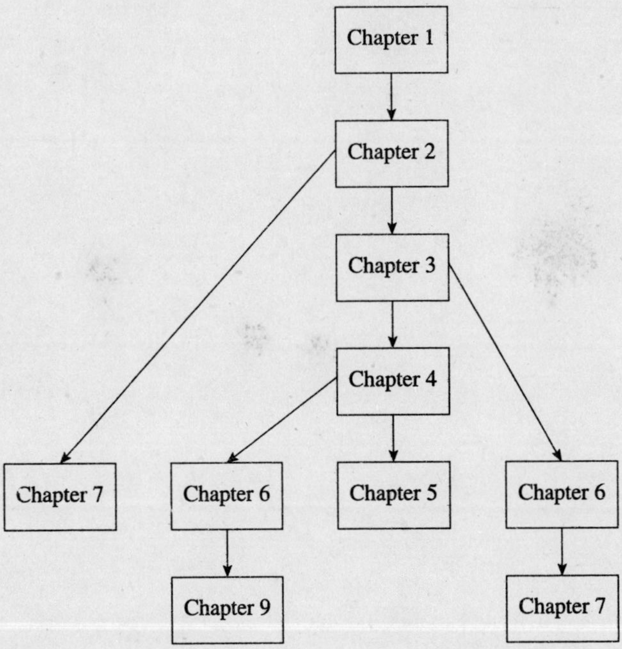

The diagram above depicts the dependency among the chapters.

Preface to the Second Edition

The second edition has greatly benefited from a sizable number of comments and suggestions I received from users of the first edition. I hope that I have corrected all the errors and misprints in the book. Important revisions were made in Chapters 1 and 4. In Chapter 1, I added two appendices (Global Stability and Periodic Solutions). In Chapter 4, I added a section on applications to mathematical biology. Influenced by a friendly and some not so friendly comments about Chapter 8 (previously Chapter 7: Asymptotic Behavior of Difference Equations), I rewrote the chapter with additional material on Birkhoff's theory. Also, due to popular demand, a new chapter (Chapter 9) under the title "Applications to Continued Fractions and Orthogonal Polynomials" has been added. This chapter gives a rather thorough presentation of continued fractions and orthogonal polynomials and their intimate connection to second-order difference equations. Chapter 8 (Oscillation Theory) has now become Chapter 7. Accordingly, the new revised suggestions for using the text are as follows.

The book may be used with considerable flexibility. For a one-semester course, one may choose one of the following options:

(i) If you want a course that emphasizes stability and control, then you may select Chapters 1, 2, and 3, and parts of Chapters 4, 5, and 6. This is perhaps appropriate for a class populated by mathematics, physics, and engineering majors.

(ii) If the focus is on the applications of difference equations to orthogonal polynomials and continued fractions, then you may select Chapters 1, 2, 3, 8, and 9.

I am indebted to K. Janglajew, who used the book several times and caught numerous glitches and typos. I am very grateful to Julio Lopez and his students, who helped me correct some mistakes and improve the exposition in Chapters 7 and 8. I am thankful to Raghib Abu-Saris, who caught some errors and typos in Chapter 4. My thanks go to Gerry Ladas, who assisted in refining Chapter 8, and to Allan Peterson, who graciously used my book and caught some mistakes in Chapter 4. I thank my brother Hatem Elaydi who read thoroughly Chapter 6 and made valuable revisions in the exercises. Many thanks to Fozi Dannan, whose comments improved Chapters 1, 4, and 9. Ronald Mickens was always there for me when I needed support, encouragement, and advice. His impact on this edition is immeasurable. My special thanks to Jenny Wolkowicki of Springer-Verlag.

I apologize in advance to all those whom I did not mention here but who have helped in one way or another to enhance the quality of this edition.

It is my pleasure to thank my former secretary, Constance Garcia, who typed the new and revised material.

San Antonio, Texas — Saber N. Elaydi
April 1999

Preface to the First Edition

This book grew out of lecture notes I used in a course on difference equations that I have taught at Trinity University for the past five years. The classes were largely populated by juniors and seniors majoring in mathematics, engineering, chemistry, computer science, and physics.

This book is intended to be used as a textbook for a course on difference equations at both the advanced undergraduate and beginning graduate levels. It may also be used as a supplement for engineering courses on discrete systems and control theory.

The main prerequisites for most of the material in this book are calculus and linear algebra. However, some topics in later chapters may require some rudiments of advanced calculus and complex analysis. Since many of the chapters in the book are independent, the instructor has great flexibility in choosing topics for a one-semester course.

This book presents the current state of affairs in many areas such as stability, Z-transform, asymptoticity, oscillations, and control theory. However, this book is by no means encyclopedic and does not contain many important topics, such as numerical analysis, combinatorics, special functions and orthogonal polynomials, boundary value problems, partial difference equations, chaos theory, and fractals. The nonselection of these topics is dictated not only by the limitations imposed by the elementary nature of this book, but also by the research interest (or lack thereof) of the author.

Great efforts were made to present even the most difficult material in an elementary format and to write in a style that makes the book accessible to students with varying backgrounds and interests. One of the main features of the book is the inclusion of a great number of applications in economics, social sciences, biology, physics, engineering, neural networks, etc. Moreover, this book contains a very extensive and carefully selected set of exercises at the end of each section. The exercises form an integral part of the text. They range from routine problems designed to build basic skills to more challenging problems that produce deeper understanding and build technique. The asterisked problems are the most challenging, and the instructor may assign them as long-term projects. Another important feature of the book is that it encourages students to make mathematical discoveries through calculator/computer experimentation.

Chapter 1 deals with first-order difference equations, or one-dimensional maps on the real line. It includes a thorough and complete analysis of stability for many popular maps (equations) such as the logistic map, the tent map, and the Baker map. The rudiments of bifurcation and chaos theory are also included in Section 1.6. This section raises more questions and gives few answers. It is intended to arouse the reader's interest in this exciting field.

In Chapter 2 we give solution methods for linear difference equations of any order. Then we apply the obtained results to investigate the stability and the oscillatory behavior of second-order difference equations. At the end of the chapter we give four applications: the propagation of annual plants, the gambler's ruin, the national income, and the transmission of information.

Chapter 3 extends the study in Chapter 2 to systems of difference equations. We introduce two methods to evaluate A^n for any matrix A. In Section 3.1 we introduce the Putzer algorithm, and in Section 3.3 the method of the Jordan form is given. Many applications are then given in Section 3.5, which include Markov chains, trade models, and the heat equation.

Chapter 4 investigates the question of stability for both scalar equations and systems. Stability of nonlinear equations is studied via linearization (Section 4.5) and by the famous method of Liapunov (Section 4.6). Our exposition here is restricted to autonomous (time-invariant) systems. I believe that the extension of the theory to nonautonomous (time-variant) systems, though technically involved, will not add much more understanding to the subject matter.

Chapter 5 delves deeply into Z-transform theory and techniques (Sections 5.1, 5.2). Then the results are applied to study the stability of Volterra difference scalar equations (Sections 5.3, 5.4) and systems (Sections 5.5, 5.6). For readers familiar with differential equations, Section 5.7 provides a comparison between the Z-transform and the Laplace transform. Most of the results on Volterra difference equations appear here for the first time in a book.

Chapter 6 takes us to the realm of control theory. Here, we cover most of the basic concepts including controllability, observability, observers, and stabilizability by feedback. Again, we restrict the presentation to autonomous (time-invariant) systems, since this is just an introduction to this vast and growing discipline. Moreover, most practitioners deal mainly with time-invariant systems.

In Chapter 7 we give a comprehensive and accessible study of asymptotic methods for difference equations. Starting from the Poincaré Theorem, the chapter covers most of the recent development in the subject. Section 7.4 (asymptotically diagonal systems) presents an extension of Levinson's Theorem to difference equations, while in Section 7.5 we carry our study to nonlinear difference equations. Several open problems are given that would serve as topics for research projects.

Finally, Chapter 8 presents a brief introduction to oscillation theory. In Section 8.1, the basic results on oscillation for three-term linear difference equations are introduced. Extension of these results to nonlinear difference equations is presented in Section 8.2. Another approach to oscillation theory, for self-adjoint equations, is presented in Section 8.3. Here we also introduce a discrete version of Sturm's Separation Theorem.

I am indebted to Gerry Ladas, who read many parts of the book and suggested many useful improvements, especially within the section on stability of scalar difference equations (Section 4.3). His influence through papers and lectures on Chapter 8 (oscillation theory) is immeasurable. My thanks go to Vlajko Kocic, who thoroughly read and made many helpful comments about Chapter 4 on Stability. Jim McDonald revised the chapters on the Z-transform and control theory (Chapters 5 and 6) and made significant improvements. I am very grateful to him for his contributions to this book. My sincere thanks go to Paul Eloe, who read the entire manuscript and offered valuable suggestions that led to many improvements in the final draft of the book. I am also grateful to Istvan Gyori for his comments on Chapter 8 and to Ronald Mickens for his review of the whole manuscript and for his advice and support. I would like to thank the following mathematicians who encouraged and helped me in numerous ways during the preparation of the book: Allan Peterson, Donald Bailey, Roberto Hasfura, Haydar Akca, and Shunian Zhang. I am grateful to my students Jeff Bator, Michelle MacArthur, and Nhung Tran, who caught misprints and mistakes in the earlier drafts of this book. My special thanks are due to my student Julie Lundquist, who proofread most of the book and made improvements in the presentation of many topics. My thanks go to Constance Garcia, who skillfully typed the entire manuscript with its many, many revised versions. And finally, it is a pleasure to thank Ina Lindemann and Robert Wexler from Springer-Verlag for their enthusiastic support of this project.

San Antonio, Texas
December 1995

Saber N. Elaydi

Contents

List of Symbols

$B(x_0, \delta)$	Ball centered at x_0 with radius δ
$B(\delta)$	Ball centered at origin with radius δ
Δ	The difference operator
$\mathcal{L}$	Moment functional
$K(a_n/b_n)$	Continued fraction
$\mathbb{R}$	The set of real numbers
$\mathbb{R}^+$	The set of nonnegative real numbers
$\mathbb{Z}$	The set of integers
$\mathbb{Z}^+$	The set of nonnegative integers
$\mathbb{C}$	The set of complex numbers
Γ	The gamma function
$F(a, b; c; z)$	The hypergeometric function
$(\nu)_n$	The Pochhammer symbol
$P_n^{(\alpha,\beta)}(x)$	Jacobi polynomials
$P_n(x)$	Legendre polynomials
$P_n^{\nu}(x)$	Gegenbauer polynomials
$L_n^{\alpha}(x)$	Laguerre polynomials
$H_n(x)$	Hermite polynomials
$O(x)$	The orbit of x
Δ^n	$\Delta^{n-1}(\Delta)$
$\prod_{i=n_0}^{n-1}$	Product
Sf	The Schwarzian derivative of f
E	Shift operator
f^n	The nth iterate of f
$x^{(k)}$	Factorial polynomial
Δ^{-1}	The antidifference operator

$\det A$	The determinant of a matrix A
$W(n)$	The Casoration
A^T	Transpose of a matrix A
diag	Diagonal matrix
$\rho(A)$	Spectral radius of A
$\|\|A\|\|$	Norm of a matrix A
$\overline{G}$	Closure of G
$\Omega(x_0)$	Limit set of x_0
$\tilde{x}(z)$	z-transform of $x(n)$
$Z(x(n))$	z-transform of $x(n)$
o	Little "oh"
O	Big "oh"
$f \sim g$	f is asymptotic to g

1

Dynamics of First-Order Difference Equations

1.1 Introduction

Difference equations usually describe the evolution of certain phenomena over the course of time. For example, if a certain population has discrete generations, the size of the $(n+1)$st generation $x(n+1)$ is a function of the nth generation $x(n)$. This relation expresses itself in the *difference equation*

$$x(n+1) = f(x(n)). \tag{1.1.1}$$

We may look at this problem from another point of view. Starting from a point x_0, one may generate the sequence

$$x_0, f(x_0), f(f(x_0)), f(f(f(x_0))), \ldots.$$

For convenience we adopt the notation

$$f^2(x_0) = f(f(x_0)), \quad f^3(x_0) = f(f(f(x_0))), \quad \text{etc.}$$

$f(x_0)$ is called the *first iterate* of x_0 under f; $f^2(x_0)$ is called the second iterate of x_0 under f; more generally, $f^n(x_0)$ is the nth iterate of x_0 under f. The set of all (positive) iterates $\{f^n(x_0) : n \geq 0\}$ where $f^0(x_0) = x_0$ by definition, is called the (*positive*) *orbit* of x_0 and will be denoted by $O(x_0)$. This iterative procedure is an example of a *discrete dynamical system.* Letting $x(n) = f^n(x_0)$, we have

$$x(n+1) = f^{n+1}(x_0) = f[f^n(x_0)] = f(x(n)),$$

and hence we recapture (1.1.1). Observe that $x(0) = f^0(x_0) = x_0$. For example, let $f(x) = x^2$ and $x_0 = 0.6$. To find the sequence of iterates

$\{f^n(x_0)\}$, we key 0.6 into a calculator and then repeatedly depress the x^2 button. We obtain the numbers

$$0.6, 0.36, 0.1296, 0.01679616, \ldots.$$

A few more key strokes on the calculator will be enough to convince the reader that the iterates $f^n(0.6)$ tend to 0. The reader is invited to verify that for all $x_0 \in (0,1)$, $f^n(x_0)$ tends to 0 as n tends to ∞, and that $f^n(x_0)$ tends to ∞ if $x_0 \notin [-1,1]$. Obviously, $f^n(0) = 0$, $f^n(1) = 1$ for all positive integers n, and $f^n(-1) = 1$ for $n = 1, 2, 3, \ldots$.

After this discussion one may conclude correctly that difference equations and discrete dynamical systems represent two sides of the same coin. For instance, when mathematicians talk about difference equations, they usually refer to the analytic theory of the subject, and when they talk about discrete dynamical systems, they generally refer to its geometrical and topological aspects.

If the function f in (1.1.1) is replaced by a function g of two variables, that is, $g : \mathbb{Z}^+ \times \mathbb{R} \to \mathbb{R}$, where $\mathbb{Z}^+$ is the set of nonnegative integers and $\mathbb{R}$ is the set of real numbers, then we have

$$x(n+1) = g(n, x(n)). \tag{1.1.2}$$

Equation (1.1.2) is called *nonautonomous* or time-variant, whereas (1.1.1) is called *autonomous* or time-invariant. The study of (1.1.2) is much more complicated and does not lend itself to the discrete dynamical system theory of first-order equations. If an initial condition $x(n_0) = x_0$ is given, then for $n \geq n_0$ there is a *unique solution* $x(n) \equiv x(n, n_0, x_0)$ of (1.1.2) such that $x(n_0, n_0, x_0) = x_0$. This may be shown easily by iteration. Now,

$$\begin{aligned}
x(n_0+1, n_0, x_0) &= g(n_0, x(n_0)) = g(n_0, x_0),\\
x(n_0+2, n_0, x_0) &= g(n_0+1, x(n_0+1)) = g(n_0+1, g(n_0, x_0)),\\
x(n_0+3, n_0, x_0) &= g(n_0+2, x(n_0+2)) = g[n_0+2, g(n_0+1, g(n_0, x_0))].
\end{aligned}$$

And, inductively, we get $x(n, n_0, x_0) = g[n-1, x(n-1, n_0, x_0)]$.

1.2 Linear First-Order Difference Equations

In this section we study the simplest special cases of (1.1.1) and (1.1.2), namely, linear equations. A typical linear *homogeneous* first-order equation is given by

$$x(n+1) = a(n)x(n), \qquad x(n_0) = x_0, \qquad n \geq n_0 \geq 0, \tag{1.2.1}$$

and the associated *nonhomogeneous* equation is given by

$$y(n+1) = a(n)y(n) + g(n), \qquad y(n_0) = y_0, \qquad n \geq n_0 \geq 0, \tag{1.2.2}$$

where in both equations it is assumed that $a(n) \neq 0$, and $a(n)$ and $g(n)$ are real-valued functions defined for $n \geq n_0 \geq 0$.

One may obtain the solution of (1.2.1) by a simple iteration:

$$x(n_0+1) = a(n_0)x(n_0) = a(n_0)x_0,$$
$$x(n_0+2) = a(n_0+1)x(n_0+1) = a(n_0+1)a(n_0)x_0,$$
$$x(n_0+3) = a(n_0+2)x(n_0+2) = a(n_0+2)a(n_0+1)a(n_0)x_0.$$

And, inductively, it is easy to see that

$$\begin{aligned} x(n) &= x(n_0+n-n_0)) \\ &= a(n-1)a(n-2)\cdots a(n_0)x_0, \end{aligned}$$

$$\boxed{x(n) = \left[\prod_{i=n_0}^{n-1} a(i)\right] x_0.} \tag{1.2.3}$$

The unique solution of the *nonhomogeneous* (1.2.2) may be found as follows:

$$\begin{aligned} y(n_0+1) &= a(n_0)y_0 + g(n_0), \\ y(n_0+2) &= a(n_0+1)y(n_0+1) + g(n_0+1) \\ &= a(n_0+1)a(n_0)y_0 + a(n_0+1)g(n_0) + g(n_0+1). \end{aligned}$$

Now we use mathematical induction to show that, for all $n \in \mathbb{Z}^+$,

$$\boxed{y(n) = \left[\prod_{i=n_0}^{n-1} a(i)\right] y_0 + \sum_{r=n_0}^{n-1}\left[\prod_{i=r+1}^{n-1} a(i)\right] g(r).} \tag{1.2.4}$$

To establish this, assume that formula (1.2.4) holds for $n = k$. Then from (1.2.2), $y(k+1) = a(k)y(k) + g(k)$, which by formula (1.2.4) yields

$$\begin{aligned} y(k+1) &= a(k)\left[\prod_{i=n_0}^{k-1} a(i)\right] y_0 + \sum_{r=n_0}^{k-1}\left[a(k)\prod_{i=r+1}^{k-1} a(i)\right] g(r) + g(k) \\ &= \left[\prod_{i=n_0}^{k} a(i)\right] y_0 + \sum_{r=n_0}^{k-1}\left(\prod_{i=r+1}^{k} a(i)\right) g(r) \\ &\quad + \left(\prod_{i=k+1}^{k} a(i)\right) g(k) \text{ (see footnote 1)} \\ &= \left[\prod_{i=n_0}^{k} a(i)\right] y_0 + \sum_{r=n_0}^{k}\left(\prod_{i=r+1}^{k} a(i)\right) g(r). \end{aligned}$$

Hence formula (1.2.4) holds for all $n \in \mathbb{Z}^+$.

---

[1]Notice that we have adopted the notation $\prod_{i=k+1}^{k} a(i) = 1$ and $\sum_{i=k+1}^{k} a(i) = 0$.

1.2.1 Important Special Cases

There are two special cases of (1.2.2) that are important in many applications. The first equation is given by

$$y(n+1) = ay(n) + g(n), \quad y(0) = y_0. \tag{1.2.5}$$

Using formula (1.2.4) one may establish that

$$\boxed{y(n) = a^n y_0 + \sum_{k=0}^{n-1} a^{n-k-1} g(k).} \tag{1.2.6}$$

The second equation is given by

$$y(n+1) = ay(n) + b, \qquad y(0) = y_0. \tag{1.2.7}$$

Using formula (1.2.6) we obtain

$$y(n) = \begin{cases} a^n y_0 + b\left[\dfrac{a^n - 1}{a - 1}\right] & \text{if } a \neq 1, \\ y_0 + bn & \text{if } a = 1. \end{cases} \tag{1.2.8}$$

Notice that the solution of the differential equation

$$\frac{dx}{dt} = ax(t), \quad x(0) = x_0,$$

is given by

$$x(t) = e^{at} x_0,$$

and the solution of the nonhomogeneous differential equation

$$\frac{dy}{dt} = ay(t) + g(t), \quad y(0) = y_0,$$

is given by

$$y(t) = e^{at} y_0 + \int_0^t e^{a(t-s)} g(s)\, ds.$$

Thus the exponential e^{at} in differential equations corresponds to the exponential a^n and the integral $\int_0^t e^{a(t-s)} g(s)\, ds$ corresponds to the summation $\sum\limits_{k=0}^{n-1} a^{n-k-1} g(k)$.

We now give some examples to practice the above formulas.

Example 1.1. Solve the equation

$$y(n+1) = (n+1)y(n) + 2^n (n+1)!, \qquad y(0) = 1, \qquad n > 0.$$

TABLE 1.1. Definite sum.

Number	Summation	Definite sum
1	$\sum_{k=1}^{n} k$	$\dfrac{n(n+1)}{2}$
2	$\sum_{k=1}^{n} k^2$	$\dfrac{n(n+1)(2n+1)}{6}$
3	$\sum_{k=1}^{n} k^3$	$\left[\dfrac{n(n+1)}{2}\right]^2$
4	$\sum_{k=1}^{n} k^4$	$\dfrac{n(6n^4+15n^3+10n^2-1)}{30}$
5	$\sum_{k=0}^{n-1} a^k$	$\begin{cases} (a^n-1)/(a-1) & \text{if } a \neq 1 \\ n & \text{if } a = 1 \end{cases}$
6	$\sum_{k=1}^{n-1} a^k$	$\begin{cases} (a^n-a)/(a-1) & \text{if } a \neq 1 \\ n-1 & \text{if } a = 1 \end{cases}$
7	$\sum_{k=1}^{n} ka^k,\ a \neq 1$	$\dfrac{(a-1)(n+1)a^{n+1} - a^{n+2} + a}{(a-1)^2}$

Solution

$$
\begin{aligned}
y(n) &= \prod_{i=0}^{n-1}(i+1) + \sum_{k=0}^{n-1}\left[\prod_{i=k+1}^{n-1}(i+1)\right] 2^k(k+1)! \\
&= n! + \sum_{k=0}^{n-1} n!\, 2^k \\
&= 2^n n! \text{ (from Table 1.1).}
\end{aligned}
$$

Example 1.2. Find a solution for the equation

$$x(n+1) = 2x(n) + 3^n, \qquad x(1) = 0.5.$$

Solution From (1.2.6), we have

$$
\begin{aligned}
x(n) &= \left(\frac{1}{2}\right) 2^{n-1} + \sum_{k=1}^{n-1} 2^{n-k-1}\, 3^k \\
&= 2^{n-2} + 2^{n-1} \sum_{k=1}^{n-1} \left(\frac{3}{2}\right)^k \\
&= 2^{n-2} + 2^{n-1} \frac{3}{2} \left(\frac{\left(\frac{3}{2}\right)^{n-1} - 1}{\frac{3}{2} - 1}\right) \\
&= 3^n - 5 \cdot 2^{n-2}.
\end{aligned}
$$

Example 1.3. A drug is administered once every four hours. Let $D(n)$ be the amount of the drug in the blood system at the nth interval. The body eliminates a certain fraction p of the drug during each time interval. If the amount administered is D_0, find $D(n)$ and $\lim_{n\to\infty} D(n)$.

Solution We first must create an equation to solve. Since the amount of drug in the patient's system at time $(n+1)$ is equal to the amount at time n minus the fraction p that has been eliminated from the body, plus the new dosage D_0, we arrive at the following equation:

$$D(n+1) = (1-p)D(n) + D_0.$$

Using (1.2.8), we solve the above equation, arriving at

$$D(n) = \left[D_0 - \frac{D_0}{p}\right](1-p)^n + \frac{D_0}{p}.$$

Hence,

$$\lim_{n\to\infty} D(n) = \frac{D_0}{p}. \tag{1.2.9}$$

Let $D_0 = 2$ cubic centimeters (cc), $p = 0.25$.

Then our original equation becomes

$$D(n+1) = 0.75D(n) + 2, \qquad D(0) = 2.$$

Table 1.2 gives $D(n)$ for $0 \le n \le 10$.

It follows from (1.2.9) that $\lim_{n\to\infty} D(n) = 8$, where $D^* = 8$ cc is the equilibrium amount of drug in the body. We now enter the realm of finance for our next example.

Example 1.4. Amortization

Amortization is the process by which a loan is repaid by a sequence of periodic payments, each of which is part payment of interest and part payment to reduce the outstanding principal.

Let $p(n)$ represent the outstanding principal after the nth payment $g(n)$. Suppose that interest charges compound at the rate r per payment period.

The formulation of our model here is based on the fact that the outstanding principal $p(n+1)$ after the $(n+1)$st payment is equal to the outstanding principal $p(n)$ after the nth payment plus the interest $rp(n)$ incurred during the $(n+1)$st period minus the nth payment $g(n)$. Hence

$$p(n+1) = p(n) + rp(n) - g(n),$$

TABLE 1.2. Values of $D(n)$.

n	0	1	2	3	4	5	6	7	8	9	10
$D(n)$	2	3.5	4.62	5.47	6.1	6.58	6.93	7.2	7.4	7.55	7.66

or

$$p(n+1) = (1+r)p(n) - g(n), \qquad p(0) = p_0, \tag{1.2.10}$$

where p_0 is the initial debt. By (1.2.6) we have

$$p(n) = (1+r)^n p_0 - \sum_{k=0}^{n-1} (1+r)^{n-k-1} g(k). \tag{1.2.11}$$

In practice, the payment $g(n)$ is constant and, say, equal to T. In this case,

$$p(n) = (1+r)^n p_0 - ((1+r)^n - 1)\left(\frac{T}{r}\right). \tag{1.2.12}$$

If we want to pay off the loan in exactly n payments, what would be the monthly payment T? Observe first that $p(n) = 0$. Hence from (1.2.12) we have

$$T = p_0 \left[\frac{r}{1-(1+r)^{-n}}\right].$$

Exercises 1.1 and 1.2

1. Find the solution of each difference equation:
   (a) $x(n+1) - (n+1)x(n) = 0, \qquad x(0) = c.$
   (b) $x(n+1) - 3^n x(n) = 0, \qquad x(0) = c.$
   (c) $x(n+1) - e^{2n} x(n) = 0, \qquad x(0) = c.$
   (d) $x(n+1) - \frac{n}{n+1} x(n) = 0, \qquad n \geq 1, \qquad x(1) = c.$
2. Find the general solution of each difference equation:
   (a) $y(n+1) - \frac{1}{2} y(n) = 2, \qquad y(0) = c.$
   (b) $y(n+1) - \frac{n}{n+1} y(n) = 4, \qquad y(1) = c.$
3. Find the general solution of each difference equation:
   (a) $y(n+1) - (n+1)y(n) = 2^n (n+1)!, \qquad y(0) = c.$
   (b) $y(n+1) = y(n) + e^n, \qquad y(0) = c.$
4. (a) Write a difference equation that describes the number of regions created by n lines in the plane if it is required that every pair of lines meet and no more than two lines meet at one point.
   (b) Find the number of these regions by solving the difference equation in case (a).
5. The gamma function is defined as $\Gamma(x) = \int_0^\infty t^{x-1} e^{-t}\, dt, x > 0.$
   (a) Show that $\Gamma(x+1) = x\Gamma(x), \qquad \Gamma(1) = 1.$

(b) If n is a positive integer, show that $\Gamma(n+1) = n!$.

(c) Show that $x^{(n)} = x(x-1)\cdots(x-n+1) = \dfrac{\Gamma(x+1)}{\Gamma(x-n+1)}$.

6. A space (three-dimensional) is divided by n planes, nonparallel, and no four planes having a point in common.

   (a) Write a difference equation that describes the number of regions created.

   (b) Find the number of these regions.

7. Verify (1.2.6).

8. Verify (1.2.8).

9. A debt of \$12,000 is to be amortized by equal payments of \$380 at the end of each month, plus a final partial payment one month after the last \$380 is paid. If interest is at an annual rate of 12% compounded monthly, construct an amortization schedule to show the required payments.

10. Suppose that a loan of \$80,000 is to be amortized by equal monthly payments. If the interest rate is 10% compounded monthly, find the monthly payment required to pay off the loan in 30 years.

11. Suppose the constant sum T is deposited at the end of each fixed period in a bank that pays interest at the rate r per period. Let $A(n)$ be the amount accumulated in the bank after n periods.

    (a) Write a difference equation that describes $A(n)$.

    (b) Solve the difference equation obtained in (a), when $A(0) = 0, T =$ \$200, and $r = 0.008$.

12. The temperature of a body is measured as 110° F. It is observed that the amount the temperature changes during each period of two hours is -0.3 times the difference between the previous period's temperature and the room temperature, which is 70° F.

    (a) Write a difference equation that describes the temperature $T(n)$ of the body at the end of n periods.

    (b) Find $T(n)$.

13. Suppose that you can get a 30-year mortgage at 8% interest. How much can you afford to borrow if you can afford to make a monthly payment of \$1,000?

14. Radium decreases at the rate of 0.04% per year. What is its half-life? (The half-life of a radioactive material is defined to be the time needed for half of the material to dissipate.)

15. (Carbon Dating). It has been observed that the proportion of carbon-14 in plants and animals is the same as that in the atmosphere as long as the plant or the animal is alive. When an animal or plant dies, the carbon-14 in its tissue starts decaying at the rate r.

    (a) If the half-life of carbon-14 is 5,700 years, find r.

    (b) If the amount of carbon-14 present in a bone of an animal is 70% of the original amount of carbon-14, how old is the bone?

1.3 Equilibrium Points

The notion of equilibrium points (states) is central in the study of the dynamics of any physical system. In many applications in biology, economics, physics, engineering, etc., it is desirable that all states (solutions) of a given system tend to its equilibrium state (point). This is the subject of study of stability theory, a topic of great importance to scientists and engineers. We now give the formal definition of an equilibrium point.

Definition 1.5. A point x^* in the domain of f is said to be an *equilibrium point* of (1.1.1) if it is a fixed point of f, i.e., $f(x^*) = x^*$.

In other words, x^* is a *constant solution* of (1.1.1), since if $x(0) = x^*$ is an initial point, then $x(1) = f(x^*) = x^*$, and $x(2) = f(x(1)) = f(x^*) = x^*$, and so on.

Graphically, an equilibrium point is the x-coordinate of the point where the graph of f intersects the diagonal line $y = x$ (Figures 1.1 and 1.2). For example, there are three equilibrium points for the equation

$$x(n+1) = x^3(n)$$

where $f(x) = x^3$. To find these equilibrium points, we let $f(x^*) = x^*$, or $x^3 = x$, and solve for x. Hence there are three equilibrium points, $-1, 0, 1$ (Figure 1.1). Figure 1.2 illustrates another example, where $f(x) = x^2 - x + 1$ and the difference equation is given by

$$x(n+1) = x^2(n) - x(n) + 1.$$

Letting $x^2 - x + 1 = x$, we find that 1 is the only equilibrium point.

There is a phenomenon that is unique to difference equations and cannot possibly occur in differential equations. It is possible in difference equations that a solution may not be an equilibrium point but may reach one after finitely many iterations. In other words, a nonequilibrium state may go to an equilibrium state in a finite time. This leads to the following definition.

Definition 1.6. Let x be a point in the domain of f. If there exists a positive integer r and an equilibrium point x^* of (1.1.1) such that $f^r(x) = x^*$, $f^{r-1}(x) \neq x^*$, then x is an *eventually equilibrium (fixed) point.*

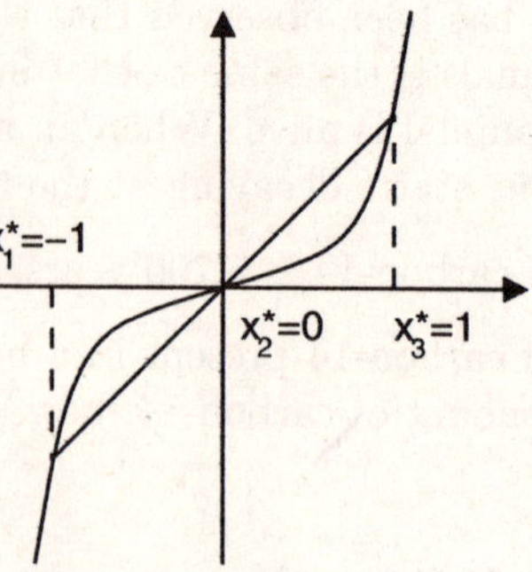

FIGURE 1.1. Fixed points of $f(x) = x^3$.

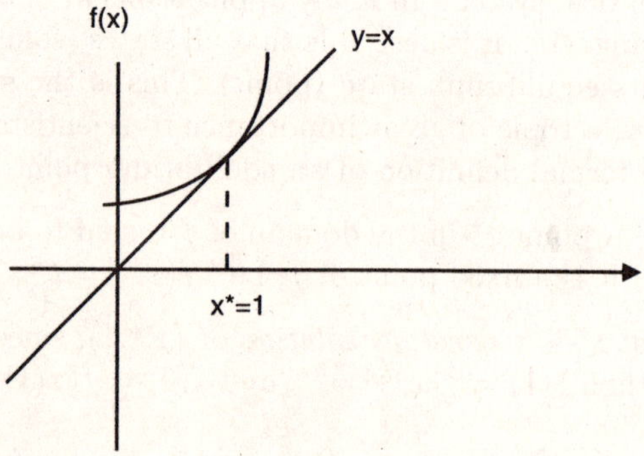

FIGURE 1.2. Fixed points of $f(x) = x^2 - x + 1$.

Example 1.7. The Tent Map

Consider the equation (Figure 1.3)

$$x(n+1) = T(x(n)),$$

where

$$T(x) = \begin{cases} 2x & \text{for } 0 \le x \le \dfrac{1}{2}, \\ 2(1-x) & \text{for } \dfrac{1}{2} < x \le 1. \end{cases}$$

There are two equilibrium points, 0 and $\frac{2}{3}$ (see Figure 1.3). The search for eventually equilibrium points is not as simple algebraically. If $x(0) = \frac{1}{4}$, then $x(1) = \frac{1}{2}$, $x(2) = 1$, and $x(3) = 0$. Thus $\frac{1}{4}$ is an eventually equilibrium point. The reader is asked to show that if $x = k/2^n$, where k and n are positive integers with $0 < k/2^n \le 1$, then x is an eventually equilibrium point (Exercises 1.3, Problem 15).

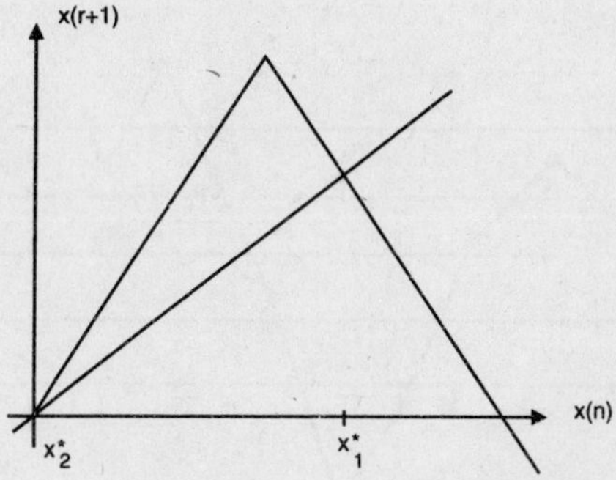

FIGURE 1.3. Equilibrium points of the tent map.

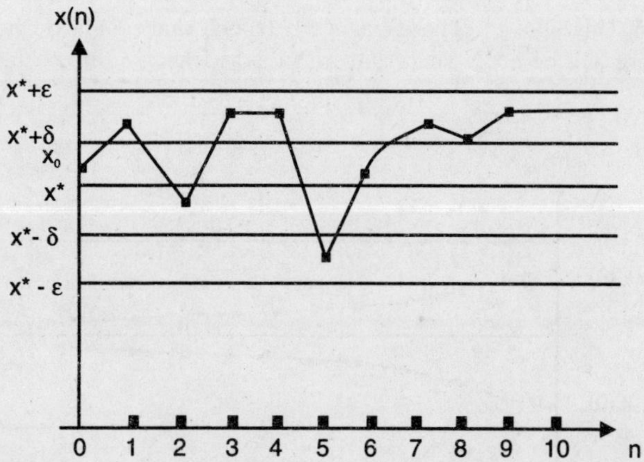

FIGURE 1.4. Stable x^*. If $x(0)$ is within δ from x^*, then $x(n)$ is within ε from $x(n)$ for all $n > 0$.

One of the main objectives in the study of a dynamical system is to analyze the behavior of its solutions near an equilibrium point. This study constitutes the stability theory. Next we introduce the basic definitions of stability.

Definition 1.8. (a) The equilibrium point x^* of (1.1.1) is *stable* (Figure 1.4) if given $\varepsilon > 0$ there exists $\delta > 0$ such that $|x_0 - x^*| < \delta$ implies $|f^n(x_0) - x^*| < \varepsilon$ for all $n > 0$. If x^* is not stable, then it is called *unstable* (Figure 1.5).

(b) The point x^* is said to be *attracting* if there exists $\eta > 0$ such that

$$|x(0) - x^*| < \eta \quad \text{implies} \quad \lim_{n\to\infty} x(n) = x^*.$$

If $\eta = \infty$, x^* is called a *global attractor* or *globally attracting*.

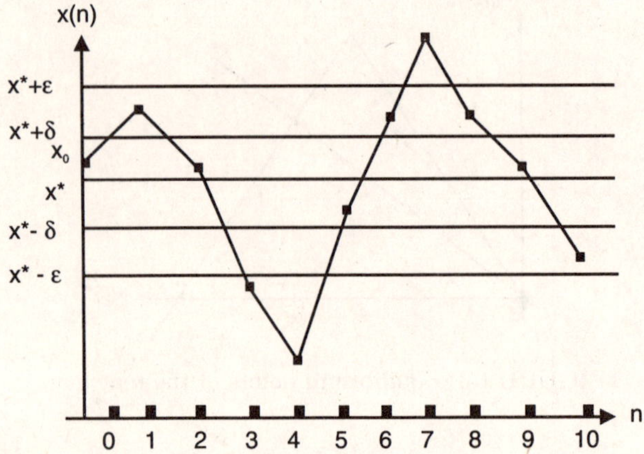

FIGURE 1.5. Unstable x^*. There exists $\varepsilon > 0$ such that no matter how close $x(0)$ is to x^*, there will be an N such that $x(N)$ is at least ε from x^*.

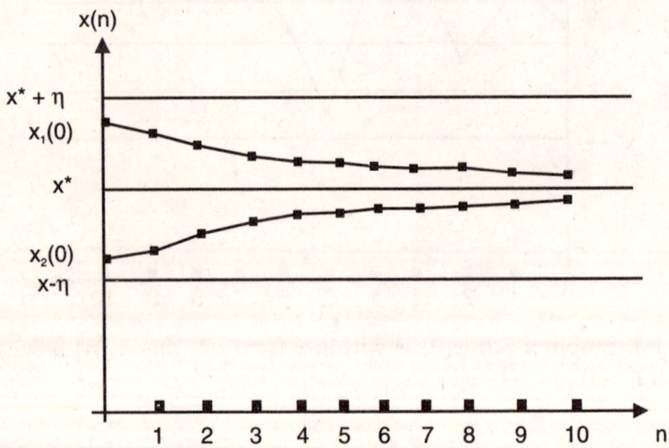

FIGURE 1.6. Asymptotically stable x^*. Stable if $x(0)$ is within η of x^*; then $\lim_{n\to\infty} x(n) = x^*$.

(c) The point x^* is an *asymptotically stable equilibrium point* if it is stable and attracting.

If $\eta = \infty$, x^* is said to be *globally asymptotically stable* (Figure 1.7).

To determine the stability of an equilibrium point from the above definitions may prove to be a mission impossible in many cases. This is due to the fact that we may not be able to find the solution in a closed form even for the deceptively simple-looking equation (1.1.1). In this section we present some of the simplest but most powerful tools of the trade to help us understand the behavior of solutions of (1.1.1) in the vicinity of equilib-

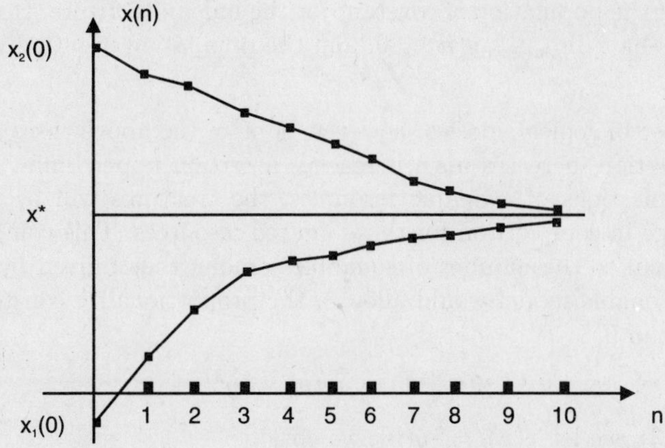

FIGURE 1.7. Globally asymptotically stable x^*. Stable and $\lim_{n\to\infty} x(n) = x^*$ for all $x(0)$.

rium points, namely, the graphical techniques. A hand-held calculator may fulfill all your graphical needs in this section.

1.3.1 The Stair Step (Cobweb) Diagrams

We now give, in excruciating detail, another important graphical method for analyzing the stability of equilibrium (and periodic) points for (1.1.1). Since $x(n+1) = f(x(n))$, we may draw a graph of f in the $(x(n), x(n+1))$ plane. Then, given $x(0) = x_0$, we pinpoint the value $x(1)$ by drawing a vertical line through x_0 so that it also intersects the graph of f at $(x_0, x(1))$. Next, draw a horizontal line from $(x_0, x(1))$ to meet the diagonal line $y = x$ at the point $(x(1), x(1))$. A vertical line drawn from the point $(x(1), x(1))$ will meet the graph of f at the point $(x(1), x(2))$. Continuing this process, one may find $x(n)$ for all $n > 0$.

Example 1.9. The Logistic Equation

Let $y(n)$ be the size of a population at time n. If μ is the rate of growth of the population from one generation to another, then we may consider a mathematical model in the form

$$y(n+1) = \mu y(n), \qquad \mu > 0. \tag{1.3.1}$$

If the initial population is given by $y(0) = y_0$, then by simple iteration we find that

$$y(n) = \mu^n y_0 \tag{1.3.2}$$

is the solution of (1.3.1). If $\mu > 1$, then $y(n)$ increases indefinitely, and $\lim_{n\to\infty} y(n) = \infty$. If $\mu = 1$, then $y(n) = y_0$ for all $n > 0$, which means that the size of the population is constant for the indefinite future. However, for $\mu < 1$, we have $\lim_{n\to\infty} y(n) = 0$, and the population eventually becomes extinct.

For most biological species, however, none of the above cases is valid as the population increases until it reaches a certain upper limit. Then, due to the limitations of available resources, the creatures will become testy and engage in competition for those limited resources. This competition is proportional to the number of squabbles among them, given by $y^2(n)$. A more reasonable model would allow b, the proportionality constant, to be greater than 0,

$$y(n+1) = \mu y(n) - by^2(n). \tag{1.3.3}$$

If in (1.3.3), we let $x(n) = \frac{b}{\mu} y(n)$, we obtain

$$x(n+1) = \mu x(n)(1 - x(n)) = f(x(n)). \tag{1.3.4}$$

This equation is the simplest nonlinear first-order difference equation, commonly referred to as the (discrete) logistic equation. However, a closed-form solution of (1.3.4) is not available (except for certain values of μ). In spite of its simplicity, this equation exhibits rather rich and complicated dynamics. To find the equilibrium points of (1.3.4) we let $f(x^*) = \mu x^*(1 - x^*) = x^*$. Thus, we pinpoint two equilibrium points: $x^* = 0$ and $x^* = (\mu - 1)/\mu$.

Figure 1.8 gives the stair step diagram of $(x(n), x(n+1))$ when $\mu = 2.5$ and $x(0) = 0.1$. In this case, we also have two equilibrium points. One, $x^* = 0$, is unstable, and the other, $x^* = 0.6$, is asymptotically stable.

Example 1.10. The Cobweb Phenomenon (Economics Application)

Here we study the pricing of a certain commodity. Let $S(n)$ be the number of units supplied in period n, $D(n)$ the number of units demanded in period n, and $p(n)$ the price per unit in period n.

For simplicity, we assume that $D(n)$ depends only linearly on $p(n)$ and is denoted by

$$D(n) = -m_d p(n) + b_d, \qquad m_d > 0, \qquad b_d > 0. \tag{1.3.5}$$

This equation is referred to as the price–demand curve. The constant m_d represents the sensitivity of consumers to price. We also assume that the price–supply curve relates the supply in any period to the price one period before, i.e.,

$$S(n+1) = m_s p(n) + b_s, \qquad m_s > 0, \qquad b_s > 0. \tag{1.3.6}$$

The constant m_s is the sensitivity of suppliers to price. The slope of the demand curve is negative because an increase of one unit in price produces a decrease of m_d units in demand. Correspondingly, an increase of one unit

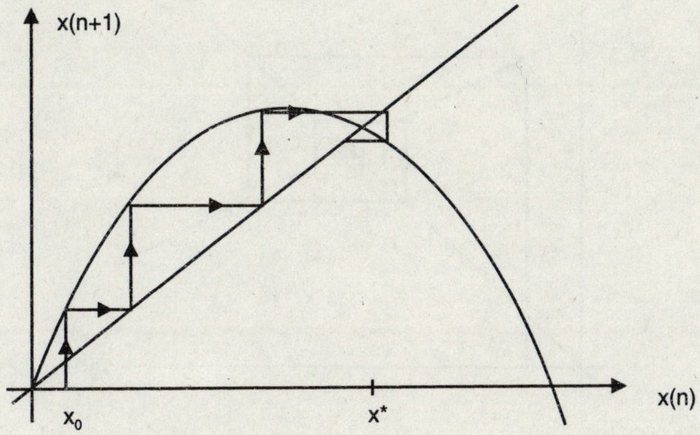

FIGURE 1.8. Stair step diagram for $\mu = 2.5$.

in price causes an increase of m_s units in supply, creating a positive slope for that curve.

A third assumption we make here is that the market price is the price at which the quantity demanded and the quantity supplied are equal, that is, at which $D(n+1) = S(n+1)$.

Thus

$$-m_d p(n+1) + b_d = m_s p(n) + b_s,$$

or

$$p(n+1) = Ap(n) + B = f(p(n)), \tag{1.3.7}$$

where

$$A = -\frac{m_s}{m_d}, \qquad B = \frac{b_d - b_s}{m_d}. \tag{1.3.8}$$

This equation is a first-order linear difference equation. The equilibrium price p^* is defined in economics as the price that results in an intersection of the supply $S(n+1)$ and demand $D(n)$ curves. Also, since p^* is the unique fixed point of $f(p)$ in (1.3.7), $p^* = B/(1-A)$. (This proof arises later as Exercises 1.3, Problem 6.) Because A is the ratio of the slopes of the supply and demand curves, this ratio determines the behavior of the price sequence. There are three cases to be considered:

(a) $-1 < A < 0$,

(b) $A = -1$,

(c) $A < -1$.

The three cases are now depicted graphically using our old standby, the stair step diagram.

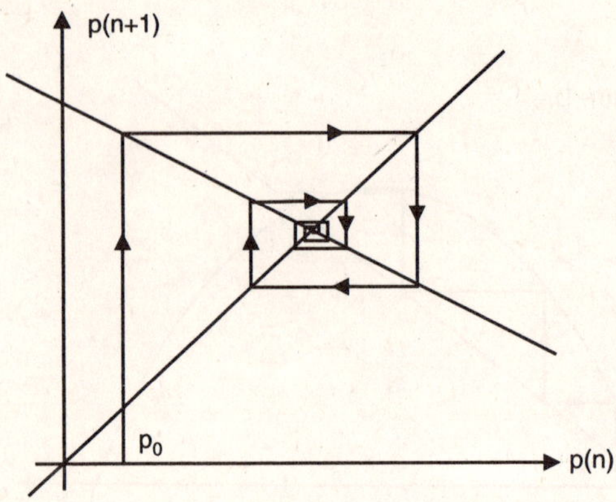

FIGURE 1.9. Asymptotically stable equilibrium price.

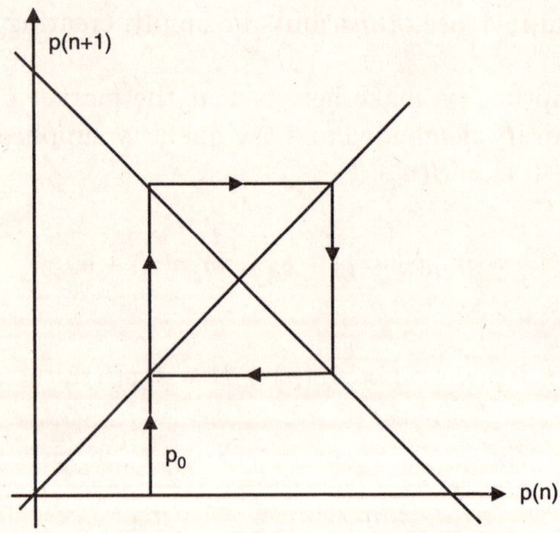

FIGURE 1.10. Stable equilibrium price.

(i) In case (a), prices alternate above and below but converge to the equilibrium price p^*. In economics lingo, the price p^* is considered "stable"; in mathematics, we refer to it as "asymptotically stable" (Figure 1.9).

(ii) In case (b), prices oscillate between two values only. If $p(0) = p_0$, then $p(1) = -p_0 + B$ and $p(2) = p_0$. Hence the equilibrium point p^* is stable (Figure 1.10).

(iii) In case (c), prices oscillate infinitely about the equilibrium point p^* but progressively move further away from it. Thus, the equilibrium point is considered unstable (Figure 1.11).

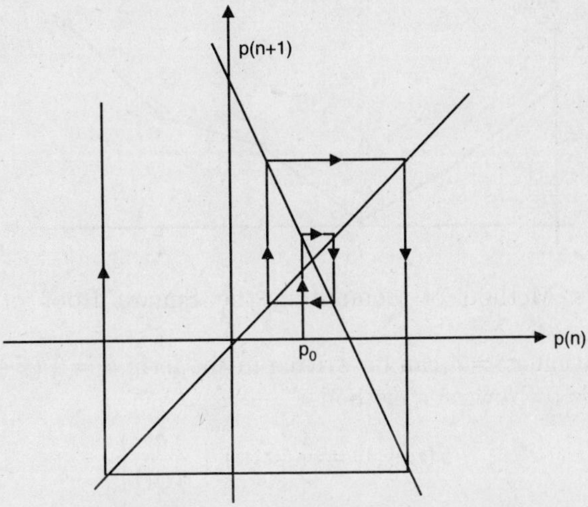

FIGURE 1.11. Unstable equilibrium price.

An explicit solution of (1.3.7) with $p(0) = p_0$ is given by

$$p(n) = \left(p_0 - \frac{B}{1-A}\right) A^n + \frac{B}{1-A} \text{ (Exercises 1.3, Problem 9).} \quad (1.3.9)$$

This explicit solution allows us to restate cases (a) and (b) as follows.

1.3.2 The Cobweb Theorem of Economics

If the suppliers are less sensitive to price than the consumers (i.e., $m_s < m_d$), the market will then be stable. If the suppliers are more sensitive than the consumers, the market will be unstable.

One might also find the closed-form solution (1.3.9) by using a computer algebra program, such as Maple. One would enter this program:

$$\text{rsolve}(\{p(n+1) = a * p(n) + b, p(0) = p_0\}, p(n)).$$

Exercises 1.3

1. Contemplate the equation $x(n+1) = f(x(n))$, where $f(0) = 0$.

   (a) Prove that $x(n) \equiv 0$ is a solution of the equation.

   (b) Show that the function depicted in the following $(n, x(n))$ diagram cannot possibly be a solution of the equation:

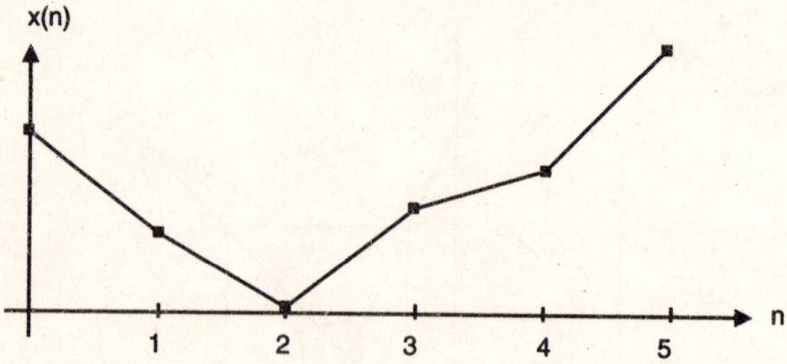

2. (Newton's Method of Computing the Square Root of a Positive Number)
The equation $x^2 = a$ can be written in the form $x = \frac{1}{2}(x + a/x)$. This form leads to Newton's method

$$x(n+1) = \frac{1}{2}\left[x(n) + \frac{a}{x(n)}\right].$$

(a) Show that this difference equation has two equilibrium points, $-\sqrt{a}$ and $\sqrt{a}$.

(b) Sketch a stair step diagram for $a = 3, x(0) = 1$, and $x(0) = -1$.

(c) What can you conclude from (b)?

3. (Pielou's Logistic Equation)
E.C. Pielou [119] referred to the following equation as the discrete logistic equation:

$$x(n+1) = \frac{\alpha x(n)}{1 + \beta x(n)}, \qquad \alpha > 1, \qquad \beta > 0.$$

(a) Find the positive equilibrium point.

(b) Demonstrate, using the stair step diagram, that the positive equilibrium point is asymptotically stable, taking $\alpha = 2$ and $\beta = 1$.

4. Find the equilibrium points and determine their stability for the equation

$$x(n+1) = 5 - \frac{6}{x(n)}.$$

5. (a) Draw a stair step diagram for (1.3.4) for $\mu = 0.5$, 3, and 3.3. What can you conclude from these diagrams?

(b) Determine whether these values for μ give rise to periodic solutions of period 2.

6. (The Cobweb Phenomenon [equation (1.3.7)]). Economists define the equilibrium price p^* of a commodity as the price at which the demand function $D(n)$ is equal to the supply function $S(n+1)$. These are defined in (1.3.5) and (1.3.6), respectively.

(a) Show that $p^* = \frac{B}{1-A}$, where A and B are defined as in (1.3.8).

(b) Let $m_s = 2$, $b_s = 3$, $m_d = 1$, and $b_d = 15$. Find the equilibrium price p^*. Then draw a stair step diagram, for $p(0) = 2$.

7. Continuation of Problem 6:
Economists use a different stair step diagram, as we will explain in the following steps:

(i) Let the x-axis represent the price $p(n)$ and the y-axis represent $S(n+1)$ or $D(n)$. Draw the supply line and the demand line and find their point of intersection p^*.

(ii) Starting with $p(0) = 2$ we find $s(1)$ by moving vertically to the supply line, then moving horizontally to find $D(1)$ (since $D(1) = S(1)$), which determines $p(1)$ on the price axis. The supply $S(2)$ is found on the supply line directly above $p(1)$, and then $D(2)$ ($= S(2)$) is found by moving horizontally to the demand line, etc.

(iii) Is p^* stable?

8. Repeat Exercises 6 and 7 for:

(a) $m_s = m_d = 2, b_d = 10$, and $b_s = 2$.

(b) $m_s = 1, m_d = 2, b_d = 14$, and $b_s = 2$.

9. Verify that formula (1.3.9) is a solution of (1.3.7).

10. Use formula (1.3.9) to show that:

(a) If $-1 < A < 0$, then $\lim_{n\to\infty} p(n) = B/1 - A$.

(b) If $A < -1$, then $p(n)$ is unbounded.

(c) If $A = -1$, then $p(n)$ takes only two values:

$$p(n) = \begin{cases} p(0) & \text{if } n \text{ is even,} \\ p(1) = B - p_0 & \text{if } n \text{ is odd.} \end{cases}$$

11. Suppose that the supply and demand equations are given by $D(n) = -2p(n) + 3$ and $S(n+1) = p^2(n) + 1$.

(a) Assuming that the market price is the price at which supply equals demand, find a difference equation that relates $p(n+1)$ to $p(n)$.

(b) Find the positive equilibrium value of this equation.

(c) Use the stair step diagrams to determine the stability of the positive equilibrium value.

12. Consider Baker's map defined by

$$B(x) = \begin{cases} 2x & \text{for } 0 \le x \le \frac{1}{2}, \\ 2x - 1 & \text{for } \frac{1}{2} < x \le 1. \end{cases}$$

    (i) Draw the function $B(x)$ on [0,1].

    (ii) Show that $x \in [0,1]$ is an eventually fixed point if and only if it is of the form $x = k/2^n$, where k and n are positive integers,[2] with $0 \le k \le 2^n - 1$.

13. Find the fixed points and the eventually fixed points of $x(n+1) = f(x(n))$, where $f(x) = x^2$.

14. Find an eventually fixed point of the tent map of Example 1.7 that is not in the form $k/2^n$.

15. Consider the tent map of Example 1.7. Show that if $x = k/2^n$, where k and n are positive integers with $0 < k/2^n \le 1$, then x is an eventually fixed point.

1.4 Numerical Solutions of Differential Equations

Differential equations have been extensively used as mathematical models for a wide variety of physical and artificial phenomena. Such models describe populations or objects that evolve continuously in which time (or the independent variable) is a subset of the set of real numbers. In contrast, difference equations describe populations or objects that evolve discretely in which time (or the independent variable) is a subset of the set of integers. In many instances, one is unable to solve a given differential equation. In this case, we need to use a numerical scheme to approximate the solutions of the differential equations. A numerical scheme leads to the construction of an associated difference equation that is more amenable to computation either by a graphing-held calculator or by a computer. Here we present a couple of simple numerical schemes. We begin by *Euler's method*, one of the oldest numerical methods.

1.4.1 *Euler's Method*

Consider the first-order differential equation

$$x'(t) = g(t, x(t)), \qquad x(t_0) = x_0, \qquad t_0 \le t \le b. \tag{1.4.1}$$

---

[2]A number $x \in [0,1]$ is called a dyadic rational if it has the form $k/2^n$ for some nonnegative integers k and n, with $0 \le k \le 2^n - 1$.

Let us divide the interval $[t_0, b]$ into N equal subintervals. The size of each subinterval is called the *step size* of the method and is denoted by $h = (b - t_0)/N$. This step size defines the *nodes* $t_0, t_1, t_2, \ldots, t_N$, where $t_j = t_0 + jh$. Euler's method approximates $x'(t)$ by $(x(t+h) - x(t))/h$.

Substituting this value into (1.4.1) gives

$$x(t+h) = x(t) + hg(t, x(t)),$$

and for $t = t_0 + nh$, we obtain

$$x[t_0 + (n+1)h] = x(t_0 + nh) + hg[t_0 + nh, x(t_0 + nh)] \tag{1.4.2}$$

for $n = 0, 1, 2, \ldots, N-1$.

Adapting the difference equation notation and replacing $x(t_0 + nh)$ by $x(n)$ gives

$$x(n+1) = x(n) + hg[n, x(n)]. \tag{1.4.3}$$

Equation (1.4.3) defines *Euler's algorithm*, which approximates the solutions of the differential equation (1.4.1) at the node points.

Note that x^* is an equilibrium point of (1.4.3) if and only if $g(x^*) = 0$. Thus the differential equation (1.4.1) and the difference equation (1.4.3) have the same equilibrium points.

Example 1.11. Let us now apply Euler's method to the differential equation:

$$x'(t) = 0.7x^2(t) + 0.7, \qquad x(0) = 1, \qquad t \in [0, 1] \qquad (DE) \text{ (see footnote 3)}.$$

Using the separation of variable method, we obtain

$$\frac{1}{0.7} \int \frac{dx}{x^2 + 1} = \int dt.$$

Hence

$$\tan^{-1}(x(t)) = 0.7t + c.$$

Letting $x(0) = 1$, we get $c = \frac{\pi}{4}$. Thus, the exact solution of this equation is given by $x(t) = \tan\left(0.7t + \frac{\pi}{4}\right)$.

The corresponding difference equation using Euler's method is

$$x(n+1) = x(n) + 0.7h(x^2(n) + 1), \qquad x(0) = 1 \qquad (\Delta E) \text{ (see footnote 4)}$$

Table 1.3 shows the Euler approximations for $h = 0.2$ and 0.1, as well as the exact values. Figure 1.12 depicts the $(n, x(n))$ diagram. Notice that the smaller the step size we use, the better the approximation we have.

---

[3] $DE \equiv$ differential equation.

[4] $\Delta E \equiv$ difference equation.

TABLE 1.3.

n	t	(ΔE) Euler ($h = 0.2$) $x(n)$	(ΔE) Euler ($h = 0.1$) $x(n)$	Exact (DE) $x(t)$
0	0	1	1	1
1	0.1		1.14	1.150
2	0.2	1.28	1.301	1.328
3	0.3		1.489	1.542
4	0.4	1.649	1.715	1.807
5	0.5		1.991	2.150
6	0.6	2.170	2.338	2.614
7	0.7		2.791	3.286
8	0.8	2.969	3.406	4.361
9	0.9		4.288	6.383
10	1	4.343	5.645	11.681

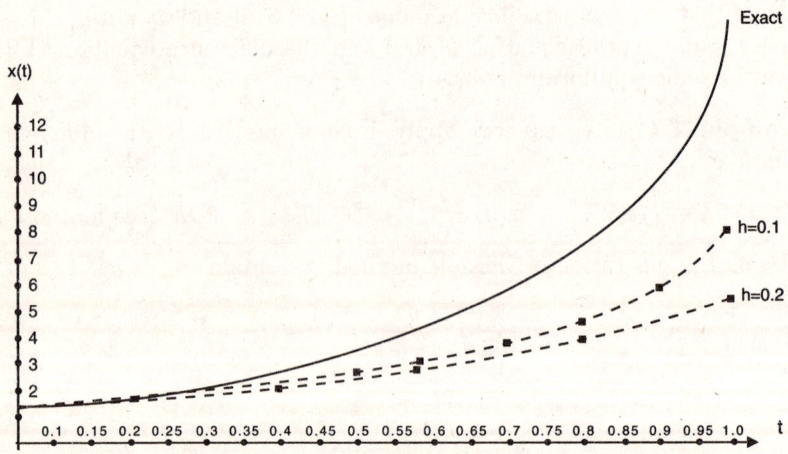

FIGURE 1.12. The $(n, x(n))$ diagram.

Example 1.12. Consider the logistic differential equation

$$x'(t) = ax(t)(1 - x(t)), \quad x(0) = x_0.$$

The equilibrium points (or constant solutions) are obtained by letting $x'(t) = 0$. Hence $ax(1 - x) = 0$ and we then have two equilibrium points $x_1^* = 0$ and $x_2^* = 1$. The exact solution of the equation is obtained by separation of variables,

$$\frac{dx}{x(1-x)} = a\,dt,$$

$$\int \frac{dx}{x} + \int \frac{dx}{1-x} = \int a\,dt,$$

$$\ln\left(\frac{x}{1-x}\right) = at + c,$$

$$\frac{x}{1-x} = e^{at+c} = be^{at}, \quad b = e^c.$$

Hence

$$x(t) = \frac{be^{at}}{1+be^{at}}.$$

Now $x(0) = x_0 = \frac{b}{1+b}$ gives $b = \frac{x_0}{1-x_0}$. Substituting in $x(t)$ yields

$$x(t) = \frac{x_0 e^{at}}{1 - x_0 + x_0 e^{at}} = \frac{x_0 e^{at}}{1 + x_0(e^{at}-1)}.$$

If $a > 0$, $\lim_{t\to\infty} x(t) = 1$, and thus all solutions converge to the equilibrium point $x_2^* = 1$. On the other hand, if $a < 0$, $\lim_{t\to\infty} x(t) = 0$, and thus all solutions converge to the equilibrium point $x_1^* = 0$.

Let us now apply Euler's method to the logistic differential equation. The corresponding difference equation is given by

$$x(n+1) = x(n) + h\,ax(n)(1 - x(n)), \quad x(0) = x_0.$$

This equation has two equilibrium points $x_1^* = 0$, $x_2^* = 1$ as in the differential equation case.

Let $y(n) = \frac{ha}{1+ha}x(n)$. Then we have

$$y(n+1) = (1+ha)y(n)(1-y(n))$$

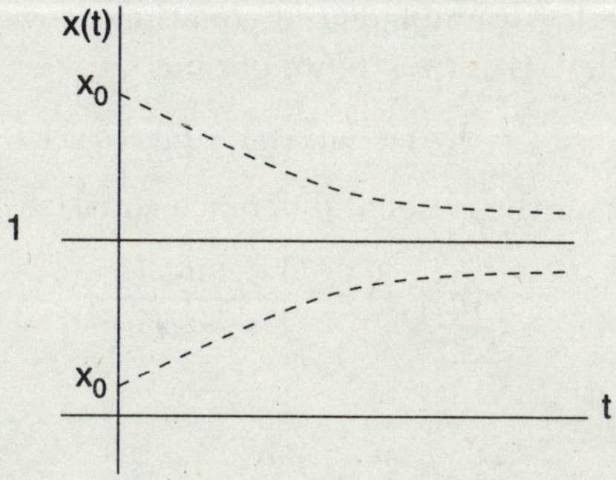

FIGURE 1.13. If $a > 0$, all solutions with $x_0 > 0$ converge to $x_2^* = 1$.

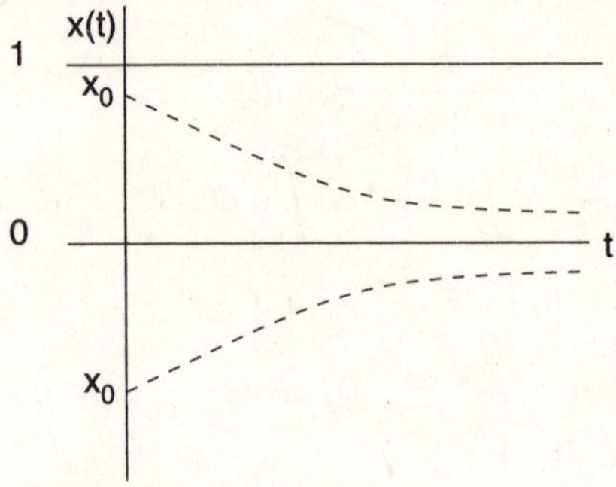

FIGURE 1.14. If $a < 0$, all solutions with $x_0 < 1$ converge to $x_1^* = 0$.

or

$$y(n+1) = \mu y(n)(1-y(n)), \quad y(0) = \frac{ha}{1+ha}x(0), \quad \text{and} \quad \mu = 1+ha.$$

The corresponding equilibrium points are $y_1^* = 0$ and $y_2^* = \frac{\mu-1}{\mu} = \frac{ha}{1+ha}$ which correspond to $x_1^* = 0$ and $x_2^* = 1$, respectively. Using the Cobweb diagram, we observe that for $1 < \mu < 3$ $(0 < ha < 2)$, all solutions whose initial point y_0 in the interval $(0,1)$ converge to the equilibrium point $y_2^* = \frac{ha}{1+ha}$ (Figure 1.15) and for $0 < \mu < 1$ $(-1 < ha < 0)$, all solutions whose initial point y_0 in the interval $(0,1)$ converge to the equilibrium point $y_2^* = 0$ (Figure 1.16). However, for $\mu > 3$ $(ha > 2)$, almost all solutions where initial points are in the interval $(0,1)$ do not converge to either equilibrium point y_1^* or y_2^*. In fact, we will see in later sections that for $\mu > 3.57$ $(ha > 2.57)$, solutions of the difference equation behave in a "chaotic" manner (Figure 1.17). In the next section we will explore another numerical scheme that has been proven effective in a lot of cases [100].

1.4.2 A Nonstandard Scheme

Consider again the logistic differential equation. Now if we replace $x^2(n)$ in Euler's method by $x(n)x(n+1)$ we obtain

$$x(n+1) = x(n) + hax(n) - hax(n)x(n+1).$$

Simplifying we obtain the rational difference equation

$$x(n+1) = \frac{(1+ha)x(n)}{1+hax(n)}$$

or

$$x(n+1) = \frac{\alpha x(n)}{1+\beta x(n)}$$

with $\alpha = 1+ha$, $\beta = \alpha - 1 = ha$.

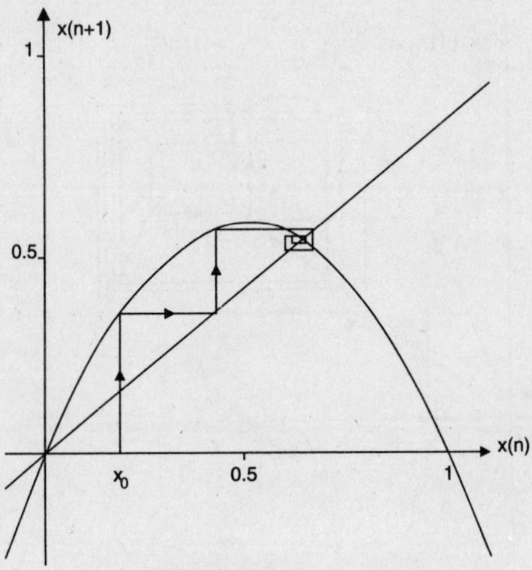

FIGURE 1.15. $0 < ha < 2$.

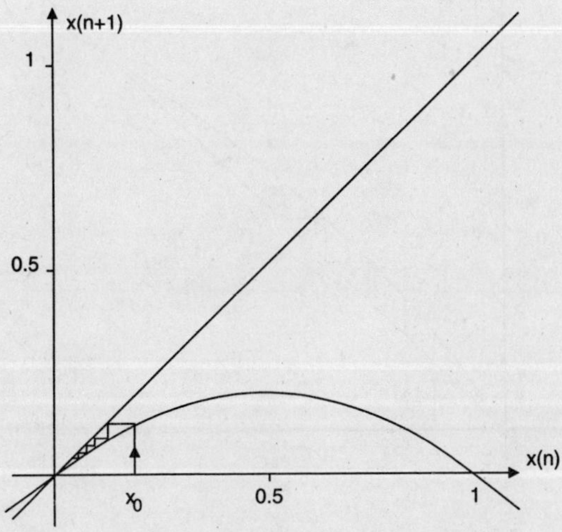

FIGURE 1.16. $-1 < ha < 0$.

This equation has two equilibrium points $x_1^* = 0$ and $x_2^* = 1$. From the Cobweb diagram (Figure 1.18) we conclude that $\lim_{n\to\infty} x(n) = 1$ if $\alpha > 1$.

Since $h > 0$, $\alpha > 1$ if and only if $a > 0$. Thus all solutions converge to the equilibrium point $x_2^* = 1$ if $a > 0$ as in the differential equation case regardless of the size of h.

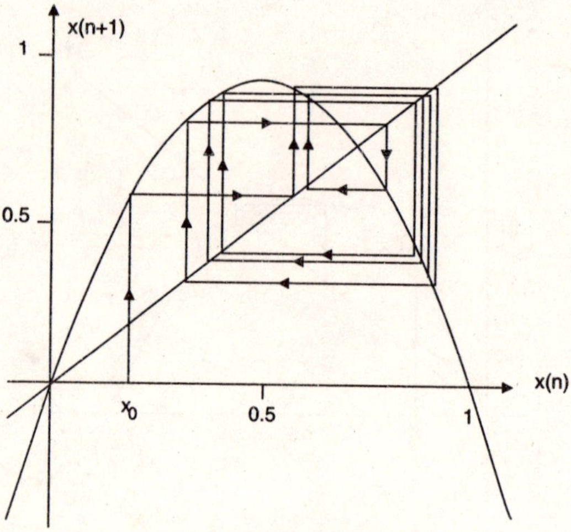

FIGURE 1.17. $ha > 2.57$.

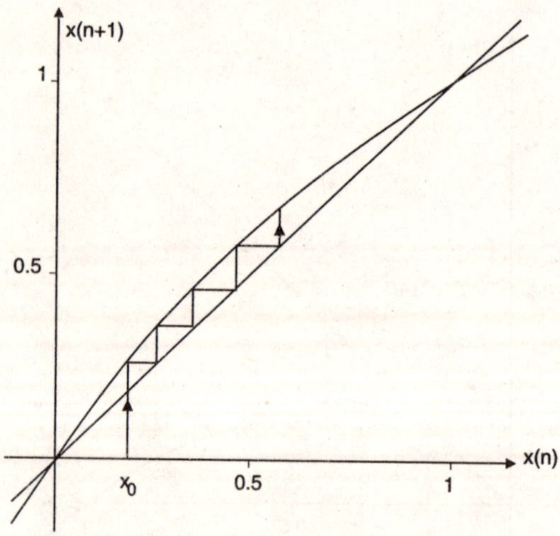

FIGURE 1.18. $\alpha = 1 + ha, \quad \beta = \alpha - 1 = ha$.

Exercises 1.4

In Problems 1–5

(a) Find the associated difference equation.

(b) Draw an $(n, y(n))$ diagram.

(c) Find, if possible, the exact solution of the differential equation and draw its graph on the same plot as that drawn in part in (b).

1. $y' = -y^2$, $\quad y(0) = 1$, $\quad 0 \le t \le 1$, $\quad h = 0.2, 0.1$.
2. $y' = -y + \frac{4}{y}$, $\quad y(0) = 1$, $\quad 0 \le t \le 1$, $\quad h = 0.25$.
3. $y' = -y + 1$, $\quad y(0) = 2$, $\quad 0 \le t \le 1$, $\quad h = 0.25$.
4. $y' = y(1 - y)$, $\quad y(0) = 0.1$, $\quad 0 \le t \le 1$, $\quad h = 0.25$.
5. $y' = y^2 + 2$, $\quad y(0) = \frac{1}{4}$, $\quad 0 \le t \le 1$, $\quad h = 0.25$.
6. Use a nonstandard numerical method to find the associated difference equation of the differential equation in Problem 1.
7. Do Problem 4 using a nonstandard numerical method and compare your results with Euler's method.
8. Do Problem 5 using a nonstandard numerical method and compare your result with Euler's method.
9. Use both Euler's method and a nonstandard method to discretize the differential equation

$$y'(t) = y^2 + t, \qquad y(0) = 1, \qquad 0 \le t \le 1, \qquad h = 0.2.$$

Draw the $n - y(n)$ diagram for both methods. Guess which method gives a better approximation to the differential equation.

10. (a) Use the Euler method with $h = 0.25$ on $[0, 1]$ to find the value of y corresponding to $t = 0.5$ for the differential equation

$$\frac{dy}{dt} = 2t + y, \qquad y(0) = 1.$$

(b) Compare the result obtained in (a) with the exact value.

11. Given the differential equation of Problem 10, show that a better approximation is given by the difference equation

$$y(n+1) = y(n) + \frac{1}{2}h(y'(n) + y'(n+1)).$$

This method is sometimes called the *modified Euler method.*

1.5 Criterion for the Asymptotic Stability of Equilibrium Points

In this section we give a simple but powerful criterion for the asymptotic stability of equilibrium points. The following theorem is our main tool in this section.

Theorem 1.13. *Let x^* be an equilibrium point of the difference equation*

$$x(n+1) = f(x(n)), \tag{1.5.1}$$

where f is continuously differentiable at x^*. The following statements then hold true:

(i) *If $|f'(x^*)| < 1$, then x^* is asymptotically stable.*

(ii) *If $|f'(x^*)| > 1$, then x^* is unstable.*

PROOF.

(i) Suppose that $|f'(x^*)| < M < 1$. Then there is an interval $J = (x^*-\gamma, x^*+\gamma)$ containing x^* such that $|f'(x)| \leq M < 1$ for all $x \in J$. For if not, then for each open interval $I_n = (x^* - \frac{1}{n}, x^* + \frac{1}{n})$ (for large n) there is a point $x_n \in I_n$ such that $|f'(x_n)| > M$. As $n \to \infty$, $x_n \to x^*$. Since f' is a continuous function, it follows that

$$\lim_{n\to\infty} f'(x_n) = f'(x^*).$$

Consequently,

$$M \leq \lim_{n\to\infty} |f'(x_n)| = |f'(x^*)| < M,$$

which is a contradiction. This proves our statement. For $x(0) \in J$, we have

$$|x(1) - x^*| = |f(x(0)) - f(x^*)|.$$

By the Mean Value Theorem, there exists ξ between $x(0)$ and x^* such that

$$|f(x(0)) - f(x^*)| = |f'(\xi)|\,|x(0) - x^*|.$$

Thus

$$|f(x(0)) - x^*| \leq M|x(0) - x^*|.$$

Hence

$$|x(1) - x^*| \leq M|x(0) - x^*|. \tag{1.5.2}$$

Since $M < 1$, inequality (1.5.2) shows that $x(1)$ is closer to x^* than $x(0)$. Consequently, $x(1) \in J$.

By induction we conclude that

$$|x(n) - x^*| \leq M^n|x(0) - x^*|.$$

For $\varepsilon > 0$ we let $\delta = \frac{\varepsilon}{2M}$. Thus $|x(0) - x^*| < \delta$ implies that $|x(n) - x^*| < \varepsilon$ for all $n > 0$. This conclusion suggests stability. Furthermore, $\lim_{n\to\infty} |x(n) - x^*| = 0$, and thus $\lim_{n\to\infty} x(n) = x^*$; we conclude asymptotic stability. □

The proof of part (ii) is left as Exercises 1.5, Problem 11.

Remark: In the literature of dynamical systems, the equilibrium point x^* is said to be *hyperbolic* if $|f'(x^*)| \neq 1$.

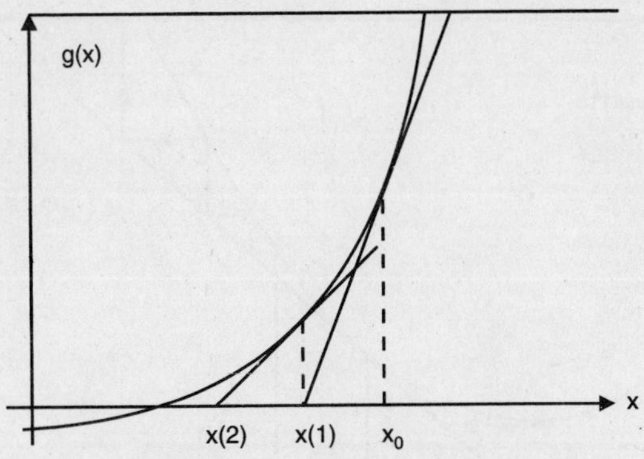

FIGURE 1.19. Newton's method.

Example 1.14. The Newton–Raphson Method

The Newton–Raphson method is one of the most famous numerical methods for finding the roots of the equation $g(x) = 0$, where $g(x)$ is continually differentiable (i.e., its derivative exists and is continuous).

Newton's algorithm for finding a zero x^* of $g(x)$ is given by the difference equation

$$\boxed{x(n+1) = x(n) - \frac{g(x(n))}{g'(x(n))},} \tag{1.5.3}$$

where $x(0) = x_0$ is your initial guess of the root x^*. Here $f(x) = x - \frac{g(x)}{g'(x)}$.

Note first that the zero x^* of $g(x)$ is also an equilibrium point of (1.5.3). To determine whether Newton's algorithm provides a sequence $\{x(n)\}$ that converges to x^* we use Theorem 1.13:

$$|f'(x^*)| = \left|1 - \frac{[g'(x^*)]^2 - g(x^*)g''(x^*)}{[g'(x^*)]^2}\right| = 0,$$

since $g(x^*) = 0$. By Theorem 1.13, $\lim_{n\to\infty} x(n) = x^*$ if $x(0) = x_0$ is close enough to x^* and $g'(x^*) \neq 0$.

Observe that Theorem 1.13 does not address the nonhyperbolic case where $|f'(x^*)| = 1$. Further analysis is needed here to determine the stability of the equilibrium point x^*. Our first discussion will address the case where $f'(x^*) = 1$.

Theorem 1.15. *Suppose that for an equilibrium point x^* of (1.5.1), $f'(x^*) = 1$. The following statements then hold:*

(i) *If $f''(x^*) \neq 0$, then x^* is unstable.*

(ii) *If $f''(x^*) = 0$ and $f'''(x^*) > 0$, then x^* is unstable.*

(iii) *If $f''(x^*) = 0$ and $f'''(x^*) < 0$, then x^* is asymptotically stable.*

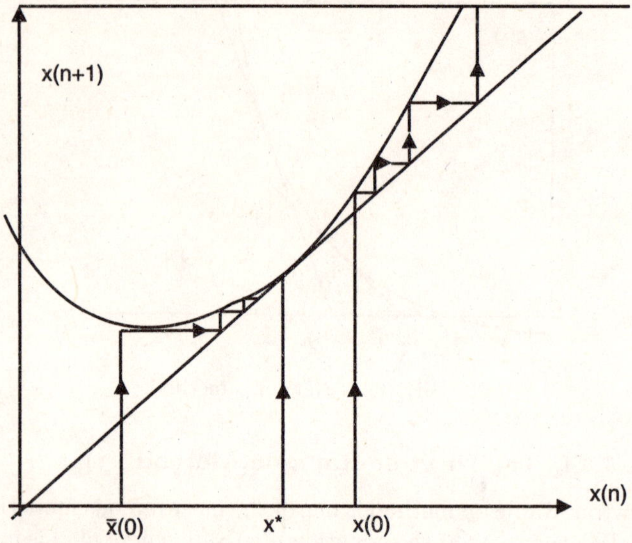

FIGURE 1.20. Unstable. $f''(x^*) > 0$ (semistable from the left).

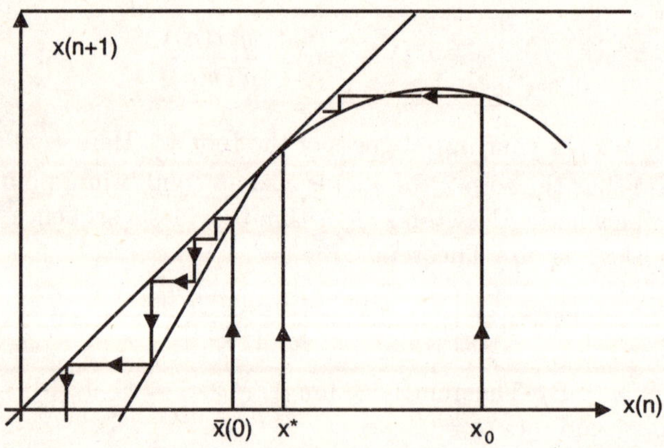

FIGURE 1.21. Unstable. $f''(x^*) < 0$ (semistable from the right).

PROOF.

(i) If $f''(x^*) \neq 0$, then the curve $y = f(x)$ is either concave upward if $f''(x^*) > 0$ or concave downward if $f''(x^*) < 0$, as shown in Figures 1.20, 1.21, 1.22, 1.23. If $f''(x^*) > 0$, then $f'(x) > 1$ for all x in a small interval $I = (x^*, x^*+\varepsilon)$. Using the same proof as in Theorem 1.13, it is easy to show that x^* is unstable. On the other hand, if $f''(x^*) < 0$, then $f'(x) > 1$ for all x in a small interval $I = (x^* - \varepsilon, x^*)$. Hence x^* is again unstable. □

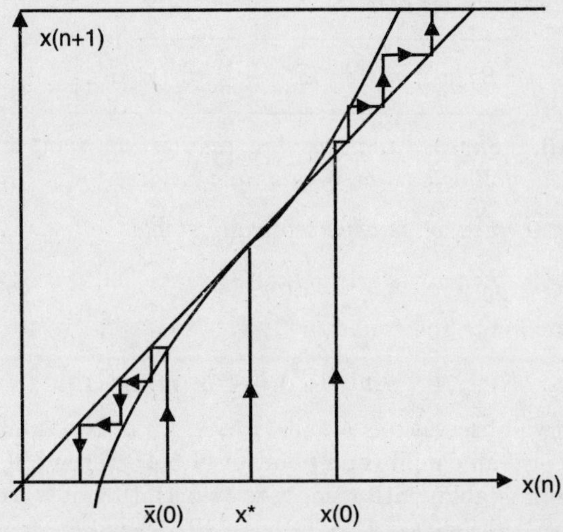

FIGURE 1.22. Unstable. $f'(x^*) = 1, f''(x^*) = 0$, and $f'''(x^*) > 0$.

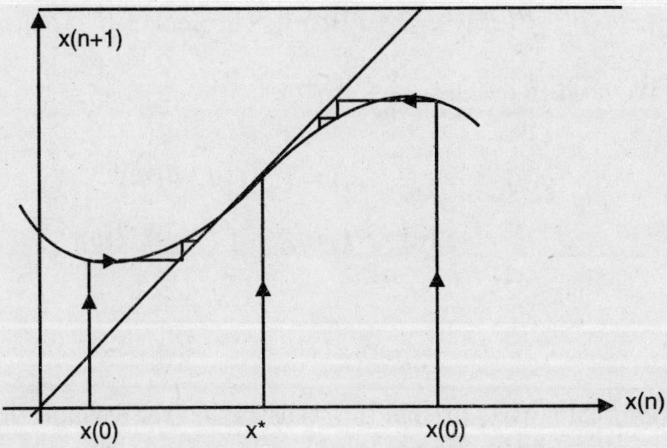

FIGURE 1.23. Asymptotically stable. $f'(x^*) = 1,\ f''(x^*) = 0$, and $f'''(x^*) < 0$.

Proofs of parts (ii) and (iii) remain for the student's pleasure as Exercises 1.5, Problem 14.

We now use the preceding result to investigate the case $f'(x^*) = -1$.

But before doing so, we need to introduce the notion of the Schwarzian derivative of a function f:

$$\boxed{Sf(x) = \frac{f'''(x)}{f'(x)} - \frac{3}{2}\left[\frac{f''(x)}{f'(x)}\right]^2}$$

Note that if $f'(x^*) = -1$, then

$$\boxed{Sf(x^*) = -f'''(x^*) - \frac{3}{2}\left(f''(x^*)\right)^2.}$$

Theorem 1.16. *Suppose that for the equilibrium point x^* of (1.1.1), $f'(x^*) = -1$. The following statements then hold:*

(i) *If $Sf(x^*) < 0$, then x^* is asymptotically stable.*

(ii) *If $Sf(x^*) > 0$, then x^* is unstable.*

PROOF. Contemplate the equation

$$y(n+1) = g(y(n)), \text{ where } g(y) = f^2(y). \tag{1.5.4}$$

We will make two observations about (1.5.4). First, the equilibrium point x^* of (1.1.1) is also an equilibrium point of (1.5.4). Second, if x^* is asymptotically stable (unstable) with respect to (1.5.4), then it is so with respect to (1.1.1). (Why?) (Exercises 1.5, Problem 12.) Now,

$$\frac{d}{dy}g(y) = \frac{d}{dy}f(f(y)) = f'(f(y))f'(y).$$

Thus $\frac{d}{dy}g(x^*) = [f'(x^*)]^2 = 1$. Hence Theorem 1.15 applies to this situation. We need to evaluate $\frac{d^2}{dy^2}g(x^*)$:

$$\begin{aligned}\frac{d^2}{dy^2}g(y) &= \frac{d^2}{dy^2}f(f(y)) = [f'(f(y))f'(y)]' \\ &= [f'(y)]^2 f''(f(y)) + f'(f(y))f''(y).\end{aligned}$$

Hence

$$\frac{d^2}{dy^2}g(x^*) = 0.$$

Now, Theorem 1.15 [parts (ii) and (iii)] tells us that the asymptotic stability of x^* is determined by the sign of $[g(x^*)]'''$. Using the chain rule again, one may show that

$$[g(x^*)]''' = -2f'''(x^*) - 3[f''(x^*)]^2. \tag{1.5.5}$$

(The explicit proof with the chain rule remains as Exercises 1.5, Problem 13.) This step rewards us with parts (i) and (ii), and the proof of the theorem is now complete. □

Example 1.17. Consider the difference equation $x(n+1) = x^2(n)+3x(n)$. Find the equilibrium points and determine their stability.

Solution The equilibrium points are 0 and -2. Now, $f'(x) = 2x+3$. Since $f'(0) = 3$, it follows from Theorem 1.13 that 0 is unstable. Now, $f'(-2) = -1$, so Theorem 1.16 applies. Using (1.5.5) we obtain $-2f'''(-2)-$

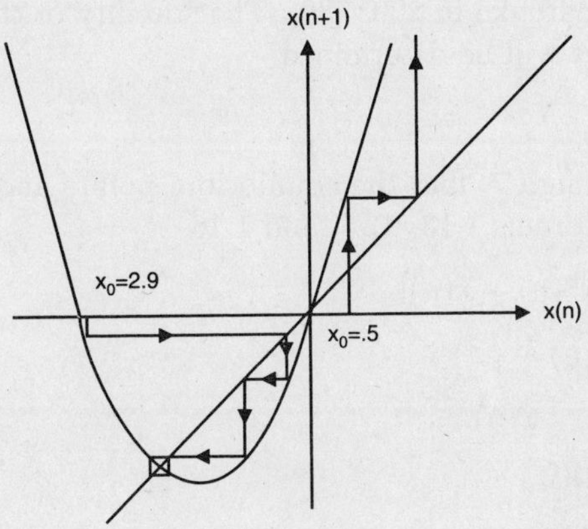

FIGURE 1.24. Stair step diagram for $x(n+1) = x^2(n) + 3x(n)$.

$3[f''(-2)]^2 = -12 < 0$. Theorem 1.16 then declares that the equilibrium point -2 is asymptotically stable. Figure 1.24 illustrates the stair step diagram of the equation.

Remark: One may generalize the result in the preceding example to a general quadratic map $Q(x) = ax^2 + bx + c, a \neq 0$. Let x^* be an equilibrium point of $Q(x)$, i.e., $Q(x^*) = x^*$. Then the following statements hold true.

(i) If $Q'(x^*) = -1$, then by Theorem 1.16, the equilibrium point x^* is asymptotically stable. In fact, there are two equilibrium points for $Q(x)$;

$$x_1^* = [(1-b) - \sqrt{(b-1)^2 - 4ac}]/2a;$$
$$x_2^* = [(1-b) + \sqrt{(b-1)^2 - 4ac}]/2a.$$

It is easy to see that $Q'(x_1^*) = -1$, if $(b-1)^2 = 4ac+4$ and $Q'(x_2^*) \neq -1$. Thus x_1^* is asymptotically stable if $(b-1)^2 = 4ac+4$ (Exercises 1.5, Problem 8).

(ii) If $Q'(x^*) = 1$, then by Theorem 1.15, x^* is unstable. In this case, we have only one equilibrium point $x^* = (1-b)/2a$. Thus, x^* is unstable if $(b-1)^2 = 4ac$.

Remark:

(i) Theorem 1.15 fails if for a fixed point x^*, $f'(x^*) = 1$, $f''(x^*) = f'''(x^*) = 0$. For example, for the map $f(x) = x + (x-1)^4$ and its fixed point $x^* = 1$, $f'(x^*) = 1$, $f''(x^*) = f'''(x^*) = 0$, and $f^{(4)}(x^*) = 24 > 0$.

(ii) Theorem 1.16 fails if $f'(x^*) = -1$, and $Sf(x^*) = 0$. This may be illustrated by the function $f(x) = -x + 2x^2 - 4x^3$. For the fixed $x^* = 0$, $f'(x^*) = -1$, and $Sf(x^*) = 0$.

In Appendix A, we present the general theory developed by Dannan, Elaydi, and Ponomarenko in 2003 [30]. The stability of the fixed points in the above examples will be determined.

Exercises 1.5

In Problems 1 through 7, find the equilibrium points and determine their stability using Theorems 1.13, 1.15, and 1.16.

1. $x(n+1) = \frac{1}{2}[x^3(n) + x(n)]$.
2. $x(n+1) = x^2(n) + \frac{1}{8}$.
3. $x(n+1) = \tan^{-1} x(n)$.
4. $x(n+1) = x^2(n)$.
5. $x(n+1) = x^3(n) + x(n)$.
6. $x(n+1) = \dfrac{\alpha x(n)}{1+\beta x(n)}, \qquad \alpha > 1$ and $\beta > 0$.
7. $x(n+1) = -x^3(n) - x(n)$.
8. Let $Q(x) = ax^2 + bx + c$, $a \neq 0$, and let x^* be a fixed point of Q. Prove the following statements:

   (i) If $Q'(x^*) = -1$, then x^* is asymptotically stable. Then prove the rest of Remark (i).

   (ii) If $Q'(x^*) = 1$, then x^* is unstable. Then prove the rest of Remark (ii).
9. Suppose that in (1.5.3), $g(x^*) = g'(x^*) = 0$ and $g''(x^*) \neq 0$. Prove that x^* is an equilibrium point of (1.5.3).
10. Prove Theorem 1.13, part (ii).
11. Prove that if x^* is an equilibrium point of (1.5.1), then it is an equilibrium point of (1.5.1). Show also that the converse is false in general. For what class of maps $f(x)$ does the converse hold?
12. Prove that if an equilibrium point x^* of (1.5.1) is asymptotically stable with respect to (1.5.4) (or unstable, as the case may be), it is also so with respect to (1.1.1).
13. Verify formula (1.5.5).
14. Prove Theorem 1.15, parts (ii) and (iii).
15. Definition of *Semistability*. An equilibrium point x^* of $x(n+1) = f(x(n))$ is semistable (from the right) if given $\varepsilon > 0$ there exists $\delta > 0$ such that if $x(0) > x^*$, $x(0) - x^* < \delta$, then $x(n) - x^* < \varepsilon$. Semistability from the left is defined similarly. If in addition, $\lim_{n\to\infty} x(n) = x^*$

whenever $x(0) - x^* < \eta\{x^* - x(0) < \eta\}$, then x^* is said to be semi-asymptotically stable from the right {or from the left, whatever the case may be}.
Suppose that if $f'(x^*) = 1$, then $f''(x^*) \neq 0$. Prove that x^* is:

(i) semiasymptotically stable from the right from the right if $f''(x^*) < 0$;

(ii) semiasymptotically stable from the left from the left if $f''(x^*) > 0$.

16. Determine whether the equilibrium point $x^* = 0$ is semiasymptotically stable from the left or from the right.

(a) $x(n+1) = x^3(n) + x^2(n) + x(n)$.

(b) $x(n+1) = x^3(n) - x^2(n) + x(n)$.

1.6 Periodic Points and Cycles

The second most important notion in the study of dynamical systems is the notion of periodicity. For example, the motion of a pendulum is periodic. We have seen in Example 1.10 that if the sensitivity m_s of the suppliers to price is equal to the sensitivity of consumers to price, then prices oscillate between two values only.

Definition 1.18. Let b be in the domain of f. Then:

(i) b is called a *periodic* point of f (or of (1.5.1)) if for some positive integer k, $f^k(b) = b$. Hence a point is *k-periodic* if it is a fixed point of f^k, that is, if it is an equilibrium point of the difference equation

$$x(n+1) = g(x(n)), \tag{1.6.1}$$

where $g = f^k$.

The periodic orbit of b, $O(b) = \{b, f(b), f^2(b), \ldots, f^{k-1}(b)\}$, is often called a k-cycle.

(ii) b is called *eventually k-periodic* if for some positive integer m, $f^m(b)$ is a *k-periodic* point. In other words, b is eventually k-periodic if

$$f^{m+k}(b) = f^m(b).$$

Graphically, a k-periodic point is the x-coordinate of the point where the graph of f^k meets the diagonal line $y = x$. Figure 1.25 depicts the graph of f^2, where f is the logistic map, which shows that there are four fixed points of f^2, of which two are fixed points of f as shown in Figure 1.26. Hence the other two fixed points of f^2 form a 2-cycle. Notice also that the point $x_0 = 0.3$ (in Figure 1.26) goes into a 2-cycle, and thus it is an eventually

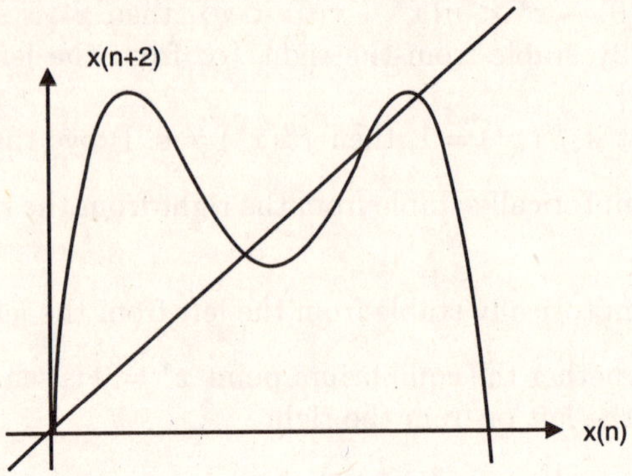

FIGURE 1.25. Graph of f^2 with four fixed points. $f(x) = 3.43x(1-x)$.

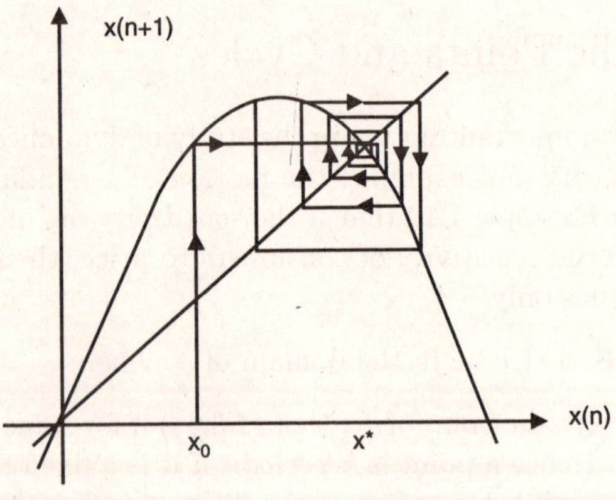

FIGURE 1.26. x_0 goes into a 2-cycle. $f(x) = 3.43x(1-x)$.

2-periodic point. Moreover, the point $x^* = 0.445$ is asymptotically stable relative to f^2 (Figure 1.27).

Observe also that if $A = -1$ in (1.3.7), then $f^2(p_0) = -(-p_0 + B) + B = p_0$. Therefore, every point is 2-periodic (see Figure 1.10). This means that in this case, if the initial price per unit of a certain commodity is p_0, then the price oscillates between p_0 and $B - p_0$.

Example 1.19. Consider again the difference equation generated by the tent function

$$T(x) = \begin{cases} 2x & \text{for } 0 \le x \le \dfrac{1}{2}, \\ 2(1-x) & \text{for } \dfrac{1}{2} < x \le 1 . \end{cases}$$

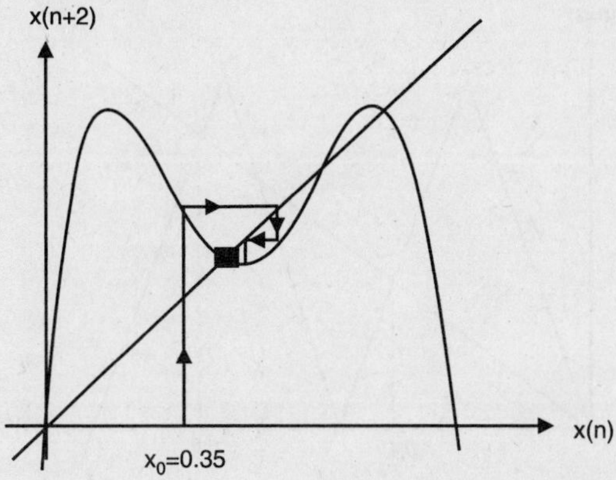

FIGURE 1.27. $x^* \approx 0.445$ is asymptotically stable relative to f^2.

This may also be written in the compact form

$$T(x) = 1 - 2\left|x - \frac{1}{2}\right|.$$

We first observe that the periodic points of period 2 are the fixed points of T^2. It is easy to verify that T^2 is given by

$$T^2(x) = \begin{cases} 4x & \text{for } 0 \le x < \frac{1}{4}, \\ 2(1-2x) & \text{for } \frac{1}{4} \le x < \frac{1}{2}, \\ 4\left(x - \frac{1}{2}\right) & \text{for } \frac{1}{2} \le x < \frac{3}{4}, \\ 4(1-x) & \text{for } \frac{3}{4} \le x \le 1. \end{cases}$$

There are four equilibrium points (Figure 1.28): 0, 0.4, $\frac{2}{3}$, and 0.8, two of which, 0 and $\frac{2}{3}$, are equilibrium points of T. Hence $\{0.4, 0.8\}$ is the only 2-cycle of T. Notice from Figure 1.29 that $x^* = 0.8$ is not stable relative to T^2.

Figure 1.30 depicts the graph of T^3. It is easy to verify that $\left\{\frac{2}{7}, \frac{4}{7}, \frac{6}{7}\right\}$ is a 3-cycle. Now,

$$T\left(\frac{2}{7}\right) = \frac{4}{7}, \quad T\left(\frac{4}{7}\right) = \frac{6}{7}, \quad T\left(\frac{6}{7}\right) = \frac{2}{7}.$$

Using a computer or hand-held calculator, one may show (using the stair step diagram) that the tent map T has periodic points of all periods. This is a phenomenon shared by all equations that possess a 3-cycle. It was discovered by Li and Yorke [92] in their celebrated paper "Period Three Implies Chaos."

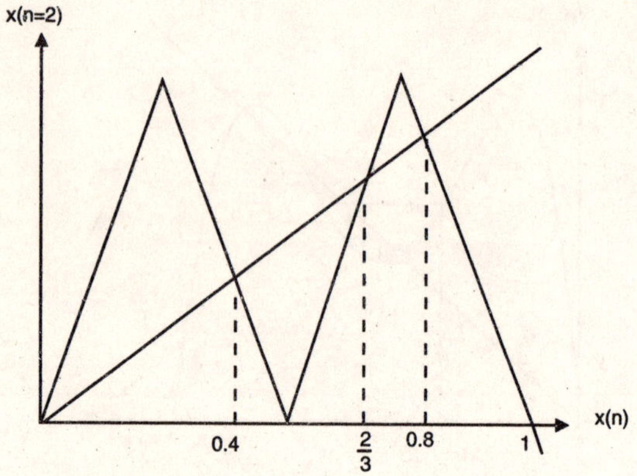

FIGURE 1.28. Fixed points of T^2.

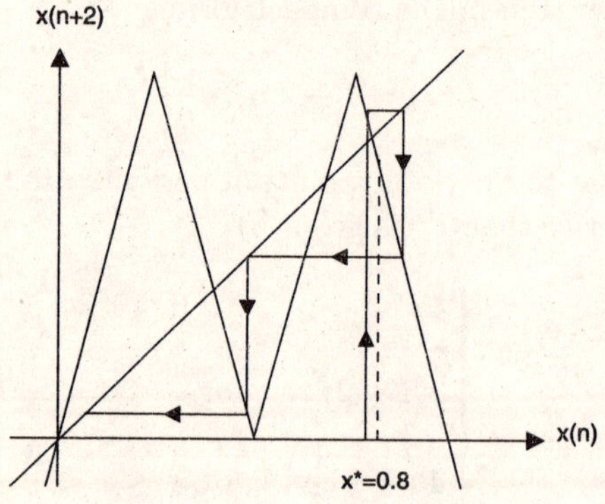

FIGURE 1.29. $x^* = 0.8$ is unstable relative to T^2.

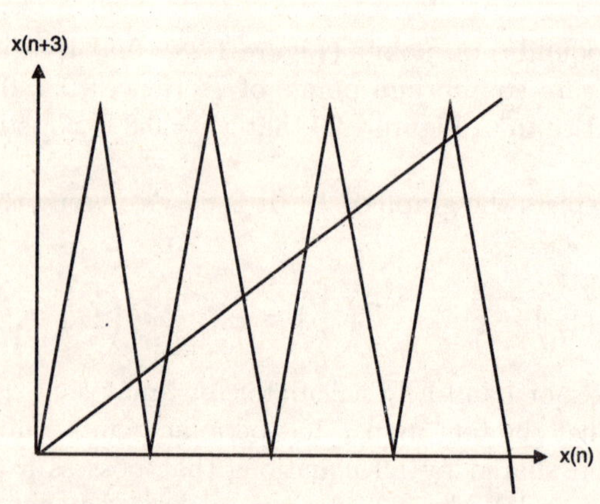

FIGURE 1.30. Fixed points of T^3.

We now turn our attention to explore the stability of periodic points.

Definition 1.20. Let b be a k-period point of f. Then b is:

(i) stable if it is a stable fixed point of f^k,

(ii) asymptotically stable if it is an asymptotically stable fixed point of f^k,

(iii) unstable if it is an unstable fixed point of f^k.

Notice that if b possesses a stability property, then so does every point in its k-cycle $\{x(0) = b, x(1) = f(b), x(2) = f^2(b), \ldots, x(k-1) = f^{k-1}(b)\}$. Hence we often speak of the stability of a k-cycle or a periodic orbit. Figure 1.29 shows that the 2-cycle in the tent map is not stable, since $x^* = 0.8$ is not stable as a fixed point of T^2, while the 2-cycle in the logistic map is asymptotically stable (see Figure 1.27).

Since the stability of a k-periodic point b of (1.1.1) reduces to the study of the stability of the point as an equilibrium point of (1.6.1), one can use all the theorems in the previous section applied to f^k. For example, Theorem 1.13 may be modified as follows.

Theorem 1.21. *Let $O(b) = \{b = x(0), x(1), \ldots, x(k-1)\}$ be a k-cycle of a continuously differentiable function f. Then the following statements hold:*

(i) *The k-cycle $O(b)$ is asymptotically stable if*

$$|f'(x(0))f'(x(1)), \ldots, f'(x(k-1))| < 1.$$

(ii) *The k-cycle $O(b)$ is unstable if*

$$|f'(x(0))f'(x(1)), \ldots, f'(x(k-1))| > 1.$$

PROOF. We apply Theorem 1.13 to (1.6.1). Notice that by using the chain rule one may show that

$$[f^k(x(r))]' = f'(x(0))f'(x(1)), \ldots, f'(x(k-1)).$$

(See Exercises 1.6, Problem 12.) □

The conclusion of the theorem now follows.

Example 1.22. Consider the map $Q(x) = x^2 - 0.85$ defined on the interval $[-2, 2]$. Find the 2-cycles and determine their stability.
Solution Observe that $Q^2(x) = (x^2 - 0.85)^2 - 0.85$. The 2-periodic points are obtained by solving the equation

$$Q^2(x) = x, \text{ or } x^4 - 1.7x^2 - x - 0.1275 = 0. \tag{1.6.2}$$

This equation has four roots, two of which are fixed points of the map $Q(x)$. These two fixed points are the roots of the equation

$$x^2 - x - 0.85 = 0. \tag{1.6.3}$$

To eliminate these fixed points of $Q(x)$ from (1.6.2) we divide the left-hand side of (1.6.2) by the left-hand side of (1.6.3) to obtain the second-degree equation

$$x^2 + x + 0.15 = 0. \tag{1.6.4}$$

The 2-periodic points are now obtained by solving (1.6.4). They are given by

$$a = \frac{-1+\sqrt{0.4}}{2}, \qquad b = \frac{-1-\sqrt{0.4}}{2}.$$

To check the stability of the cycle $\{a, b\}$ we apply Theorem 1.21. Now,

$$|Q'(a)Q'(b)| = |(-1+\sqrt{0.4})(-1-\sqrt{0.4})| = 0.6 < 1.$$

Hence by Theorem 1.21, part (i), the 2-cycle is asymptotically stable.

Exercises 1.6

1. Suppose that the difference equation $x(n+1) = f(x(n))$ has a 2-cycle whose orbit is $\{a, b\}$. Prove that:

   (i) the 2-cycle is asymptotically stable if $|f'(a)f'(b)| < 1$,

   (ii) the 2-cycle is unstable if $|f'(a)f'(b)| > 1$.

2. Let T be the tent map in Example 1.17. Show that $\left\{\frac{2}{9}, \frac{4}{9}, \frac{8}{9}\right\}$ is an unstable 3-cycle for T.

3. Let $f(x) = -\frac{1}{2}x^2 - x + \frac{1}{2}$. Show that 1 is an asymptotically stable 2-periodic point of f.

In Problems 4 through 6 find the 2-cycle and then determine its stability.

4. $x(n+1) = 3.5x(n)[1 - x(n)]$.

5. $x(n+1) = 1 - x^2$.

6. $x(n+1) = 5 - (6/x(n))$.

7. Let $f(x) = ax^3 - bx + 1$, where $a, b \in \mathbb{R}$. Find the values of a and b for which $\{0, 1\}$ is an attracting 2-cycle.

Consider Baker's function defined as follows:

$$B(x) = \begin{cases} 2x & \text{for } 0 \le x \le \dfrac{1}{2}, \\[2mm] 2x - 1 & \text{for } \dfrac{1}{2} < x \le 1. \end{cases}$$

Problems 8, 9, and 10 are concerned with Baker's function $B(x)$ on $[0, 1]$.

*8. (Hard). Draw Baker's function $B(x)$. Then find the number of n-periodic points of B.

9. Sketch the graph of B^2 and then find the 2-cycles of Baker's function B.

10. (Hard). Show that if m is an odd positive integer, then $\overline{x} = k/m$ is periodic, for $k = 1, 2, \ldots, m-1$.

11. Consider the quadratic map

$$Q(x) = ax^2 + bx + c, \quad a \neq 0.$$

   (a) If $\{d, e\}$ is a 2-cycle such that $Q'(d)Q'(e) = -1$, prove that it is asymptotically stable.

   (b) If $\{d, e\}$ is a 2-cycle with $Q'(d)Q'(e) = 1$, what can you say about the stability of the cycle?

12. (This exercise generalizes the result in Problem 1.) Let $\{x(0), x(1), \ldots, x(k-1)\}$ be a k-cycle of (1.2.1). Prove that:

   (i) if $|f'(x(0))f'(x(1)), \ldots, f'(x(k-1))| < 1$, then the k-cycle is asymptotically stable,

   (ii) if $|f'(x(0))f'(x(1)), \ldots, f'(x(k-1))| > 1$, then the k-cycle is unstable.

13. Give an example of a decreasing function that has a fixed point and a 2-cycle.

14. (i) Can a decreasing map have a k-cycle for $k > 1$?

   (ii) Can an increasing map have a k-cycle for $k > 1$?

Carvalho's Lemma. In [18] Carvalho gave a method to find periodic points of a given function. The method is based on the following lemma.

Lemma 1.23. *If k is a positive integer and $x(n)$ is a periodic sequence of period k, then the following hold true:*

(i) *If $k > 1$ is odd and $m = \frac{k-1}{2}$, then*

$$\boxed{x(n) = c_0 + \sum_{j=1}^{m} \left[c_j \cos\left(\frac{2jn\pi}{k}\right) + d_j \sin\left(\frac{2jn\pi}{k}\right) \right],}$$

for all $n \geq 1$.

(ii) *If k is even and $k = 2m$, then*

$$\boxed{x(n) = c_0 + (-1)^n c_m + \sum_{j=1}^{m-1} \left[c_j \cos\left(\frac{2jn\pi}{k}\right) + d_j \sin\left(\frac{2jn\pi}{k}\right) \right],}$$

for all $n \geq 1$.

Example 1.24 [23]. Consider the equation

$$x(n+1) = x(n)\exp(r(1-x(n))), \tag{1.6.5}$$

which describes a population with a propensity to simple exponential growth at low densities and a tendency to decrease at high densities. The quantity $\lambda = \exp(r(1-x(n)))$ could be considered the density-dependent reproductive rate of the population. This model is plausible for a single-species population that is regulated by an epidemic disease at high density.

The nontrivial fixed point of this equation is given by $x^* = 1$. Now, $f'(1) = 1 - r$. Hence $x^* = 1$ is asymptotically stable if $0 < r \le 2$ (check $r = 2$). At $r = 2$, $x^* = 1$ loses its stability and gives rise to an asymptotically stable 2-cycle. Carvalho's lemma implies

$$x(n) = a + (-1)^n b.$$

Plugging this into equation (1.6.5) yields

$$a - (-1)^n b = (a + (-1)^n b)\ \exp r(1 - a - (-1)^n b).$$

The shift $n \mapsto n+1$ gives

$$a + (-1)^n b = (a - (-1)^n b)\ \exp r(1 - a + (-1)^n b).$$

Hence

$$a^2 - b^2 = (a^2 - b^2)\ \exp 2r(1-a).$$

Thus either $a^2 = b^2$, which gives the trivial solution 0, or $a = 1$. Hence a 2-periodic solution has the form $x(n) = 1 + (-1)^n b$. Plugging this again into equation (1.6.5) yields

$$1 - (-1)^n b = (1 + (-1)^n b)\exp((-1)^{n+1} rb).$$

Let $y = (-1)^{n+1} b$.Then

$$1 + y = (1-y)e^{ry},$$

$$r = \frac{1}{y}\ln\left(\frac{1+y}{1-y}\right) = g(y).$$

The function g has its minimum at 0, where $g(0) = 2$. Thus, for $r < 2, g(y) = r$ has no solution, and we have no periodic points, as predicted earlier. However, each $r > 2$ determines values $\pm y_r$ and the corresponding coefficient $(-1)^n b$. Further analysis may show that this map undergoes bifurcation similar to that of the logistic map.

Exercises 1.6 (continued).

In Problems 15 through 20, use Carvalho's lemma (Lemma 1.23).

15. Consider Ricker's equation

$$x(n+1) = x(n)\exp(r(1 - x(n))).$$

Find the 2-period solution when $r > 2$.

16. The population of a certain species is modeled by the difference equation $x(n+1) = \mu x(n)e^{-x(n)}$, $x(n) \geq 0$, $\mu > 0$. For what values of μ does the equation have a 2-cycle?

17. Use Carvalho's lemma to find the values of c for which the map

$$Q_c(x) = x^2 + c, \quad c \in [-2, 0],$$

has a 3-cycle and then determine its stability.

18*. (Term Project). Find the values of μ where the logistic equation $x(n+1) = \mu x(n)[1 - x(n)]$ has a 3-periodic solution.

19. Use Carvalho's lemma to find the values of μ where the logistic equation $x(n+1) = \mu x(n)[1 - x(n)]$ has a 2-periodic solution.

20. Find the 3-periodic solutions of the equation $x(n+1) = ax(n)$, $a \neq 1$.

1.7 The Logistic Equation and Bifurcation

Let us now return to the most important example in this chapter: the logistic difference equation

$$x(n+1) = \mu x(n)[1 - x(n)], \tag{1.7.1}$$

which arises from iterating the function

$$F_\mu(x) = \mu x(1 - x), \qquad x \in [0, 1], \qquad \mu > 0. \tag{1.7.2}$$

1.7.1 *Equilibrium Points*

To find the equilibrium points (fixed points of F_μ) of (1.7.1) we solve the equation

$$F_\mu(x^*) = x^*.$$

Hence the fixed points are $0, x^* = (\mu - 1)/\mu$. Next we investigate the stability of each equilibrium point separately.

(a) The equilibrium point 0. (See Figures 1.31, 1.32.) Since $F'_\mu(0) = \mu$, it follows from Theorems 1.13 and 1.15 that:

  (i) 0 is an asymptotically stable fixed point for $0 < \mu < 1$,

  (ii) 0 is an unstable fixed point for $\mu > 1$.

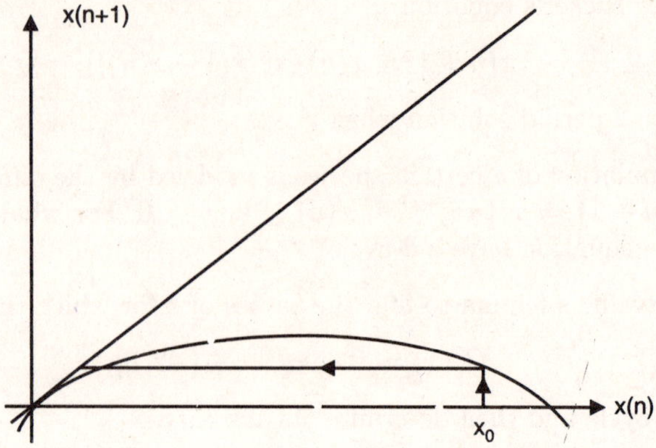

FIGURE 1.31. $0 < \mu < 1$: 0 is an asymptotically stable fixed point.

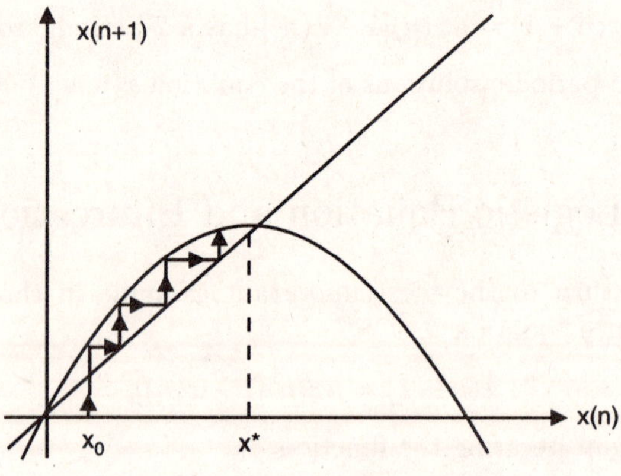

FIGURE 1.32. $\mu > 1$: 0 is an unstable fixed point, x^* is an asymptotically fixed point.

The case where $\mu = 1$ needs special attention, for we have $F_1'(0) = 1$ and $F''(0) = -2 \neq 0$. By applying Theorem 1.15 we may conclude that 0 is unstable. This is certainly true if we consider negative as well as positive initial points in the neighborhood of 0. Since negative initial points are not in the domain of F_μ, we may discard them and consider only positive initial points. Exercises 1.5, Problem 16 tells us that 0 is semiasymptotically stable from the right, i.e., $x^* = 0$ is asymptotically stable in the domain $[0, 1]$.

(b) The equilibrium point $x^* = (\mu - 1)/\mu, \mu \neq 1$. (See Figures 1.32, 1.33.)

In order to have $x^* \in (0, 1]$ we require that $\mu > 1$. Now, $F_\mu'((\mu-1)/\mu) = 2-\mu$. Thus using Theorems 1.13 and 1.16 we obtain the following conclusions:

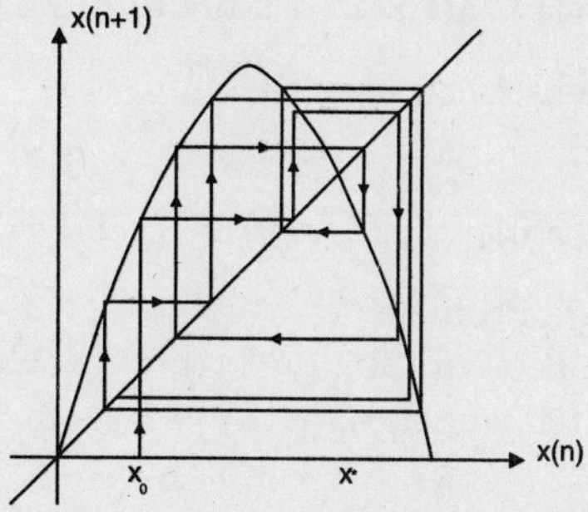

FIGURE 1.33. $\mu > 3$: x^* is an unstable fixed point.

(i) x^* is an asymptotically stable fixed point for $1 < \mu \leq 3$ (Figure 1.32).

(ii) x^* is an unstable fixed point for $\mu > 3$ (Figure 1.33).

1.7.2 2-Cycles

To find the 2-cycles we solve the equation $F_\mu^2(x) = x$ (or we solve $x_2 = \mu x_1(1 - x_1), x_1 = \mu x_2(1 - x_2)$),

$$\mu^2 x(1 - x)[1 - \mu x(1 - x)] - x = 0. \tag{1.7.3}$$

Discarding the equilibrium points 0 and $x^* = \frac{\mu-1}{\mu}$, one may then divide (1.7.3) by the factor $x(x - (\mu - 1)/\mu)$ to obtain the quadratic equation

$$\mu^2 x^2 - \mu(\mu + 1)x + \mu + 1 = 0.$$

Solving this equation produces the 2-cycle

$$x(0) = \left[(1 + \mu) - \sqrt{(\mu - 3)(\mu + 1)}\right] / 2\mu,$$

$$x(1) = \left[(1 + \mu) + \sqrt{(\mu - 3)(\mu + 1)}\right] / 2\mu. \tag{1.7.4}$$

Clearly, there are no periodic points of period 2 for $0 < \mu \leq 3$, and there is a 2-cycle for $\mu > 3$. For our reference we let $\mu_0 = 3$.

1.7.2.1 Stability of the 2-Cycle $\{x(0), x(1)\}$ for $\mu > 3$

From Theorem 1.21, this 2-cycle is asymptotically stable if

$$|F'_\mu(x(0))F'_\mu(x(1))| < 1,$$

or

$$-1 < \mu^2(1 - 2x(0))(1 - 2x(1)) < 1. \tag{1.7.5}$$

Substituting from (1.7.4) the values of $x(0)$ and $x(1)$ into (1.7.5), we obtain

$$3 < \mu < 1 + \sqrt{6} \approx 3.44949.$$

Conclusion This 2-cycle is attracting if $3 < \mu < 3.44949\ldots$.

Question What happens when $\mu = 1 + \sqrt{6}$?

In this case,

$$\left[F_\mu^2(x(0))\right]' = F_\mu'(x(0))F_\mu'(x(1)) = -1. \tag{1.7.6}$$

(Verify in Exercises 1.7, Problem 7.)

Hence we may use Theorem 1.16, part (i), to conclude that the 2-cycle is also attracting. For later reference, let $\mu_1 = 1 + \sqrt{6}$. Moreover, the 2-cycle becomes unstable when $\mu > \mu_1 = 1 + \sqrt{6}$.

1.7.3 2^2*-Cycles*

To find the 4-cycles we solve $F_\mu^4(x) = x$. The computation now becomes unbearable, and one should resort to a computer to do the work. It turns out that there is a 2^2-cycle when $\mu > 1 + \sqrt{6}$, which is attracting for $1 + \sqrt{6} < \mu < 3.544090\ldots$. This 2^2-cycle becomes unstable at $\mu > \mu_2 = 3.544090\ldots$.

When $\mu = \mu_2$, the 2^2-cycle bifurcates into a 2^3 cycle. The new 2^3 cycle is attracting for $\mu_3 < \mu \leq \mu_4$ for some number μ_4. This process of double bifurcation continues indefinitely. Thus we have a sequence $\{\mu_n\}_{n=0}^{\infty}$ where at μ_n there is a bifurcation from a 2^{n-1}-cycle to a 2^n-cycle. (See Figures 1.34, 1.35.) Table 1.4 provides some astonishing patterns.

From Table 1.4 we bring forward the following observations:

(i) The sequence $\{\mu_n\}$ seems to converge to a number $\mu_\infty = 3.57\ldots$.

(ii) The quotient $(\mu_n - \mu_{n-1})/(\mu_{n+1} - \mu_n)$ seems to tend to a number $\delta = 4.6692016\ldots$. This number is called the *Feigenbaum number* after its discoverer, the physicist Mitchell Feigenbaum [56]. In fact, Feigenbaum made a much more remarkable discovery: The number δ is universal and is independent of the form of the family of maps f_μ. However, the number μ_∞ depends on the family of functions under consideration.

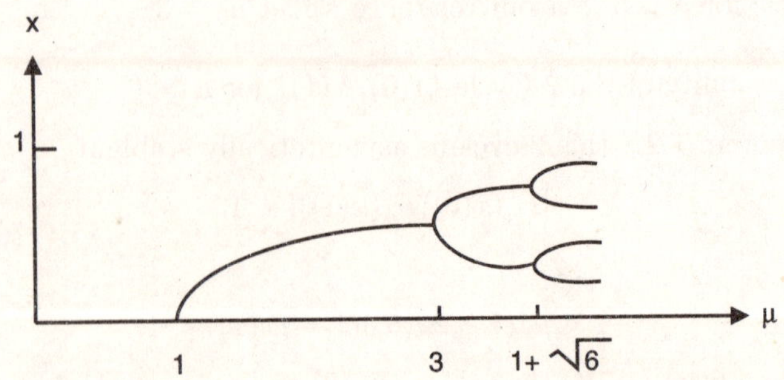

FIGURE 1.34. Partial bifurcation diagram for $\{F_\mu\}$.

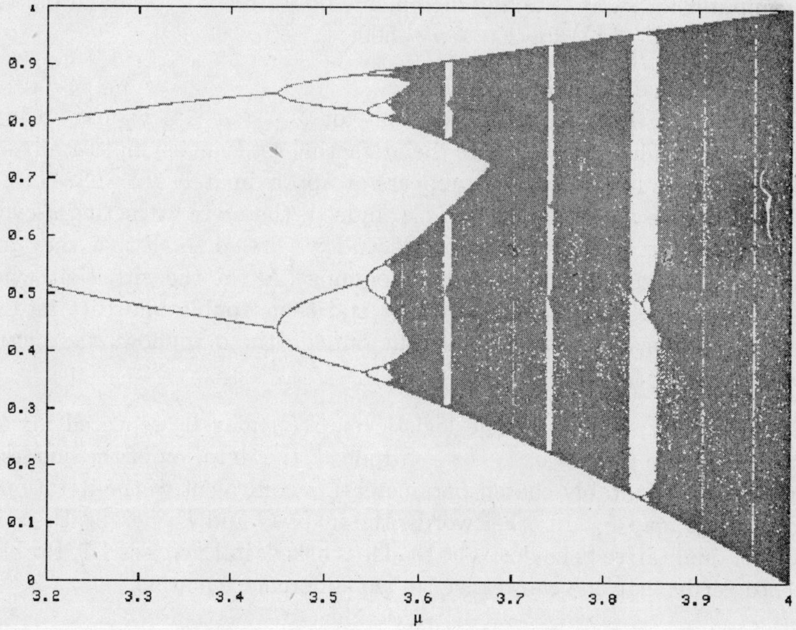

FIGURE 1.35. The bifurcation diagram of F_μ.

TABLE 1.4. Feigenbaum table.

n	μ_n	$\mu_n - \mu_{n-1}$	$\dfrac{\mu_n - \mu_{n-1}}{\mu_{n+1} - \mu_n}$
0	3	—	—
1	3.449499 ...	0.449499 ...	—
2	3.544090 ...	0.094591 ...	4.752027 ...
3	3.564407 ...	0.020313 ...	4.656673 ...
4	3.568759 ...	0.004352 ...	4.667509 ...
5	3.569692 ...	0.00093219 ...	4.668576 ...
6	3.569891 ...	0.00019964 ...	4.669354 ...

Theorem 1.25 (Feigenbaum [56] (1978)**).** *For sufficiently smooth families of maps (such as F_μ) of an interval into itself, the number $\delta =$ 4.6692016 does not in general depend on the family of maps.*

1.7.4 The Bifurcation Diagram

Here the horizontal axis represents the μ values, and the vertical axis represents higher iterates $F_\mu^n(x)$. For a fixed x_0, the diagram shows the eventual behavior of $F_\mu^n(x_0)$. The bifurcation diagram was obtained with the aid of a computer for $x_0 = \frac{1}{2}$, taking increments of $\frac{1}{500}$ for $\mu \in [0, 4]$ and plotting all points $\left(\mu, F_\mu^n\left(\frac{1}{2}\right)\right)$ for $200 \le n \le 500$.

Question What happens when $\mu > \mu_\infty$?
Answer From Figure 1.35 we see that for $\mu_\infty < \mu \le 4$ we have a large number of small windows where the attracting set is an asymptotically stable cycle. The largest window appears at approximately $\mu = 3.828427\ldots$, where we have an attracting 3-cycle. Indeed, there are attracting k-cycles for all positive integers k, but their windows are so small that they may not be noticed without sufficient zooming. As in the situation where $\mu < \mu_\infty$, these k-cycles lose stability and then double bifurcate into attracting $2^n k$-cycles. We observe that outside these windows the picture looks chaotic!

Remarks: Our analysis of the logistic map F_μ may be repeated for any quadratic map $Q(x) = ax^2 + bx + c$. Indeed, the iteration of the quadratic map Q (with suitably chosen parameters) is equivalent to the iteration of the logistic map F_μ. In other words, the maps Q and F_μ possess the same type of qualitative behavior. The reader is asked, in Exercises 1.7, Problem 11, to verify that one can transform the difference equation

$$y(n+1) = y^2(n) + c \tag{1.7.7}$$

to

$$x(n+1) = \mu x(n)[1 - x(n)] \tag{1.7.8}$$

by letting

$$y(n) = -\mu x(n) + \frac{\mu}{2}, \qquad c = \frac{\mu}{2} - \frac{\mu^2}{4}. \tag{1.7.9}$$

Note here that $\mu = 2$ corresponds to $c = 0, \mu = 3$ corresponds to $c = \frac{-3}{4}$, and $\mu = 4$ corresponds to $c = -2$. Naturally, we expect to have the same behavior of the iteration of (1.7.7) and (1.7.8) at these corresponding values of μ and c.

Comments: We are still plagued by numerous unanswered questions in connection with periodic orbits (cycles) of the difference equation

$$x(n+1) = f(x(n)). \tag{1.7.10}$$

Question A. Do all points converge to some asymptotically stable periodic orbit of (1.7.8)?
The answer is definitely no.

If $f(x) = 1 - 2x^2$ in (1.7.10), then there are no asymptotically stable (attractor) periodic orbits. Can you verify this statement? If you have some difficulty here, it is not your fault. Obviously, we need some tools to help us in verifying that there are no periodic attractors.

Question B. If there is a periodic attractor of (1.7.10), how many points converge to it?
Once again, we need more machinery to answer this question.

Question C. Can there be several distinct periodic attractors for (1.7.10)?

This question leads us to the Li–Yorke famous result "Period Three Implies Chaos" [92]. To explain this and more general results requires the introduction of the so-called *Schwarzian derivative* of $f(x)$. We will come back to these questions in Chapter 6.

Exercises 1.7

Unless otherwise stated, all the problems here refer to the logistic difference equation (1.7.1).

1. Use the stair step diagram for F_4^k on $[0,1]$, $k = 1, 2, 3, \ldots$, to demonstrate that F_4 has at least 2^k periodic points of period k (including periodic points of periods that are divisors of k).

2. Find the exact solution of $x(n+1) = 4x(n)[1-x(n)]$.

3. Let $x^* = (\mu - 1)/\mu$ be the equilibrium point of (1.7.1). Show that:

   (i) For $1 < \mu \leq 3$, x^* is an attracting fixed point.

   (ii) For $\mu > 3$, x^* is a repelling fixed point.

4. Prove that $\lim_{n\to\infty} F_2^n(x) = \frac{1}{2}$ if $0 < x < 1$.

5. Let $1 < \mu \leq 2$ and let $x^* = (\mu - 1)/\mu$ be the equilibrium point of (1.7.1). Show that if $x^* < x < \frac{1}{2}$, then $\lim_{n\to\infty} F_\mu^n(x) = x^*$.

6. Prove that the 2-cycle given by (1.7.4) is attracting if $3 < \mu < 1+\sqrt{6}$.

7. Verify formula (1.7.6). Then show that the 2-cycle in (1.7.4) is attracting when $\mu = 1 + \sqrt{6}$.

8. Verify that $\mu_2 \approx 3.54$ using a calculator or a computer.

*9. (Project). Show that the map $H_\mu(x) = \sin \mu x$ leads to the same value for the Feigenbaum number δ.

10. Show that if $|\mu - \mu_1| < \varepsilon$, then $|F\mu(x) - F\mu_1(x)| < \varepsilon$ for all $x \in [0,1]$.

11. Show that (1.7.7) can be transformed to the logistic equation (1.7.8), with $c = \frac{\mu}{2} - \frac{\mu^2}{4}$.

12. (a) Find the equilibrium points $\overset{*}{y_1}, \overset{*}{y_2}$ of (1.7.7).

    (b) Find the values of c where $\overset{*}{y_1}$ is attracting or unstable.

    (c) Find the values of c where $\overset{*}{y_2}$ is attracting or unstable.

13. Find the value of c_0 where (1.7.7) double bifurcates for $c > c_0$. Check your answer using (1.7.9).

*14. (Project). Use a calculator or a computer to develop a bifurcation diagram, as in Figures 1.34, 1.35, for (1.7.6).

*15. (Project). Develop a bifurcation diagram for the quadratic map $Q_\lambda(x) = 1 - \lambda x^2$ on the interval $[-1, 1]$, $\lambda \in (0, 2]$.

In Problems 16–19 determine the stability of the fixed points of the difference equation.

16. $x(n+1) = x(n) + \frac{1}{\pi}\sin(2\pi x(n))$.

17. $x(n+1) = 0.5\sin(\pi x(n))$.

18. $x(n+1) = 2x(n)\exp(-x(n))$.

19. A population of birds is modeled by the difference equation

$$x(n+1) = \begin{cases} 3.2x(n) & \text{for } 0 \le x(n) \le 1, \\ 0.5x(n) & \text{for } x(n) > 1, \end{cases}$$

where $x(n)$ is the number of birds in year n. Find the equilibrium points and then determine their stability.

1.8 Basin of Attraction and Global Stability (Optional)

It is customary to call an asymptotically stable fixed point or a cycle an attractor. This name makes sense since in this case all nearby points tend to the attractor. The maximal set that is attracted to an attractor M is called the *basin of attraction* of M. Our analysis applies to cycles of any period.

Definition 1.26. Let x^* be a fixed point of map f. Then the basin of attraction (or the stable set) $W^s(x^*)$ of x^* is defined as

$$W^s(x^*) = \{x : \lim_{n\to\infty} f^n(x) = x^*\}.$$

In other words, $W^s(x^*)$ consists of all points that are forward asymptotic to x^*.

Observe that if x^* is an attracting fixed point, $W^s(x^*)$ contains an open interval around x^*. The maximal interval in $W^s(x^*)$ that contains x^* is called the *immediate basin of attraction* and is denoted by $\mathcal{B}^s(x^*)$.

Example 1.27. The map $f(x) = x^2$ has one attracting fixed point $x^* = 0$. Its basin of attraction $W^s(0) = (-1, 1)$. Note that 1 is an unstable fixed point and -1 is an eventually fixed point that goes to 1 after one iteration.

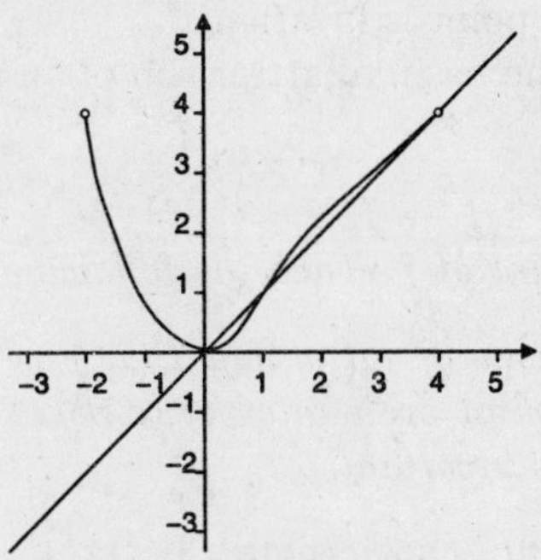

FIGURE 1.36. The basin of attraction $W^s(0) = (-1, 1)$ and $W^s(4) = [-2, -1) \cup (1, 4]$. The immediate basin of attraction $\mathcal{B}(4) = (1, 4]$.

Example 1.28. Let us now modify the map f. Consider the map $g : [-2, 4] \to [-2, 4]$ defined as

$$g(x) = \begin{cases} x^2 & \text{if } -2 \leq x \leq 1, \\ 3\sqrt{x} - 2 & \text{if } 1 < x \leq 4. \end{cases}$$

The map g has three fixed points $x_1^* = 0$, $x_2^* = 1$, $x_3^* = 4$. The basin of attraction of $x_1^* = 0$, $W^s(0) = (-1, 1)$, while the basin of attraction of $x_3^* = 4$, $W^s(4) = [-2, -1) \cup (1, 4]$. Moreover, the immediate basin of attractions of $x_1^* = 0$ is $\mathcal{B}(0) = W^s(0) = (-1, 1)$, while $\mathcal{B}(4) = (1, 4]$.

Remark: Observe that in the preceding example, the basins of attraction of the two fixed points $x_1^* = 0$ and $x_3^* = 4$ are disjoint. This is no accident and is, in fact, generally true. This is due to the uniqueness of a limit of a sequence. In other words, if the $\lim_{n \to \infty} f^n(x) = L_1$ and $\lim_{n \to \infty} f^n(x) = L_2$, then certainly $L_1 = L_2$.

It is worth noting here that finding the basin of attraction of a fixed point is in general a difficult task. But even more difficult is providing a rigorous proof. The most efficient method to determining the basin of attraction is the method of Liapunov functions, which will be developed in Chapter 4. In this section, we will develop some of the basic topological properties of the basin of attractions. Henceforth, all our maps are assumed to be *continuous.* We begin our exposition by defining the important notion of invariance.

Definition 1.29. A set M is *positively invariant* under a map f if $f(M) \subseteq M$. In other words, for every $x \in M$, $\mathcal{O}(x) \subseteq M$. Since we are only considering forward iterations of f, the prefix "positively" will henceforth be dropped.

Clearly an orbit of a point is invariant.

Next we show that the basin of attraction of an attracting fixed point is invariant and open.

Theorem 1.30. *Let $f : I \to I$, $I = [a, b]$, be a continuous map and let $x^* \in [a, b]$ be a fixed point of f. Then the following statements hold true:*

(i) *The immediate basin of attraction $\mathcal{B}(x^*)$ is an interval containing x^*, which is either an open interval (c, d) or of the form $[a, c)(c, b]$. Moreover, $\mathcal{B}(x^*)$ is invariant.*

(ii) *$W^s(x^*)$ is invariant. Furthermore, $W^s(x^*)$ is the union (maybe an infinite union) of intervals that are either open intervals or of the form $[a, c)$ or $(d, b]$.*

PROOF.

(i) We know that $\mathcal{B}(x^*)$ is a maximal interval in $W^s(x^*)$ containing x^*. Assume that $\mathcal{B}(x^*) = [c, d)$, $c \neq a$. Now for a given small $\varepsilon > 0$ there exists $m \in \mathbb{Z}^+$ such that $f^m(c) \in (x^* - \varepsilon, x^* + \varepsilon) \subset (c, d)$. Since f^m is continuous, there exists $\delta > 0$ such that if $x_0 \in (c - \delta, c + \delta)$, then $f^m(x_0) \in (x^* - \varepsilon, x^* + \varepsilon) \subset \mathcal{B}(x^*)$. Then $x_0 \in \mathcal{B}(x^*)$ and hence $(c - \delta, d) \subset W^s(x^*)$ which violates the maximality of $\mathcal{B}(x^*)$. Hence $\mathcal{B}(x^*) \neq [c, a)$, a contradiction. Analogously, one may show that $W^s(x^*) \neq (c, d]$ if $d \neq b$.

To prove the invariance of $\mathcal{B}(x^*)$, assume that there exists $y \in \mathcal{B}(x^*)$ such that $f^r(y) \notin \mathcal{B}(x^*)$ for some $r \in \mathbb{Z}^+$. Since $\mathcal{B}(x^*)$ is an interval, it follows by the Intermediate Value Theorem that $f^r(\mathcal{B}(x^*))$ is also an interval. Moreover, this interval $f^r(\mathcal{B}(x^*))$ must contain x^* since $f^r(x^*) = x^*$. Thus $f^r(\mathcal{B}(x^*)) \cap \mathcal{B}(x^*) \neq 0$, and hence $\mathcal{B}(x^*) \cup f^r(\mathcal{B}(x^*))$ is an interval in $W^s(x^*)$, which violates the maximality of $\mathcal{B}(x^*)$.

(ii) The proof of this part is analogous to the proof of part (a) and will be left to the reader to verify. □

There are several (popular) maps such as the logistic map and Ricker's map in which the basin of attraction, for the attractive fixed point, is the entire space with the exception of one or two points (fixed or eventually fixed). For the logistic map $F_\mu(x) = \mu x(1 - x)$ and $1 < \mu < 3$, the basin of attraction $W^s(x^*) = (0, 1)$ for the fixed point $x^* = \frac{\mu-1}{\mu}$. And for Ricker's map $R_p(x) = xe^{p-x}$, $0 < p < 2$, the basin of attraction $W^s(x^*) = (0, \infty)$, for $x^* = p$. Here we will consider only the logistic map and leave it to the reader to prove the statement concerning Ricker's map.

Notice that $|F'_\mu(x)| = |\mu - 2\mu x| < 1$ if and only if $-1 < \mu - 2\mu x < 1$. This implies that $\frac{\mu-1}{2\mu} < x < \frac{\mu+1}{2\mu}$. Hence $|F'_\mu(x)| < 1$ for all $x \in \left(\frac{\mu-1}{2\mu}, \frac{\mu+1}{2\mu}\right)$. Observe that $x^* = \frac{\mu-1}{\mu} \in \left(\frac{\mu-1}{2\mu}, \frac{\mu+1}{2\mu}\right)$ if and only if $1 < \mu < 3$. Now

$F_\mu\left(\frac{\mu+1}{2\mu}\right) = F_\mu\left(\frac{\mu-1}{2\mu}\right) = \frac{1}{2}\left[\frac{(\mu-1)(\mu+1)}{2\mu}\right]$. Notice that since $1 < \mu < 3$, $\frac{\mu-1}{2\mu} < \frac{1}{2}\cdot\frac{(\mu-1)(\mu+1)}{2\mu} < \frac{\mu+1}{2\mu}$. Hence $\left[\frac{\mu-1}{2\mu}, \frac{\mu+1}{2\mu}\right] \subset W^s(x^*)$.

If $z \in \left(0, \frac{\mu-1}{2\mu}\right)$, then $F'_\mu(z) > 1$. By the Mean Value Theorem, $\frac{F_\mu(z)-F_\mu(0)}{z-0} = F'_\mu(\gamma)$, for some γ with $0 < \gamma < z$. Hence

$$F_\mu(z) - F_\mu(0) = F_\mu(z) \geq \beta z$$

for some $\beta > 1$. Then for some $r \in \mathbb{Z}^+$, $F^r_\mu(z) \geq \beta^r z > \frac{\mu-1}{2\mu}$ and $F^{r-1}_\mu(z) < \frac{\mu-1}{2\mu}$. Moreover, since F is increasing on $\left[0, \frac{\mu-1}{2\mu}\right]$, $F^r_\mu(z) < F_\mu\left(\frac{\mu-1}{2\mu}\right) = \mu\left(\frac{\mu-1}{2\mu}\right)\left(1-\frac{\mu-1}{2\mu}\right) = \frac{\mu-1}{\mu}\left(\frac{\mu+1}{4}\right) \leq x^*$. Thus $z \in W^s(x^*)$. On the other hand, $F_\mu\left(\frac{\mu+1}{2\mu}, 1\right) \subset (0, x^*)$ and hence $\left(\frac{\mu+1}{2\mu}, 1\right) \subset W^s(x^*)$. This shows that $W^s(x^*) = (0,1)$.

To summarize

Lemma 1.31. *For the logistic map $F_\mu(x) = \mu x(1-x)$, $1 < \mu < 3$, $W^s(x^*) = (0,1)$ for $x^* = \frac{\mu-1}{\mu}$.*

We now turn our attention to periodic points. If $\bar{x}$ is a periodic point of period k under the map f, then its basin of attraction $W^s(\bar{x})$ is its basin of attraction as a fixed point under the map f^k. Hence $W^s(\bar{x}) = \{x : \lim_{n\to\infty}(f^k)^n(x) = \lim_{n\to\infty} f^{kn}(x) = \bar{x}\}$. Let $\{\bar{x}_1, \bar{x}_2, \ldots, \bar{x}_k\}$ be a k-cycle of a map f. Then clearly for $i \neq j$, $W^s(\bar{x}_i) \cap W^s(\bar{x}_j) = \emptyset$. (Why?) More generally, if x is a periodic point of period r and $y \neq x$ is a periodic point of period s, then $W^s(x) \cap W^s(y) = \emptyset$ (Exercises 1.8, Problem 6).

Example 1.32. Consider the function $f(x) = -x^{\frac{1}{3}}$. Then $x^* = 0$ is the only fixed point. There is a 2-cycle $\{-1, 1\}$ with $f(-1) = 1$, $f^2(-1) = -1$. The cobweb diagram (Figure 1.37) shows that $W^s(1) = (0, \infty)$, $W^s(-1) = (-\infty, 0)$.

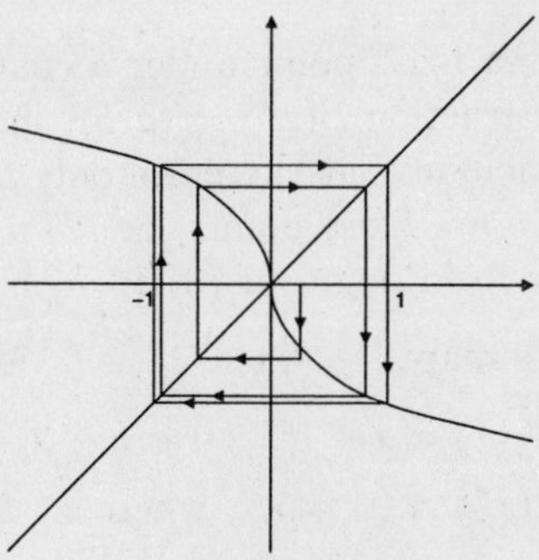

FIGURE 1.37.

Exercises 1.8

1. Investigate the basin of attraction of the fixed points of the map

$$f(x) = \begin{cases} x^2 & \text{if } -3 \le x \le 1, \\ 4\sqrt{x} - 3 & \text{if } 1 < x \le 9. \end{cases}$$

2. Let $f(x) = |x - 1|$. Find $W^s(\frac{1}{2})$.

3. Suppose that $f : I \to I$ is a continuous and onto map on an interval I. Let $\bar{x}$ be an asymptotically stable periodic point of period $k \ge 2$. Show that $W^s(f(\bar{x})) = f(W^s(\bar{x}))$.

4. Describe the basin of attraction of all fixed and periodic points of the maps:

   (i) $f(x) = x^2$,

   (ii) $g(x) = x^3$,

   (iii) $h(x) = 2xe^{-x}$,

   (iv) $q(x) = -\frac{4}{\pi} \arctan x$.

5. Investigate the basin of attraction of the origin for the map

$$f(x) = \begin{cases} \dfrac{x}{2} & \text{if } 0 \le x \le 0.2, \\ 3x - \dfrac{1}{2} & \text{if } 0.2 < x \le \dfrac{1}{2}, \\ 2 - 2x & \text{if } \dfrac{1}{2} < x \le 1. \end{cases}$$

6. Let f be a continuous map that has two periodic points x and y, $x \ne y$, with periods r and t, $r \ne t$, respectively. Prove that $W^s(x) \cap W^s(y) = \emptyset$.

7*. Suppose that a set M is invariant under a one-to-one continuous map f. A point $x \in M$ is said to be an interior point if $(x - \delta, x + \delta) \subset M$ for some $\delta > 0$. Prove that the set of all interior points of M, denoted by $\text{int}(M)$, is invariant.

8. Let x^* be an attracting fixed point under a continuous map f. If the immediate basin of attraction $\mathcal{B}(x^*) = (a, b)$, show that the set $\{a, b\}$ is invariant. Then conclude that there are only three scenarios in this case: (1) both a and b are fixed points, or (2) a or b is fixed and the other is an eventually fixed point, or (3) $\{a, b\}$ is a 2-cycle.

9. Show that for Ricker's map

$$R_p(x) = xe^{p-x}, \quad 0 < p < 2,$$
$$W^s(x^*) = (0, \infty), \quad \text{where } x^* = p.$$

10. (Term Project). Consider the logistic map $F_\mu(x) = \mu x(1 - x)$ with $3 < \mu < 1 + \sqrt{6}$. Let $c = \{\bar{x}_1, \bar{x}_2\}$ be the attracting 2-cycle. Show that

$W^s(c) = W^s(\bar{x}_1) \cup W^s(\bar{x}_2)$ is all the points in $(0,1)$ except the set of eventually fixed points (including the fixed point $\frac{\mu-1}{\mu}$).

2

Linear Difference Equations of Higher Order

In this chapter we examine linear difference equations of high order, namely, those involving a single dependent variable.[1] Such equations arise in almost every field of scientific inquiry, from population dynamics (the study of a single species) to economics (the study of a single commodity) to physics (the study of the motion of a single body). We will become acquainted with some of these applications in this chapter. We start this chapter by introducing some rudiments of difference calculus that are essential in the study of linear equations.

2.1 Difference Calculus

Difference calculus is the discrete analogue of the familiar differential and integral calculus. In this section we introduce some very basic properties of two operators that are essential in the study of difference equations. These are the *difference operator* (Section 1.2)

$$\Delta x(n) = x(n+1) - x(n)$$

and the *shift operator*

$$Ex(n) = x(n+1).$$

[1]Difference equations that involve more than one dependent variable are called systems of difference equations; we will inspect these equations in Chapter 3.

It is easy to see that

$$E^k x(n) = x(n+k).$$

However, $\Delta^k x(n)$ is not so apparent. Let I be the *identity operator*, i.e., $Ix = x$. Then, one may write $\Delta = E - I$ and $E = \Delta + I$.

Hence,

$$\Delta^k x(n) = (E-I)^k x(n)$$

$$= \sum_{i=0}^{k} (-1)^i \binom{k}{i} E^{k-i} x(n),$$

$$\boxed{\Delta^k x(n) = \sum_{i=0}^{k} (-1)^i \binom{k}{i} x(n+k-i).} \tag{2.1.1}$$

Similarly, one may show that

$$E^k x(n) = \sum_{i=0}^{k} \binom{k}{i} \Delta^{k-i} x(n). \tag{2.1.2}$$

We should point out here that the operator Δ is the counterpart of the derivative operator D in calculus. Both operators E and Δ share one of the helpful features of the derivative operator D, namely, the property of *linearity*.

"Linearity" simply means that $\Delta[ax(n)+by(n)] = a\Delta x(n) + b\Delta y(n)$ and $E[ax(n)+by(n)] = aEx(n) + bEy(n)$, for all a and $b \in \mathbb{R}$. In Exercises 2.1, Problem 1, the reader is allowed to show that both Δ and E are linear operators.

Another interesting difference, parallel to differential calculus, is the discrete analogue of the fundamental theorem of calculus. [2]

Lemma 2.1. *The following statements hold:*

(i)

$$\boxed{\sum_{k=n_0}^{n-1} \Delta x(k) = x(n) - x(n_0),} \tag{2.1.3}$$

[2]The fundamental theorem of calculus states that:

(i) $\int_a^b df(x) = f(b) - f(a)$,

(ii) $d\left(\int_a^x f(t)\,dt\right) = f(x)$.

(ii)

$$\boxed{\Delta\left(\sum_{k=n_0}^{n-1} x(k)\right) = x(n).} \tag{2.1.4}$$

PROOF. The proof remains as Exercises 2.1, Problem 3. □

We would now like to introduce a third property that the operator Δ has in common with the derivative operator D.

Let

$$p(n) = a_0 n^k + a_1 n^{k-1} + \cdots + a_k$$

be a polynomial of degree k. Then

$$\begin{aligned}\Delta p(n) &= \left[a_0(n+1)^k + a_1(n+1)^{k-1} + \cdots + a_k\right] \\ &\quad - \left[a_0 n^k + a_1 n^{k-1} + \cdots + a_k\right] \\ &= a_0 k n^{k-1} + \text{ terms of degree lower than } (k-1).\end{aligned}$$

Similarly, one may show that

$$\Delta^2 p(n) = a_0 k(k-1) n^{k-2} + \text{ terms of degree lower than } (k-2).$$

Carrying out this process k times, one obtains

$$\Delta^k p(n) = a_0 k!. \tag{2.1.5}$$

Thus,

$$\Delta^{k+i} p(n) = 0 \text{ for } i \geq 1. \tag{2.1.6}$$

2.1.1 The Power Shift

We now discuss the action of a polynomial of degree k in the shift operator E on the term b^n, for any constant b.

Let

$$p(E) = a_0 E^k + a_1 E^{k-1} + \cdots + a_k I \tag{2.1.7}$$

be a polynomial of degree k in E.

Then

$$\begin{aligned}p(E)b^n &= a_0 b^{n+k} + a_1 b^{n+k-1} + \cdots + a_k b^n \\ &= (a_0 b^k + a_1 b^{k-1} + \cdots + a_k) b^n \\ &= p(b) b^n.\end{aligned} \tag{2.1.8}$$

A generalization of formula (2.1.8) now follows.

Lemma 2.2. *Let $p(E)$ be the polynomial in (2.1.7) and let $g(n)$ be any discrete function. Then*

$$\boxed{p(E)(b^n g(n)) = b^n p(bE)g(n).} \tag{2.1.9}$$

PROOF. This is left to the reader as Exercises 2.1, Problem 4. □

2.1.2 *Factorial Polynomials*

One of the most interesting functions in difference calculus is the *factorial polynomial* $x^{(k)}$ defined as follows. Let $x \in \mathbb{R}$. Then the kth factorial of x is given by

$$x^{(k)} = x(x-1)\cdots(x-k+1), \qquad k \in \mathbb{Z}^+.$$

Thus if $x = n \in \mathbb{Z}^+$ and $n \geq k$, then

$$\boxed{n^{(k)} = \frac{n!}{(n-k)!}}$$

and

$$n^{(n)} = n!.$$

The function $x^{(k)}$ plays the same role here as that played by the polynomial x^k in differential calculus. The following Lemma 2.3 demonstrates this fact.

So far, we have defined the operators Δ and E on sequences $f(n)$. One may extend the definitions of Δ and E to continuous functions $f(t)$, $t \in \mathbb{R}$, by simply letting $\Delta f(t) = f(t+1) - f(t)$ and $Ef(t) = f(t+1)$. This extension enables us to define $\Delta f(x)$ and $Ef(x)$ where $f(x) = x^{(k)}$ by

$$\Delta x^{(k)} = (x+1)^{(k)} - x^{(k)} \quad \text{and} \quad Ex^{(k)} = (x+1)^{(k)}.$$

Using this definition one may establish the following result.

Lemma 2.3. *For fixed $k \in \mathbb{Z}^+$ and $x \in \mathbb{R}$, the following statements hold:*

(i)

$$\boxed{\Delta x^{(k)} = kx^{(k-1)};} \tag{2.1.10}$$

(ii)

$$\boxed{\Delta^n x^{(k)} = k(k-1), \ldots, (k-n+1)x^{(k-n)};} \tag{2.1.11}$$

(iii)

$$\boxed{\Delta^k x^{(k)} = k!.} \tag{2.1.12}$$

PROOF. (i)

$$\begin{aligned}\Delta x^{(k)} &= (x+1)^{(k)} - x^{(k)} \\ &= (x+1)x(x-1)\cdots(x-k+2) - x(x-1) \\ &\quad \cdots(x-k+2)(x-k+1) \\ &= x(x-1)\cdots(x-k+2)\cdot k \\ &= kx^{(k-1)}.\end{aligned}$$

The proofs of parts (ii) and (iii) are left to the reader as Exercises 2.1, Problem 5. □

If we define, for $k \in \mathbb{Z}^+$,

$$x^{(-k)} = \frac{1}{x(x+1)\cdots(x+k-1)} \tag{2.1.13}$$

and $x^{(0)} = 1$, then one may extend Lemma 2.3 to hold for all $k \in \mathbb{Z}$. In other words, parts (i), (ii), and (iii) of Lemma 2.3 hold for all $k \in \mathbb{Z}$ (Exercises 2.1, Problem 6).

The reader may wonder whether the product and quotient rules of the differential calculus have discrete counterparts. The answer is affirmative, as may be shown by the following two formulas, where proofs are left to the reader as Exercises 2.1, Problem 7.
Product Rule:

$$\boxed{\Delta[x(n)y(n)] = Ex(n)\Delta y(n) + y(n)\Delta x(n).} \tag{2.1.14}$$

Quotient Rule:

$$\boxed{\Delta\left[\frac{x(n)}{y(n)}\right] = \frac{y(n)\Delta x(n) - x(n)\Delta y(n)}{y(n)Ey(n)}.} \tag{2.1.15}$$

2.1.3 The Antidifference Operator

The discrete analogue of the indefinite integral in calculus is the antidifference operator Δ^{-1}, defined as follows. If $\Delta F(n) = 0$, then $\Delta^{-1}(0) = F(n) = c$ for some arbitrary constant c. Moreover, if $\Delta F(n) = f(n)$, then $\Delta^{-1}f(n) = F(n) + c$, for some arbitrary constant c. Hence

$$\Delta\Delta^{-1}f(n) = f(n),$$
$$\Delta^{-1}\Delta F(n) = F(n) + c,$$

and

$$\Delta\Delta^{-1} = I \quad \text{but} \quad \Delta^{-1}\Delta \neq I.$$

Using formula (2.1.4) one may readily obtain

$$\boxed{\Delta^{-1}f(n) = \sum_{i=0}^{n-1} f(i) + c.} \tag{2.1.16}$$

Formula (2.1.16) is very useful in proving that the operator Δ^{-1} is linear.

Theorem 2.4. *The operator Δ^{-1} is linear.*

PROOF. We need to show that for $a, b \in \mathbb{R}$, $\Delta^{-1}[ax(n) + by(n)] = a\Delta^{-1}x(n) + b\Delta^{-1}y(n)$. Now, from formula (2.1.16) we have

$$\begin{aligned}\Delta^{-1}[ax(n) + by(n)] = & \sum_{i=0}^{n-1} ax(i) + by(i) + c \\ &= a\sum_{i=0}^{n-1} x(i) + b\sum_{i=0}^{n-1} y(i) + c \\ &= a\Delta^{-1}x(n) + b\Delta^{-1}y(n).\end{aligned}$$

□

Next we derive the antidifference of some basic functions.

Lemma 2.5. *The following statements hold:*

(i)

$$\Delta^{-k}0 = c_1 n^{k-1} + c_2 n^{k-2} + \cdots + c_k. \tag{2.1.17}$$

(ii)

$$\Delta^{-k}1 = \frac{n^k}{k!} + c_1 n^{k-1} + c_2 n^{k-2} + \cdots + c_k. \tag{2.1.18}$$

(iii)

$$\boxed{\Delta^{-1}n^{(k)} = \frac{n^{(k+1)}}{k+1} + c, \qquad k \neq -1.} \tag{2.1.19}$$

PROOF. The proofs of parts (i) and (ii) follow by applying Δ^k to the right-hand side of formulas (2.1.17) and (2.1.18) and then applying formulas (2.1.6) and (2.1.5), respectively. The proof of part (iii) follows from formula (2.1.10).

Finally, we give the discrete analogue of the integration by parts formula, namely, the summation by parts formula:

$$\boxed{\sum_{k=0}^{n-1} y(k)\Delta x(k) = x(n)y(n) - \sum_{k=0}^{n-1} x(k+1)\Delta y(k) + c.} \tag{2.1.20}$$

To prove formula (2.1.20) we use formula (2.1.14) to obtain

$$y(n)\Delta x(n) = \Delta(x(n)y(n)) - x(n+1)\Delta y(n).$$

Applying Δ^{-1} to both sides and using formula (2.1.16), we get

$$\sum_{k=0}^{n-1} y(k)\Delta x(k) = x(n)y(n) - \sum_{k=0}^{n-1} x(k+1)\Delta y(k) + c. \qquad \square$$

Exercises 2.1

1. Show that the operators Δ and E are linear.
2. Show that $E^k x(n) = \sum_{i=0}^{k} \binom{k}{i} \Delta^{k-i} x(n)$.
3. Verify formulas (2.1.3) and (2.1.4).
4. Verify formula (2.1.9).
5. Verify formulas (2.1.11) and (2.1.12).
6. Show that Lemma 2.3 holds for $k \in \mathbb{Z}$.
7. Verify the product and quotient rules (2.1.14) and (2.1.15).
8. (Abel's Summation Formula). Prove that

$$\boxed{\sum_{k=1}^{n} x(k)y(k) = x(n+1)\sum_{k=1}^{n} y(k) - \sum_{k=1}^{n}\left(\Delta x(k)\sum_{r=1}^{k} y(r)\right).}$$

9. (Newton's Theorem). If $f(n)$ is a polynomial of degree k, show that

$$f(n) = f(0) + \frac{n^{(1)}}{1!}\Delta f(0) + \frac{n^{(2)}}{2!}\Delta^2 f(0) + \cdots + \frac{n^{(k)}}{k!}\Delta^{(k)} f(0).$$

10. (The Discrete Taylor Formula). Verify that

$$\boxed{f(n) = \sum_{i=0}^{k-1} \binom{n}{i} \Delta^i f(0) + \sum_{s=0}^{n-k} \binom{n-s-1}{k-1} \Delta^k f(s).}$$

11. (The Stirling Numbers). The Stirling numbers of the second kind $s_i(k)$ are defined by the difference equation $s_i(m+1) = s_{i-1}(m) + is_i(m)$ with $s_i(i) = s_1(i) = 1$ and $1 \le i \le m$, $s_1(k) = 0$ for $1 > k$. Prove that

$$x^m = \sum_{i=1}^{m} s_i(m) x^{(i)}. \tag{2.1.21}$$

12. Use (2.1.21) to verify Table 2.1 which gives the Stirling numbers $s_i(k)$ for $1 \le i, k \le 7$.
13. Use Table 2.1 and formula (2.1.21) to write x^3, x^4, and x^5 in terms of the factorial polynomials $x^{(k)}$ (e.g., $x^2 = x^{(1)} + x^{(2)}$).
14. Use Problem 13 to find

TABLE 2.1. Stirling numbers $s_i(k)$.

$i\backslash k$	1	2	3	4	5	6	7
1	1	1	1	1	1	1	1
2		1	3	7	15	31	63
3			1	6	25	90	301
4				1	10	65	350
5					1	15	140
6						1	21
7							1

(i) $\Delta^{-1}(n^3+1)$.

(ii) $\Delta^{-1}\left(\dfrac{5}{n(n+3)}\right)$.

15. Use Problem 13 to solve the difference equation $y(n+1) = y(n) + n^3$.

16. Use Problem 13 to solve the difference equation $y(n+1) = y(n) - 5n^2$.

17. Consider the difference equation[3]

$$y(n+1) = a(n)y(n) + g(n). \tag{2.1.22}$$

(a) Put $y(n) = \left(\prod_{i=0}^{n-1} a(i)\right) u(n)$ in (2.1.22). Then show that $\Delta u(n) = g(n)/\prod_{i=0}^{n} a(i)$.

(b) Prove that

$$y(n) = \left(\prod_{i=0}^{n-1} a(i)\right) y_0 + \sum_{r=0}^{n-1}\left(\prod_{i=r+1}^{n-1} a(i)\right) g(r), \qquad y_0 = y(0).$$

(Compare with Section 1.2.)

2.2 General Theory of Linear Difference Equations

The normal form of a kth-order *nonhomogeneous linear* difference equation is given by

$$y(n+k) + p_1(n)\,y(n+k-1) + \cdots + p_k(n)\,y(n) = g(n), \tag{2.2.1}$$

where $p_i(n)$ and $g(n)$ are real-valued functions defined for $n \geq n_0$ and $p_k(n) \neq 0$ for all $n \geq n_0$. If $g(n)$ is identically zero, then (2.2.1) is said to be a homogeneous equation. Equation (2.2.1) may be written in the form

$$y(n+k) = -p_1(n)\,y(n+k-1) - \cdots - p_k(n)\,y(n) + g(n). \tag{2.2.2}$$

[3]This method of solving a nonhomogeneous equation is called the method of variation of constants.

By letting $n = 0$ in (2.2.2), we obtain $y(k)$ in terms of $y(k-1), y(k-2), \cdots, y(0)$. Explicitly, we have

$$y(k) = -p_1(0)y(k-1) - p_2(0)y(k-2) - \cdots - p_k(0)y(0) + g(0).$$

Once $y(k)$ is computed, we can go to the next step and evaluate $y(k+1)$ by letting $n = 1$ in (2.2.2). This yields

$$y(k+1) = -p_1(1)y(k) - p_2(1)y(k-1) - \cdots - p_k(1)y(1) + g(1).$$

By repeating the above process, it is possible to evaluate all $y(n)$ for $n \geq k$. Let us now illustrate the above procedure by an example.

Example 2.6. Consider the third-order difference equation

$$y(n+3) - \frac{n}{n+1}y(n+2) + ny(n+1) - 3y(n) = n, \tag{2.2.3}$$

where $y(1) = 0, y(2) = -1$, and $y(3) = 1$. Find the values of $y(4)$, $y(5)$, $y(6)$, and $y(7)$.

Solution First we rewrite (2.2.3) in the convenient form

$$y(n+3) = \frac{n}{n+1}y(n+2) - ny(n+1) + 3y(n) + n. \tag{2.2.4}$$

Letting $n = 1$ in (2.2.4), we have

$$y(4) = \frac{1}{2}y(3) - y(2) + 3y(1) + 1 = \frac{5}{2}.$$

For $n = 2$,

$$y(5) = \frac{2}{3}y(4) - 2y(3) + 3y(2) + 2 = -\frac{4}{3}.$$

For $n = 3$,

$$y(6) = \frac{3}{4}y(5) - 3y(4) + 3y(3) + 3 = -\frac{3}{2}.$$

For $n = 4$,

$$y(7) = \frac{4}{5}y(6) - 4y(5) + 3y(4) + 4 = 20.9.$$

Now let us go back to (2.2.1) and formally define its solution. A sequence $\{y(n)\}_{n_0}^{\infty}$ or simply $y(n)$ is said to be a *solution* of (2.2.1) if it satisfies the equation. Observe that if we specify the initial data of the equation, we are led to the corresponding initial value problem

$$y(k+n) + p_1(n)y(n+k-1) + \cdots + p_k(n)y(n) = g(n), \tag{2.2.5}$$

$$y(n_0) = a_0, \quad y(n_0+1) = a_1, \ldots, y(n_0+k-1) = a_{k-1}, \tag{2.2.6}$$

where the a_i's are real numbers. In view of the above discussion, we conclude with the following result.

Theorem 2.7. *The initial value problems* (2.2.5) *and* (2.2.6) *have a* unique *solution* $y(n)$.

PROOF. The proof follows by using (2.2.5) for $n = n_0, n_0 + 1, n_0 + 2, \ldots$. Notice that any $n \geq n_0 + k$ may be written in the form $n = n_0 + k + (n - n_0 - k)$. By *uniqueness* of the solution $y(n)$ we mean that if there is another solution $\tilde{y}(n)$ of the initial value problems (2.2.5) and (2.2.6), then $\tilde{y}(n)$ must be identical to $y(n)$. This is again easy to see from (2.2.5). □

The question still remains whether we can find a closed-form solution for (2.2.1) or (2.2.5) and (2.2.6). Unlike our amiable first-order equations, obtaining a closed-form solution of (2.2.1) is a formidable task. However, if the coefficients p_i in (2.2.1) are constants, then a solution of the equation may be easily obtained, as we see in the next section.

In this section we are going to develop the general theory of kth-order linear *homogeneous* difference equations of the form

$$x(n+k) + p_1(n)x(n+k-1) + \cdots + p_k(n)x(n) = 0. \qquad (2.2.7)$$

We start our exposition by introducing three important definitions.

Definition 2.8. The functions $f_1(n), f_2(n), \ldots, f_r(n)$ are said to be *linearly dependent* for $n \geq n_0$ if there are constants $a_1, a_2, \ldots, a_r$, not all zero, such that

$$a_1 f_1(n) + a_2 f_2(n) + \cdots + a_r f_r(n) = 0, \qquad n \geq n_0.$$

If $a_j \neq 0$, then we may divide (2.2.7) by a_j to obtain

$$\begin{aligned} f_j(n) &= -\frac{a_1}{a_j} f_1(n) - \frac{a_2}{a_j} f_2(n) \cdots - \frac{a_r}{a_j} f_r(n) \\ &= -\sum_{i \neq j} \frac{a_i}{a_j} f_i(n). \end{aligned} \qquad (2.2.8)$$

Equation (2.2.8) simply says that each f_j with nonzero coefficient is a *linear combination* of the other f_i's. Thus two functions $f_1(n)$ and $f_2(n)$ are linearly dependent if one is a multiple of the other, i.e., $f_1(n) = af_2(n)$, for some constant a.

The negation of linear dependence is *linear independence.* Explicitly put, the functions $f_1(n), f_2(n), \ldots, f_r(n)$ are said to be *linearly independent* for $n \geq n_0$ if whenever

$$a_1 f_1(n) + a_2 f_2(n) + \cdots + a_r f_r(n) = 0$$

for all $n \geq n_0$, then we must have $a_1 = a_2 = \cdots = a_r = 0$.

Let us illustrate this new concept by an example.

Example 2.9. Show that the functions 3^n, $n3^n$, and $n^2 3^n$ are linearly independent on $n \geq 1$.

Solution Suppose that for constants a_1, a_2, and a_3 we have

$$a_1 3^n + a_2 n 3^n + a_3 n^2 3^n = 0, \qquad \text{for all } n \geq 1.$$

Then by dividing by 3^n we get

$$a_1 + a_2 n + a_3 n^2 = 0, \qquad \text{for all } n \geq 1.$$

This is impossible unless $a_3 = 0$, since a second-degree equation in n possesses at most two solutions $n \geq 1$. Hence $a_1 = a_2 = a_3 = 0$. Similarly, $a_2 = 0$, whence $a_1 = 0$, which establishes the linear independence of our functions.

Definition 2.10. A set of k linearly independent solutions of (2.2.7) is called a *fundamental set* of solutions.

As you may have noticed from Example 2.9, it is not practical to check the linear independence of a set of solutions using the definition. Fortunately, there is a simple method to check the linear independence of solutions using the so-called Casoratian $W(n)$, which we now define for the eager reader.

Definition 2.11. The Casoratian[4] $W(n)$ of the solutions $x_1(n), x_2(n), \ldots, x_r(n)$ is given by

$$W(n) = \det \begin{pmatrix} x_1(n) & x_2(n) & \ldots & x_r(n) \\ x_1(n+1) & x_2(n+1) & \ldots & x_r(n+1) \\ \vdots & & & \\ x_1(n+r-1) & x_2(n+r-1) & \ldots & x_r(n+r-1) \end{pmatrix}. \tag{2.2.9}$$

Example 2.12. Consider the difference equation

$$x(n+3) - 7x(n+1) + 6x(n) = 0.$$

(a) Show that the sequences 1, $(-3)^n$, and 2^n are solutions of the equation.

(b) Find the Casoratian of the sequences in part (a).

Solution

(a) Note that $x(n) = 1$ is a solution, since $1 - 7 + 6 = 0$. Furthermore, $x(n) = (-3)^n$ is a solution, since

$$(-3)^{n+3} - 7(-3)^{n+1} + 6(-3)^n = (-3)^n[-27 + 21 + 6] = 0.$$

Finally, $x(n) = 2^n$ is a solution, since

$$(2)^{n+3} - 7(2)^{n+1} + 6(2)^n = 2^n[8 - 14 + 6] = 0.$$

[4]This is the discrete analogue of the Wronskian in differential equations.

(b) Now,

$$\begin{aligned}
W(n) &= \det\begin{pmatrix} 1 & (-3)^n & 2^n \\ 1 & (-3)^{n+1} & 2^{n+1} \\ 1 & (-3)^{n+2} & 2^{n+2} \end{pmatrix} \\
&= \begin{vmatrix} (-3)^{n+1} & (2)^{n+1} \\ (-3)^{n+2} & (2)^{n+2} \end{vmatrix} - (-3)^n \begin{vmatrix} 1 & (2)^{n+1} \\ 1 & (2)^{n+2} \end{vmatrix} \\
&\quad + (2)^n \begin{vmatrix} 1 & (-3)^{n+1} \\ 1 & (-3)^{n+2} \end{vmatrix} \\
&= (2)^{n+2}(-3)^{n+1} - (2)^{n+1}(-3)^{n+2} - (-3)^n((2)^{n+2} - (2)^{n+1}) \\
&\quad + (2)^n((-3)^{n+2} - (-3)^{n+1}) \\
&= -12(2)^n(-3)^n - 18(2)^n(-3)^n - 4(2)^n(-3)^n \\
&\quad + 2(2)^n(-3)^n + 9(2)^n(-3)^n + 3(2)^n(-3)^n \\
&= -20(2)^n(-3)^n.
\end{aligned}$$

Next we give a formula, called Abel's formula, to compute the Casoratian $W(n)$. The significance of Abel's formula is its effectiveness in the verification of the linear independence of solutions.

Lemma 2.13 (Abel's Lemma). *Let $x_1(n), x_2(n), \ldots, x_k(n)$ be solutions of (2.2.7) and let $W(n)$ be their Casoratian. Then, for $n \geq n_0$,*

$$\boxed{W(n) = (-1)^{k(n-n_0)} \left(\prod_{i=n_0}^{n-1} p_k(i) \right) W(n_0).} \tag{2.2.10}$$

Proof. We will prove the lemma for $k = 3$, since the general case may be established in a similar fashion. So let $x_1(n)$, $x_2(n)$, and $x_3(n)$ be three independent solutions of (2.2.7). Then from formula (2.2.9) we have

$$W(n+1) = \det\begin{pmatrix} x_1(n+1) & x_2(n+1) & x_3(n+1) \\ x_1(n+2) & x_2(n+2) & x_3(n+2) \\ x_1(n+3) & x_2(n+3) & x_3(n+3) \end{pmatrix}. \tag{2.2.11}$$

From (2.2.7) we have, for $1 \leq i \leq 3$,

$$x_i(n+3) = -p_3(n)x_i(n) - [p_1(n)x_i(n+2) + p_2(n)x_i(n+1)]. \tag{2.2.12}$$

Now, if we use formula (2.2.12) to substitute for $x_1(n+3)$, $x_2(n+3)$, and $x_3(n+3)$ in the last row of formula (2.2.11), we obtain

$$W(n+1) = \det \begin{pmatrix} x_1(n+1) & x_2(n+1) & x_3(n+1) \\ x_1(n+2) & x_2(n+2) & x_3(n+2) \\ -p_3x_1(n) & -p_3x_2(n) & -p_3x_3(n) \\ -(p_2x_1(n+1) & -(p_2x_2(n+1) & -(p_2x_3(n+1) \\ +p_1x_1(n+2)) & +p_1x_2(n+2)) & +p_1x_3(n+2)) \end{pmatrix}. \tag{2.2.13}$$

Using the properties of determinants, it follows from (2.2.13) that

$$W(n+1) = \det \begin{pmatrix} x_1(n+1) & x_2(n+1) & x_3(n+1) \\ x_1(n+2) & x_2(n+2) & x_3(n+2) \\ -p_3(n)x_1(n) & -p_3(n)x_2(n) & -p_3(n)x_3(n) \end{pmatrix} \tag{2.2.14}$$

$$= -p_3(n) \det \begin{pmatrix} x_1(n+1) & x_2(n+1) & x_3(n+1) \\ x_1(n+2) & x_2(n+2) & x_3(n_2) \\ x_1(n) & x_2(n) & x_3(n) \end{pmatrix}$$

$$= -p_3(n)(-1)^2 \det \begin{pmatrix} x_1(n) & x_2(n) & x_3(n) \\ x_1(n+2) & x_2(n+2) & x_3(n+2) \\ x_1(n+1) & x_2(n+1) & x_3(n+1) \end{pmatrix}.$$

Thus

$$W(n+1) = (-1)^3 p_3(n) W(n). \tag{2.2.15}$$

Using formula (1.2.3), the solution of (2.2.15) is given by

$$\boxed{W(n) = \left[\prod_{i=n_0}^{n-1} (-1)^3 p_3(i)\right] W(n_0) = (-1)^{3(n-n_0)} \prod_{i=n_0}^{n-1} p_3(i) W(n_0).}$$

□

This completes the proof of the lemma for $k = 3$. The general case is left to the reader as Exercises 2.2, Problem 6.

We now examine and treat one of the special cases that arises as we try to apply this Casoratian. For example, if (2.2.7) has constant coefficients $p_1, p_2, \ldots, p_k$, then we have

$$\boxed{W(n) = (-1)^{k(n-n_0)} p_k^{(n-n_0)} W(n_0).} \tag{2.2.16}$$

Formula (2.2.10) has the following important correspondence.

Corollary 2.14. *Suppose that $p_k(n) \neq 0$ for all $n \geq n_0$. Then the Casoratian $W(n) \neq 0$ for all $n \geq n_0$ if and only if $W(n_0) \neq 0$.*

PROOF. This corollary follows immediately from formula (2.2.10) (Exercises 2.2, Problem 7). □

Let us have a close look at Corollary 2.14 and examine what it really says. The main point in the corollary is that either the Casoratian is identically zero (i.e., zero for all $n \geq n_0$, for some n_0) or never zero for any $n \geq n_0$. Thus to check whether $W(n) \neq 0$ for all $n \in \mathbb{Z}^+$, we need only to check whether $W(0) \neq 0$. Note that we can always choose the most suitable n_0 and compute $W(n_0)$ there.

Next we examine the relationship between the linear independence of solutions and their Casoratian. Basically, we will show that a set of k solutions is a *fundamental set* (i.e., linearly independent) if their Casoratian $W(n)$ is never zero.

To determine the preceding statement we contemplate k solutions $x_1(n), x_2(n), \ldots, x_k(n)$ of (2.2.7). Suppose that for some constants a_1, $a_2, \ldots, a_k$ and $n_0 \in \mathbb{Z}^+$,

$$a_1 x_1(n) + a_2 x_2(n) + \cdots + a_k(n)\, x_k(n) = 0, \qquad \text{for all } n \geq n_0.$$

Then we can generate the following $k-1$ equations:

$$\begin{aligned} a_1 x_1(n+1) + a_2 x_2(n+1) + \cdots + a_k x_k(n+1) &= 0, \\ &\vdots \\ a_1 x_1(n+k-1) + a_2 x_2(n+k-1) + \cdots + a_k x_k(n+k-1) &= 0. \end{aligned}$$

This assemblage may be transcribed as

$$X(n)\xi = 0, \tag{2.2.17}$$

where

$$X(n) = \begin{pmatrix} x_1(n) & x_2(n) & \ldots & x_k(n) \\ x_1(n+1) & x_2(n+1) & \ldots & x_k(n+1) \\ \vdots & \vdots & & \vdots \\ x_1(n+k-1) & x_2(n+k-1) & \ldots & x_k(n+k-1) \end{pmatrix},$$

$$\xi = \begin{pmatrix} a_1 \\ a_2 \\ \vdots \\ a_k \end{pmatrix}.$$

Observe that $W(n) = \det X(n)$.

Linear algebra tells us that the vector (2.2.17) has only the trivial (or zero) solution (i.e., $a_1 = a_2 = \cdots = a_k = 0$) if and only if the matrix $X(n)$ is nonsingular (invertible) (i.e., $\det X(n) = W(n) \neq 0$ for all $n \geq n_0$). This deduction leads us to the following conclusion.

Theorem 2.15. *The set of solutions $x_1(n), x_2(n), \ldots, x_k(n)$ of* (2.2.7) *is a fundamental set if and only if for some $n_0 \in Z^+$, the Casoratian $W(n_0) \neq 0$.*

PROOF. Exercises 2.2, Problem 8. □

Example 2.16. Verify that $\{n, 2^n\}$ is a fundamental set of solutions of the equation

$$x(n+2) - \frac{3n-2}{n-1}x(n+1) + \frac{2n}{n-1}x(n) = 0.$$

Solution We leave it to the reader to verify that n and 2^n are solutions of the equation. Now, the Casoratian of the solutions n, 2^n is given by

$$W(n) = \det \begin{pmatrix} n & 2^n \\ n+1 & 2^{n+1} \end{pmatrix}.$$

Thus

$$W(0) = \det \begin{pmatrix} 0 & 1 \\ 1 & 2 \end{pmatrix} = -1 \neq 0.$$

Hence by Theorem 2.15, the solutions n, 2^n are linearly independent and thus form a fundamental set.

Example 2.17. Consider the third-order difference equation

$$x(n+3) + 3x(n+2) - 4x(n+1) - 12x(n) = 0.$$

Show that the functions 2^n, $(-2)^n$, and $(-3)^n$ form a fundamental set of solutions of the equation.

Solution

(i) Let us verify that 2^n is a legitimate solution by substituting $x(n) = 2^n$ into the equation:

$$2^{n+3} + (3)(2^{n+1}) - (4)(2^{n+1}) - (12)(2^n) = 2^n[8 + 12 - 8 - 12] = 0.$$

We leave it to the reader to verify that $(-2)^n$ and $(-3)^n$ are solutions of the equation.

(ii) To affirm the linear independence of these solutions we construct the Casoratian

$$W(n) = \det \begin{pmatrix} 2^n & (-2)^n & (-3)^n \\ 2^{n+1} & (-2)^{n+1} & (-3)^{n+1} \\ 2^{n+2} & (-2)^{n+2} & (-3)^{n+2} \end{pmatrix}.$$

Thus

$$W(0) = \det \begin{pmatrix} 1 & 1 & 1 \\ 2 & -2 & 3 \\ 4 & 4 & 9 \end{pmatrix} = -20 \neq 0.$$

By Theorem 2.15, the solutions 2^n, $(-2)^n$, and 3^n are linearly independent, and thus form a fundamental set.

We are now ready to discuss the fundamental theorem of homogeneous linear difference equations.

Theorem 2.18 (The Fundamental Theorem). *If $p_k(n) \neq 0$ for all $n \geq n_0$, then (2.2.7) has a fundamental set of solutions for $n \geq n_0$.*

PROOF. By Theorem 2.7, there are solutions $x_1(n), x_2(n), \ldots, x_k(n)$ such that $x_i(n_0+i-1) = 1$, $x_i(n_0) = x_i(n_0+1) = \cdots = x_i(n_0+i-2) = x_i(n_0+i) = \cdots = x_i(n_0+k-1) = 0, 1 \leq i \leq k$. Hence $x_1(n_0) = 1, x_2(n_0+1) = 1, x_3(n_0+2) = 1, \ldots, x_k(n_0+k-1) = 1$. It follows that $W(n_0) = \det I = 1$. This implies by Theorem 2.15 that the set $\{x_1(n), x_2(n), \ldots, x_k(n)\}$ is a fundamental set of solutions of (2.2.7). □

We remark that there are infinitely many fundamental sets of solutions of (2.2.7). The next result presents a method of generating fundamental sets starting from a known set.

Lemma 2.19. *Let $x_1(n)$ and $x_2(n)$ be two solutions of (2.2.7). Then the following statements hold:*

(i) *$x(n) = x_1(n) + x_2(n)$ is a solution of (2.2.7).*

(ii) *$\tilde{x}(n) = ax_1(n)$ is a solution of (2.2.7) for any constant a.*

PROOF. (Exercises 2.2, Problem 9.) □

From the preceding lemma we conclude the following principle.

Superposition Principle. If $x_1(n), x_2(n), \ldots, x_r(n)$ are solutions of (2.2.7), then

$$x(n) = a_1x_1(n) + a_2x_2(n) + \cdots + a_rx_r(n)$$

is also a solution of (2.2.7) (Exercises 2.2, Problem 12).

Now let $\{x_1(n), x_2(n), \ldots, x_k(n)\}$ be a fundamental set of solutions of (2.2.7) and let $x(n)$ be any given solution of (2.2.7). Then there are constants $a_1, a_2, \ldots, a_k$ such that $x(n) = \sum_{i=1}^{k} a_ix_i(n)$. To show this we use the notation (2.2.17) to write $X(n)\xi = \hat{x}(n)$, where

$$\hat{x}(n) = \begin{pmatrix} x(n) \\ x(n+1) \\ \vdots \\ x(n+k-1) \end{pmatrix}.$$

Since $X(n)$ is invertible (Why?), it follows that

$$\xi = X^{-1}(n)\hat{x}(n),$$

and, for $n = n_0$,

$$\xi = X^{-1}(n_0)\hat{x}(n_0).$$

The above discussion leads us to define the general solution of (2.2.7).

Definition 2.20. Let $\{x_1(n), x_2(n), \ldots, x_k(n)\}$ be a fundamental set of solutions of (2.2.7). Then the *general solution* of (2.2.7) is given by $x(n) = \sum_{i=1}^{k} a_i x_i(n)$, for arbitrary constants a_i.

It is worth noting that any solution of (2.2.7) may be obtained from the general solution by a suitable choice of the constants a_i.

The preceding results may be restated using the elegant language of linear algebra as follows: Let S be the set of all solutions of (2.2.7) with the operations $+$, $\cdot$ defined as follows:

(i) $(x+y)(n) = x(n) + y(n)$, for $x, y \in S$, $n \in Z^+$,

(ii) $(ax)(n) = ax(n)$, for $x \in S$, a constant.

Equipped with linear algebra we now summarize the results of this section in a compact form.

Theorem 2.21. *The space $(S, +, \cdot)$ is a linear (vector) space of dimension k.*

PROOF. Use Lemma 2.19. To construct a basis of S we can use the fundamental set in Theorem 2.18 (Exercises 2.2, Problem 11). □

Exercises 2.2

1. Find the Casoratian of the following functions and determine whether they are linearly dependent or independent:

   (a) $5^n, 3 \cdot 5^{n+2}, e^n$.

   (b) $5^n, n\,5^n, n^2\,5^n$.

   (c) $(-2)^n, 2^n, 3$.

   (d) $0, 3^n, 7^n$.

2. Find the Casoratian $W(n)$ of the solutions of the difference equations:

   (a) $x(n+3) - 10x(n+2) + 31x(n+1) - 30x(n) = 0$, if $W(0) = 6$.

   (b) $x(n+3) - 3x(n+2) + 4x(n+1) - 12x(n) = 0$, if $W(0) = 26$.

3. For the following difference equations and their accompanied solutions:

   (i) determine whether these solutions are linearly independent, and

(ii) find, if possible, using only the given solutions, the general solution:

(a) $x(n+3) - 3x(n+2) + 3x(n+1) - x(n) = 0; \quad 1, n, n^2,$

(b) $x(n+2) + x(n) = 0; \quad \cos\left(\frac{n\pi}{2}\right), \sin\left(\frac{n\pi}{2}\right),$

(c) $x(n+3)+x(n+2)-8x(n+1)-12x(n) = 0; \quad 3^n, (-2)^n, (-2)^{n+3},$

(d) $x(n+4) - 16x(n) = 0; \quad 2^n, n2^n, n^2 2^n.$

4. Verify formula (2.2.10) for the general case.

5. Show that the Casoratian $W(n)$ in formula (2.2.9) may be given by the formula

$$W(n) = \det \begin{pmatrix} x_1(n) & x_2(n) & \dots & x_k(n) \\ \Delta x_1(n) & \Delta x_2(n) & \dots & \Delta x_k(n) \\ \vdots & \vdots & & \vdots \\ \Delta^{k-1} x_1(n) & \Delta^{k-1} x_2(n) & \dots & \Delta^{k-1} x_k(n) \end{pmatrix}.$$

6. Verify formula (2.2.16).

7. Prove Corollary 2.14.

8. Prove Theorem 2.15.

9. Prove Lemma 2.19.

10. Prove the superposition principle: If $x_1(n), x_2(n), \dots, x_r$ are solutions of (2.2.7), then any linear combination of them is also a solution of (2.2.7).

11. Prove Theorem 2.21.

12. Suppose that for some integer $m \geq n_0, p_k(m) = 0$ in (2.2.1).

    (a) What is the value of the Casoratian for $n \geq m$?

    (b) Does Corollary 2.14 still hold? (Why?)

*13. Show that the equation $\Delta^2 y(n) = p(n) y(n+1)$ has a fundamental set of solutions whose Casoratian $W(n) = -1$.

14. Contemplate the second-order difference equation $u(n+2)+p_1(n)u(n+1) + p_2(n)u(n) = 0$. If $u_1(n)$ and $u_2(n)$ are solutions of the equation and $W(n)$ is their Casoratian, prove that

$$\boxed{u_2(n) = u_1(n) \left[\sum_{r=0}^{n-1} W(r)/u_1(r)u_1(r+1)\right].} \tag{2.2.18}$$

15. Contemplate the second-order difference equation $u(n+2) - \frac{(n+3)}{(n+2)} u(n+1) + \frac{2}{(n+2)} u(n) = 0$.

(a) Verify that $u_1(n) = \frac{2^n}{n!}$ is a solution of the equation.

(b) Use formula (2.2.18) to find another solution $u_2(n)$ of the equation.

16. Show that $u(n) = (n+1)$ is a solution of the equation $u(n+2) - u(n+1) - 1/(n+1)u(n) = 0$ and then find a second solution of the equation by using the method of Exercises 2.2, Problem 15.

2.3 Linear Homogeneous Equations with Constant Coefficients

Consider the kth-order difference equation

$$x(n+k) + p_1 x(n+k-1) + p_2 x(n+k-2) + \cdots + p_k x(n) = 0, \quad (2.3.1)$$

where the p_i's are constants and $p_k \neq 0$. Our objective now is to find a fundamental set of solutions and, consequently, the general solution of (2.3.1). The procedure is rather simple. We suppose that solutions of (2.3.1) are in the form λ^n, where λ is a complex number. Substituting this value into (2.3.1), we obtain

$$\lambda^k + p_1 \lambda^{k-1} + \cdots + p_k = 0. \quad (2.3.2)$$

This is called the *characteristic equation* of (2.3.1), and its roots λ are called the *characteristic roots*. Notice that since $p_k \neq 0$, none of the characteristic roots is equal to zero. (Why?) (Exercises 2.3, Problem 19.)

We have two situations to contemplate:

Case (a). Suppose that the characteristic roots $\lambda_1, \lambda_2, \ldots, \lambda_k$ are distinct. We are now going to show that the set $\{\lambda_1^n, \lambda_2^n, \ldots, \lambda_k^n\}$ is a fundamental set of solutions. To prove this, by virtue of Theorem 2.15 it suffices to show that $W(0) \neq 0$, where $W(n)$ is the Casoratian of the solutions. That is,

$$W(0) = \det \begin{pmatrix} 1 & 1 & \cdots & 1 \\ \lambda_1 & \lambda_2 & \cdots & \lambda_k \\ \lambda_1^2 & \lambda_2^2 & \cdots & \lambda_k^2 \\ \vdots & \vdots & & \vdots \\ \lambda_1^{k-1} & \lambda_2^{k-1} & \cdots & \lambda_k^{k-1} \end{pmatrix}. \quad (2.3.3)$$

This determinant is called the *Vandermonde determinant*.

It may be shown by mathematical induction that

$$\boxed{W(0) = \prod_{1 \le i < j \le k} (\lambda_j - \lambda_i).} \quad (2.3.4)$$

The reader will prove this conclusion in Exercises 2.3, Problem 20.

Since all the λ_i's are distinct, it follows from (2.3.4) that $W(0) \neq 0$. This fact proves that $\{\lambda_1^n, \lambda_2^n, \ldots, \lambda_k^n\}$ is a fundamental set of solutions of (2.3.1). Consequently, the general solution of (2.3.1) is

$$x(n) = \sum_{i=1}^{k} a_i \lambda_i^n, \quad a_i \text{ a complex number.} \tag{2.3.5}$$

Case (b). Suppose that the distinct characteristic roots are $\lambda_1, \lambda_2, \ldots, \lambda_r$ with multiplicities $m_1, m_2, \ldots, m_r$ with $\sum_{i=1}^r m_i = k$, respectively. In this case, (2.3.1) may be written as

$$(E - \lambda_1)^{m_1}(E - \lambda_2)^{m_2} \cdots (E - \lambda_r)^{m_r} x(n) = 0. \tag{2.3.6}$$

A vital observation here is that if $\psi_1(n), \psi_2(n), \ldots, \psi_{m_i}(n)$ are solutions of

$$(E - \lambda_i)^{m_i} x(n) = 0, \tag{2.3.7}$$

then they are also solutions of (2.3.6). For if $\Psi_s(n)$ is a solution of (2.3.7), then $(E - \lambda_i)^{m_i}\Psi_s(n) = 0$. Now

$$\begin{aligned}&(E - \lambda_1)^{m_1} \cdots (E - \lambda_i)^{m_i} \cdots (E - \lambda_r)^{m_r}\Psi_s(n)\\ &= (E - \lambda_1)^{m_1} \cdots (E - \lambda_{i-1})^{m_{i-1}}(E - \lambda_{i+1})^{m_{i+1}} \cdots\\ &(E - \lambda_r)^{m_r}(E - \lambda_i)^{m_i}\Psi_s(n) = 0.\end{aligned}$$

Suppose we are able to find a fundamental set of solutions for each (2.3.7), $1 \leq i \leq r$. It is not unreasonable to expect, then, that the union of these r fundamental sets would be a fundamental set of solutions of (2.3.6). In the following lemma we will show that this is indeed the case.

Lemma 2.22. *The set* $G_i = \left\{\lambda_i^n, \binom{n}{1}\lambda_i^{n-1}, \binom{n}{2}\lambda_2^{n-2}, \ldots, \binom{n}{m_i-1}\lambda_i^{n-m_i+1}\right\}$ *is a fundamental set of solutions of* (2.3.7) *where* $\binom{n}{1} = n$, $\binom{n}{2} = \frac{n(n-1)}{2!}, \ldots, \binom{n}{r} = \frac{n(n-1)\cdots(n-r+1)}{r!}$.

PROOF. To show that G_i is a fundamental set of solutions of (2.3.7), it suffices, by virtue of Corollary 2.14, to show that $W(0) \neq 0$. But

$$W(0) = \begin{vmatrix} 1 & 0 & \ldots & 0 \\ \lambda_i & 1 & \ldots & 0 \\ \lambda_i^2 & 2\lambda_i & \ldots & 0 \\ \vdots & \vdots & & \vdots \\ \lambda_i^{m_i-1} & \frac{(m_i-1)}{1!}\lambda_i^{m_i-2} & \ldots & \frac{1}{2!3!\cdots(m_i-2)!} \end{vmatrix}.$$

Hence

$$W(0) = \frac{1}{(2!3!\ldots(m_i-2)!} \neq 0.$$

It remains to show that $\binom{n}{r}\lambda_i^{n-r}$ is a solution of (2.3.7). From equation (2.1.9) it follows that

$$\begin{aligned}(E-\lambda_i)^{m_i}\binom{n}{r}\lambda_i^{n-r} &= \lambda_i^{n-r}(\lambda_i E-\lambda_i)^{m_i}\binom{n}{r}\\ &= \lambda_i^{n+m_i-r}(E-I)^{m_i}\binom{n}{r}\\ &= \lambda_i^{n+m_i-r}\Delta^{m_i}\binom{n}{r}\\ &= 0 \quad \text{using (2.1.6).}\end{aligned}$$

□

Now we are finally able to find a fundamental set of solutions.

Theorem 2.23. *The set $G = \bigcup_{i=1}^{r} G_i$ is a fundamental set of solutions of* (2.3.6).

PROOF. By Lemma 2.22, the functions in G are solutions of (2.3.6). Now

$$W(0) = \det\begin{pmatrix} 1 & 0 & \cdots & 1 & 0 & \cdots\\ \lambda_1 & 1 & \cdots & \lambda_r & 1 & \cdots\\ \lambda_1^2 & 2\lambda_1 & \cdots & \lambda_r^2 & 2\lambda_r & \cdots\\ \vdots & \vdots & & \vdots & \vdots & \\ \lambda_1^{k-1} & (k-1)\lambda_1^{k-2} & \cdots & \lambda_r^{k-1} & (k-1)\lambda_r^{k-2} & \cdots \end{pmatrix}. \tag{2.3.8}$$

This determinant is called the generalized Vandermonde determinant. (See Appendix B.) It may be shown [76] that

$$\boxed{W(0) = \prod_{1\le i<j\le k} (\lambda_j-\lambda_i)^{m_j m_i}.} \tag{2.3.9}$$

As $\lambda_i \neq \lambda_j$, $W(0) \neq 0$. Hence by Corollary 2.14 the Casoratian $W(n) \neq 0$ for all $n \geq 0$. Thus by Theorem 2.15, G is a fundamental set of solutions. □

Corollary 2.24. *The general solution of* (2.3.6) *is given by*

$$x(n) = \sum_{i=1}^{r} \lambda_i^n \left(a_{i0} + a_{i1}n + a_{i2}n^2 + \cdots + a_{i,m_i-1}n^{m_i-1}\right). \tag{2.3.10}$$

PROOF. Use Lemma 2.22 and Theorem 2.23. □

Example 2.25. Solve the equation

$$\begin{aligned} x(n+3) - 7x(n+2) + 16x(n+1) - 12x(n) = 0,\\ x(0) = 0, \quad x(1) = 1, \quad x(2) = 1.\end{aligned}$$

Solution The characteristic equation is

$$r^3 - 7r^2 + 16r - 12 = 0.$$

Thus, the characteristic roots are $\lambda_1 = 2 = \lambda_2, \lambda_3 = 3$.

The characteristic roots give us the general solution

$$x(n) = a_0 2^n + a_1 n 2^n + b_1 3^n.$$

To find the constants a_0, a_1, and b_1, we use the initial data

$$\begin{aligned} x(0) &= a_0 + b_1 = 0, \\ x(1) &= 2a_0 + 2a_1 + 3b_1 = 1, \\ x(2) &= 4a_0 + 8a_1 + 9b_1 = 1. \end{aligned}$$

Finally, after solving the above system of equations, we obtain

$$a_0 = 3, \quad a_1 = 2, \quad b_1 = -3.$$

Hence the solution of the equation is given by $x(n) = 3(2^n) + 2n(2^n) - 3^{n+1}$.

Example 2.26. Complex Characteristic Roots

Suppose that the equation $x(n+2) + p_1 x(n+1) + p_2 x(n) = 0$ has the complex roots $\lambda_1 = \alpha + i\beta, \lambda_2 = \alpha - i\beta$. Its general solution would then be

$$x(n) = c_1(\alpha + i\beta)^n + c_2(\alpha - i\beta)^n.$$

Recall that the point (α, β) in the complex plane corresponds to the complex number $\alpha + i\beta$. In polar coordinates,

$$\alpha = r\cos\theta, \qquad \beta = r\sin\theta, \qquad r = \sqrt{\alpha^2 + \beta^2}, \qquad \theta = \tan^{-1}\left(\frac{\beta}{\alpha}\right).$$

Hence,[5]

$$\begin{aligned} x(n) &= c_1(r\cos\theta + ir\sin\theta)^n + c_2(r\cos\theta - ir\sin\theta)^n \\ &= r^n[(c_1 + c_2)\cos(n\theta) + i(c_1 - c_2)\sin(n\theta)] \\ &= r^n[a_1\cos(n\theta) + a_2\sin(n\theta)], \end{aligned} \tag{2.3.11}$$

where $a_1 = c_1 + c_2$ and $a_2 = i(c_1 - c_2)$.

Let

$$\cos\omega = \frac{a_1}{\sqrt{a_1^2 + a_2^2}}, \qquad \sin\omega = \frac{a_2}{\sqrt{a_1^2 + a_2^2}}, \qquad \omega = \tan^{-1}\left(\frac{a_2}{a_1}\right).$$

[5] We used De Moivre's Theorem: $[r(\cos\theta + i\sin\theta)]^n = r^n(\cos n\theta + i\sin n\theta)$.

Then (2.3.11) becomes

$$\begin{aligned} x(n) &= r^n\sqrt{a_1^2 + a_2^2}[\cos\omega\cos(n\theta) + \sin\omega\sin(n\theta)] \\ &= r^n\sqrt{a_1^2 + a_2^2}\cos(n\theta - \omega), \\ x(n) &= Ar^n\cos(n\theta - \omega). \end{aligned} \tag{2.3.12}$$

Example 2.27. The Fibonacci Sequence (The Rabbit Problem)

This problem first appeared in 1202, in *Liber abaci*, a book about the abacus, written by the famous Italian mathematician Leonardo di Pisa, better known as Fibonacci. The problem may be stated as follows: How many pairs of rabbits will there be after one year if starting with one pair of mature rabbits, if each pair of rabbits gives birth to a new pair each month starting when it reaches its maturity age of two months? (See Figure 2.1.)

Table 2.2 shows the number of pairs of rabbits at the end of each month. The first pair has offspring at the end of the first month, and thus we have two pairs. At the end of the second month only the first pair has offspring, and thus we have three pairs. At the end of the third month, the first and second pairs will have offspring, and hence we have five pairs. Continuing this procedure, we arrive at Table 2.2. If $F(n)$ is the number of pairs of rabbits at the end of n months, then the recurrence relation that represents this model is given by the second-order linear difference equation

$$F(n+2) = F(n+1) + F(n), \qquad F(0) = 1, \qquad F(1) = 2, \qquad 0 \le n \le 10.$$

This example is a special case of the Fibonacci sequence, given by

$$F(n+2) = F(n+1) + F(n), \qquad F(0) = 0, \qquad F(1) = 1, \qquad n \ge 0. \tag{2.3.13}$$

The first 14 terms are given by 1, 2, 3, 5, 8, 13, 21, 34, 55, 89, 144, 233, and 377, as already noted in the rabbit problem.

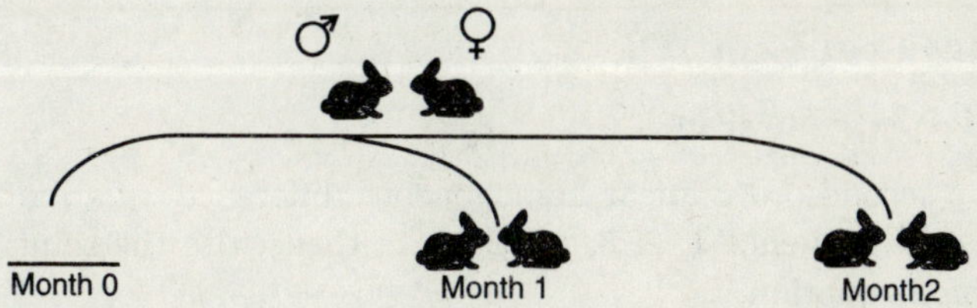

FIGURE 2.1.

TABLE 2.2. Rabbits' population size.

Month	0	1	2	3	4	5	6	7	8	9	10	11	12
Pairs	1	2	3	5	8	13	21	34	55	89	144	233	377

The characteristic equation of (2.3.13) is

$$\lambda^2 - \lambda - 1 = 0.$$

Hence the characteristic roots are $\alpha = \frac{1+\sqrt{5}}{2}$ and $\beta = \frac{1-\sqrt{5}}{2}$.
The general solution of (2.3.13) is

$$F(n) = a_1 \left(\frac{1+\sqrt{5}}{2}\right)^n + a_2 \left(\frac{1-\sqrt{5}}{2}\right)^n, \qquad n \geq 1. \tag{2.3.14}$$

Using the initial values $F(1) = 1$ and $F(2) = 1$, one obtains

$$a_1 = \frac{1}{\sqrt{5}}, \qquad a_2 = -\frac{1}{\sqrt{5}}.$$

Consequently,

$$F(n) = \frac{1}{\sqrt{5}}\left[\left(\frac{1+\sqrt{5}}{2}\right)^n - \left(\frac{1-\sqrt{5}}{2}\right)^n\right] = \frac{1}{\sqrt{5}}(\alpha^n - \beta^n). \tag{2.3.15}$$

It is interesting to note that $\lim_{n\to\infty} \frac{F(n+1)}{F(n)} = \alpha \approx 1.618$ (Exercises 2.3, Problem 15). This number is called the *golden mean*, which supposedly represents the ratio of the sides of a rectangle that is most pleasing to the eye. This Fibonacci sequence is very interesting to mathematicians; in fact, an entire publication, *The Fibonacci Quarterly*, dwells on the intricacies of this fascinating sequence.

Exercises 2.3

1. Find homogeneous difference equations whose solutions are:
   (a) $2^{n-1} - 5^{n+1}$.
   (b) $3\cos\left(\frac{n\pi}{2}\right) - \sin\left(\frac{n\pi}{2}\right)$.
   (c) $(n+2)5^n \sin\left(\frac{n\pi}{4}\right)$.
   (d) $(c_1 + c_2 n + c_3 n^2)7^n$.
   (e) $1 + 3n - 5n^2 + 6n^3$.

2. Find a second-order linear homogeneous difference equation that generates the sequence 1, 2, 5, 12, 29, ...; then write the solution of the obtained equation.

In each of Problems 3 through 8, write the general solution of the difference equation.

3. $x(n+2) - 16x(n) = 0$.
4. $x(n+2) + 16x(n) = 0$.
5. $(E-3)^2(E^2+4)x(n) = 0$.

6. $\Delta^3 x(n) = 0$.

7. $(E^2+2)^2 x(n) = 0$.

8. $x(n+2) - 6x(n+1) + 14x(n) = 0$.

9. Consider Example 2.26. Verify that $x_1(n) = r^n \cos n\theta$ and $x_2(n) = r^n \sin n\theta$ are two linearly independent solutions of the given equation.

10. Consider the integral defined by

$$I_k(\varphi) = \int_0^{\pi} \frac{\cos(k\theta) - \cos(k\varphi)}{\cos\theta - \cos\varphi} d\theta, \quad k = 0, 1, 2, \ldots, \quad \varphi \in \mathbb{R}.$$

(a) Show that $I_k(\varphi)$ satisfies the difference equation

$$I_{n+2}(\varphi) - 2\cos\varphi I_{n+1}(\varphi) + I_n(\varphi) = 0, \qquad I_0(\varphi) = 0, \qquad I_1(\varphi) = \pi.$$

(b) Solve the difference equation in part (a) to find $I_n(\varphi)$.

11. The Chebyshev polynomials of the first and second kinds are defined, respectively, as follows:

$$T_n(x) = \cos(n\cos^{-1}(x)), \qquad U_n(x) = \frac{1}{\sqrt{1-x^2}} \sin[(n+1)\cos^{-1}(x)],$$

for $|x| < 1$.

(a) Show that $T_n(x)$ obeys the difference equation

$$T_{n+2}(x) - 2xT_{n+1}(x) + T_n(x) = 0, \qquad T_0(x) = 1, \qquad T_1(x) = x.$$

(b) Solve the difference equation in part (a) to find $T_n(x)$.

(c) Show that $U_n(x)$ satisfies the difference equation

$$U_{n+2}(x) - 2xU_{n+1}(x) + U_n(x) = 0, \qquad U_0(x) = 1, \qquad U_1(x) = 2x.$$

(d) Write down the first three terms of $T_n(x)$ and $U_n(x)$.

(e) Show that $T_n(\cos\theta) = \cos n\theta$ and that

$$U_n(\cos\theta) = (\sin[(n+1)\theta])/\sin\theta.$$

12. Show that the general solution of

$$x(n+2) - 2sx(n+1) + x(n) = 0, \qquad |s| < 1,$$

is given by

$$x(n) = c_1 T_n(s) + c_2 U_n(s).$$

13. Show that the general solution of $x(n+2) + p_1 x(n+1) + p_2 x(n) = 0$, $p_2 > 0$, $p_1^2 < 4p_2$, is given by $x(n) = r^n[c_1 T_n(s) + c_2 U_{n-1}(s)]$, where $r = \sqrt{p_2}$ and $s = P_1/(2\sqrt{p_2})$.

14. The Lucas numbers L_n are defined by the difference equation

$$L_{n+2} = L_{n+1} + L_n, \quad n \geq 0, \quad L_0 = 2, \quad L_1 = 1.$$

Solve the difference equation to find L_n.

15. Show that $\lim_{n\to\infty}(F(n+1))/F(n) = \alpha$, where $\alpha = (1+\sqrt{5})/2$.

16. Prove that consecutive Fibonacci numbers $F(n)$ and $F(n+1)$ are relatively prime.

17. (a) Prove that $F(n)$ is the nearest integer to $1/\sqrt{5}((1+\sqrt{5})/2)^n$.

    (b) Find $F(17)$, $F(18)$, and $F(19)$, applying part (a).

*18. Define $x = a \bmod p$ if $x = mp + a$. Let p be a prime number with $p > 5$.

    (a) Show that $F(p) = 5^{(p-1)/2} \bmod p$.

    (b) Show that $F(p) = \pm 1 \bmod p$.

19. Show that if $p_k \neq 0$ in (2.3.1), then none of its characteristic roots is equal to zero.

20. Show that the Vandermonde determinant (2.3.3) is equal to

$$\prod_{1\leq i<j\leq k} (\lambda_j - \lambda_i).$$

21. Find the value of the $n \times n$ tridiagonal determinant

$$D(n) = \begin{vmatrix} b & a & 0 & \dots & 0 & 0 \\ a & b & a & \dots & 0 & 0 \\ 0 & a & b & \dots & 0 & 0 \\ \vdots & \vdots & & & & \vdots \\ 0 & 0 & 0 & \dots & b & a \\ 0 & 0 & 0 & \dots & a & b \end{vmatrix}.$$

22. Find the value of the $n \times n$ tridiagonal determinant

$$D(n) = \begin{vmatrix} a & b & 0 & \dots & 0 & 0 \\ c & a & b & \dots & 0 & 0 \\ 0 & c & a & \dots & 0 & 0 \\ \vdots & \vdots & \vdots & & \vdots & \vdots \\ 0 & 0 & 0 & \dots & a & b \\ 0 & 0 & 0 & \dots & c & a \end{vmatrix}.$$

2.4 Linear Nonhomogeneous Equations: Method of Undetermined Coefficients

In the last two sections we developed the theory of linear homogeneous difference equations. Moreover, in the case of equations with constant coefficients we have shown how to construct their solutions. In this section we focus our attention on solving the kth-order linear nonhomogeneous equation

$$y(n+k) + p_1(n)y(n+k-1) + \cdots + p_k(n)y(n) = g(n), \tag{2.4.1}$$

where $p_k(n) \neq 0$ for all $n \geq n_0$. The sequence $g(n)$ is called the *forcing term*, the *external force*, the *control*, or the *input* of the system. As we will discuss later in Chapter 6, equation (2.4.1) represents a physical system in which $g(n)$ is the input and $y(n)$ is the output (Figure 2.2). Thus solving (2.4.1) amounts to determining the output $y(n)$ given the input $g(n)$. We may look at $g(n)$ as a control term that the designing engineer uses to force the system to behave in a specified way.

Before proceeding to present general results concerning (2.4.1) we would like to raise the following question: Do solutions of (2.4.1) form a vector space? In other words, is the sum of two solutions of (2.4.1) a solution of (2.4.1)? And is a multiple of a solution of (2.4.1) a solution of (2.4.1)? Let us answer these questions through the following example.

Example 2.28. Contemplate the equation

$$y(n+2) - y(n+1) - 6y(n) = 5(3^n).$$

(a) Show that $y_1(n) = n(3^{n-1})$ and $y_2(n) = (1+n)3^{n-1}$ are solutions of the equation.

(b) Show that $y(n) = y_2(n) - y_1(n)$ is not a solution of the equation.

(c) Show that $\varphi(n) = cn(3^{n-1})$ is not a solution of the equation, where c is a constant.

Solution

(a) The verification that y_1 and y_2 are solutions is left to the reader.

(b) $y(n) = y_2(n) - y_1(n) = 3^{n-1}$. Substituting this into the equation yields

$$3^{n+1} - 3^n - 63^{n-1} = 3^n[3-1-2] = 0 \neq 5(3^n).$$

(c) By substituting for $\varphi(n)$ into the equation we see easily that $\varphi(n)$ is not a solution.

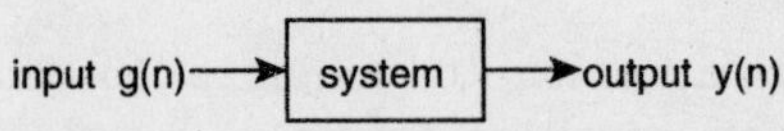

FIGURE 2.2. Input–output system.

Conclusion

(i) From the above example we conclude that in contrast to the situation for homogeneous equations, solutions of the nonhomogeneous equation (2.4.1) do not form a vector space. In particular, neither the sum (difference) of two solutions nor a multiple of a solution is a solution.

(ii) From part (b) in Example 2.28 we found that the difference of the solutions $y_2(n)$ and $y_1(n)$ of the nonhomogeneous equation is actually a solution of the associated homogeneous equation. This is indeed true for the general nth-order equation, as demonstrated by the following result.

Theorem 2.29. *If $y_1(n)$ and $y_2(n)$ are solutions of* (2.4.1), *then $x(n) = y_1(n) - y_2(n)$ is a solution of the corresponding homogeneous equation*

$$x(n+k) + p_1(n)x(n+k-1) + \cdots + p_k(n)x(n) = 0. \tag{2.4.2}$$

PROOF. The reader will undertake the justification of this theorem in Exercises 2.4, Problem 12. □

It is customary to refer to the general solution of the homogeneous equation (2.4.2) as the *complementary solution* of the nonhomogeneous equation (2.4.1), and it will be denoted by $y_c(n)$. A solution of the nonhomogeneous equation (2.4.1) is called a *particular solution* and will be denoted by $y_p(n)$. The next result gives us an algorithm to generate all solutions of the nonhomogeneous equation (2.4.1).

Theorem 2.30. *Any solution $y(n)$ of* (2.4.1) *may be written as*

$$\boxed{y(n) = y_p(n) + \sum_{i=1}^{k} a_i x_i(n),}$$

where $\{x_1(n), x_2(n), \ldots, x_k(n)\}$ is a fundamental set of solutions of the homogeneous equation (2.4.2).

PROOF. Observe that according to Theorem 2.29, $y(n) - y_p(n)$ is a solution of the homogeneous equation (2.4.2). Thus $y(n) - y_p(n) = \sum_{i=1}^{k} a_i x_i(n)$, for some constants a_i.

The preceding theorem leads to the definition of the *general solution* of the nonhomogeneous equation (2.4.1) as

$$y(n) = y_c(n) + y_p(n). \tag{2.4.3}$$

□

We now turn our attention to finding a particular solution y_p of nonhomogeneous equations with constant coefficients such as

$$y(n+k) + p_1 y(n+k-1) + \cdots + p_k y(n) = g(n). \tag{2.4.4}$$

Because of its simplicity, we use the method of *undetermined coefficients* to compute y_p.

Basically, the method consists in making an intelligent guess as to the form of the particular solution and then substituting this function into the difference equation. For a completely arbitrary nonhomogeneous term $g(n)$, this method is not effective. However, definite rules can be established for the determination of a particular solution by this method if $g(n)$ is a linear combination of terms, each having one of the forms

$$a^n, \quad \sin(bn), \quad \cos(bn), \quad \text{or } n^k, \tag{2.4.5}$$

or products of these forms such as

$$a^n \sin(bn), \qquad a^n n^k, \qquad a^n n^k \cos(bn), \ldots . \tag{2.4.6}$$

Definition 2.31. A polynomial operator $N(E)$, where E is the shift operator, is said to be an *annihilator* of $g(n)$ if

$$N(E)g(n) = 0. \tag{2.4.7}$$

In other words, $N(E)$ is an annihilator of $g(n)$ if $g(n)$ is a solution of (2.4.7). For example, an annihilator of $g(n) = 3^n$ is $N(E) = E - 3$, since $(E-3)y(n) = 0$ has a solution $y(n) = 3^n$. An annihilator of $g(n) = \cos\frac{n\pi}{2}$ is $N(E) = E^2 + 1$, since $(E^2+1)y(n) = 0$ has a solution $y(n) = \cos\frac{n\pi}{2}$. Let us now rewrite (2.4.4) using the shift operator E as

$$p(E)y(n) = g(n), \tag{2.4.8}$$

where $p(E) = E^k + p_1E^{k-1} + p_2E^{k-2} + \cdots + p_kI$.

Assume now that $N(E)$ is an annihilator of $g(n)$ in (2.4.8). Applying $N(E)$ on both sides of (2.4.8) yields

$$N(E)p(E)y(n) = 0. \tag{2.4.9}$$

Let $\lambda_1, \lambda_2, \ldots, \lambda_k$ be the characteristic roots of the homogeneous equation

$$p(E)y(n) = 0, \tag{2.4.10}$$

and let $\mu_1, \mu_2, \ldots, \mu_k$ be the characteristic roots of

$$N(E)y(n) = 0. \tag{2.4.11}$$

We must consider two separate cases.

Case 1. None of the λ_i's equals any of the μ_i's. In this case, write $y_p(n)$ as the general solution of (2.4.11) with undetermined constants. Substituting back this "guesstimated" particular solution into (2.4.4), we find the values of the constants. Table 2.3 contains several types of functions $g(n)$ and their corresponding particular solutions.

Case 2. $\lambda_i = \mu_j$ for some i, j. In this case, the set of characteristic roots of (2.4.9) is equal to the union of the sets $\{\lambda_i\}$, $\{\mu_j\}$ and, consequently,

TABLE 2.3. Particular solutions $y_p(n)$.

$g(n)$	$y_p(n)$
a^n	$c_1 a^n$
n^k	$c_0 + c_1 n + \cdots + c_k n^k$
$n^k a^n$	$c_0 a^n + c_1 n a^n + \cdots + c_k n^k a^n$
$\sin bn, \cos bn$	$c_1 \sin bn + c_2 \cos bn$
$a^n \sin bn, a^n \cos bn$	$(c_1 \sin bn + c_2 \cos bn)a^n$
$a^n n^k \sin bn, a^n n^k \cos bn$	$(c_0 + c_1 n + \cdots + c_k n^k)a^n \sin(bn)$ $+ (d_0 + d_1 n + \cdots d_k n^k)a^n \cos(bn)$

contains roots of higher multiplicity than the two individual sets of characteristic roots. To determine a particular solution $y_p(n)$, we first find the general solution of (2.4.9) and then drop all the terms that appear in $y_c(n)$. Then proceed as in Case 1 to evaluate the constants.

Example 2.32. Solve the difference equation

$$y(n+2) + y(n+1) - 12y(n) = n2^n. \tag{2.4.12}$$

Solution The characteristic roots of the homogeneous equation are $\lambda_1 = 3$ and $\lambda_2 = -4$.

Hence,

$$y_c(n) = c_1 3^n + c_2(-4)^n.$$

Since the annihilator of $g(n) = n2^n$ is given by $N(E) = (E-2)^2$ (Why?), we know that $\mu_1 = \mu_2 = 2$. This equation falls in the realm of Case 1, since $\lambda_i \neq \mu_j$, for any i, j. So we let

$$y_p(n) = a_1 2^n + a_2 n 2^n.$$

Substituting this relation into equation (2.4.12) gives

$$a_1 2^{n+2} + a_2(n+2)2^{n+2} + a_1 2^{n+1} + a_2(n+1)2^{n+1} - 12a_1 2^n - 12a_2 n 2^n = n2^n,$$

$$(10a_2 - 6a_1)2^n - 6a_2 n 2^n = n2^n.$$

Hence

$$10a_2 - 6a_1 = 0 \quad \text{and} \quad -6a_2 = 1,$$

or

$$a_1 = \frac{-5}{18}, \qquad a_2 = \frac{-1}{6}.$$

The particular solution is

$$y_p(n) = \frac{-5}{18}2^n - \frac{1}{6}n2^n,$$

and the general solution is

$$y(n) = c_1 3^n + c_2(-4)^n - \frac{5}{18}2^n - \frac{1}{6}n2^n.$$

Example 2.33. Solve the difference equation

$$(E-3)(E+2)y(n) = 5(3^n). \tag{2.4.13}$$

Solution The annihilator of $5(3^n)$ is $N(E) = (E-3)$. Hence, $\mu_1 = 3$. The characteristic roots of the homogeneous equation are $\lambda_1 = 3$ and $\lambda_2 = -2$. Since $\lambda_1 = \mu_1$, we apply the procedure for Case 2.

Thus,

$$(E-3)^2(E+2)y(n) = 0. \tag{2.4.14}$$

Now $y_c(n) = c_1 3^n + c_2(-2)^n$.

We now know that the general solution of (2.4.14) is given by

$$\tilde{y}(n) = (a_1 + a_2 n)3^n + a_3(-2)^n.$$

Omitting from $\tilde{y}(n)$ the terms 3^n and $(-2)^n$ that appeared in $y_c(n)$, we set $y_p(n) = a_2 n 3^n$. Substituting this $y_p(n)$ into (2.4.13) gives

$$a_2(n+2)3^{n+2} - a_2(n+1)3^{n+1} + 6a_2 n 3^n = 5.3^n,$$

or

$$a_2 = \frac{1}{3}.$$

Thus $y_p(n) = n3^{n-1}$, and the general solution of (2.4.13) is

$$y(n) = c_1 3^n + c_2(-2)^n + n3^{n-1}.$$

Example 2.34. Solve the difference equation

$$y(n+2) + 4y(n) = 8(2^n)\cos\left(\frac{n\pi}{2}\right). \tag{2.4.15}$$

Solution The characteristic equation of the homogeneous equation is

$$\lambda^2 + 4 = 0.$$

The characteristic roots are

$$\lambda_1 = 2i, \qquad \lambda_2 = -2i.$$

Thus $r = 2, \theta = \pi/2$, and

$$y_c(n) = 2^n\left(c_1\cos\left(\frac{n\pi}{2}\right) + c_2\sin\left(\frac{n\pi}{2}\right)\right).$$

Notice that $g(n) = 2^n\cos\left(\frac{n\pi}{2}\right)$ appears in $y_c(n)$. Using Table 2.3, we set

$$y_p(n) = 2^n\left(an\cos\left(\frac{n\pi}{2}\right) + bn\sin\left(\frac{n\pi}{2}\right)\right). \tag{2.4.16}$$

Substituting (2.4.16) into (2.4.15) gives

$$\begin{aligned} &2^{n+2}\left[a(n+2)\cos\left(\frac{n\pi}{2}+\pi\right) + b(n+2)\sin\left(\frac{n\pi}{2}+\pi\right)\right] \\ &+ (4)2^n\left[an\cos\left(\frac{n\pi}{2}\right) + bn\sin\left(\frac{n\pi}{2}\right)\right] = 8(2^n)\cos\left(\frac{n\pi}{2}\right). \end{aligned}$$

Replacing $\cos((n\pi)/2 + \pi)$ by $-\cos((n\pi)/2)$, and $\sin((n\pi)/2 + \pi)$ by $-\sin((n\pi)/2)$ and then comparing the coefficients of the cosine terms leads us to $a = -1$. Then by comparing the coefficients of the sine terms, we realize that $b = 0$.

By substituting these values back into (2.4.16), we know that

$$y_p(n) = -2^n n \cos\left(\frac{n\pi}{2}\right),$$

and the general solution of (2.4.15), arrived at by adding $y_c(n)$ and $y_p(n)$, is

$$y(n) = 2^n \left(c_1 \cos\frac{n\pi}{2} + c_2 \sin\left(\frac{n\pi}{2}\right) - n\cos\left(\frac{n\pi}{2}\right)\right).$$

Exercises 2.4.

For Problems 1 through 6, find a particular solution of the difference equation.

1. $y(n+2) - 5y(n+1) + 6y(n) = 1 + n.$
2. $y(n+2) + 8y(n+1) + 12y(n) = e^n.$
3. $y(n+2) - 5y(n+1) + 4y(n) = 4^n - n^2.$
4. $y(n+2) + 8y(n+1) + 7y(n) = ne^n.$
5. $y(n+2) - y(n) = n\cos\left(\frac{n\pi}{2}\right).$
6. $(E^2 + 9)^2 y(n) = \sin\left(\frac{n\pi}{2}\right) - \cos\left(\frac{n\pi}{2}\right).$

For Problems 7 through 9 find the solution of the difference equation.

7. $\Delta^2 y(n) = 16, \quad y(0) = 2, \quad y(1) = 3.$
8. $\Delta^2 y(n) + 7y(n) = 2\sin\left(\frac{n\pi}{4}\right), \quad y(0) = 0, \quad y(1) = 1.$
9. $(E - 3)(E^2 + 1)y(n) = 3^n, \quad y(0) = 0, \quad y(1) = 1, \quad y(2) = 3.$

For Problems 10 and 11 find the general solution of the difference equation.

10. $y(n+2) - y(n) = n2^n \sin\left(\frac{n\pi}{2}\right).$
11. $y(n+2) + 8y(n+1) + 7y(n) = n2^n.$
12. Prove Theorem 2.29.
13. Consider the difference equation $y(n+2) + p_1 y(n+1) + p_2 y(n) = g(n)$, where $p_1^2 < 4p_2$ and $0 < p_2 < 1$. Show that if y_1 and y_2 are two solutions of the equation, then $y_1(n) - y_2(n) \to 0$ as $n \to \infty$.
14. Determine the general solution of $y(n+2) + \lambda^2 y(n) = \sum_{m=1}^{N} a_m \sin(m\pi n)$, where $\lambda > 0$ and $\lambda \neq m\pi, m = 1, 2, \ldots, N$.

15. Solve the difference equation

$$y(n+2)+y(n)=\begin{cases}1 & \text{if } 0\le n\le 2,\\ -1 & \text{if } n>2,\end{cases}$$

with $y(0)=0$ and $y(1)=1$.

2.4.1 The Method of Variation of Constants (Parameters)

Contemplate the second-order nonhomogeneous difference equation

$$y(n+2)+p_1(n)y(n+1)+p_2(n)y(n)=g(n) \tag{2.4.17}$$

and the corresponding homogeneous difference equation

$$y(n+2)+p_1(n)y(n+1)+p_2(n)y(n)=0. \tag{2.4.18}$$

The method of variation of constants is commonly used to find a particular solution $y_p(n)$ of (2.4.17) when the coefficients $p_1(n)$ and $p_2(n)$ are not constants. The method assumes that a particular solution of (2.4.17) may be written in the form

$$y(n)=u_1(n)y_1(n)+u_2(n)y_2(n), \tag{2.4.19}$$

where $y_1(n)$ and $y_2(n)$ are two linearly independent solutions of the homogeneous equation (2.4.18), and $u_1(n), u_2(n)$ are sequences to be determined later.

16. (a) Show that

$$\begin{aligned}y(n+1)&=u_1(n)y_1(n+1)+u_2(n)y_2(n+1)\\ &\quad+\Delta u_1(n)y_1(n+1)+\Delta u_2(n)y_2(n+1).\end{aligned} \tag{2.4.20}$$

(b) The method stipulates that

$$\Delta u_1(n)y_1(n+1)+\Delta u_2(n)y_2(n+1)=0. \tag{2.4.21}$$

Use (2.4.20) and (2.4.21) to show that

$$\begin{aligned}y(n+2)&=u_1(n)y_1(n+2)+u_2(n)y_2(n+2)\\ &\quad+\Delta u_1(n)y_1(n+2)+\Delta u_2(n)y_2(n+2).\end{aligned}$$

(c) By substituting the above expressions for $y(n)$, $y(n+1)$, and $y(n+2)$ into (2.4.17), show that

$$\Delta u_1(n)y_1(n+2)+\Delta u_2(n)y_2(n+2)=g(n). \tag{2.4.22}$$

(d) Using expressions (2.4.21) and (2.4.22), show that

$$\Delta u_1(n)=\frac{-g(n)y_2(n+1)}{W(n+1)},\qquad u_1(n)=\sum_{r=0}^{n-1}\frac{-g(r)y_2(r+1)}{W(r+1)}, \tag{2.4.23}$$

$$\Delta u_2(n) = \frac{g(n)y_1(n+1)}{W(n+1)}, \qquad u_2(n) = \sum_{r=0}^{n-1} \frac{g(r)y_1(r+1)}{W(r+1)}, \tag{2.4.24}$$

where $W(n)$ is the Casoratian of $y_1(n)$ and $y_2(n)$.

17. Use formulas (2.4.23) and (2.4.24) to solve the equation

$$y(n+2) - 7y(n+1) + 6y(n) = n.$$

18. Use the variation of constants method to solve the initial value problem

$$y(n+2) - 5y(n+1) + 6y(n) = 2^n, \qquad y(1) = y(2) = 0.$$

19. Use Problem 16(d) to show that the unique solution of (2.4.17) with $y(0) = y(1) = 0$ is given by

$$y(n) = \sum_{r=0}^{n-1} \frac{y_1(r+1)y_2(n) - y_2(r+1)y_1(n)}{W(r+1)}.$$

20. Consider the equation

$$x(n+1) = ax(n) + f(n). \tag{2.4.25}$$

(a) Show that

$$x(n) = a^n \left[x(0) + \frac{f(0)}{a} + \frac{f(1)}{a^2} + \cdots + \frac{f(n-1)}{a^n} \right] \tag{2.4.26}$$

is a solution of (2.4.25).

(b) Show that if $|a| < 1$ and $\{f(n)\}$ is a bounded sequence, i.e., $|f(n)| \leq M$, for some $M > 0$, $n \in \mathbb{Z}^+$, then all solutions of (2.4.25) are bounded.

(c) Suppose that $a > 1$ and $\{f(n)\}$ is bounded on $\mathbb{Z}^+$. Show that if we choose

$$x(0) = -\left(\frac{f(0)}{a} + \frac{f(1)}{a^2} + \cdots + \frac{f(n)}{a^{n+1}} + \cdots \right) = -\sum_{i=0}^{\infty} \frac{f(i)}{a^{i+1}}, \tag{2.4.27}$$

then the solution $x(n)$ given by (2.4.26) is bounded on $\mathbb{Z}^+$. Give an explicit expression for $x(n)$ in this case.

(d) Under the assumptions of part (c), show that for any choice of $x(0)$, excepting that value given by (2.4.27), the solution of (2.4.25) is unbounded.

2.5 Limiting Behavior of Solutions

To simplify our exposition we restrict our discussion to the second-order difference equation

$$y(n+2) + p_1 y(n+1) + p_2 y(n) = 0. \tag{2.5.1}$$

Suppose that λ_1 and λ_2 are the characteristic roots of the equation. Then we have the following three cases:

(a) λ_1 and λ_2 are distinct real roots. Then $y_1(n) = \lambda_1^n$ and $y_2(n) = \lambda_2^n$ are two linearly independent solutions of (2.5.1). If $|\lambda_1| > |\lambda_2|$, then we call $y_1(n)$ the *dominant solution*, and λ_1 the *dominant characteristic root*. Otherwise, $y_2(n)$ is the dominant solution, and λ_2 is the dominant characteristic root. We will now show that the limiting behavior of the general solution $y(n) = a_1\lambda_1^n + a_2\lambda_2^n$ is determined by the behavior of the dominant solution. So assume, without loss of generality, that $|\lambda_1| > |\lambda_2|$. Then

$$y(n) = \lambda_1^n \left[a_1 + a_2 \left(\frac{\lambda_2}{\lambda_1} \right)^n \right].$$

Since

$$\left| \frac{\lambda_2}{\lambda_1} \right| < 1,$$

it follows that

$$\left(\frac{\lambda_2}{\lambda_1} \right)^n \to 0 \quad \text{as } n \to \infty.$$

Consequently, $\lim_{n\to\infty} y(n) = \lim_{n\to\infty} a_1\lambda_1^n$. There are six different situations that may arise here depending on the value of λ_1 (see Figure 2.3).

1. $\lambda_1 > 1$: The sequence $\{a_1\lambda_1^n\}$ diverges to ∞ (unstable system).

2. $\lambda_1 = 1$: The sequence $\{a_1\lambda_1^n\}$ is a constant sequence.

3. $0 < \lambda_1 < 1$: The sequence $\{a_1\lambda_1^n\}$ is monotonically decreasing to zero (stable system).

4. $-1 < \lambda_1 < 0$: The sequence $\{a_1\lambda_1^n\}$ is oscillating around zero (i.e., alternating in sign) and converging to zero (stable system).

5. $\lambda_1 = -1$: The sequence $\{a_1\lambda_1^n\}$ is oscillating between two values a_1 and $-a_1$.

6. $\lambda_1 < -1$: The sequence $\{a_1\lambda_1^n\}$ is oscillating but increasing in magnitude (unstable system).

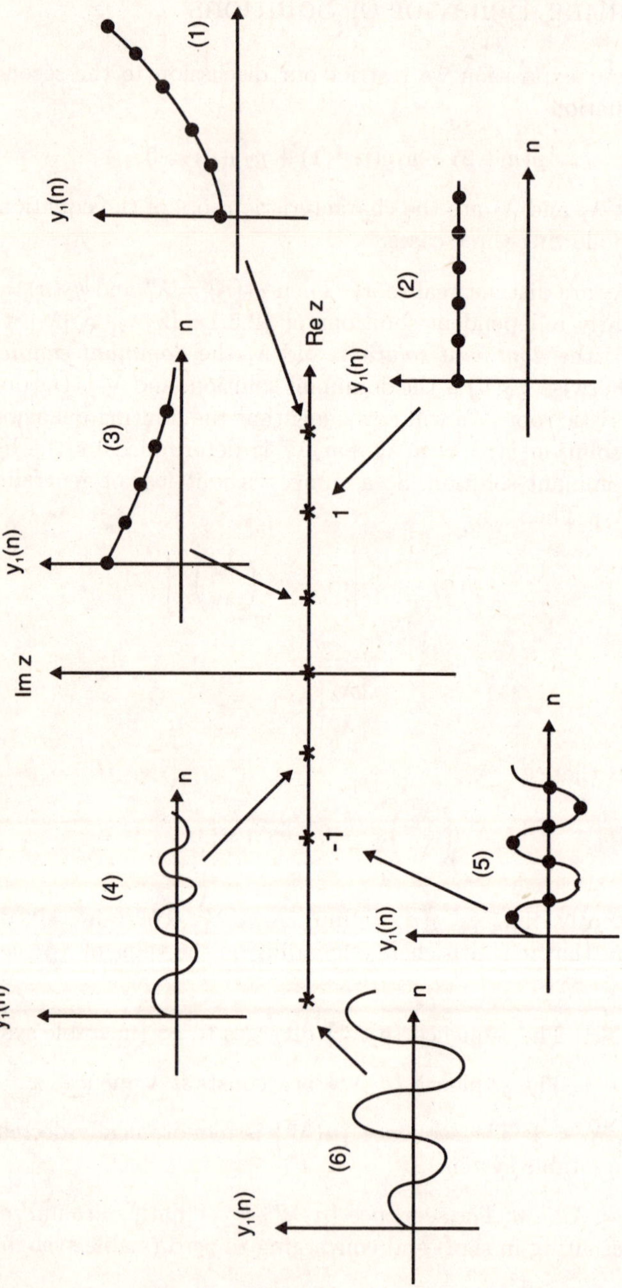

FIGURE 2.3. $(n, y(n))$ diagrams for real roots.

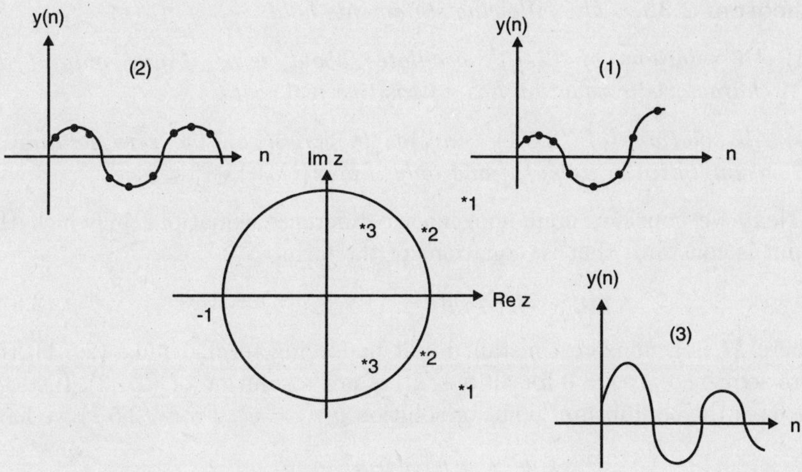

FIGURE 2.4. $(n, y(n))$ diagrams for comples roots.

(b) $\lambda_1 = \lambda_2 = \lambda$.
The general solution of (2.5.1) is given by $y(n) = (a_1 + a_2 n)\lambda^n$. Clearly, if $|\lambda| \geq 1$, the solution $y(n)$ diverges either monotonically if $\lambda \geq 1$ or by oscillating if $\lambda \leq -1$. However, if $|\lambda| < 1$, then the solution converges to zero, since $\lim_{n\to\infty} n\lambda^n = 0$ (Why?).

(c) Complex roots: $\lambda_1 = \alpha + i\beta$ and $\lambda_2 = \alpha - i\beta$, where $\beta \neq 0$.
As we have seen in Section 2.3, formula (2.3.12), the solution of (2.5.1) is given by $y(n) = ar^n \cos(n\theta - \omega)$, where

$$r = \sqrt{\alpha^2 + \beta^2}, \qquad \theta = \tan^{-1}\left(\frac{\beta}{\alpha}\right).$$

The solution $y(n)$ clearly oscillates, since the cosine function oscillates. However, $y(n)$ oscillates in three different ways depending on the location of the conjugate characteristic roots, as may be seen in Figure 2.4.

1. $r > 1$: Here λ_1 and $\lambda_2 = \overline{\lambda}_1$ are outside the unit circle. Hence $y(n)$ is oscillating but increasing in magnitude (unstable system).

2. $r = 1$: Here λ_1 and $\lambda_2 = \overline{\lambda}_1$ lie on the unit circle. In this case $y(n)$ is oscillating but constant in magnitude.

3. $r < 1$: Here λ_1 and $\lambda_2 = \overline{\lambda}_1$ lie inside the unit disk. The solution $y(n)$ oscillates but converges to zero as $n \to \infty$ (stable system).

Finally, we summarize the above discussion in the following theorem.

Theorem 2.35. *The following statements hold:*

(i) *All solutions of* (2.5.1) *oscillate (about zero) if and only if the characteristic equation has no positive real roots.*

(ii) *All solutions of* (2.5.1) *converge to zero (i.e., the zero solution is asymptotically stable) if and only if* $\max\{|\lambda_1|, |\lambda_2|\} < 1$.

Next we consider nonhomogeneous difference equations in which the input is constant, that is, equations of the form

$$y(n+2) + p_1 y(n+1) + p_2 y(n) = M, \tag{2.5.2}$$

where M is a nonzero constant input or forcing term. Unlike (2.5.1), the zero sequence $y(n) = 0$ for all $n \in \mathbb{Z}^+$ is not a solution of (2.5.2). Instead, we have the equilibrium point or solution $y(n) = y^*$. From (2.5.2) we have

$$y^* + p_1 y^* + p_2 y^* = M,$$

or

$$y^* = \frac{M}{1 + p_1 + p_2}. \tag{2.5.3}$$

Thus $y_p(n) = y^*$ is a particular solution of (2.5.2). Consequently, the general solution of (2.5.2) is given by

$$y(n) = y^* + y_c(n). \tag{2.5.4}$$

It is clear that $y(n) \to y^*$ if and only if $y_c(n) \to 0$ as $n \to \infty$. Furthermore, $y(n)$ oscillates[6] about y^* if and only if $y_c(n)$ oscillates about zero. These observations are summarized in the following theorem.

Theorem 2.36. *The following statements hold:*

(i) *All solutions of the nonhomogeneous equation* (2.5.2) *oscillate about the equilibrium solution* y^* *if and only if none of the characteristic roots of the homogeneous equation* (2.5.1) *is a positive real number.*

(ii) *All solutions of* (2.5.2) *converge to* y^* *as* $n \to \infty$ *if and only if* $\max\{|\lambda_1|, |\lambda_2|\} < 1$, *where* λ_1 *and* λ_2 *are the characteristic roots of the homogeneous equation* (2.5.1).

Theorems 2.35 and 2.36 give necessary and sufficient conditions under which a second-order difference equation is asymptotically stable. In many applications, however, one needs to have explicit criteria for stability based on the values of the coefficients p_1 and p_2 of (2.5.2) or (2.5.1). The following result provides us with such needed criteria.

---

[6]We say $y(n)$ oscillates about y^* if $y(n) - y^*$ alternates sign, i.e., if $y(n) > y^*$, then $y(n+1) < y^*$.

Theorem 2.37. *The conditions*

$$1 + p_1 + p_2 > 0, \qquad 1 - p_1 + p_2 > 0, \qquad 1 - p_2 > 0 \tag{2.5.5}$$

are necessary and sufficient for the equilibrium point (solution) of equations (2.5.1) *and* (2.5.2) *to be asymptotically stable (i.e., all solutions converge to* y^**).*

PROOF. Assume that the equilibrium point of (2.5.1) or (2.5.2) is asymptotically stable. In virtue of Theorems 2.35 and 2.36, the roots λ_1, λ_2 of the characteristic equation $\lambda^2 + p_1\lambda + p_2 = 0$ lie inside the unit disk, i.e., $|\lambda_1| < 1$ and $|\lambda_2| < 1$. By the quadratic formula, we have

$$\lambda_1 = \frac{-p_1 + \sqrt{p_1^2 - 4p_2}}{2} \quad \text{and} \quad \lambda_2 = \frac{-p_1 - \sqrt{p_1^2 - 4p_2}}{2}. \tag{2.5.6}$$

Then we have two cases to consider.

Case 1. λ_1, λ_2 are real roots, i.e., $p_1^2 - 4p_2 \geq 0$. From formula (2.5.6) we have

$$-2 < -p_1 + \sqrt{p_1^2 - 4p_2} < 2,$$

or

$$-2 + p_1 < \sqrt{p_1^2 - 4p_2} < 2 + p_1. \tag{2.5.7}$$

Similarly, one obtains

$$-2 + p_1 < -\sqrt{p_1^2 - 4p_2} < 2 + p_1. \tag{2.5.8}$$

Squaring the second inequality in expression (2.5.7) yields

$$1 + p_1 + p_2 > 0. \tag{2.5.9}$$

Similarly, if we square the first inequality in expression (2.5.8) we obtain

$$1 - p_1 + p_2 > 0. \tag{2.5.10}$$

Now from the second inequality of (2.5.7) and the first inequality of (2.5.8) we obtain

$$2 + p_1 > 0 \quad \text{and} \quad 2 - p_1 > 0 \quad \text{or} \quad |p_1| < 2$$

since $p_1^2 - 4p_2 \geq 0$, $p_2 \leq p_1^2/4 < 1$. This completes the proof of (2.5.5) in this case.

Case 2. λ_1 and λ_2 are complex conjugates, i.e., $p_1^2 - 4p_2 < 0$. In this case we have

$$\lambda_{1,2} = \frac{-p_1}{2} \pm \frac{i}{2}\sqrt{4p_2 - p_1^2}.$$

Moreover, since $p_1^2 < 4p_2$, it follows that $-2\sqrt{p_2} < p_1 < 2\sqrt{p_2}$. Now $|\lambda_1|^2 = \frac{p_1^2}{4} + \frac{4p_2}{4} - \frac{p_1^2}{4} = p_2$. Since $|\lambda_1| < 1$, it follows that $0 < p_2 < 1$.

Hence to show that the first two inequalities of (2.5.5) hold we need to show that the function $f(x) = 1 + x - 2\sqrt{x} > 0$ for $x \in (0, 1)$. Observe that $f(0) = 1$, and $f'(x) = 1 - \frac{1}{\sqrt{x}}$. Thus $x = 1$ is a local minimum as $f(x)$ decreases for $x \in (0, 1)$. Hence $f(x) > 0$ for all $x \in (0, 1)$.

This completes the proof of the necessary conditions. The converse is left to the reader as Exercises 2.5, Problem 8. □

Example 2.38. Find conditions under which the solutions of the equation

$$y(n+2) - \alpha(1+\beta)y(n+1) + \alpha\beta y(n) = 1, \qquad \alpha, \beta > 0,$$

(a) converge to the equilibrium point y^*, and

(b) oscillate about y^*.

Solution Let us first find the equilibrium point y^*. Be letting $y(n) = y^*$ in the equation, we obtain

$$y^* = \frac{1}{1-\alpha}, \qquad \alpha \neq 1.$$

(a) Applying condition (2.5.5) to our equation yields

$$\alpha < 1, \qquad 1 + \alpha + 2\alpha\beta > 0, \qquad \alpha\beta < 1.$$

Clearly, the second inequality $1+\alpha+2\alpha\beta > 0$ is always satisfied, since α, β are both positive numbers.

(b) The solutions are oscillatory about y^* if either λ_1, λ_2 are negative real numbers or complex conjugates. In the first case we have

$$\alpha^2(1+\beta)^2 > 4\alpha\beta, \quad \text{or} \quad \alpha > \frac{4\beta}{(1+\beta)^2},$$

and

$$\alpha(1+\beta) < 0,$$

which is impossible. Thus if $\alpha > 4\beta/(1+\beta)^2$, we have no oscillatory solutions.

Now, λ_1 and λ_2 are complex conjugates if

$$\alpha^2(1+\beta)^2 < 4\alpha\beta \quad \text{or} \quad \alpha < \frac{4\beta}{(1+\beta)^2}.$$

Hence all solutions are oscillatory if

$$\alpha < \frac{4\beta}{(1+\beta)^2}.$$

For the treatment of the general kth-order scalar difference equations, the reader is referred to Chapter 4, on stability, and Chapter 8, on oscillation.

Exercises 2.5.

In Problems 1 through 4:

(a) Determine the stability of the equilibrium point by using Theorem 2.35 or Theorem 2.36.

(b) Determine the oscillatory behavior of the solutions of the equation.

1. $y(n+2) - 2y(n+1) + 2y(n) = 0$.
2. $y(n+2) + \frac{1}{4}y(n) = \frac{5}{4}$.
3. $y(n+2) + y(n+1) + \dfrac{1}{2}y(n) = -5$.
4. $y(n+2) - 5y(n+1) + 6y(n) = 0$.
5. Determine the stability of the equilibrium point of the equations in Problems 1 through 4 by using Theorem 2.37.
6. Show that the stability conditions (2.5.5) for the equation $y(n+2) - \alpha y(n+1) + \beta y(n) = 0$, where α, β are constants, may be written as
$$-1-\beta < \alpha < 1+\beta, \qquad \beta < 1.$$
7. Contemplate the equation $y(n+2) - p_1 y(n+1) - p_2 y(n) = 0$. Show that if $|p_1| + |p_2| < 1$, then all solutions of the equation converge to zero.
8. Prove that conditions (2.5.5) imply that all solutions of (2.5.2) converge to the equilibrium point y^*.
9. Determine conditions under which all solutions of the difference equation in Problem 7 oscillate.
10. Determine conditions under which all solutions of the difference equation in Problem 6 oscillate.
11. Suppose that p is a real number. Prove that every solution of the difference equation $y(n+2) - y(n+1) + py(n) = 0$ oscillates if and only if $p > \frac{1}{4}$.

*12. Prove that a necessary and sufficient condition for the asymptotic stability of the zero solution of the equation
$$y(n+2) + p_1 y(n+1) + p_2 y(n) = 0$$
is
$$|p_1| < 1 + p_2 < 2.$$

13. Determine the limiting behavior of solutions of the equation

$$y(n+2) = \alpha c + \alpha\beta(y(n+1) - y(n))$$

if:

(i) $\alpha\beta = 1$,

(ii) $\alpha\beta = 2$,

(iii) $\alpha\beta = \frac{1}{2}$,

provided that α, β, and c are positive constants.

14. If $p_1 > 0$ and $p_2 > 0$, show that all solutions of the equation

$$y(n+2) + p_1 y(n+1) + p_2 y(n) = 0$$

are oscillatory.

15. Determine the limiting behavior of solutions of the equation

$$y(n+2) - \frac{\beta}{\alpha} y(n+1) + \frac{\beta}{\alpha} y(n) = 0,$$

where α and β are constants, if:

(i) $\beta > 4\alpha$,

(ii) $\beta < 4\alpha$.

2.6 Nonlinear Equations Transformable to Linear Equations

In general, most nonlinear difference equations cannot be solved explicitly. However, a few types of nonlinear equations can be solved, usually by transforming them into linear equations. In this section we discuss some tricks of the trade.

Type I. Equations of Riccati type:

$$x(n+1)x(n) + p(n)x(n+1) + q(n)x(n) = 0. \tag{2.6.1}$$

To solve the Riccati equation, we let

$$z(n) = \frac{1}{x(n)}$$

in (2.6.1) to give us

$$q(n)z(n+1) + p(n)z(n) + 1 = 0. \tag{2.6.2}$$

The nonhomogeneous equation requires a different transformation

$$y(n+1)y(n) + p(n)y(n+1) + q(n)y(n) = g(n). \tag{2.6.3}$$

If we let $y(n) = (z(n+1)/z(n)) - p(n)$ in (2.6.3) we obtain

$$z(n+2) + (q(n) - p(n+1))z(n+1) - (g(n) + p(n)q(n))z(n) = 0.$$

Example 2.39. The Pielou Logistic Equation

The most popular continuous model of the growth of a population is the well-known Verhulst–Pearl equation given by

$$x'(t) = x(t)[a - bx(t)], \qquad a, b > 0, \tag{2.6.4}$$

where $x(t)$ is the size of the population at time t; a is the rate of the growth of the population if the resources were unlimited and the individuals did not affect one another, and $-bx^2(t)$ represents the negative effect on the growth due to crowdedness and limited resources. The solution of (2.6.4) is given by

$$x(t) = \frac{a/b}{1 + (e^{-at}/cb)}.$$

Now,

$$\begin{aligned} x(t+1) &= \frac{a/b}{1 + \left(e^{-a(t+1)}/cb\right)} \\ &= \frac{e^a(a/b)}{1 + (e^{-at}/cb) + (e^a - 1)}. \end{aligned}$$

Dividing by $[1 + (e^{-at}/cb)]$, we obtain

$$x(t+1) = \frac{e^a x(t)}{\left[1 + \frac{b}{a}(e^a - 1)x(t)\right]},$$

or

$$x(n+1) = \frac{\alpha x(n)}{[1 + \beta x(n)]}, \tag{2.6.5}$$

where $\alpha = e^a$ and $\beta = \frac{b}{a}(e^a - 1)$.

This equation is titled the *Pielou logistic equation.*

Equation (2.6.5) is of Riccati type and may be solved by letting $x(n) = 1/z(n)$. This gives us the equation

$$z(n+1) = \frac{1}{\alpha} z(n) + \frac{\beta}{\alpha},$$

whose solution is given by

$$z(n) = \begin{cases} \left[c - \dfrac{\beta}{\alpha - 1}\right] \alpha^{-n} + (\beta/(\alpha - 1)) & \text{if } \alpha \neq 1, \\ c + \beta n & \text{if } \alpha = 1. \end{cases}$$

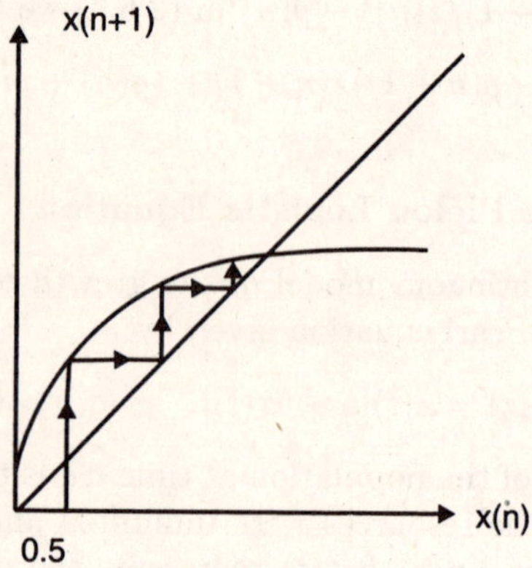

FIGURE 2.5. Asymptotically stable equilibrium points.

Thus

$$x(n) = \begin{cases} \alpha^n(\alpha-1)/[\beta\alpha^n + c(\alpha-1) - \beta] & \text{if } \alpha \neq 1, \\ \dfrac{1}{c+\beta n} & \text{if } \alpha = 1. \end{cases}$$

Hence

$$\lim_{n\to\infty} x(n) = \begin{cases} (\alpha-1)/\beta & \text{if } \alpha \neq 1, \\ 0 & \text{if } \alpha = 1. \end{cases}$$

This conclusion shows that the equilibrium point $(\alpha - 1)/\beta$ is globally asymptotically stable if $\alpha \neq 1$. Figure 2.5 illustrates this for $\alpha = 3$, $\beta = 1$, and $x(0) = 0.5$.

Type II. Equations of general Riccati type:

$$x(n+1) = \frac{a(n)x(n) + b(n)}{c(n)x(n) + d(n)} \tag{2.6.6}$$

such that $c(n) \neq 0, a(n)d(n) - b(n)c(n) \neq 0$ for all $n \geq 0$.

To solve this equation we let

$$c(n)x(n) + d(n) = \frac{y(n+1)}{y(n)}. \tag{2.6.7}$$

Then by substituting

$$x(n) = \frac{y(n+1)}{c(n)y(n)} - \frac{d(n)}{c(n)}$$

into (2.6.6) we obtain

$$\frac{y(n+2)}{c(n+1)y(n+1)} - \frac{d(n+1)}{c(n+1)} = \frac{a(n)\left[\dfrac{y(n+1)}{c(n)y(n)} - \dfrac{d(n)}{c(n)}\right] + b(n)}{\dfrac{y(n+1)}{y(n)}}.$$

This equation simplifies to

$$\begin{aligned} &y(n+2) + p_1(n)y(n+1) + p_2(n)y(n) = 0, \\ &y(0) = 1, \quad y(1) = c(0)x(0) + d(0), \end{aligned} \tag{2.6.8}$$

where

$$p_1(n) = -\frac{c(n)d(n+1) + a(n)c(n+1)}{c(n)},$$

$$p_2(n) = (a(n)d(n) - b(n)c(n))\frac{c(n+1)}{c(n)}.$$

Example 2.40. Solve the difference equation

$$x(n+1) = \frac{2x(n)+3}{3x(n)+2}.$$

Solution Here $a = 2, b = 3, c = 3$, and $d = 2$. Hence $ad - bc \neq 0$. Using the transformation

$$3x(n) + 2 = \frac{y(n+1)}{y(n)}, \tag{2.6.9}$$

we obtain, as in (2.6.8),

$$y(n+2) - 4y(n+1) - 5y(n) = 0, \qquad y(0) = 1, \qquad y(1) = 3x(0) + 2,$$

with characteristic roots $\lambda_1 = 5, \lambda_2 = -1$.

Hence

$$y(n) = c_1 5^n + c_2(-1)^n. \tag{2.6.10}$$

From formula (2.6.9) we have

$$\begin{aligned} x(n) &= \frac{1}{3}\frac{y(n+1)}{y(n)} - \frac{2}{3} = \frac{1}{3}\frac{c_1 5^{n+1} + c_2(-1)^{n+1}}{c_1 5^n + c_2(-1)^n} - \frac{2}{3} \\ &= \frac{(c_1 5^n - c_2(-1)^n)}{(c_1 5^n + c_2(-1)^n)} = \frac{5^n - c(-1)^n}{5^n + c(-1)^n}, \end{aligned}$$

where

$$c = \frac{c_1}{c_2}.$$

Type III. Homogeneous difference equations of the type

$$f\left(\frac{x(n+1)}{x(n)}, n\right) = 0.$$

Use the transformation $z(n) = \frac{x(n+1)}{x(n)}$ to convert such an equation to a linear equation in $z(n)$, thus allowing it to be solved.

Example 2.41. Solve the difference equation

$$x^2(n+1) - 3x(n+1)x(n) + 2x^2(n) = 0. \tag{2.6.11}$$

Solution Dividing by $x^2(n)$, equation (2.6.11) becomes

$$\left[\frac{x(n+1)}{x(n)}\right]^2 - 3\left[\frac{x(n+1)}{x(n)}\right] + 2 = 0, \tag{2.6.12}$$

which is of Type III.

Letting $z(n) = \frac{x(n+1)}{x(n)}$ in (2.6.12) creates

$$z^2(n) - 3z(n) + 2 = 0.$$

We can factor this down to

$$[z(n) - 2][z(n) - 1] = 0,$$

and thus either $z(n) = 2$ or $z(n) = 1$.

This leads to

$$x(n+1) = 2x(n) \quad \text{or} \quad x(n+1) = x(n).$$

Starting with $x(0) = x_0$, there are infinitely many solutions $x(n)$ of (2.6.11) of the form

$$x_0, \ldots, x_0; 2x_0, \ldots, 2x_0; 2^2x_0, \ldots, 2^2x_0; \ldots .$$[7]

Type IV. Consider the difference equation of the form

$$(y(n+k))^{r_1}(y(n+k-1))^{r_2} \cdots (y(n))^{r_{k+1}} = g(n). \tag{2.6.13}$$

Let $z(n) = \ln y(n)$, and rearrange to obtain

$$r_1 z(n+k) + r_2 z(n+k-1) + \cdots + r_{k+1} z(n) = \ln g(n). \tag{2.6.14}$$

Example 2.42. Solve the difference equation

$$x(n+2) = \frac{x^2(n+1)}{x^2(n)}. \tag{2.6.15}$$

Solution Let $z(n) = \ln x(n)$ in (2.6.15). Then as in (2.6.12) we obtain

$$z(n+2) - 2z(n+1) + 2z(n) = 0.$$

The characteristic roots are $\lambda_1 = 1 + i$, $\lambda_2 = 1 - i$.

Thus,

$$z(n) = (2)^{n/2}\left[c_1 \cos\left(\frac{n\pi}{4}\right) + c_2 \sin\left(\frac{n\pi}{4}\right)\right].$$

---

[7]This solution was given by Sebastian Pancratz of the Technical University of Munich.

Therefore,

$$x(n) = \exp\left[(2)^{n/2}\left\{c_1 \cos\left(\frac{n\pi}{4}\right) + c_2 \sin\left(\frac{n\pi}{4}\right)\right\}\right].$$

Exercises 2.6

1. Find the general solution of the difference equation
$$y^2(n+1) - 2y(n+1)y(n) - 3y^2(n) = 0.$$
2. Solve the difference equation
$$y^2(n+1) - (2+n)y(n+1)y(n) + 2ny^2(n) = 0.$$
3. Solve $y(n+1)y(n) - y(n+1) + y(n) = 0$.
4. Solve $y(n+1)y(n) - \frac{2}{3}y(n+1) + \frac{1}{6}y(n) = \frac{5}{18}$.
5. Solve $y(n+1) = 5 - \frac{6}{y(n)}$.
6. Solve $x(n+1) = \frac{x(n)+a}{x(n)+1}, 1 \neq a > 0$.
7. Solve $x(n+1) = x^2(n)$.
8. Solve the logistic difference equation
$$x(n+1) = 2x(n)(1 - x(n)).$$
9. Solve the logistic equation
$$x(n+1) = 4x(n)[1 - x(n)].$$
10. Solve $x(n+1) = \frac{1}{2}\left(x(n) - \frac{a}{x(n)}\right), a > 0$.
11. Solve $y(n+2) = y^3(n+1)/y^2(n)$.
12. Solve $x(n+1) = \frac{2x(n)+4}{x(n)-1}$.
13. Solve $y(n+1) = \frac{2 - y^2(n)}{2(1 - y(n))}$.
14. Solve $x(n+1) = \frac{2x(n)}{x(n)+3}$.
15. Solve $y(n+1) = 2y(n)\sqrt{1 - y^2(n)}$.
16. The "regular falsi" method for finding the roots of $f(x) = 0$ is given by
$$x(n+1) = \frac{x(n-1)f(x(n)) - x(n)f(x(n-1))}{f(x(n)) - f(x(n-1))}.$$

(a) Show that for $f(x) = x^2$, this difference equation becomes

$$x(n+1) = \frac{x(n-1)x(n)}{x(n-1)+x(n)}.$$

(b) Let $x(1) = 1, x(2) = 1$ for the equation in part (a). Show that the solution of the equation is $x(n) = 1/F(n)$, where $F(n)$ is the nth Fibonacci number.

2.7 Applications

2.7.1 Propagation of Annual Plants

The material of this section comes from Edelstein–Keshet [37] of plant propagation. Our objective here is to develop a mathematical model that describes the number of plants in any desired generation. It is known that plants produce seeds at the end of their growth season (say August), after which they die. Furthermore, only a fraction of these seeds survive the winter, and those that survive germinate at the beginning of the season (say May), giving rise to a new generation of plants.

Let

γ = number of seeds produced per plant in August,

α = fraction of one-year-old seeds that germinate in May,

β = fraction of two-year-old seeds that germinate in May,

σ = fraction of seeds that survive a given winter.

If $p(n)$ denotes the number of plants in generation n, then

$$p(n) = \begin{pmatrix} \text{plants from} \\ \text{one-year-old seeds} \end{pmatrix} + \begin{pmatrix} \text{plants from} \\ \text{two-year-old seeds} \end{pmatrix},$$

$$p(n) = \alpha s_1(n) + \beta s_2(n), \tag{2.7.1}$$

where $s_1(n)$ (respectively, $s_2(n)$) is the number of one-year-old (two-year-old) seeds in April (before germination). Observe that the number of seeds left after germination may be written as

$$\text{seeds left} = \begin{pmatrix} \text{fraction} \\ \text{not germinated} \end{pmatrix} \times \begin{pmatrix} \text{original number} \\ \text{of seeds in April} \end{pmatrix}.$$

This gives rise to two equations:

$$\tilde{s}_1(n) = (1-\alpha)s_1(n), \tag{2.7.2}$$

$$\tilde{s}_2(n) = (1-\beta)s_2(n), \tag{2.7.3}$$

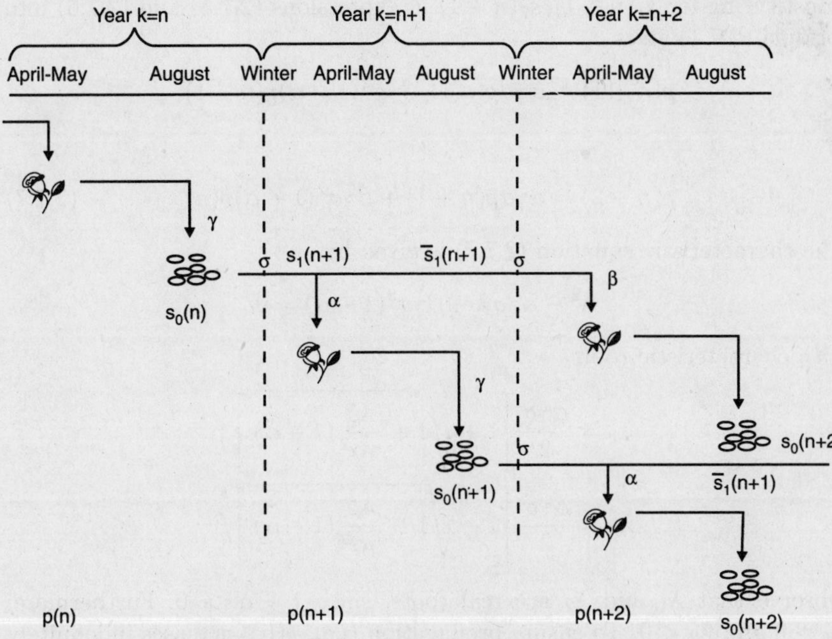

FIGURE 2.6. Propogation of annual plants.

where $\tilde{s}_1(n)$ (respectively, $\tilde{s}_2(n)$) is the number of one-year (two-year-old) seeds left in May after some have germinated. New seeds $s_0(n)$ (0-year-old) are produced in August (Figure 2.6) at the rate of γ per plant,

$$s_0(n) = \gamma p(n). \tag{2.7.4}$$

After winter, seeds $s_0(n)$ that were new in generation n will be one year old in the next generation $n+1$, and a fraction $\sigma s_0(n)$ of them will survive. Hence

$$s_1(n+1) = \sigma s_0(n),$$

or, by using formula (2.7.4), we have

$$s_1(n+1) = \sigma\gamma p(n). \tag{2.7.5}$$

Similarly,

$$s_2(n+1) = \sigma\tilde{s}_1(n),$$

which yields, by formula (2.7.2),

$$\begin{aligned} s_2(n+1) &= \sigma(1-\alpha)s_1(n), \\ s_2(n+1) &= \sigma^2\gamma(1-\alpha)p(n-1). \end{aligned} \tag{2.7.6}$$

Substituting for $s_1(n+1), s_2(n+1)$ in expressions (2.7.5) and (2.7.6) into formula (2.7.1) gives

$$p(n+1) = \alpha\gamma\sigma p(n) + \beta\gamma\sigma^2(1-\alpha)p(n-1),$$

or

$$p(n+2) = \alpha\gamma\sigma p(n+1) + \beta\gamma\sigma^2(1-\alpha)p(n). \tag{2.7.7}$$

The characteristic equation (2.7.7) is given by

$$\lambda^2 - \alpha\gamma\sigma\lambda - \beta\gamma\sigma^2(1-\alpha) = 0$$

with characteristic roots

$$\lambda_1 = \frac{\alpha\gamma\sigma}{2}\left[1+\sqrt{1+\frac{4\beta}{\gamma\alpha^2}(1-\alpha)}\right],$$

$$\lambda_2 = \frac{\alpha\gamma\sigma}{2}\left[1-\sqrt{1+\frac{4\beta}{\gamma\alpha^2}(1-\alpha)}\right].$$

Observe that λ_1 and λ_2 are real roots, since $1-\alpha > 0$. Furthermore, $\lambda_1 > 0$ and $\lambda_2 < 0$. To ensure propagation (i.e., $p(n)$ increases indefinitely as $n \to \infty$) we need to have $\lambda_1 > 1$. We are not going to do the same with λ_2, since it is negative and leads to undesired fluctuation (oscillation) in the size of the plant population. Hence

$$\frac{\alpha\gamma\sigma}{2}\left[1+\sqrt{1+\frac{4\beta}{\gamma\alpha^2}(1-\alpha)}\right] > 1,$$

or

$$\frac{\alpha\gamma\sigma}{2}\sqrt{1+\frac{4\beta(1-\alpha)}{\gamma\alpha^2}} > 1 - \frac{\alpha\gamma\sigma}{2}.$$

Squaring both sides and simplifying yields

$$\gamma > \frac{1}{\alpha\sigma + \beta\sigma^2(1-\alpha)}. \tag{2.7.8}$$

If $\beta = 0$, that is, if no two-year-old seeds germinate in May, then condition (2.7.8) becomes

$$\gamma > \frac{1}{\alpha\sigma}. \tag{2.7.9}$$

Condition (2.7.9) says that plant propagation occurs if the product of the fraction of seeds produced per plant in August, the fraction of one-year-old seeds that germinate in May, and the fraction of seeds that survive a given winter exceeds 1.

2.7.2 Gambler's Ruin

A gambler plays a sequence of games against an adversary in which the probability that the gambler wins \$1.00 in any given game is a known value q, and the probability of his losing \$1.00 is $1-q$, where $0 \leq q \leq 1$. He quits gambling if he either loses all his money or reaches his goal of acquiring N dollars. If the gambler runs out of money first, we say that the gambler has been ruined. Let $p(n)$ denote the probability that the gambler will be ruined if he possesses n dollars. He may be ruined in two ways. First, winning the next game; the probability of this event is q; then his fortune will be $n+1$, and the probability of being ruined will become $p(n+1)$. Second, losing the next game; the probability of this event is $1-q$, and the probability of being ruined is $p(n-1)$. Hence applying the theorem of total probabilities, we have

$$p(n) = qp(n+1) + (1-q)p(n-1).$$

Replacing n by $n+1$, we get

$$p(n+2) - \frac{1}{q}p(n+1) + \frac{(1-q)}{q}p(n) = 0, \quad n = 0, 1, \ldots, N, \tag{2.7.10}$$

with $p(0) = 1$ and $p(N) = 0$. The characteristic equation is given by

$$\lambda^2 - \frac{1}{q}\lambda + \frac{1-q}{q} = 0,$$

and the characteristic roots are given by

$$\lambda_1 = \frac{1}{2q} + \frac{1-2q}{2q} = \frac{1-q}{q},$$

$$\lambda_2 = \frac{1}{2q} - \frac{1-2q}{2q} = 1.$$

Hence the general solution may be written as

$$p(n) = c_1 + c_2\left(\frac{1-q}{q}\right)^n, \qquad \text{if } q \neq \frac{1}{2}.$$

Now using the initial conditions $p(0) = 1, P(N) = 0$ we obtain

$$c_1 + c_2 = 1, \qquad c_1 + c_2\left(\frac{1-q}{q}\right)^N = 0,$$

which gives

$$c_1 = \frac{-\left(\frac{1-q}{q}\right)^N}{1 - \left(\frac{1-q}{q}\right)^N}, \qquad c_2 = \frac{1}{1 - \left(\frac{1-q}{q}\right)^N}.$$

Thus

$$p(n) = \frac{\left(\frac{1-q}{q}\right)^n - \left(\frac{1-q}{q}\right)^N}{1 - \left(\frac{1-q}{q}\right)^N}. \qquad (2.7.11)$$

The special case $q = \frac{1}{2}$ must be treated separately, since in this case we have repeated roots $\lambda_1 = \lambda_2 = 1$. This is certainly the case when we have a fair game. The general solution in this case may be given by

$$p(n) = a_1 + a_2 n,$$

which with the initial conditions yields

$$p(n) = 1 - \frac{n}{N} = \frac{N-n}{N}. \qquad (2.7.12)$$

For example, suppose you start with \$4, the probability that you win a dollar is 0.3, and you will quit if you run out of money or have a total of \$10. Then $n = 4$, $q = 0.3$, and $N = 10$, and the probability of being ruined is given by

$$p(4) = \frac{\left(\frac{7}{3}\right)^4 - \left(\frac{7}{3}\right)^{10}}{1 - \left(\frac{7}{3}\right)^{10}} = 0.994.$$

On the other hand, if $q = 0.5$, $N = \$100.00$, and $n = 20$, then from formula (2.7.12) we have

$$p(20) = 1 - \frac{20}{100} = 0.8.$$

Observe that if $q \le 0.5$ and $N \to \infty$, $p(n)$ tends to 1 in both formulas (2.7.11) and (2.7.12), and the gambler's ruin is certain.

The probability that the gambler wins is given by

$$\tilde{p}(n) = 1 - p(n) = \begin{cases} \dfrac{1 - \left(\frac{1-q}{q}\right)^n}{1 - \left(\frac{1-q}{q}\right)^N}, & \text{if } q \ne 0.5, \\[2ex] \dfrac{n}{N}, & \text{if } q = 0.5. \end{cases} \qquad (2.7.13)$$

2.7.3 National Income

In a capitalist country the national income $Y(n)$ in a given period n may be written as

$$Y(n) = C(n) + I(n) + G(n), \qquad (2.7.14)$$

where

$C(n) =$ consumer expenditure for purchase of consumer goods,

$I(n) =$ induced private investment for buying capital equipment, and

$G(n) =$ government expenditure,

where n is usually measured in years.

We now make some assumptions that are widely accepted by economists (see, for example, Samuelson [129]).

(a) Consumer expenditure $C(n)$ is proportional to the national income $Y(n-1)$ in the preceding year $n-1$, that is,

$$C(n) = \alpha Y(n-1), \tag{2.7.15}$$

where $\alpha > 0$ is commonly called the *marginal propensity to consume.*

(b) Induced private investment $I(n)$ is proportional to the increase in consumption $C(n) - C(n-1)$, that is,

$$I(n) = \beta[C(n) - C(n-1)], \tag{2.7.16}$$

where $\beta > 0$ is called the *relation.*

(c) Finally, the government expenditure $G(n)$ is constant over the years, and we may choose our units such that

$$G(n) = 1. \tag{2.7.17}$$

Employing formulas (2.7.15), (2.7.16), and (2.7.17) in formula (2.7.14) produces the second-order difference equation

$$Y(n+2) - \alpha(1+\beta)Y(n+1) + \alpha\beta Y(n) = 1, \qquad n \in \mathbb{Z}^+. \tag{2.7.18}$$

Observe that this is the same equation we have already studied, in detail, in Example 2.38. As we have seen there, the equilibrium state of the national income $Y^* = 1/(1-\alpha)$ is asymptotically stable (or just stable in the theory of economics) if and only if the following conditions hold:

$$\alpha < 1, \qquad 1 + \alpha + 2\alpha\beta > 0, \qquad \alpha\beta < 1. \tag{2.7.19}$$

Furthermore, the national income $Y(n)$ fluctuates (oscillates) around the equilibrium state Y^* if and only if

$$\alpha < \frac{4\beta}{(1+\beta)^2}. \tag{2.7.20}$$

Now consider a concrete example where $\alpha = \frac{1}{2}, \beta = 1$. Then $Y^* = 2$, i.e., $Y^* =$ twice the government expenditure. Then clearly, conditions (2.7.19) and (2.7.12) are satisfied. Hence the national income $Y(n)$ always converges in an oscillatory fashion to $Y^* = 2$, regardless of what the initial national income $Y(0)$ and $Y(1)$ are. (See Figure 2.7.)

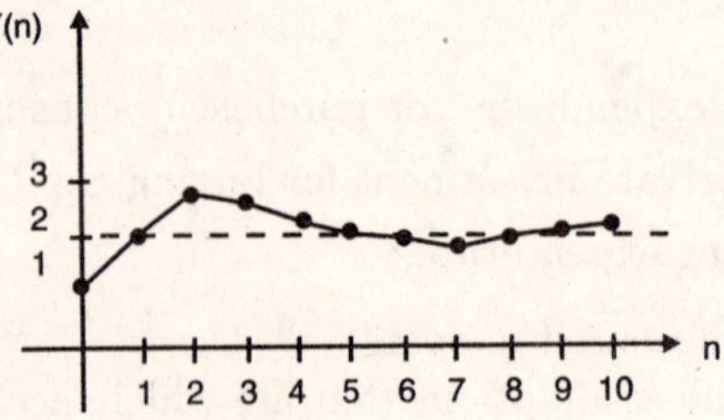

FIGURE 2.7. Solution of $Y(n+2) - Y(n+1) + Y(n) = 1, Y(0) = 1, Y(1) = 2$.

The actual solution may be given by

$$Y(n) = A\left(\frac{1}{\sqrt{2}}\right)^n \cos\left(\frac{n\pi}{4} - \omega\right) + 2.$$

Figure 2.7 depicts the solution $Y(n)$ if $Y(0) = 1$ and $Y(1) = 2$. Here we find that $A = -\sqrt{2}$ and $\omega = \pi/4$ and, consequently, the solution is

$$Y(n) = -\left(\frac{1}{\sqrt{2}}\right)^{n-1} \cos\left[\frac{(n+1)}{4}\pi\right] + 2.$$

Finally, Figure 2.8 depicts the parameter diagram $(\beta - \alpha)$, which shows regions of stability and regions of instability.

2.7.4 The Transmission of Information

Suppose that a signaling system has two signals s_1 and s_2 such as dots and dashes in telegraphy. Messages are transmitted by first encoding them into a string, or sequence, of these two signals. Suppose that s_1 requires exactly n_1 units of time, and s_2 exactly n_2 units of time, to be transmitted. Let $M(n)$ be the number of possible message sequences of duration n. Now, a signal of duration time n either ends with an s_1 signal or with an s_2 signal.

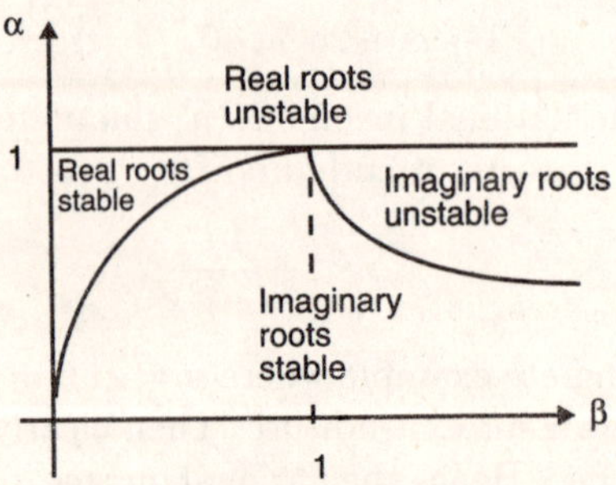

FIGURE 2.8. Parametric diagram $(\beta - \alpha)$.

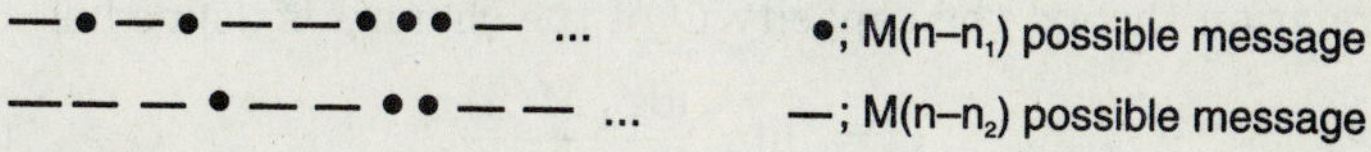

FIGURE 2.9. Two signals, one ends with s_1 and the other with s_2.

If the message ends with s_1, the last signal must start at $n - n_1$ (since s_1 takes n_1 units of time). Hence there are $M(n - n_1)$ possible messages to which the last s_1 may be appended. Hence there are $M(n - n_1)$ messages of duration n that end with s_1. By a similar argument, one may conclude that there are $M(n - n_2)$ messages of duration n that end with s_2. (See Figure 2.9.) Consequently, the total number of messages $x(n)$ of duration n may be given by

$$M(n) = M(n - n_1) + M(n - n_2).$$

If $n_1 \geq n_2$, then the above equation may be written in the familiar form of an n_1th-order equation

$$M(n + n_1) - M(n + n_1 - n_2) - M(n) = 0. \tag{2.7.21}$$

On the other hand, if $n_1 \leq n_2$, then we obtain the n_2th-order equation

$$M(n + n_2) - M(n + n_2 - n_1) - M(n) = 0. \tag{2.7.22}$$

An interesting special case is that in which $n_1 = 1$ and $n_2 = 2$. In this case we have

$$M(n + 2) - M(n + 1) - M(n) = 0,$$

or

$$M(n + 2) = M(n + 1) + M(n),$$

which is nothing but our Fibonacci sequence $\{0, 1, 1, 2, 3, 5, 8, \ldots\}$, which we encountered in Example 2.27. The general solution (see formula (2.3.14)) is given by

$$M(n) = a_1 \left(\frac{1 + \sqrt{5}}{2}\right)^n + a_2 \left(\frac{1 - \sqrt{5}}{2}\right)^n, \qquad n = 0, 1, 2, \ldots. \tag{2.7.23}$$

To find a_1 and a_2 we need to specify $M(0)$ and $M(1)$. Here a sensible assumption is to let $M(0) = 0$ and $M(1) = 1$. Using these initial data in (2.7.23) yields

$$a_1 = \frac{1}{\sqrt{5}}, \qquad a_2 = -\frac{1}{\sqrt{5}},$$

and the solution of our problem now becomes

$$M(n) = \frac{1}{\sqrt{5}} \left(\frac{1 + \sqrt{5}}{2}\right)^n - \frac{1}{\sqrt{5}} \left(\frac{1 - \sqrt{5}}{2}\right)^n. \tag{2.7.24}$$

In information theory, the capacity C of the channel is defined as

$$C = \lim_{n\to\infty} \frac{\log_2 M(n)}{n}, \tag{2.7.25}$$

where $\log_2$ denotes the logarithm base 2.

From (2.7.24) we have

$$C = \lim_{n\to\infty} \frac{\log_2 \frac{1}{\sqrt{5}}}{n} + \lim_{n\to\infty} \frac{1}{n} \log_2 \left[\left(\frac{1+\sqrt{5}}{2}\right) - \left(\frac{1-\sqrt{5}}{2}\right)^n \right]. \tag{2.7.26}$$

Since $\left(\frac{1-\sqrt{5}}{2}\right) \approx 0.6 < 1$, it follows that $\left(\frac{1-\sqrt{5}}{2}\right)^n \to 0$ as $n \to \infty$.

Observe also that the first term on the right-hand side of (2.7.26) goes to zero as $n \to \infty$.

Thus

$$C = \lim_{n\to\infty} \frac{1}{n} \log_2 \left(\frac{1+\sqrt{5}}{2}\right)^n,$$

$$C = \log_2 \left(\frac{1+\sqrt{5}}{2}\right) \approx 0.7. \tag{2.7.27}$$

Exercises 2.7

1. The model for annual plants was given by (2.7.7) in terms of the plant population $p(n)$.

   (a) Write the model in terms of $s_1(n)$.

   (b) Let $\alpha = \beta = 0.01$ and $\sigma = 1$. How big should γ be to ensure that the plant population increases in size?

2. An alternative formulation for the annual plant model is that in which we define the beginning of a generation as the time when seeds are produced. Figure 2.10 shows the new method.

   Write the difference equation in $p(n)$ that represents this model. Then find conditions on γ under which plant propagation occurs.

3. A planted seed produces a flower with one seed at the end of the first year and a flower with two seeds at the end of two years and each year thereafter. Suppose that each seed is planted as soon as it is produced.

   (a) Write the difference equation that describes the number of flowers $F(n)$ at the end of the nth year.

   (b) Compute the number of flowers at the end of 3, 4, and 5 years.

4. Suppose that the probability of winning any particular bet is 0.49. If you start with \$50 and will quit when you have \$100, what is the probability of ruin (i.e., losing all your money):

(i) if you make \$1 bets?

(ii) if you make \$10 bets?

(iii) if you make \$50 bets?

5. John has m chips and Robert has $(N - m)$ chips. Suppose that John has a probability p of winning each game, where one chip is bet on in each play. If $G(m)$ is the expected value of the number of games that will be played before either John or Robert is ruined:

   (a) Show that $G(m)$ satisfies the second-order equation
   $$G(m+2) + pG(m+1) + (1-p)G(m) = 0. \tag{2.7.28}$$

   (b) What are the values of $G(0)$ and $G(N)$?

   (c) Solve the difference equation (2.7.28) with the boundary conditions in part (b).

6. Suppose that in a game we have the following situation: On each play, the probability that you will win \$2 is 0.1, the probability that you will win \$1 is 0.3, and the probability that you will lose \$1 is 0.6. Suppose you quit when either you are broke or when you have at least N dollars. Write a third-order difference equation that describes the probability $p(n)$ of eventually going broke if you have n dollars. Then find the solution of the equation.

7. Suppose that Becky plays a roulette wheel that has 37 divisions: 18 are red, 18 are black, and one is green. Becky can bet on either the red or black, and she wins a sum equal to her bet if the outcome is a division of that color; otherwise, she loses the bet. If the bank has one

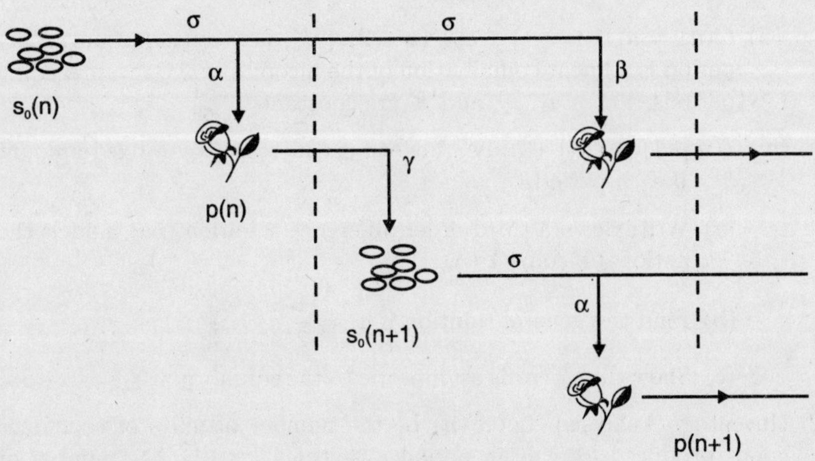

FIGURE 2.10. Annual plant model.

million dollars and she has \$5000, what is the probability that Becky can break the bank, assuming that she bets \$100 on either red or black for each spin of the wheel?

8. In the national income model (2.7.14), assume that the government expenditure $G(n)$ is proportional to the national income $Y(n-2)$ two periods past, i.e., $G(n) = \gamma Y(n-2)$, $0 < \gamma < 1$. Derive the difference equation for the national income $Y(n)$. Find the conditions for stability and oscillations of solutions.

9. Determine the behavior (stability, oscillations) of solutions of (2.7.18) for the cases:

   (a) $\alpha = \dfrac{4\beta}{(1+\beta)^2}$.

   (b) $\alpha > \dfrac{4\beta}{(1+\beta)^2}$.

10. Modify the national income model such that instead of the government having fixed expenditures, it increases its expenditures by 5% each time period, that is, $G(n) = (1.05)^n$.

    (a) Write down the second-order difference equation that describes this model.

    (b) Find the equilibrium value.

    (c) If $\alpha = 0.5, \beta = 1$, find the general solution of the equation.

11. Suppose that in the national income we make the following assumptions:

    (i) $Y(n) = C(n) + I(n)$, i.e., there is no government expenditure.

    (ii) $C(n) = a_1 Y(n-1) + a_2 Y(n-2) + K$, i.e., consumption in any period is a linear combination of the incomes of the two preceding periods, where a_1, a_2, and K are constants.

    (iii) $I(n+1) = I(n) + h$, i.e., investment increases by a fixed amount $h > 0$ each period.

    (a) Write down a third-order difference equation that models the national income $Y(n)$.

    (b) Find the general solution if $a_1 = \dfrac{1}{2}, a_2 = \dfrac{1}{4}$.

    (c) Show that $Y(n)$ is asymptotic to the equilibrium $Y^* = \alpha + \beta n$.

12. (Inventory Analysis). Let $S(n)$ be the number of units of consumer goods produced for sale in period n, and let $T(n)$ be the number of units of consumer goods produced for inventories in period n. Assume that there is a constant noninduced net investment V_0 in each period.

Then the total income $Y(n)$ produced in time n is given by $Y(n) = T(n) + S(n) + V_0$.

(a) Develop a difference equation that models the total income $Y(n)$, under the assumptions:

(i) $S(n) = \beta Y(n-1)$,

(ii) $T(n) = \beta Y(n-1) - \beta Y(n-2)$.

(b) Obtain conditions under which:

(i) solutions converge to the equilibrium,

(ii) solutions are oscillatory.

(c) Interpret your results in part (b).

13. Let $I(n)$ denote the level of inventories at the close of period n.

(a) Show that $I(n) = I(n-1) + S(n) + T(n) - \beta Y(n)$ where $S(n), T(n), Y(n)$ are as in Problem 12.

(b) Assuming that $S(n) = 0$ (passive inventory adjustment), show that

$$I(n) - I(n-1) = (1-\beta)Y(n) - V_0$$

where V_0 is as in Problem 12.

(c) Suppose as in part (b) that $s(n) = 0$. Show that

$$I(n+2) - (\beta+1)I(n+1) + \beta I(n) = 0.$$

(d) With $\beta \neq 1$, show that

$$I(n) = \left(I(0) - \frac{c}{1-\beta}\right)\beta^n + \frac{c}{1-\beta},$$

where $(E-\beta)I(n) = c$.

14. Consider (2.7.21) with $n_1 = n_2 = 2$ (i.e., both signals s_1 and s_2 take two units of time for transmission).

(a) Solve the obtained difference equation with the initial conditions $M(2) = M(3) = 2$.

(b) Find the channel capacity c.

15. Consider (2.7.21) with $n_1 = n_2 = 1$ (i.e., both signals take one unit of time for transmission).

(a) Solve the obtained difference equation.

(b) Find the channel capacity c.

16. (Euler's method for solving a second-order differential equation.) Recall from Section 1.4.1 that one may approximate $x'(t)$ by $(x(n+1)-x(n))/h$, where h is the step size of the approximation and $x(n)=x(t_0+nh)$.

    (a) Show that $x''(t)$ may be approximated by
    $$\frac{x(n+2)-2x(n+1)+x(n)}{h^2}.$$

    (b) Write down the corresponding difference equation of the differential equation
    $$x''(t)=f(x(t),x'(t)).$$

17. Use Euler's method described in Problem 16 to write the corresponding difference equation of
$$x''(t)-4x(t)=0, \quad x(0)=0, \quad x'(0)=1.$$
Solve both differential and difference equations and compare the results.

18. (The Midpoint Method). The midpoint method stipulates that one may approximate $x'(t)$ by $(x(n+1)-x(n-1))/h$, where h is the step size of the approximation and $t=t_0+nh$.

    (a) Use the method to write the corresponding difference equation of the differential equation $x'(t)=g(t,x(t))$.

    (b) Use the method to write the corresponding difference equation of $x'(t)=0.7x^2+0.7, x(0)=1, t\in[0,1]$. Then solve the obtained difference equation.

    (c) Compare your findings in part (b) with the results in Section 1.4.1. Determine which of the two methods, Euler or midpoint, is more accurate.

3

Systems of Linear Difference Equations

In the last chapter we concerned ourselves with linear difference equations, namely, those equations with only one independent and one dependent variable. Since not every situation that we will encounter will be this simple, we must be prepared to deal with systems of more than one dependent variable.

Thus, in this chapter we deal with those equations of two or more dependent variables known as first-order difference equations. These equations naturally apply to various fields of scientific endeavor, like biology (the study of competitive species in population dynamics), physics (the study of the motions of interacting bodies), the study of control systems, neurology, and electricity. Furthermore, we will also transform those high-order linear difference equations that we investigated in Chapter 2 into systems of first-order equations. This transformation will probably prove to be of little practical use in the realm of boundary value problems and oscillations, but will be substantiated as an immensely helpful tool in the study of stability theory later on, see [3], [79], [87].

3.1 Autonomous (Time-Invariant) Systems

In this section we are interested in finding solutions of the following system of k linear equations:

$$x_1(n+1) = a_{11}x_1(n) + a_{12}x_2(n) + \cdots + a_{1k}x_k(n),$$
$$x_2(n+1) = a_{21}x_1(n) + a_{22}x_2(n) + \cdots + a_{2k}x_k(n),$$

$$\vdots \qquad \vdots \qquad \vdots \qquad \vdots$$

$$x_k(n+1) = a_{k1}x_1(n) + a_{k2}x_2(n) + \cdots + a_{kk}x_k(n).$$

This system may be written in the vector form

$$x(n+1) = Ax(n), \tag{3.1.1}$$

where $x(n) = (x_1(n), x_2(n), \ldots, x_k(n))^T \in \mathbb{R}^k$, and $A = (a_{ij})$ is a $k \times k$ real nonsingular matrix. Here T indicates the transpose of a vector. System (3.1.1) is considered autonomous, or time-invariant, since the values of A are all constants. Nonautonomous, or time-variant, systems will be considered later in Section 3.3.

If for some $n_0 \geq 0, x(n_0) = x_0$ is specified, then system (3.1.1) is called an initial value problem. Furthermore, by simple iteration (or by direct substitution into the equation), one may show that the solution is given by

$$\boxed{x(n, n_0, x_0) = A^{n-n_0}x_0,} \tag{3.1.2}$$

where $A^0 = I$, the $k \times k$ identity matrix. Notice that $x(n_0, n_0, x_0) = x_0$. If $n_0 = 0$, then the solution in formula (3.1.2) may be written as $x(n, x_0)$, or simply $x(n)$. We now show that we may assume that $n_0 = 0$ without loss of generality.

Let $y(n - n_0) = x(n)$. Then (3.1.1) becomes

$$y(n+1) = Ay(n), \tag{3.1.3}$$

with $y(0) = x(n_0)$ and

$$y(n) = A^n y(0). \tag{3.1.4}$$

A parallel theory exists for systems of linear differential equations. The solution of the initial value problem

$$\frac{dx}{dt} = Ax(t), \quad x(t_0) = x_0,$$

where A is a $k \times k$ matrix, $x \in \mathbb{R}^k$, is given by

$$x(t) = e^{A(t-t_0)}x_0.$$

3.1.1 *The Discrete Analogue of the Putzer Algorithm*

In differential equations the Putzer algorithm is used to compute e^{At}. Here, we introduce an analogous algorithm to compute A^n. First, let us review the rudiments of matrix theory that are vital in the development of this algorithm. In what follows $\mathbb{C}$ denotes the set of complex numbers.

Recall that for a real $k \times k$ matrix $A = (a_{ij})$, an eigenvalue of A is a real or complex number λ such that $A\xi = \lambda\xi$ for some nonzero $\xi \in \mathbb{C}^k$. Equivalently, this relation may be written as

$$(A - \lambda I)\xi = 0. \tag{3.1.5}$$

Equation (3.1.5) has a nonzero solution if and only if

$$\det(A - \lambda I) = 0,$$

or

$$\lambda^k + a_1\lambda^{k-1} + a_2\lambda^{k-2} + \cdots + a_{k-1}\lambda + a_k = 0. \tag{3.1.6}$$

Equation (3.1.6) is called the *characteristic equation* of A, whose roots λ are called the *eigenvalues* of A. If $\lambda_1, \lambda_2, \ldots, \lambda_k$ are the eigenvalues of A (some of them may be repeated), then one may write (3.1.6) as

$$p(\lambda) = \prod_{j=1}^{k} (\lambda - \lambda_j). \tag{3.1.7}$$

We are now ready to state the Cayley–Hamilton theorem, one of the fundamental results of matrix theory.

Theorem 3.1. *Every matrix satisfies its characteristic equation. That is,*

$$p(A) = \prod_{j=1}^{k} (A - \lambda_j I) = 0, \tag{3.1.8}$$

or

$$A^k + a_1 A^{k-1} + a_2 A^{k-2} + \cdots + a_k I = 0. \tag{3.1.9}$$

3.1.2 The Development of the Algorithm for A^n

Let A be a $k \times k$ real matrix. We look for a representation of A^n in the form

$$\boxed{A^n = \sum_{j=1}^{s} u_j(n) M(j-1),} \tag{3.1.10}$$

where the $u_j(n)$'s are scalar functions to be determined later, and

$$M(j) = (A - \lambda_j I) M(j-1), \quad M(0) = I, \tag{3.1.11}$$

or

$$M(j+1) = (A - \lambda_{j+1} I) M(j), \quad M(0) = I.$$

By iteration, one may show that

$$M(n) = (A - \lambda_n I)(A - \lambda_{n-1} I) \cdots (A - \lambda_1 I),$$

or, in compact form,

$$M(n) = \prod_{j=1}^{n} (A - \lambda_j I). \tag{3.1.12}$$

Notice that by the Cayley–Hamilton theorem we have

$$M(k) = \prod_{j=1}^{k}(A - \lambda_j I) = 0.$$

Consequently, $M(n) = 0$ for all $n \geq k$. In light of this observation, we may rewrite formula (3.1.10) as

$$A^n = \sum_{j=1}^{k} u_j(n)M(j-1). \tag{3.1.13}$$

If we let $n = 0$ in formula (3.1.13) we obtain

$$A^0 = I = u_1(0)I + u_2(0)M(1) + \cdots + u_k(0)M(k-1). \tag{3.1.14}$$

Equation (3.1.14) is satisfied if

$$u_1(0) = 1 \quad \text{and} \quad u_2(0) = u_3(0) = \cdots = u_k(0) = 0. \tag{3.1.15}$$

From formula (3.1.13) we have

$$\begin{aligned}\sum_{j=1}^{k} u_j(n+1)M(j-1) = AA^n &= A\left[\sum_{j=1}^{k} u_j(n)M(j-1)\right] \\ &= \sum_{j=1}^{k} u_j(n)AM(j-1).\end{aligned}$$

Substituting for $AM(j-1)$ from (3.1.11) yields

$$\sum_{j=1}^{k} u_j(n+1)M(j-1) = \sum_{j=1}^{k} u_j(n)[M(j) + \lambda_j M(j-1)]. \tag{3.1.16}$$

Comparing the coefficients of $M(j), 1 \leq j \leq k$, in (3.1.16), and applying condition (3.1.15), we obtain

$$\begin{aligned}&u_1(n+1) = \lambda_1 u_1(n), && u_1(0) = 1, \\ &u_j(n+1) = \lambda_j u_j(n) + u_{j-1}(n), && u_j(0) = 0, \quad j = 2, 3, \ldots, k.\end{aligned} \tag{3.1.17}$$

The solutions of (3.1.17) are given by

$$u_1(n) = \lambda_1^n, \qquad u_j(n) = \sum_{i=0}^{n-1} \lambda_j^{n-1-i} u_{j-1}(i), \qquad j = 2, 3, \ldots, k. \tag{3.1.18}$$

Equations (3.1.12) and (3.1.18) together constitute an algorithm for computing A^n, which henceforth will be called the *Putzer algorithm.* For more details and other algorithms, the interested reader may consult the paper by Elaydi and Harris [46].

Example 3.2. Find A^n if

$$A = \begin{pmatrix} 0 & 1 & 1 \\ -2 & 3 & 1 \\ -3 & 1 & 4 \end{pmatrix}.$$

Solution The eigenvalues of A are obtained by solving the characteristic equation

$$\det(A - \lambda I) = \det \begin{pmatrix} -\lambda & 1 & 1 \\ -2 & 3-\lambda & 1 \\ -3 & 1 & 4-\lambda \end{pmatrix} = 0.$$

Hence

$$\begin{aligned} p(\lambda) &= \lambda^3 - 7\lambda^2 + 16\lambda - 12 \\ &= (\lambda - 2)^2(\lambda - 3) = 0. \end{aligned}$$

The eigenvalues of A are $\lambda_1 = \lambda_2 = 2$, $\lambda_3 = 3$,

$$M(0) = I, \quad M(1) = A - 2I = \begin{pmatrix} -2 & 1 & 1 \\ -2 & 1 & 1 \\ -3 & 1 & 2 \end{pmatrix},$$

$$M(2) = (A - 2I), \quad M(1) = (A - 2I)^2 = \begin{pmatrix} -1 & 0 & 1 \\ -1 & 0 & 1 \\ -2 & 0 & 2 \end{pmatrix}.$$

Now

$$\begin{aligned} u_1(n) &= 4^n, \\ u_2(n) &= \sum_{i=0}^{n-1} 2^{(n-1-i)} \cdot 2^i = n2^{n-1}, \\ u_3(n) &= \sum_{i=0}^{n-1} 3^{(n-1-i)} (i2^{i-1}) \\ &= \frac{3^{n-1}}{2} \sum_{i=0}^{n-1} i \left(\frac{2}{3}\right)^i \\ &= \frac{3^{n-1}}{2} \left[\frac{\left(\frac{2}{3} - 1\right)^n \left(\frac{2}{3}\right)^n - \left(\frac{2}{3}\right)^{n+1} + \frac{2}{3}}{\left(\frac{2}{3} - 1\right)^2} \right] \quad \text{(from Table 1.1)} \\ &= -2^n + 3^n - n2^{n-1}. \end{aligned}$$

Thus

$$A^n = \sum_{j=1}^{n} u_j(n)M(j-1)$$

$$= \begin{pmatrix} 2^{n-1} - 3^n - n2^{n-1} & n2^{n-1} & -2^n + 3^n \\ 2^n - 3^n - n2^{n-1} & (n+2)2^{n-1} & -2^n + 3^n \\ 2^{n+1} - 2\cdot 3^n - n2^{n-1} & n2^{n-1} & -2^n + 2\cdot 3^n \end{pmatrix}.$$

Example 3.3. Find the solution of the difference system $x(n+1) = Ax(n)$, where

$$A = \begin{pmatrix} 4 & 1 & 2 \\ 0 & 2 & -4 \\ 0 & 1 & 6 \end{pmatrix}.$$

Solution The eigenvalues of A may be obtained by solving the characteristic equation $\det(A - \lambda I) = 0$. Now,

$$\det \begin{pmatrix} 4-\lambda & 1 & 2 \\ 0 & 2-\lambda & -4 \\ 0 & 1 & 6-\lambda \end{pmatrix} = (4-\lambda)(\lambda-4)^2 = 0.$$

Hence, the eigenvalues of A are $\lambda_1 = \lambda_2 = \lambda_3 = 4$. So

$$M(0) = I, \quad M(1) = A - 4I = \begin{pmatrix} 0 & 1 & 2 \\ 0 & -2 & -4 \\ 0 & 1 & 2 \end{pmatrix},$$

$$M(2) = (A-4I)M(1) = \begin{pmatrix} 0 & 0 & 0 \\ 0 & 0 & 0 \\ 0 & 0 & 0 \end{pmatrix}.$$

Now,

$$u_1(n) = 4^n,$$

$$u_2(n) = \sum_{i=0}^{n-1} (4^{n-1-i})(4^i) = n(4^{n-1}),$$

$$u_3(n) = \sum_{i=0}^{n-1} 4^{n-1-i}(i4^{i-1})$$

$$= 4^{n-2} \sum_{i=0}^{n-1} i$$

$$= \frac{n(n-1)}{2} 4^{n-2}.$$

Using (3.1.13), we have

$$A^n = 4^n \begin{pmatrix} 1 & 0 & 0 \\ 0 & 1 & 0 \\ 0 & 0 & 1 \end{pmatrix} + n4^{n-1} \begin{pmatrix} 0 & 1 & 2 \\ 0 & -2 & -4 \\ 0 & 1 & 2 \end{pmatrix}$$

$$+ \frac{n(n-1)}{2} 4^{n-2} \begin{pmatrix} 0 & 0 & 0 \\ 0 & 0 & 0 \\ 0 & 0 & 0 \end{pmatrix}$$

$$= \begin{pmatrix} 4^n & n4^{n-1} & 2n4^{n-1} \\ 0 & 4^n - 2n4^{n-1} & -n4^n \\ 0 & n4^{n-1} & 4^n + 2n4^{n-1} \end{pmatrix}.$$

The solution of the difference equation is given by

$$x(n) = A^n x(0) = \begin{pmatrix} 4^n x_1(0) + n4^{n-1} x_2(0) + 2n4^{n-1} x_3(0) \\ (4^n - 2n4^{n-1}) x_2(0) - n4^n x_3(0) \\ n4^{n-1} x_2(0) + (4^n + 2n4^{n-1}) x_3(0) \end{pmatrix},$$

where $x(0) = (x_1(0), x_2(0), x_3(0))^T$.

Exercises 3.1

In Problems 1 through 4, use the discrete Putzer algorithm to evaluate A^n.

1. $A = \begin{bmatrix} 1 & 1 \\ -2 & 4 \end{bmatrix}$.

2. $A = \begin{bmatrix} -1 & 2 \\ 3 & 0 \end{bmatrix}$.

3. $A = \begin{bmatrix} 1 & 2 & -1 \\ 0 & 1 & 0 \\ 4 & -4 & 5 \end{bmatrix}$.

4. $A = \begin{bmatrix} 2 & 1 & 0 \\ 0 & 2 & 1 \\ 0 & 0 & 2 \end{bmatrix}$.

5. Solve the system
$x_1(n+1) = -x_1(n) + x_2(n), \qquad x_1(0) = 1,$
$x_2(n+1) = 2x_2(n), \qquad x_2(0) = 2.$

6. Solve the system
$x_1(n+1) = x_2(n),$
$x_2(n+1) = x_3(n),$
$x_3(n+1) = 2x_1(n) - x_2(n) + x_3(n).$

7. Solve the system

$$x(n+1) = \begin{bmatrix} 1 & -2 & -2 \\ 0 & 0 & -1 \\ 0 & 2 & 3 \end{bmatrix} x(n), \qquad x(0) = \begin{pmatrix} 1 \\ 1 \\ 0 \end{pmatrix}.$$

8. Solve the system

$$x(n+1) = \begin{pmatrix} 1 & 3 & 0 & 0 \\ 0 & 2 & 1 & -1 \\ 0 & 0 & 2 & 0 \\ 0 & 0 & 0 & 3 \end{pmatrix} x(n).$$

9. Verify that the matrix $A = \begin{pmatrix} 2 & -1 \\ 1 & 3 \end{pmatrix}$ satisfies its characteristic equation (the Cayley Hamilton Theorem).

10. Let $\rho(A) = \max\{|\lambda| : \lambda \text{ is an eigenvalue of } A\}$. Suppose that $\rho(A) = \rho_0 < \beta$.

    (a) Show that $|u_j(n)| \leq \dfrac{\beta^n}{(\beta - \rho_0)}$, $j = 1, 2, \ldots, k$.

    (b) Show that if $\rho_0 < 1$, then $u_j(n) \to 0$ as $n \to \infty$. Conclude that $A^n \to 0$ as $n \to \infty$.

    (c) If $\alpha < \min\{|\lambda| : \lambda \text{ is an eigenvalue of } A\}$, establish a lower bound for $|u_j(n)|$.

11. If a $k \times k$ matrix A has distinct eigenvalues $\lambda_1, \lambda_2, \ldots, \lambda_k$, then one may compute $A^n, n \geq k$, using the following method. Let $p(\lambda)$ be the characteristic polynomial of A. Divide λ^n by $p(\lambda)$ to obtain $\lambda^n = p(\lambda)q(\lambda) + r)(\lambda)$, where the remainder $r(\lambda)$ is a polynomial of degree at most $(k-1)$. Thus one may write $A^n = p(A)q(A) + r(A)$.

    (a) Show that $A^n = r(A) = a_0 I + a_1 A + a_2 A^2 + \cdots + a_{k-1} A^{k-1}$.

    (b) Show that $\lambda_1^n = r(\lambda_1), \lambda_2^n = r(\lambda_2), \ldots, \lambda_k^n = r(\lambda_k)$.

    (c) Use part (b) to find $a_0, a_1, \ldots, a_{k-1}$.

12. Extend the method of Problem 11 to the case of repeated roots.

13. Apply the method of Problem 12 to find A^n for:

    (i) $A = \begin{pmatrix} 1 & 1 \\ -2 & 4 \end{pmatrix}$.

    (ii) $A = \begin{pmatrix} 1 & 2 & -1 \\ 1 & 0 & 1 \\ 4 & -4 & 5 \end{pmatrix}$.

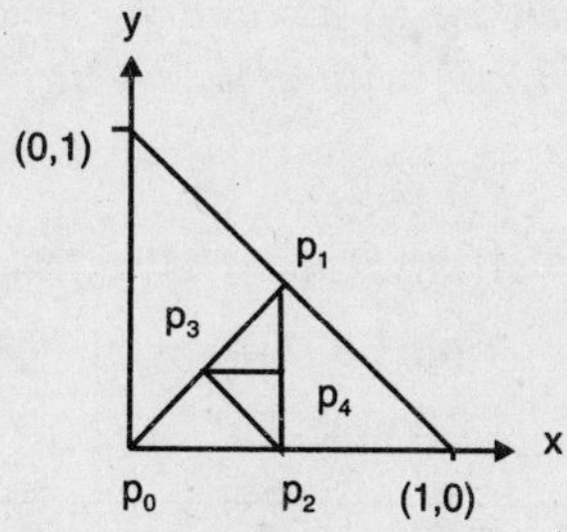

FIGURE 3.1.

14. Apply the method of Problem 12 to find A^n for

$$A = \begin{pmatrix} 4 & 1 & 2 \\ 0 & 2 & -4 \\ 0 & 1 & 6 \end{pmatrix}.$$

15.[1] Consider the right triangle in Figure 3.1 where $p(0) = (0,0)$, $p(1) = \left(\frac{1}{2}, \frac{1}{2}\right)$, and $p(2) = \left(\frac{1}{2}, 0\right)$. For $p(n) = (x(n), y(n))$ with $n \geq 3$, we have

$$x(n+3) = \frac{1}{2}(x(n) + x(n+1)),$$
$$y(n+3) = \frac{1}{2}(y(n) + y(n+1)).$$

(a) Write each equation as a system $z(n+1) = Az(n)$.

(b) Find $\lim_{n\to\infty} p(n)$.

3.2 The Basic Theory

Now contemplate the system

$$x(n+1) = A(n)x(n), \tag{3.2.1}$$

where $A(n) = (a_{ij}(n))$ is a $k \times k$ nonsingular matrix function. This is a *homogeneous linear difference system* that is *nonautonomous*, or *time-variant.*

The corresponding nonhomogeneous system is given by

$$y(n+1) = A(n)y(n) + g(n), \tag{3.2.2}$$

where $g(n) \in R^k$.

We now establish the existence and uniqueness of solutions of (3.2.1).

---

[1]Proposed by C.V. Eynden and solved by Trinity University Problem Solving Group (1994).

Theorem 3.4. *For each $x_0 \in \mathbb{R}^k$ and $n_0 \in \mathbb{Z}^+$ there exists a unique solution $x(n, n_0, x_0)$ of (3.2.1) with $x(n_0, n_0, x_0) = x_0$.*

PROOF. From (3.2.1),

$$x(n_0 + 1, n_0, x_0) = A(n_0)x(n_0) = A(n_0)x_0,$$
$$x(n_0 + 2, n_0, x_0) = A(n_0 + 1)x(n_0 + 1) = A(n_0 + 1)A(n_0)x_0.$$

Inductively, one may conclude that

$$\boxed{x(n, n_0, x_0) = \left[\prod_{i=n_0}^{n-1} A(i)\right] x_0,} \tag{3.2.3}$$

where

$$\prod_{i=n_0}^{n-1} A(i) = \begin{cases} A(n-1)A(n-2)\cdots A(n_0) & \text{if } n > n_0, \\ I & \text{if } n = n_0. \end{cases}$$

Formula (3.2.3) gives the unique solution with the desired properties. □

We will now develop the notion of a fundamental matrix, a central building block in the theory of linear systems.

Definition 3.5. The solutions $x_1(n), x_2(n), \ldots, x_k(n)$ of (3.2.1) are said to be linearly independent for $n \geq n_0 \geq 0$ if whenever $c_1x_1(n) + c_2x_2(n) + \cdots + c_kx_k(n) = 0$ for all $n \geq n_0$, then $c_i = 0, 1 \leq i \leq k$.

Let $\Phi(n)$ be a $k \times k$ matrix whose columns are solutions of (3.2.1). We write

$$\Phi(n) = [x_1(n), x_2(n), \ldots, x_k(n)].$$

Now,

$$\begin{aligned} \Phi(n+1) &= [A(n)x_1(n), A(n)x_2(n), \ldots, A(n)x_k(n)] \\ &= A(n)[x_1(n), x_2(n), \ldots, x_k(n)] \\ &= A(n)\Phi(n). \end{aligned}$$

Hence, $\Phi(n)$ satisfies the matrix difference equation

$$\Phi(n+1) = A(n)\Phi(n). \tag{3.2.4}$$

Furthermore, the solutions $x_1(n), x_2(n), \ldots, x_k(n)$ are linearly independent for $n \geq n_0$ if and only if the matrix $\Phi(n)$ is nonsingular ($\det \Phi(n) \neq 0$) for all $n \geq n_0$. (Why?) This actually leads to the next definition.

Definition 3.6. If $\Phi(n)$ is a matrix that is nonsingular for all $n \geq n_0$ and satisfies (3.2.4), then it is said to be a *fundamental matrix* for system equation (3.2.1).

Note that if $\Phi(n)$ is a fundamental matrix and C is any nonsingular matrix, then $\Phi(n)C$ is also a fundamental matrix (Exercises 3.2, Problem

6). Thus there are infinitely many fundamental matrices for a given system. However, there is one fundamental matrix that we already know, namely,

$$\Phi(n) = \prod_{i=n_0}^{n-1} A(i), \quad \text{with } \Phi(n_0) = I$$

(Exercises 3.2, Problem 5). In the autonomous case when A is a constant matrix, $\Phi(n) = A^{n-n_0}$, and if $n_0 = 0$, then $\Phi(n) = A^n$. Consequently, it would be much more suitable to use the Putzer algorithm to compute the fundamental matrix for an autonomous system.

Theorem 3.7. *There is a unique solution* $\Psi(n)$ *of the matrix* (3.2.4) *with* $\Psi(n_0) = I$.

PROOF. One may think of the matrix difference equation (3.2.4) as a system of k^2 first-order difference equations. Thus, to complete the point, we may apply the "existence and uniqueness" Theorem 3.4 to obtain a k^2-vector solution ν such that $\nu(n_0) = (1, 0, \ldots, 1, 0, \ldots)^T$, where 1's appear at the first, $(k+2)$th, $(2k+3)$th, ... slots and 0's everywhere else. The vector ν is then converted to the $k \times k$ matrix $\Psi(n)$ by grouping the components into sets of k elements in which each set will be a column. Clearly, $\Psi(n_0) = I$. □

We may add here that starting with any fundamental matrix $\Phi(n)$, the fundamental matrix $\Phi(n)\Phi^{-1}(n_0)$ is such a matrix. This special fundamental matrix is denoted by $\Phi(n, n_0)$ and is referred to as the *state transition* matrix.

One may, in general, write $\Phi(n, m) = \Phi(n)\Phi^{-1}(m)$ for any two positive integers n, m with $n \geq m$. The fundamental matrix $\Phi(n, m)$ has some agreeable properties that we ought to list here. Observe first that $\Phi(n, m)$ is a solution of the matrix difference equation $\Phi(n+1, m) = A(n)\Phi(n, m)$ (Exercises 3.2, Problem 2). The reader is asked to prove the following statements:

(i) $\Phi^{-1}(n, m) = \Phi(m, n)$ (Exercises 3.2, Problem 3).

(ii) $\Phi(n, m) = \Phi(n, r)\Phi(r, m)$ (Exercises 3.2, Problem 3).

(iii) $\Phi(n, m) = \prod_{i=m}^{n-1} A(i)$ (Exercises 3.2, Problem 3).

Corollary 3.8. *The unique solution of* $x(n, n_0, x_0)$ *of* (3.2.1) *with* $x(n, n_0, x_0) = x_0$ *is given by*

$$\boxed{x(n, n_0, x_0) = \Phi(n, n_0)x_0.} \tag{3.2.5}$$

Checking the linear independence of a fundamental matrix $\Phi(n)$ for $n \geq n_0$ is a formidable task. We will instead show that it suffices to establish linear independence at n_0.

Lemma 3.9 Abel's Formula. *For any $n \geq n_0 \geq 0$,*

$$\boxed{\det \Phi(n) = \left(\prod_{i=n_0}^{n-1} [\det \ A(i)] \right) \det \Phi(n_0).} \tag{3.2.6}$$

PROOF. Taking the determinant of both sides of (3.2.4) we obtain the scalar difference equation

$$\det \Phi(n+1) = \det \ A(n) \ \det \Phi(n)$$

whose solution is given by (3.2.6). □

Corollary 3.10. *If in* (3.2.1) *A is a constant matrix, then*

$$\boxed{\det \Phi(n) = [\det \ A]^{n-n_0} \ \det \Phi(n_0).} \tag{3.2.7}$$

PROOF. The proof follows from formula (3.2.6). □

Corollary 3.11. *The fundamental matrix $\Phi(n)$ is nonsingular for all $n \geq n_0$ if and only if $\Phi(n_0)$ is nonsingular.*

PROOF. This follows from formula (3.2.6), having noted that $\det \ A(i) \neq 0$ for $i \geq n_0$. □

Corollary 3.12. *The solutions $x_1(n), x_2(n), \ldots, x_k(n)$ of* (3.2.1) *are linearly independent for $n \geq n_0$ if and only if $\Phi(n_0)$ is nonsingular.*

PROOF. This follows immediately from Corollary 3.11. □

The following theorem establishes the existence of k linearly independent solutions of (3.2.1).

Theorem 3.13. *There are k linearly independent solutions of system* (3.2.1) *for $n \geq n_0$.*

PROOF. For each $i = 1, 2, \ldots, \ k$, let $e_i = (0, 0, \ldots, \ 1, \ldots, \ 0)^T$ be the standard unit vector in R^k where all the components are zero except the ith component, which is equal to 1. By Theorem 3.4, for each e_i, $1 \leq i \leq k$, there exists a solution $x(n, n_0, e_i)$ of (3.2.1) with $x(n_0, n_0, e_i) = e_i$. To prove that the set $\{x(n, n_0, e_i) | 1 \leq i \leq k\}$ is linearly independent, according to Corollary 3.11 it suffices to show that $\Phi(n_0)$ is nonsingular. But this fact is obvious, since $\Phi(n_0) = I$. The proof of the theorem is now complete. □

Linearity Principle. An important feature of the solutions of system (3.2.1) is that they are closed under addition and scalar multiplication. That is to say, if $x_1(n)$ and $x_2(n)$ are solutions of (3.2.1) and $c \in \mathbb{R}$, then:

(1) $x_1(n) + x_2(n)$ is a solution of (3.2.1),

(2) $cx_1(n)$ is a solution of (3.2.1).

This is called the *linearity principle.*

PROOF. Statement (1) can be proved as follows. Let $x(n) = x_1(n)+x_2(n)$. Then

$$\begin{aligned} x(n+1) &= x_1(n+1) + x_2(n+1) \\ &= Ax_1(n) + Ax_2(n) \\ &= A[x_1(n) + x_2(n)] \\ &= Ax(n). \end{aligned}$$

The proof of (2) is similar. □

An immediate consequence of the linearity principle is that if $x_1(n), x_2(n), \ldots, x_k(n)$ are also solutions of system (3.2.1), then so is any linear combination of the form

$$x(n) = c_1x_1(n) + c_2x_2(n) + \cdots + c_kx_k(n).$$

This leads to the following definition.

Definition 3.14. Assuming that $\{x_i(n) | 1 \leq i \leq k\}$ is any linearly independent set of solutions of (3.2.1), the general solution of (3.2.1) is defined to be

$$\boxed{x(n) = \sum_{i=1}^{k} c_i x_i(n),} \tag{3.2.8}$$

where $c_i \in \mathbb{R}$ and at least one $c_i \neq 0$.

Formula (3.2.8) may be written as

$$x(n) = \Phi(n)c, \tag{3.2.9}$$

where $\Phi(n) = (x_1(n), x_2(n), \ldots, x_k(n))$ is a fundamental matrix, and $c = (c_1, c_2, \ldots, c_k)^T \in \mathbb{R}^k$.

Remark: The set S of all solutions of system (3.2.1) forms a linear (vector) space under addition and scalar multiplication. Its basis is any fundamental set of solutions and hence its dimension is k. The basis $\{x_1(n), x_2(n), \ldots, x_k(n)\}$ spans all solutions of equation (3.2.1). Hence any solution $x(n)$ of equation (3.2.1) can be written in the form (3.2.8) or equivalently (3.2.9). This is why we call $x(n)$ in (3.2.8) a general solution.

Let us now focus our attention on the nonhomogeneous system (3.2.2). We define a particular solution $y_p(n)$ of (3.2.2) as any k-vector function that satisfies the nonhomogeneous difference system. The following result gives us a mechanism to find the general solution of system (3.2.2).

Theorem 3.15. *Any solution $y(n)$ of (3.2.2) can be written as*

$$y(n) = \Phi(n)c + y_p(n) \tag{3.2.10}$$

for an appropriate choice of the constant vector c, and a particular solution $y_p(n)$.

PROOF. Let $y(n)$ be a solution of (3.2.2) and let $y_p(n)$ be any particular solution of (3.2.2). If $x(n) = y(n) - y_p(n)$, then

$$\begin{aligned} x(n+1) &= y(n+1) - y_p(n+1) \\ &= A(n)y(n) - A(n)y_p(n) \\ &= A(n)[y(n) - y_p(n)] \\ &= A(n)x(n). \end{aligned}$$

Thus $x(n)$ is a solution of the homogeneous equation (3.2.1). Hence $x(n) = \Phi(n)c$ for some vector constant c. Thus

$$y(n) - y_p(n) = \Phi(n)c$$

which proves (3.2.10). □

We now give a formula to evaluate $y_p(n)$.

Lemma 3.16. *A particular solution of (3.2.2) may be given by*

$$y_p(n) = \sum_{r=n_0}^{n-1} \Phi(n, r+1)g(r)$$

with $y_p(n_0) = 0$.

PROOF.

$$\begin{aligned} y_p(n+1) &= \sum_{r=n_0}^{n} \Phi(n+1, r+1)g(r) \\ &= \sum_{r=n_0}^{n-1} A(n)\Phi(n, r+1)g(r) + \Phi(n+1, n+1)g(n) \\ &= A(n)y_p(n) + g(n). \end{aligned}$$

Hence, $y_p(n)$ is a solution of (3.2.2). Furthermore, $y_p(n_0) = 0$. □

Theorem 3.17 (Variation of Constants Formula). *The unique solution of the initial value problem*

$$y(n+1) = A(n)y(n) + g(n), \quad y(n_0) = y_0, \tag{3.2.11}$$

is given by

$$\boxed{y(n, n_0, y_0) = \Phi(n, n_0)y_0 + \sum_{r=n_0}^{n-1} \Phi(n, r+1)g(r),} \tag{3.2.12}$$

or, more explicitly, by

$$y(n, n_0, y_0) = \left(\prod_{i=n_0}^{n-1} A(i)\right) y_0 + \sum_{r=n_0}^{n-1} \left(\prod_{i=r+1}^{n-1} A(i)\right) g(r). \tag{3.2.13}$$

PROOF. This theorem follows immediately from Theorem 3.15 and Lemma 3.16. □

Corollary 3.18. *For autonomous systems when A is a constant matrix, the solution of* (3.2.11) *is given by*

$$\boxed{y(n, n_0, y_0) = A^{n-n_0} y_0 + \sum_{r=n_0}^{n-1} A^{n-r-1} g(r).} \tag{3.2.14}$$

Example 3.19. Solve the system $y(n+1) = Ay(n) + g(n)$, where

$$A = \begin{pmatrix} 2 & 1 \\ 0 & 2 \end{pmatrix}, \quad g(n) = \begin{pmatrix} n \\ 1 \end{pmatrix}, \quad y(0) = \begin{pmatrix} 1 \\ 0 \end{pmatrix}.$$

Solution Using the Putzer algorithm, one may show that

$$A^n = \begin{pmatrix} 2^n & n2^{n-1} \\ 0 & 2^n \end{pmatrix}.$$

Hence,

$$\begin{aligned} y(n) &= \begin{pmatrix} 2^n & n2^{n-1} \\ 0 & 2^n \end{pmatrix} \begin{pmatrix} 1 \\ 0 \end{pmatrix} \\ &\quad + \sum_{r=0}^{n-1} \begin{pmatrix} 2^{n-r-1} & (n-r-1)2^{n-r-2} \\ 0 & 2^{n-r-1} \end{pmatrix} \begin{pmatrix} r \\ 1 \end{pmatrix} \\ &= \begin{pmatrix} 2^n \\ 0 \end{pmatrix} + \sum_{r=0}^{n-1} \begin{pmatrix} r2^{n-r-1} + (n-r-1)2^{n-r-2} \\ 2^{n-r-1} \end{pmatrix} \\ &= \begin{pmatrix} 2^n \\ 0 \end{pmatrix} + 2^n \begin{pmatrix} \frac{1}{4} \sum_{r=1}^{n-1} r \left(\frac{1}{2}\right)^r + \frac{n-1}{4} \sum_{r=0}^{n-1} \left(\frac{1}{2}\right)^r \\ \frac{1}{2} \sum_{r=0}^{n-1} \left(\frac{1}{2}\right)^r \end{pmatrix}^{*} \end{aligned}$$

$*$ $\displaystyle\sum_{r=1}^{n-1} ra^r = \frac{a(1-a^n) - na^{n+1}(1-a)}{(1-a)^2}.$

$$= \begin{pmatrix} 2^n \\ 0 \end{pmatrix} + 2^n \begin{pmatrix} \dfrac{1}{2}\left[1-\left(\dfrac{1}{2}\right)^n\right] - n\left(\dfrac{1}{2}\right)^{n+2} + \dfrac{n-1}{2}\left[1-\left(\dfrac{1}{2}\right)^n\right] \\ 1-\left(\dfrac{1}{2}\right)^n \end{pmatrix}$$

$$= \begin{pmatrix} 2^n \\ 0 \end{pmatrix} + 2^n \begin{pmatrix} -\dfrac{n}{4}\left(\dfrac{1}{2}\right)^n + \dfrac{n}{2} - \dfrac{n}{2}\left(\dfrac{1}{2}\right) \\ 1-\left(\dfrac{1}{2}\right)^n \end{pmatrix}$$

$$= \begin{pmatrix} 2^n \\ 0 \end{pmatrix} + \begin{pmatrix} n2^{n-1} - \dfrac{3}{4}n \\ 2^n - 1 \end{pmatrix}$$

$$= \begin{pmatrix} 2^n + n2^{n-1} - \dfrac{3}{4}n \\ 2^n - 1 \end{pmatrix}.$$

We now revisit scalar equations of order k and demonstrate how to transform them into a k-dimensional system of first-order equations. Consider again the equation

$$y(n+k) + p_1(n)y(n+k-1) + \cdots + p_k(n)y(n) = g(n). \tag{3.2.15}$$

This relation may be written as a system of first-order equations of dimension k. We let

$$\begin{aligned} z_1(n) &= y(n), \\ z_2(n) &= y(n+1) = z_1(n+1), \\ z_3(n) &= y(n+2) = z_2(n+1), \\ &\vdots \\ z_k(n) &= y(n+k-1) = z_{k-1}(n+1). \end{aligned}$$

Let $z(n) = (z_1(n), z_2(n), \ldots, z_k(n))$.

Hence,

$$\begin{aligned} z_1(n+1) &= z_2(n), \\ z_2(n+1) &= z_3(n), \\ &\vdots \\ z_{k-1}(n+1) &= z_k(n), \\ z_k(n+1) &= -p_k(n)z_1(n) - p_{k-1}(n)z_2(n), \ldots, \\ &\quad - p_1(n)z_k(n) + g(n). \end{aligned}$$

In vector notation, we transcribe this system as

$$z(n+1) = A(n)z(n) + h(n), \tag{3.2.16}$$

where

$$A(n) = \begin{pmatrix} 0 & 1 & 0 & \dots & 0 \\ 0 & 0 & 1 & \dots & 0 \\ 0 & 0 & 0 & \dots & 0 \\ \vdots & \vdots & \vdots & & \vdots \\ 0 & 0 & 0 & & 1 \\ -p_k(n) & -p_{k-1}(n) & -p_{k-2}(n) & \dots & -p_1(n) \end{pmatrix} \tag{3.2.17}$$

and

$$h(n) = \begin{pmatrix} 0 \\ 0 \\ 0 \\ \vdots \\ g(n) \end{pmatrix}.$$

If $g(n) = 0$, we arrive at the homogeneous system

$$z(n+1) = A(n)z(n). \tag{3.2.18}$$

The matrix $A(n)$ is called the *companion matrix* of (3.2.15).

Consider now the kth-order homogeneous equation with constant coefficients

$$x(n+k) + p_1 x(n+k-1) + p_2 x(n+k-2) + \cdots + p_k x(n) = 0, \tag{3.2.19}$$

which is equivalent to the system where A is the companion matrix defined in formula (3.2.17) with all p_i's constant,

$$z(n+1) = Az(n). \tag{3.2.20}$$

We first observe that the Casoratian of (3.2.19) is denoted by $C(n) = \det \Phi(n)$, where $\Phi(n)$ is a fundamental matrix of (3.2.20). (Why?) (Exercises 3.2, Problem 14.) The characteristic equation of A is given by

$$\lambda^k + p_1\lambda^{k-1} + p_2\lambda^{k-2} + \cdots + p_{k-1}\lambda + p_k = 0 \quad \text{(Exercises 3.2, Problem 13)},$$

which correlates with (2.3.2). Hence, the eigenvalues of A are the roots of the characteristic equation of (2.3.1).

Exercises 3.2

1. Let $\Phi_1(n)$ and $\Phi_2(n)$ be two fundamental matrices of system (3.2.1). Prove that $\Phi_1(n)\Phi_1^{-1}(n_0) = \Phi_2(n)\Phi_2^{-1}(n_0)$ for any $n_0 \geq 0$.

2. Let $\Phi(n,m)$ be a fundamental matrix of (3.2.1). Show that:
   (i) $\Phi(n,m)$ is a solution of $\Phi(n+1,m) = A(n)\Phi(n,m)$.
   (ii) $\Phi(n,m)$ is a solution of $\Phi(n,m+1) = \Phi(n,m)A^{-1}(m)$.

3. Let $\Phi(n,m)$ be a fundamental matrix of (3.2.1). Show that:
   (a) $\Phi(n,m) = A^{n-m}$ if $A(n) \equiv A$ is a constant matrix.
   (b) $\Phi(n,m) = \Phi(n,r)\Phi(r,m)$.
   (c) $\Phi^{-1}(n,m) = \Phi(m,n)$.
   (d) $\Phi(n,m) = \prod_{i=m}^{n-1} A(i)$.

4. Let $\Phi(n)$ be a fundamental matrix of (3.2.1). Show that each column of $\Phi(n)$ is a solution of (3.2.1).

5. Show that $\Phi(n) = \prod_{i=n_0}^{n-1} A(i)$ is a fundamental matrix of (3.2.1).

6. Show that if $\Phi(n)$ is a fundamental matrix of (3.2.1) and C is any nonsingular matrix, then $\Phi(n)C$ is also a fundamental matrix of (3.2.1).

7. Show that if $\Phi_1(n), \Phi_2(n)$ are two fundamental matrices of (3.2.1), then there exists a nonsingular matrix C such that $\Phi_2(n) = \Phi_1(n)C$.

8. Solve the system:
   $y_1(n+1) = y_2(n)$,
   $y_2(n+1) = y_3(n) + 2$,
   $y_3(n+1) = y_1(n) + 2y_3(n) + n^2$.

9. Solve the system:
   $y_1(n+1) = 2y_1(n) + 3y_2(n) + 1$,
   $y_2(n+1) = y_1(n) + 4y_2(n)$,
   $y_1(0) = 0, y_2(0) = -1$.

10. Solve the system $y(n+1) = Ay(n) + g(n)$ if

$$A = \begin{pmatrix} 2 & 2 & -2 \\ 0 & 3 & 1 \\ 0 & 1 & 3 \end{pmatrix}, \qquad g(n) = \begin{pmatrix} 1 \\ n \\ n^2 \end{pmatrix}.$$

11. For system equation (3.2.18) show that

$$\det A(n) = (-1)^k p_k(n).$$

12. If $\Phi(n)$ is a fundamental matrix of (3.2.18), prove that

$$\det \Phi(n) = (-1)^{k(n-n_0)} \left(\prod_{i=n_0}^{n-1} p_k(i) \right) \det \Phi(n_0).$$

13. Prove by induction that the characteristic equation of

$$A = \begin{pmatrix} 0 & 1 & 0 & \dots & 0 \\ 0 & 0 & 1 & \dots & 0 \\ \vdots & \vdots & \vdots & & \\ 0 & 0 & 0 & \dots & 1 \\ -p_k & -p_{k-1} & -p_{k-2} & \dots & -p_1 \end{pmatrix}$$

is

$$\lambda^k + p_1\lambda^{k-1} + p_2\lambda^{k-2} + \cdots + p_{k-1}\lambda + p_k = 0.$$

14. Let $W(n)$ be the Casoratian of (3.2.15) with $g(n) = 0$. Prove that there exists a fundamental matrix $\Phi(n)$ of (3.2.18) such that $W(n) = \det \Phi(n)$.

Use the methods of systems to solve the difference equation for Problems 15 through 19.

15. $x(n+2) + 8x(n+1) + 12x(n) = 0$.

16. $x(n+2) - 16x(n) = 0$.

17. $y(n+2) - 5y(n+1) + 4y(n) = 4^n$.

18. $\Delta^2 y(n) = 16$.

19. $\Delta^2 x(n) + \Delta x(n) - x(n) = 0$.

3.3 The Jordan Form: Autonomous (Time-Invariant) Systems Revisited

The Jordan form of a matrix is vital for both theoretical and computational purposes in autonomous systems. In this section we will briefly describe the Jordan form and derive a new method for computing fundamental matrices.

3.3.1 Diagonalizable Matrices

We say that the two $k \times k$ matrices A and B are *similar* if there exists a nonsingular matrix P such that $P^{-1}AP = B$. It may be shown in this case that A and B have the same eigenvalues and, in fact, the eager student will prove this supposition in Exercises 3.3, Problem 15. If a matrix A is similar to a diagonal matrix $D = \operatorname{diag}(\lambda_1, \lambda_2, \dots, \lambda_k)$, then A is said to be *diagonalizable*. Notice here that the diagonal elements of D, namely, $\lambda_1, \lambda_2, \dots, \lambda_k$, are the eigenvalues of A. We remark here that only special types of matrices are diagonalizable. For those particular diagonalizable matrices, computing A^n is simple. For if

$$P^{-1}AP = D = \text{diag}[\lambda_1, \lambda_2, \ldots, \lambda_k],$$

then

$$A = PDP^{-1},$$

and, consequently,

$$A^n = (PDP^{-1})^n = PD^nP^{-1}.$$

Explicitly,

$$A^n = P \begin{bmatrix} \lambda_1^n & & & 0 \\ & \lambda_2^n & & \\ & & \ddots & \\ 0 & & & \lambda_k^n \end{bmatrix} P^{-1}. \tag{3.3.1}$$

If we are interested in finding another (but simpler) fundamental matrix of the equation

$$x(n+1) = Ax(n), \tag{3.3.2}$$

then we let

$$\Phi(n) = A^nP = P \begin{bmatrix} \lambda_1^n & & & 0 \\ & \lambda_2^n & & \\ & & \ddots & \\ 0 & & & \lambda_k^n \end{bmatrix}. \tag{3.3.3}$$

From formula (3.3.3) we have $\Phi(0) = P$ and, consequently,

$$A^n = \Phi(n)\Phi^{-1}(0). \tag{3.3.4}$$

Now, formula (3.3.3) is useful only if one can pinpoint the matrix P. Fortunately, this is an easy task. We will now reveal how to compute P.

Let $P = (\xi_1, \xi_2, \ldots, \xi_k)$, where ξ_i is the ith column of P. Since $P^{-1}AP = D$, then $AP = PD$. This implies that $A\xi_i = \lambda_i \xi_i, i = 1, 2, \ldots, k$ (Exercises 3.3, Problem 15). Thus, $\xi_i, 1 \le i \le k$, is the eigenvector of A corresponding to the eigenvalue λ_i, and hence the ith column of P is the eigenvector of A corresponding to the ith eigenvalue λ_i of A. Since P is nonsingular, its columns (and hence the eigenvectors $\xi_1, \xi_2, \ldots, \xi_k$ of A) are linearly independent. Reversing the above steps, one may show that the converse of the above statement is true. Namely, if there are k linearly independent eigenvectors of a $k \times k$ matrix A, then it is diagonalizable. The following theorem summarizes the above discussion.

Theorem 3.20. *A $k \times k$ matrix is diagonalizable if and only if it has k linearly independent eigenvectors.*

Let us revert back to formula (3.3.3), which gives us a computational method to find a fundamental matrix $\Phi(n)$. Let $\lambda_1, \lambda_2, \ldots, \lambda_k$ be the eigenvalues of A and let $\xi_1, \xi_2, \ldots, \xi_k$ be the corresponding linearly independent eigenvectors of A. Then from formula (3.3.3) we have

$$\begin{aligned}\Phi(n) &= [\xi_1, \xi_2, \ldots, \xi_k]\begin{bmatrix}\lambda_1^n & & & 0\\ & \lambda_2^n & & \\ & & \ddots & \\ 0 & & & \lambda_k^n\end{bmatrix}\\ &= [\lambda_1^n\xi_1, \lambda_2^n\xi_2, \ldots, \lambda_k^n\xi_k]. \end{aligned} \tag{3.3.5}$$

Notice that since columns of $\Phi(n)$ are solutions of (3.3.2), it follows that for each $i, 1 \le i \le k, x(n) = \lambda_i^n\xi_i$ is a solution of (3.3.2).

Hence, the general solution of (3.3.2) may be given by

$$x(n) = c_1\lambda_1^n\xi_1 + c_2\lambda_2^n\xi_2 + \cdots + c_k\lambda_k^n\xi_k. \tag{3.3.6}$$

The following example illustrates the above method.

Example 3.21. Find the general solution of $x(n+1) = Ax(n)$, where

$$A = \begin{pmatrix}2 & 2 & 1\\ 1 & 3 & 1\\ 1 & 2 & 2\end{pmatrix}.$$

Solution The eigenvalues of A may be obtained by solving the characteristic equation

$$\det(A - \lambda I) = \det\begin{pmatrix}2-\lambda & 2 & 1\\ 1 & 3-\lambda & 1\\ 1 & 2 & 2-\lambda\end{pmatrix} = 0.$$

This determinant produces $(\lambda - 1)^2(\lambda - 5) = 0$. Thus, $\lambda_1 = 5$, and $\lambda_2 = \lambda_3 = 1$. To find the corresponding eigenvectors, we solve the equation $(A - \lambda I)x = 0$. Hence, for $\lambda_1 = 5$,

$$\begin{pmatrix}-3 & 2 & 1\\ 1 & -2 & 1\\ 1 & 2 & -3\end{pmatrix}\begin{pmatrix}x_1\\ x_2\\ x_3\end{pmatrix} = \begin{pmatrix}0\\ 0\\ 0\end{pmatrix}.$$

Solving this system gives us the first eigenvector

$$\xi_1 = \begin{pmatrix}1\\ 1\\ 1\end{pmatrix}.$$

For $\lambda_2 = \lambda_3 = 1$, we have

$$\begin{pmatrix} 1 & 2 & 1 \\ 1 & 2 & 1 \\ 1 & 2 & 1 \end{pmatrix} \begin{pmatrix} x_1 \\ x_2 \\ x_3 \end{pmatrix} = \begin{pmatrix} 0 \\ 0 \\ 0 \end{pmatrix}.$$

Consequently, $x_1 + 2x_2 + x_3 = 0$ is the only equation obtained from this algebraic system. To solve the system, two of the three unknown terms $x_1, x_2,$ and x_3 must be arbitrarily chosen. So if we let $x_1 = 1$ and $x_2 = 0$, then $x_3 = -1$, and we obtain the eigenvector

$$\xi_2 = \begin{pmatrix} 1 \\ 0 \\ -1 \end{pmatrix}.$$

On the other hand, if we let $x_1 = 0$ and $x_2 = 1$, then $x_3 = -2$, and we obtain the third eigenvector

$$\xi_3 = \begin{pmatrix} 0 \\ 1 \\ -2 \end{pmatrix}.$$

Obviously, there are infinitely many choices for ξ_2, ξ_3. Using formula (3.3.6), we see that the general solution is

$$x(n) = c_1 5^n \begin{pmatrix} 1 \\ 1 \\ 1 \end{pmatrix} + c_2 \begin{pmatrix} 1 \\ 0 \\ -1 \end{pmatrix} + c_3 \begin{pmatrix} 0 \\ 1 \\ -2 \end{pmatrix},$$

or

$$x(n) = \begin{pmatrix} c_1 5^n + c_2 \\ c_1 5^n + c_3 \\ c_1 5^n - c_2 - 2c_3 \end{pmatrix}. \tag{3.3.7}$$

Suppose that in the above problem we are given an initial value

$$x(0) = \begin{pmatrix} 0 \\ 1 \\ 0 \end{pmatrix}$$

and must find the solution $x(n)$ with this initial value. One way of doing this is by letting $n = 0$ in the solution given by formula (3.3.7) and evaluating the constants $c_1, c_2,$ and c_3.

Thus

$$\begin{aligned} c_1 + c_2 &= 0, \\ c_1 + c_3 &= 1, \\ c_1 - c_2 - 2c_3 &= 0. \end{aligned}$$

Solving this system gives $c_1 = \frac{1}{2}, c_2 = -\frac{1}{2}$, and $c_3 = \frac{1}{2}$, leading us to the solution

$$x(n) = \begin{pmatrix} \frac{1}{2}5^n - \frac{1}{2} \\ \frac{1}{2}5^n + \frac{1}{2} \\ \frac{1}{2}5^n - \frac{1}{2} \end{pmatrix}.$$

We now introduce yet another method to find the solution. Let

$$x(n) = \Phi(n)\Phi^{-1}(0)x(0),$$

where

$$\Phi(n) = (\lambda_1^n \xi_1, \lambda_2^n \xi_2, \lambda_3^n \xi_3) = \begin{pmatrix} 5^n & 1 & 0 \\ 5^n & 0 & 1 \\ 5^n & -1 & -2 \end{pmatrix}$$

and

$$\Phi(0) = \begin{pmatrix} 1 & 1 & 0 \\ 1 & 0 & 1 \\ 1 & -1 & -2 \end{pmatrix}.$$

Thus,

$$\Phi^{-1}(0) = \begin{pmatrix} \frac{1}{4} & \frac{1}{2} & \frac{1}{4} \\ \frac{3}{4} & -\frac{1}{2} & -\frac{1}{4} \\ -\frac{1}{4} & \frac{1}{2} & -\frac{1}{4} \end{pmatrix}.$$

This gives

$$x(n) = \begin{pmatrix} 5^n & 1 & 0 \\ 5^n & 0 & 1 \\ 5^n & -1 & -2 \end{pmatrix} \begin{pmatrix} \frac{1}{4} & \frac{1}{2} & \frac{1}{4} \\ \frac{3}{4} & -\frac{1}{2} & -\frac{1}{4} \\ -\frac{1}{4} & \frac{1}{2} & -\frac{1}{4} \end{pmatrix} \begin{pmatrix} 0 \\ 1 \\ 0 \end{pmatrix}$$

$$= \begin{pmatrix} \frac{1}{2}5^n - \frac{1}{2} \\ \frac{1}{2}5^n + \frac{1}{2} \\ \frac{1}{2}5^n - \frac{1}{2} \end{pmatrix}.$$

In the next example we will examine the case where the matrix A has complex eigenvalues. Notice that if A is a real matrix (which we are assuming here) and if $\lambda = \alpha + i\beta$ is an eigenvalue of A, then $\overline{\lambda} = \alpha - i\beta$ is also an eigenvalue of A. Moreover, if ξ is the eigenvector of A corresponding to the eigenvalue $\lambda = \alpha + i\beta$, then $\overline{\xi}$ is the eigenvector of A corresponding to the eigenvalue $\overline{\lambda} = \alpha - i\beta$. Taking advantage of these observations, one may be able to simplify considerably the computation involved in finding a fundamental matrix of the system of equations (3.3.2).

Suppose that $\xi = \xi_1 + i\xi_2$. A solution of system (3.3.2) may then be given by $x(n) = (\alpha + i\beta)^n(\xi_1 + i\xi_2)$. Also, if

$$r = \sqrt{\alpha^2 + \beta^2},$$

then

$$\theta = \tan^{-1}\left(\frac{\beta}{\alpha}\right).$$

This solution may now be written as

$$\begin{aligned} x(n) &= [r\cos\theta + i\sin\theta)]^n(\xi_1 + i\xi_2) \\ &= r^n(\cos n\theta + i\sin n\theta)(\xi_1 + i\xi_2) \\ &= r^n[(\cos n\theta)\xi_1 - (\sin n\theta)\xi_2] + ir^n[(\cos n\theta)\xi_2 + (\sin n\theta)\xi_1] \\ &= u(n) + i\,v(n), \end{aligned}$$

where $u(n) = r^n[(\cos n\theta)\xi_1 - (\sin n\theta)\xi_2]$ and $v(n) = r^n[(\cos n\theta)\xi_2 + (\sin n\theta)\xi_1]$. One might show (Exercises 3.3, Problem 7) that $u(n)$ and $v(n)$ are linearly independent solutions of system (3.3.2). Hence, we do not need to consider the solution generated by $\overline{\lambda}$ and $\overline{\xi}$.

Example 3.22. Find a general solution of the system $x(n+1) = Ax(n)$, where

$$A = \begin{pmatrix} 1 & -5 \\ 1 & -1 \end{pmatrix}.$$

Solution The eigenvalues of A are $\lambda_1 = 2i$, $\lambda_2 = -2i$, and the corresponding eigenvectors are

$$\xi_1 = \begin{pmatrix} \frac{1}{5} - \frac{2}{5}i \\ 1 \end{pmatrix}, \quad \xi_2 = \begin{pmatrix} \frac{1}{5} + \frac{2}{5}i \\ 1 \end{pmatrix}.$$

Hence,

$$x(n) = (2i)^n \begin{pmatrix} \frac{1}{5} - \frac{2}{5}i \\ 1 \end{pmatrix}$$

is a solution. Since

$$i = \cos\frac{\pi}{2} + i\sin\frac{\pi}{2},$$
$$i^n = \cos\frac{n\pi}{2} + i\sin\frac{n\pi}{2},$$

this solution may be written as

$$\begin{aligned}
x(n) &= 2^n\left(\cos\left(\frac{n\pi}{2}\right) + i\sin\left(\frac{n\pi}{2}\right)\right)\begin{pmatrix}\frac{1}{5} - \frac{2}{5}i \\ 1\end{pmatrix} \\
&= 2^n\begin{pmatrix}\frac{1}{5}\cos\left(\frac{n\pi}{2}\right) + \frac{2}{5}\sin\frac{n\pi}{2} \\ \cos\left(\frac{n\pi}{2}\right)\end{pmatrix} \\
&\quad + i2^n\begin{pmatrix}\frac{-2}{5}\cos\left(\frac{n\pi}{2}\right) + \frac{1}{5}\sin\left(\frac{n\pi}{2}\right) \\ \sin\left(\frac{n\pi}{2}\right)\end{pmatrix}.
\end{aligned}$$

Thus,

$$u(n) = 2^n\begin{pmatrix}\frac{1}{5}\cos\left(\frac{n\pi}{2}\right) + \frac{2}{5}\sin\left(\frac{n\pi}{2}\right) \\ \cos\left(\frac{n\pi}{2}\right)\end{pmatrix}$$

and

$$v(n) = 2^n\begin{pmatrix}\frac{-2}{5}\cos\left(\frac{n\pi}{2}\right) + \frac{1}{5}\sin\left(\frac{n\pi}{2}\right) \\ \sin\left(\frac{n\pi}{2}\right)\end{pmatrix}$$

are two linearly independent solutions. A general solution may be given as

$$\begin{aligned}
x(n) &= c_1 2^n\begin{pmatrix}\frac{1}{5}\cos\left(\frac{n\pi}{2}\right) + \frac{2}{5}\sin\left(\frac{n\pi}{2}\right) \\ \cos\left(\frac{n\pi}{2}\right)\end{pmatrix} \\
&\quad + c_2 2^n\begin{pmatrix}\frac{-2}{5}\cos\left(\frac{n\pi}{2}\right) + \frac{1}{5}\sin\left(\frac{n\pi}{2}\right) \\ \sin\left(\frac{n\pi}{2}\right)\end{pmatrix} \\
&= 2^n\begin{bmatrix}\left(\frac{1}{5}c_1 - \frac{2}{5}c_2\right)\cos\left(\frac{n\pi}{2}\right) + \left(\frac{2}{5}c_1 + \frac{1}{5}c_2\right)\sin\left(\frac{n\pi}{2}\right) \\ c_1\cos\left(\frac{n\pi}{2}\right) + c_2\sin\left(\frac{n\pi}{2}\right)\end{bmatrix}.
\end{aligned}$$

So far, we have discussed the solution of system (3.3.2) if the matrix A is diagonalizable. We remark here that a sufficient condition for a $k\times k$ matrix A to be diagonalizable is that it have k distinct eigenvalues (Exercises 3.3,

Problem 20). If the matrix A has repeated roots, then it is diagonalizable if it is *normal*, that is to say, if $A^T A = AA^T$. (For a proof see [111].) Examples of normal matrices are:

(i) symmetric matrices ($A^T = A$),

(ii) skew symmetric matrices ($A^T = -A$),

(iii) unitary matrices ($A^T A = AA^T = I$).

3.3.2 The Jordan Form

We now turn our attention to the general case where the matrix A is not diagonalizable. This happens when A has repeated eigenvalues, and one is not able to generate k linearly independent eigenvectors. For example, the following matrices are not diagonalizable:

$$\begin{bmatrix} 2 & 1 \\ 0 & 2 \end{bmatrix}, \quad \begin{bmatrix} 2 & 0 & 0 \\ 0 & 2 & 1 \\ 0 & 0 & 2 \end{bmatrix}, \quad \begin{bmatrix} 2 & 0 & 0 & 0 \\ 0 & 3 & 0 & 0 \\ 0 & 0 & 4 & 1 \\ 0 & 0 & 0 & 4 \end{bmatrix}.$$

If a $k \times k$ matrix A is not diagonalizable, then it is akin to the so-called *Jordan form*, i.e., $P^{-1}AP = J$, where

$$J = \operatorname{diag}(J_1, J_2, \ldots, J_r), \qquad 1 \le r \le k, \tag{3.3.8}$$

and

$$J_i = \begin{pmatrix} \lambda_i & 1 & 0 & \ldots & 0 \\ 0 & \lambda_i & 1 & & 0 \\ & & \ddots & & \\ 0 & 0 & \ddots & \ddots & \vdots \\ \vdots & \vdots & & \ddots & 1 \\ 0 & 0 & & & \lambda_i \end{pmatrix}. \tag{3.3.9}$$

The matrix J_i is called a *Jordan block.*

These remarks are formalized in the following theorem.

Theorem 3.23 (The Jordan Canonical Form). *Any $k \times k$ matrix A is similar to a Jordan form given by the formula* (3.3.8), *where each J_i is an $s_i \times s_i$ matrix of the form* (3.3.9), *and $\sum_{i=1}^{r} s_i = k$.*

The number of Jordan blocks corresponding to one eigenvalue λ is called the *geometric multiplicity* of λ, and this number, in turn, equals the number of linearly independent eigenvectors corresponding to λ.

The *algebraic multiplicity* of an eigenvalue λ is the number of times it is repeated. If the algebraic multiplicity of λ is 1 (i.e., λ is not repeated), then we refer to λ as *simple.* If the geometric multiplicity of λ is equal to its algebraic multiplicity (i.e., only 1×1 Jordan blocks correspond to λ), then it is called *semisimple.* For example, the matrix

$$\begin{bmatrix} 3 & 0 & 0 & 0 & 0 \\ 0 & 2 & 0 & 0 & 0 \\ 0 & 0 & 2 & 0 & 0 \\ 0 & 0 & 0 & 5 & 1 \\ 0 & 0 & 0 & 0 & 5 \end{bmatrix}$$

has one simple eigenvalue 3, one semisimple eigenvalue 2, and one eigenvalue 5, which is neither simple nor semisimple.

To illustrate the theorem, we list below the possible Jordan forms of a 3×3 matrix with an eigenvalue $\lambda = 5$, of multiplicity 3. In the matrix, different Jordan blocks are indicated by squares.

$$\underset{(a)}{\begin{pmatrix} \boxed{5} & 0 & 0 \\ 0 & \boxed{5} & 0 \\ 0 & 0 & \boxed{5} \end{pmatrix}} \quad \underset{(b)}{\begin{pmatrix} \boxed{\begin{matrix} 5 & 1 \\ 0 & 5 \end{matrix}} & \begin{matrix} 0 \\ 0 \end{matrix} \\ \begin{matrix} 0 & 0 \end{matrix} & \boxed{5} \end{pmatrix}} \quad \underset{(c)}{\begin{pmatrix} \boxed{5} & \begin{matrix} 0 & 0 \end{matrix} \\ \begin{matrix} 0 \\ 0 \end{matrix} & \boxed{\begin{matrix} 5 & 1 \\ 0 & 5 \end{matrix}} \end{pmatrix}} \quad \underset{(d)}{\begin{pmatrix} \boxed{\begin{matrix} 5 & 1 & 0 \\ 0 & 5 & 1 \\ 0 & 0 & 5 \end{matrix}} \end{pmatrix}}.$$

Recall that s_i is the order of the ith Jordan block and r is the number of Jordan blocks in a Jordan form. In (a) the matrix is diagonalizable, and we have three Jordan blocks of order 1. Thus, $s_1 = s_2 = s_3 = 1, r = 3$, and the geometric multiplicity of λ is 3.

In (b) there are two Jordan blocks with $s_1 = 2, s_2 = 1, r = 2$, and the geometric multiplicity of λ is 2.

In (c) there are also two Jordan blocks with $s_1 = 1, s_2 = 2, r = 2$, and the geometric multiplicity of λ is 2. In (d) there is only one Jordan block with $s_1 = 3, r = 1$, and the geometric multiplicity of λ is 1. The linearly independent eigenvectors corresponding to $\lambda = 5$ in (a), (b), (c), (d) are, respectively,

$$\underbrace{\begin{pmatrix} 1 \\ 0 \\ 0 \end{pmatrix}, \begin{pmatrix} 0 \\ 1 \\ 0 \end{pmatrix}, \begin{pmatrix} 0 \\ 0 \\ 1 \end{pmatrix}}_{(a)} \quad \underbrace{\begin{pmatrix} 1 \\ 0 \\ 0 \end{pmatrix}, \begin{pmatrix} 0 \\ 0 \\ 1 \end{pmatrix}}_{(b)} \quad \underbrace{\begin{pmatrix} 1 \\ 0 \\ 0 \end{pmatrix}, \begin{pmatrix} 0 \\ 1 \\ 0 \end{pmatrix}}_{(c)} \quad \underset{(d)}{\begin{pmatrix} 1 \\ 0 \\ 0 \end{pmatrix}}.$$

Note that a matrix of the form

$$\begin{pmatrix} \lambda & 1 & 0 & \dots & 0 \\ 0 & \lambda & 1 & \dots & 0 \\ \vdots & \vdots & \vdots & & \vdots \\ 0 & 0 & 0 & \dots & 1 \\ 0 & 0 & 0 & \dots & \lambda \end{pmatrix}$$

has only one eigenvector, namely, the unit vector $e_1 = (1, 0, \dots, 0)^T$. This shows us that the linearly independent eigenvectors of the Jordan form J given by formula (3.3.8) are

$$e_1, e_{s_1+1}, e_{s_1+s_2+1}, \dots, e_{s_1+s_2+\dots+s_{r-1}+1}.$$

Now, since $P^{-1}AP = J$, then

$$AP = PJ. \tag{3.3.10}$$

Let $P = (\xi_1, \xi_2, \dots, \xi_k)$. Equating the first s_1 columns of both sides in formula (3.3.10), we obtain

$$A\xi_1 = \lambda_1 \xi_1, \dots, A\xi_i = \lambda_1 \xi_i + \xi_{i-1}, \qquad i = 2, 3, \dots, s_1. \tag{3.3.11}$$

Clearly, ξ_1 is the only eigenvector of A in the *Jordan chain* $\xi_1, \xi_2, \dots, \xi_{s_1}$. The other vectors $\xi_2, \xi_3, \dots, \xi_{s_1}$ are called *generalized eigenvectors* of A, and they may be obtained by using the difference equation

$$(A - \lambda_1 I)\xi_i = \xi_{i-1}, \qquad i = 2, 3, \dots, s_1. \tag{3.3.12}$$

Repeating this process for the remainder of the Jordan blocks, one may find the generalized eigenvectors corresponding to the mth Jordan block using the difference equation

$$(A - \lambda_m I)\xi_{m_i} = \xi_{m_i - 1}, \qquad i = 2, 3, \dots, s_m. \tag{3.3.13}$$

Now we know that $A^n = (PJP^{-1})^n = PJ^nP^{-1}$, where

$$J^n = \begin{bmatrix} J_1^n & & & 0 \\ & J_2^n & & \\ & & \ddots & \\ 0 & & & J_k^n \end{bmatrix}.$$

Notice that for any $J_i, i = 1, 2, \dots, r$, we have $J_i = \lambda_i I + N_i$, where

$$N_i = \begin{pmatrix} 0 & 1 & 0 & \dots & 0 \\ 0 & 0 & 1 & & 0 \\ \vdots & \vdots & & & 1 \\ 0 & 0 & & \dots & 0 \end{pmatrix}$$

is an $s_i \times s_i$ nilpotent matrix (i.e., $N_i^r = 0$ for all $r \geq s_i$). Hence,

$$J_i^n = (\lambda_i I + N_i)^n = \lambda_i^n I + \binom{n}{1} \lambda_i^{n-1} N_i + \binom{n}{2} \lambda_i^{n-2} N_i^2 + \cdots + \binom{n}{s_i - 1} \lambda_i^{n-s_i+1} N_i^{s_i-1}$$

$$= \begin{pmatrix} \lambda_i^n & \binom{n}{1} \lambda_i^{n-1} & \binom{n}{2} \lambda_i^{n-2} & \cdots & \binom{n}{s_i - 1} \lambda_i^{n-s_i+1} \\ 0 & \lambda_i^n & \binom{n}{1} \lambda_i^{n-1} & \cdots & \binom{n}{s_i - 2} \lambda_i^{n-s_i+2} \\ \vdots & \vdots & & \ddots & \vdots \\ & & & & \binom{n}{1} \lambda_i^{n-1} \\ 0 & 0 & & \cdots & \lambda_i^n \end{pmatrix}. \tag{3.3.14}$$

The lines inside J_i^n indicate that the entries in each diagonal are identical.

We can now substantiate that the general solution of system (3.3.2) is

$$x(n) = A^n c = P J^n P^{-1} c,$$

or

$$x(n) = P J^n \hat{c}, \tag{3.3.15}$$

where

$$\hat{c} = P^{-1} c.$$

Hence, a fundamental matrix of system (3.3.2) may be given by $\Phi(n) = PJ^n$. Also, the state transition matrix may be given by $\Phi(n, n_0) = PJ^{n-n_0}P^{-1}$ and thus $x(n, n_0, x_0) = PJ^{n-n_0}P^{-1}x_0$.

The following corollary arises directly from an immediate consequence of formula (3.3.14).

Corollary 3.24. *Assuming that A is any $k \times k$ matrix, then* $\lim_{n\to\infty} A^n = 0$ *if and only if* $|\lambda| < 1$ *for all eigenvalues* λ *of* A.

PROOF. (Exercises 3.3, Problem 21.) □

The importance of the preceding corollary lies in the fact that if $\lim_{n\to\infty} A^n = 0$, then $\lim_{n\to\infty} x^n = \lim_{n\to\infty} A^n x(0) = 0$. This fact reminds us that if $|\lambda| < 1$ for all eigenvalues of A, then all solutions $x(n)$ of (3.3.1) tend toward the zero vector as $n \to \infty$.

Example 3.25. Find the general solution of $x(n+1) = Ax(n)$ with

$$A = \begin{pmatrix} 4 & 1 & 2 \\ 0 & 2 & -4 \\ 0 & 1 & 6 \end{pmatrix}.$$

Solution Note that this example uses conclusions from Example 3.3. The eigenvalues are $\lambda_1 = \lambda_2 = \lambda_3 = 4$. To find the eigenvectors, we solve the equation $(A - \lambda I)\xi = 0$, or

$$\begin{pmatrix} 0 & 1 & 2 \\ 0 & -2 & -4 \\ 0 & 1 & 2 \end{pmatrix} \begin{pmatrix} d_1 \\ d_2 \\ d_3 \end{pmatrix} = \begin{pmatrix} 0 \\ 0 \\ 0 \end{pmatrix}.$$

Hence,

$$\begin{aligned} d_2 + 2d_3 &= 0, \\ -2d_2 - 4d_3 &= 0, \\ d_2 + 2d_3 &= 0. \end{aligned}$$

These equations imply that $d_2 = -2d_3$, thus generating two eigenvectors,

$$\xi_1 = \begin{pmatrix} 0 \\ -2 \\ 1 \end{pmatrix} \quad \text{and} \quad \xi_2 = \begin{pmatrix} 1 \\ -2 \\ 1 \end{pmatrix}.$$

We must now find one generalized eigenvector ξ_3. Applying formula (3.3.11), let us test $(A - 4I)\xi_3 = \xi_1$:

$$\begin{pmatrix} 0 & 1 & 2 \\ 0 & -2 & -4 \\ 0 & 1 & 2 \end{pmatrix} \begin{pmatrix} a_1 \\ a_2 \\ a_3 \end{pmatrix} = \begin{pmatrix} 0 \\ -2 \\ 1 \end{pmatrix}.$$

This system is an inconsistent system that has no solution. The second attempt will use

$$(A - 4I)\xi_3 = \xi_2,$$

or

$$\begin{pmatrix} 0 & 1 & 2 \\ 0 & -2 & -4 \\ 0 & 1 & 2 \end{pmatrix} \begin{pmatrix} a_1 \\ a_2 \\ a_3 \end{pmatrix} = \begin{pmatrix} 1 \\ -2 \\ 1 \end{pmatrix}.$$

Hence, $a_2 + 2a_3 = 1$. One may now set

$$\xi_3 = \begin{pmatrix} 0 \\ -1 \\ 1 \end{pmatrix}.$$

Thus,

$$P = \begin{pmatrix} 0 & 1 & 0 \\ -2 & -2 & -1 \\ 1 & 1 & 1 \end{pmatrix},$$

$$J = \begin{pmatrix} 4 & 0 & 0 \\ 0 & 4 & 1 \\ 0 & 0 & 4 \end{pmatrix},$$

and

$$J^n = \begin{pmatrix} 4^n & 0 & 0 \\ 0 & 4^n & n4^{n-1} \\ 0 & 0 & 4^n \end{pmatrix}.$$

Hence,

$$x(n) = PJ^n\hat{c} = \begin{pmatrix} 0 & 4^n & n4^{n-1} \\ -2 \cdot 4^n & -2 \cdot 4^n & -2n4^{n-1} - 4^n \\ -4^n & 4^n & n4^{n-1} + 4^n \end{pmatrix} \begin{pmatrix} \hat{c}_1 \\ \hat{c}_2 \\ \hat{c}_3 \end{pmatrix}.$$

Example 3.26. Solve the system

$$x(n+1) = Ax(n), \quad x(0) = \begin{pmatrix} 1 \\ 1 \\ 1 \end{pmatrix},$$

where

$$A = \begin{pmatrix} 3/2 & 1/2 & 1/2 \\ 1/2 & 5/2 & -1/2 \\ 0 & 1 & 2 \end{pmatrix}.$$

Solution The eigenvalues of A are $\lambda_1 = \lambda_2 = \lambda_3 = 2$. We have a sole eigenvector,

$$\xi_1 = \begin{pmatrix} 1 \\ 0 \\ 1 \end{pmatrix}.$$

We now need to compose two generalized eigenvectors, using (3.3.13):

$$(A - 2I)\xi_2 = \xi_1 \text{ gives } \xi_2 = \begin{pmatrix} 1 \\ 1 \\ 2 \end{pmatrix}$$

and

$$(A-2I)\xi_3=\xi_2 \text{ gives } \xi_3=\begin{pmatrix}1\\2\\1\end{pmatrix}.$$

So

$$P=\begin{pmatrix}1&1&1\\0&1&2\\1&2&1\end{pmatrix},\quad J=\begin{pmatrix}2&1&0\\0&2&1\\0&0&2\end{pmatrix},$$

$$J^n=\begin{pmatrix}2^n & n2^{n-1} & \dfrac{n(n-1)}{2}2^{n-2}\\ 0 & 2^n & n2^{n-1}\\ 0 & 0 & 2^n\end{pmatrix}.$$

Now,

$$\begin{aligned}x(n,x_0)&=PJ^nP^{-1}x_0\\ &=2^{n-4}\begin{pmatrix}n^2-5n+16 & n^2+3n & -n^2+5n\\ 4n & 4n+16 & -4n\\ n^2-n & n^2+7n & -n^2+n+16\end{pmatrix}\begin{pmatrix}1\\1\\1\end{pmatrix}\\ &=2^{n-4}\begin{pmatrix}n^2+3n+16\\4n+16\\n^2+7n+16\end{pmatrix}.\end{aligned}$$

3.3.3 Block-Diagonal Matrices

In general, the generalized eigenvectors corresponding to an eigenvalue λ of algebraic multiplicity m are the solutions of the equation

$$(A-\lambda I)^m\xi=0. \tag{3.3.16}$$

The first eigenvector ξ_1 corresponding to λ is obtained by solving the equation

$$(A-\lambda I)\xi=0.$$

The second eigenvector or generalized eigenvector ξ_2 is obtained by solving the equation

$$(A-\lambda I)^2\xi=0.$$

And so on.

Now if J is the Jordan form of A, that is, $P^{-1}AP=J$ or $A=PJP^{-1}$, then λ is an eigenvalue of A if and only if it is an eigenvalue of J. Moreover, if ξ is an eigenvector of A, then $\tilde{\xi}=P^{-1}\xi$ is an eigenvector of J.

We would like to know the structure of the eigenvectors $\tilde{\xi}$ of J. For this we appeal to the following simple lemma from Linear Algebra.

Lemma 3.27. *Let* $C = \begin{pmatrix} A & 0 \\ 0 & B \end{pmatrix}$ *be a* $k \times k$ *block-diagonal matrix such that* A *is an* $r \times r$ *matrix and* B *is an* $s \times s$ *matrix, with* $r + s = k$. *Then the following statements hold true:*

(i) *If* λ *is an eigenvalue of* A, *then it is an eigenvalue of* C. *Moreover, the eigenvector and the generalized eigenvectors corresponding to* λ *are of the form* $\xi = (a_1, a_2, \ldots, a_r, 0, \ldots, 0)^T$ *for some* $a_i \in \mathbb{R}$.

(ii) *If* λ *is an eigenvalue of* B, *then it is an eigenvalue of* C. *Moreover, the eigenvector and the generalized eigenvectors corresponding to* λ *are of the form* $\xi = (0, \ldots, 0, a_{r+1}, a_{r+2}, \ldots, a_s)$ *for some* $a_{r+i} \in \mathbb{R}$.

PROOF.

(i) Suppose that λ is an eigenvalue of A, and $V = (a_1, a_2, \ldots, a_r)^T$ is the corresponding eigenvector. Define $\xi = (a_1, \ldots, a_r, 0, \ldots, 0) \in \mathbb{R}^k$. Then clearly $C\xi = \lambda\xi$, and thus λ is an eigenvalue of C. Let the $k \times k$ identity matrix I be written in the form $I = \begin{pmatrix} I_r & 0 \\ 0 & I_s \end{pmatrix}$, where I_r and I_s are, respectively, the $r \times r$ and $s \times s$ identity matrices. Let λ be an eigenvalue of A with algebraic multiplicity m. Then

$$(C - \lambda I)\xi = \begin{pmatrix} A - \lambda I_r & 0 \\ 0 & B - \lambda I_S \end{pmatrix} \begin{pmatrix} \xi_1 \\ \vdots \\ \xi_r \\ \vdots \\ \xi_s \end{pmatrix} = \begin{pmatrix} 0 \\ \vdots \\ 0 \\ \vdots \\ 0 \end{pmatrix}.$$

Hence

$$(A - \lambda I_r) \begin{pmatrix} \xi_1 \\ \vdots \\ \xi_r \end{pmatrix} = \begin{pmatrix} 0 \\ \vdots \\ 0 \end{pmatrix}$$

has a nontrivial solution

$$\tilde{\xi} = \begin{pmatrix} a_1 \\ \vdots \\ a_r \end{pmatrix}.$$

However

$$(B - \lambda I_s)\begin{pmatrix}\xi_{r+1}\\ \vdots\\ \xi_s\end{pmatrix} = \begin{pmatrix}0\\ \vdots\\ 0\end{pmatrix}$$

has only the trivial solution

$$\begin{pmatrix}0\\ \vdots\\ 0\end{pmatrix}.$$

Then $\xi = (a_1, \ldots, a_r, 0, \ldots, 0)^T$ is an eigenvector of C corresponding to λ. The same analysis can be done for generalized eigenvectors by solving $(C - \lambda I)^i \xi = 0$, $1 \le i \le m$.

(ii) The proof of the second part is analogous and will be omitted. □

Exercises 3.3

In Problems 1 through 6, use formula (3.3.6) to find the solution of $x(n+1) = Ax(n)$, where A is given in the exercise.

1. $A = \begin{pmatrix}2 & -1\\ 0 & 4\end{pmatrix}, \quad x(0) = \begin{pmatrix}1\\ 2\end{pmatrix}.$

2. $A = \begin{pmatrix}1 & 0\\ 1 & 2\end{pmatrix}.$

3. $A = \begin{pmatrix}2 & 3 & 0\\ 4 & 3 & 0\\ 0 & 0 & 6\end{pmatrix}, \quad x(0) = \begin{pmatrix}0\\ 1\\ 0\end{pmatrix}.$

4. $A = \begin{pmatrix}2 & -1 & 0\\ 0 & 4 & 0\\ 2 & 5 & 3\end{pmatrix}.$

5. $A = \begin{pmatrix}1 & 0 & 1\\ 1 & 2 & 3\\ 0 & 0 & 3\end{pmatrix}.$

6. $A = \begin{pmatrix}1 & 1 & 0\\ -1 & 1 & 0\\ 1 & 0 & 1\end{pmatrix}, \quad x(0) = \begin{pmatrix}1\\ 0\\ 1\end{pmatrix}.$

7. Suppose that $x(n) = u(n) + iv(n)$ is a solution of (3.3.2), where $u(n)$ and $v(n)$ are real vectors. Prove that $u(n)$ and $v(n)$ are linearly independent solutions of (3.3.2).

8. Utilize Problem 7 to find a fundamental matrix of $x(n+1) = Ax(n)$ with

$$A = \begin{pmatrix} 1 & 1 & 0 & 1 \\ -1 & 1 & 0 & 1 \\ 0 & 0 & 2 & 1 \\ 0 & 0 & -1 & 2 \end{pmatrix}.$$

9. Apply Problem 7 to find a fundamental matrix of $x(n+1) = Ax(n)$ with

$$A = \begin{pmatrix} 1 & 1 & 0 \\ -1 & 1 & 0 \\ 1 & 0 & 1 \end{pmatrix}.$$

10. Find the eigenvalues and the corresponding eigenvectors and generalized eigenvectors for the matrix A.

(a) $A = \begin{pmatrix} 3 & 1 \\ 0 & 3 \end{pmatrix}$. (b) $A = \begin{pmatrix} 2 & 1 & 0 \\ 0 & 2 & 1 \\ 0 & 0 & 2 \end{pmatrix}$.

(c) $A = \begin{pmatrix} 4 & 2 & 3 \\ -\frac{1}{2} & 2 & 0 \\ 0 & 0 & 3 \end{pmatrix}$. (d) $A = \begin{pmatrix} 2 & 0 & 0 & 0 \\ 0 & 2 & 1 & 0 \\ 0 & 0 & 2 & 1 \\ 0 & 0 & 0 & 2 \end{pmatrix}$.

11. Find A^n for the matrices in Problem 10 using the Jordan form.

12. Use the Jordan form to solve $x(n+1) = Ax(n)$ with

$$A = \begin{pmatrix} 3 & 2 & 1 \\ -1 & 3 & 2 \\ 1 & -3 & -2 \end{pmatrix}.$$

13. Use the Jordan form to solve $x(n+1) = Ax(n)$ with

$$A = \begin{pmatrix} 3 & 2 & 3 \\ -1/2 & 1 & 0 \\ 0 & 0 & 2 \end{pmatrix}.$$

14. Let A and B be two similar matrices with $P^{-1}AP = B$.

    (i) Show that A and B have the same eigenvalues.

    (ii) Show that if ξ is an eigenvector of B, then $P\xi$ is an eigenvector of A.

15. Suppose that $P^{-1}AP = D = \text{diag}(\lambda_1, \lambda_2, \ldots, \lambda_k)$, where $P = [\xi_1, \xi_2, \ldots, \xi_k]$ is a nonsingular $k \times k$ matrix. Show that $\xi_1, \xi_2, \ldots, \xi_k$ are the eigenvectors of A that correspond to the eigenvalues $\lambda_1, \lambda_2, \ldots, \lambda_k$, respectively.

16. Let A be a 4×4 matrix with an eigenvalue $\lambda = 3$ of multiplicity 4. Write all possible Jordan forms of A.

17. Show that $(PJP^{-1})^n = PJ^nP^{-1}$.

18. If λ is an eigenvalue of A, and ξ is the corresponding eigenvector of A, show that $\lambda^n \xi$ is a solution of (3.3.2).

19. Let

$$A = \begin{pmatrix} \lambda & 1 & \ldots & 0 \\ 0 & \lambda & \ldots & 0 \\ \vdots & \vdots & & \vdots \\ 0 & 0 & \ldots & 1 \\ 0 & 0 & \ldots & \lambda \end{pmatrix}.$$

Then one may write $A = \lambda I + N$, where

$$N = \begin{pmatrix} 0 & 1 & \ldots & 0 \\ 0 & 0 & & \vdots \\ \vdots & \vdots & & 1 \\ 0 & 0 & & 0 \end{pmatrix}.$$

Show that for any $\alpha > 0$, A is similar to a matrix

$$B = \lambda I + \alpha N = \begin{pmatrix} \lambda & \alpha & \ldots & 0 \\ 0 & \lambda & & \\ \vdots & \vdots & & \alpha \\ 0 & 0 & & \lambda \end{pmatrix}.$$

20. Prove that if a $k \times k$ matrix A has k distinct eigenvalues, then:

    (i) A has k linearly independent eigenvectors.

    (ii) A is diagonalizable. (Use mathematical induction.)

21. Prove Corollary 3.24.

22. Consider the companion matrix A of (3.2.17) with the coefficients p_i constant. Assume that the eigenvalues of A are real and distinct. Let V denote the Vandermonde matrix

$$V = \begin{pmatrix} 1 & 1 & \dots & 1 \\ \lambda_1 & \lambda_2 & \dots & \lambda_k \\ \vdots & \vdots & & \vdots \\ \lambda_1^{k-1} & \lambda_2^{k-1} & \dots & \lambda_k^{k-1} \end{pmatrix}.$$

Show that $V^{-1}AV$ is a diagonal matrix.

23. Consider the companion matrix A (3.3.16) with $p_i(n)$ constants. Suppose $\lambda_1, \lambda_2, \dots, \lambda_r$ are the distinct eigenvalues of A with multiplicities $m_1, m_2, \dots, m_r$ and $\sum_{i=1}^{r} m_i = k$. Let V be the generalized Vandermonde matrix (2.3.9). Show that $V^{-1}AV = J$, where J is in the Jordan form (3.3.8).

3.4 Linear Periodic Systems

In this section we regard the linear periodic system

$$x(n+1) = A(n)x(n), \tag{3.4.1}$$

where for all $n \in \mathbb{Z}, \mathbb{A}(\ltimes + \mathbb{N}) = \mathbb{A}(\ltimes)$, for some positive integer N.

We now show that the study of the periodic system (3.4.1) simplifies to the study of an associated autonomous system. This inference is the analogue of Floquet theory in differential equations. But before we prove that analogue, we need the following theorem.

Lemma 3.28. *Let B be a $k \times k$ nonsingular matrix and let m be any positive integer. Then there exists some $k \times k$ matrix C such that $C^m = B$.*

Proof. Let

$$P^{-1}BP = J = \begin{pmatrix} J_1 & & & \\ & J_2 & & \\ & & \ddots & \\ & & & J_r \end{pmatrix}$$

be the Jordan form of B. Let us write

$$J_i = \lambda_i \left(I_i + \frac{1}{\lambda_i} N_i \right),$$

where I_i is the $s_i \times s_i$ identity matrix and

$$N_i = \begin{pmatrix} 0 & 1 & & & 0 \\ & 0 & 1 & & \\ & & \ddots & \ddots & \\ & & & & 1 \\ 0 & & & & 0 \end{pmatrix}.$$

Observe that

$$N_i^{s_i} = 0. \tag{3.4.2}$$

To motivate our construction, we formally write

$$\begin{aligned} H_i &= \exp\left[\frac{1}{m}\ln J_i\right] \\ &= \exp\left[\frac{1}{m}\left\{\ln \lambda_i I_i + \ln\left(I_i + \frac{1}{\lambda_i}N_i\right)\right\}\right] \\ &= \exp\left[\frac{1}{m}\left\{\ln \lambda_i I_i + \sum_{s=1}^{\infty}\frac{(-1)^{s+1}}{s}\left(\frac{N_i}{\lambda_i}\right)^s\right\}\right]. \end{aligned}$$

Applying formula (3.4.2), we obtain

$$H_i = \exp\left[\frac{1}{m}\left\{\ln \lambda_i I_i + \sum_{s=1}^{s_i-1}\frac{(-1)^{s+1}}{s}\left(\frac{N_i}{\lambda_i}\right)^s\right\}\right]. \tag{3.4.3}$$

Hence, H_i is a well-defined matrix. Furthermore, $H_i^m = J_i$. □

Now, if we let

$$H = \begin{pmatrix} H_1 & & 0 \\ & H_2 & \\ & & \ddots \\ 0 & & & H_r \end{pmatrix},$$

where H_i is defined in formula (3.4.3), then

$$H^m = \begin{bmatrix} H_1^m & & & 0 \\ & H_2^m & & \\ & & \ddots & \\ 0 & & & H_r^m \end{bmatrix} = J.$$

Define $C = PHP^{-1}$. Then $C^m = PH^mP^{-1} = PJP^{-1} = B$.

Armed with this lemma, we are now prepared to introduce the primary result for this section.

Lemma 3.29. *For system* (3.4.1), *the following statements hold:*

(i) *If* $\Phi(n)$ *is a fundamental matrix, then so is* $\Phi(n+N)$.

(ii) $\Phi(n+N) = \Phi(n)C$, *for some nonsingular matrix* C.

(iii) $\Phi(n+N, N) = \Phi(n, 0)$.

PROOF.

(i) Let $\Phi(n)$ be a fundamental matrix of system (3.4.1). Then $\Phi(n+1) = A(n)\Phi(n)$. Now

$$\begin{aligned}\Phi(n+N+1) &= A(n+N)\Phi(n+N) \\ &= A(n)\Phi(n+N)\end{aligned}$$

Hence $\Phi(n+N)$ is also a fundamental matrix of system (3.4.1).

(ii) Observe that $\Psi_1(n, n_0) = \Phi(n+N)\Phi^{-1}(n_0+N)$ and $\Psi_2(n, n_0) = \Phi(n)\Phi^{-1}(n_0)$ are fundamental matrices of system (3.4.1) with the same initial condition $\Psi_1(n_0, n_0) = \Psi_2(n_0, n_0) = I$. By the uniqueness of fundamental matrices (Theorem 3.7) $\Psi_1(n, n_0) = \Psi_2(n, n_0)$. This implies that

$$\begin{aligned}\Phi(n+N) &= \Phi(n)\Phi^{-1}(n_0)\Phi(n_0+N) \\ &= \Phi(n)C\end{aligned}$$

(iii) This is left to the reader as Problem 1. □

There are many consequences of this lemma, including the following theorem.

Theorem 3.30. *For every fundamental matrix* $\Phi(n)$ *of system* (3.4.1), *there exists a nonsingular periodic matrix* $P(n)$ *of period* N *such that*

$$\boxed{\Phi(n) = P(n)B^n.} \tag{3.4.4}$$

PROOF. By Lemma 3.28, there exists some matrix B such that $B^N = C$, where C is the matrix specifed in Lemma 3.29(ii). Define $P(n) = \Phi(n)B^{-n}$, where $B^{-n} = (B^n)^{-1}$. Then

$$\begin{aligned}P(n+N) &= \Phi(n+N)B^{-N}B^{-n} \\ &= \Phi(n)CB^{-N}B^{-n} \quad \text{[using part (ii) of Lemma 3.29]} \\ &= \Phi(n)B^{-n} \\ &= P(n).\end{aligned}$$

We now know that $P(n)$ has period N and is clearly nonsingular. (Why?) From the definition of $P(n)$ it thus follows that $\Phi(n) = P(n)B^n$. □

Remark: If $z(n)$ is a solution of the system

$$z(n+1) = Bz(n), \tag{3.4.5}$$

then

$$x(n) = \Phi(n)c = P(n)B^n c,$$

or

$$x(n) = P(n)z(n). \tag{3.4.6}$$

The value of this remark lies in the fact that the qualitative study of the periodic system of equations (3.4.1) reduces to the study of the autonomous system (3.4.5).

The matrix $C = B^N$, which may be found using Lemma 3.29 part (ii), is referred to as a *monodromy matrix* of (3.4.1). The eigenvalues λ of B are called the *Floquet exponents* of (3.4.1); the corresponding eigenvalues λ^N of B^N are called the *Floquet multipliers* of (3.4.1). The reason we call λ^N a multiplier is that there exists a solution $x(n)$ of (3.4.1) such that $x(n+N) = \lambda^N x(n)$. (See Exercises 3.4, Problem 9.) Notice that the Floquet exponents (multipliers) do not depend upon the monodromy matrix chosen, that is, they do not hinge upon the particular fundamental matrix $\Phi(n)$ used to define the monodromy matrix. The following lemma explicitly states this truth.

Lemma 3.31. *If $\Phi(n)$ and $\Psi(n)$ are two fundamental matrices of* (3.4.1) *such that*

$$\Phi(n+N) = \Phi(n)C,$$
$$\Psi(n+N) = \Psi(n)E,$$

then C and E are similar (and thus they have the same eigenvalues).

PROOF. The reader will prove this lemma in Exercises 3.4, Problem 2. □

Lemma 3.32. *A complex number λ is a Floquet exponent of* (3.4.1) *if and only if there is a nontrivial solution of* (3.4.1) *of the form $\lambda^n q(n)$, where $q(n)$ is a vector function with $q(n+N) = q(n)$ for all n.*

PROOF. First, we assume that λ is a Floquet exponent of (3.4.1). Then, we also know that $\det(B^n - \lambda^n I) = 0$. Now choose $x_0 \in R^k, x_0 \neq 0$, such that $(B^n - \lambda^n I)x_0 = 0$ for all n. (Why?) (See Exercises 3.4, Problem 4.) Hence, we have the equation $B^n x_0 = \lambda^n x_0$.

Thus, $P(n)B^n x_0 = \lambda^n P(n)x_0$, where $P(n)$ is the periodic matrix defined in formula (3.4.4). By formula (3.4.4) now,

$$x(n, n_0, y_0) = \Phi(n, n_0)x_0 = P(n)B^n x_0 = \lambda^n P(n)x_0 = \lambda^n q(n),$$

and we have the desired periodic solution of (3.4.1), where $q(n) = P(n)x_0$. Conversely, if $\lambda^n q(n), q(n+N) = q(n) \neq 0$ is a solution of (3.4.1), Theorem 3.30 then implies that

$$\lambda^n q(n) = P(n)B^n x_0 \tag{3.4.7}$$

for some nonzero vector x_0. This implies that

$$\lambda^{n+N} q(n) = P(n) B^{n+N} x_0. \tag{3.4.8}$$

But, from (3.4.7),

$$\lambda^{n+N} q(n) = \lambda^N P(n) B^n x_0. \tag{3.4.9}$$

Equating the right-hand sides of formulas (3.4.8) and (3.4.9), we obtain

$$P(n) B^n [B^N - \lambda^N I] x_0 = 0,$$

and thus

$$\det[B^N - \lambda^N I] = 0.$$

This manipulation shows that λ is a Floquet exponent of (3.4.1). □

Using the preceding theorem, one may easily conclude the following results.

Corollary 3.33. *The following statements hold:*

(i) *System* (3.4.1) *has a periodic solution of period N if and only if it has a Floquet multiplier equal to* 1.

(ii) *There is a Floquet multiplier equal to* -1 *if and only if system* (3.4.1) *has a periodic solution of period* $2N$.

PROOF. Use Lemma 3.32 as you prove Corollary 3.33 in Exercises 3.4, Problem 3. □

Remark: Lemma 3.29, part (ii), gives us a formula to find the monodromy matrix $C = B^N$, whose eigenvalues happen to be the Floquet multipliers of (3.4.1). From Lemma 3.29,

$$C = \Phi^{-1}(n)\Phi(n+N).$$

By letting $n = 0$, we have

$$C = \Phi^{-1}(0)\Phi(N). \tag{3.4.10}$$

If we take $\Phi(N) = A(N-1)A(N-2)\cdots A(0)$, then $\Phi(0) = I$. Thus, formula (3.4.10) becomes

$$C = \Phi(N),$$

or

$$C = A(N-1)A(N-2)\cdots A(0). \tag{3.4.11}$$

We now give an example to illustrate the above results.

Example 3.34. Consider the planar system

$$x(n+1) = A(n)x(n),$$

$$A(n) = \begin{pmatrix} 0 & (-1)^n \\ (-1)^n & 0 \end{pmatrix}.$$

Clearly, $A(n+2) = A(n)$ for all $n \in \mathbb{Z}$.

Applying formula (3.4.10),

$$B^2 = C = A(1)A(0) = \begin{pmatrix} -1 & 0 \\ 0 & -1 \end{pmatrix}.$$

Thus the Floquet multipliers are -1, -1. By virtue of Corollary 3.33, the system has a 4-periodic solution. Note that since $A(n)$ has the constant eigenvalues -1, 1, $\rho(A(n)) = 1$.

The above example may suggest that there is some kind of relationship between the eigenvalues of $A(n)$ and its Floquet multipliers. To dispel any such thoughts we offer the following example.

Example 3.35. Consider system (3.2.1) with

$$A(n) = \begin{pmatrix} 0 & \dfrac{2+(-1)^n}{2} \\ \dfrac{2-(-1)^n}{2} & 0 \end{pmatrix}.$$

This is a system of period 2. The eigenvalues of A are $\pm\frac{\sqrt{3}}{2}$, and hence $\rho(A) = \frac{\sqrt{3}}{2} < 1$. Now,

$$B^2 = C = A(1)A(0) = \begin{pmatrix} \dfrac{1}{4} & 0 \\ 0 & \dfrac{9}{4} \end{pmatrix}.$$

Thus, the Floquet multipliers are $\frac{1}{4}$ and $\frac{9}{4}$. Hence, $\rho(B) = \frac{3}{2}$.

Exercises 3.4

1. Prove Lemma 3.29 (iii).
2. Prove Lemma 3.31.
3. Prove Corollary 3.33.
4. Suppose that $(B - \lambda I)x_0 = 0$ for some $x_0 \in R^k, x_0 \neq 0$. Prove that $(B^n - \lambda^n I)x_0 = 0$ for all $n \in Z^t$.
5. Let $a_1(n), a_2(n)$ be N-periodic functions and let $\Psi_1(n), \Psi_2(n)$ be solutions of

$$x(n+2) + a_1(n)x(n+1) + a_2(n)x(n) = 0 \qquad (3.4.12)$$

such that $\Psi_1(0) = 1, \Psi_1(1) = 0, \Psi_2(0) = 0$, and $\Psi_2(1) = 1$. Show that the Floquet multipliers satisfy the equation $\lambda^2 + b\lambda + c = 0$, where

$$b = -[\Psi_1(N) + \Psi_2(N+1)], \qquad c = \prod_{i=0}^{N-1} a_2(i).$$

6. In Problem 5, let $a_2(n) \equiv 1$. Show that the product of the Floquet multipliers is equal to 1.

7. In Problem 5, let $a_2(n) \equiv 1$. Show that if $b = 2$, there is at least one solution of period $2N$ while for $b = -2$ there is at least one solution of period N.

8. In Problem 5 it is clear that if $\lambda = 1$, then (3.4.12) has a periodic solution of period N. Show that $x(n+2)+a_1(n)x(n+1)+a_2(n)x(n) = 0$ has a periodic solution of period $2N$ if and only if $\lambda = -1$.

9. Show that there exists a solution $x(n)$ of (3.4.1) that satisfies $x(n+N) = \lambda x(n)$ if and only if λ is a Floquet multiplier.

3.5 Applications

3.5.1 Markov Chains

In 1906 the Russian mathematician A.A. Markov developed the concept of Markov chains. We can describe a Markov chain as follows: Suppose that we conduct some experiment with a set of k outcomes, or states, $S = \{s_1, s_2, \ldots, s_k\}$. The experiment is repeated such that the probability (p_{ij}) of the state $s_i, 1 \le i \le k$, occurring on the $(n+1)$th repetition depends only on the state s_j occurring on the nth repetition of the experiment. In other words, the system has no memory: The future state depends only on the present state. In probability theory language, $p_{ij} = p(s_i|s_j)$ is the probability of s_i occurring on the next repetition, given that s_j occurred on the last repetition. Given that s_j has occurred in the last repetition, one of $s_1, s_2, \ldots, s_k$ must occur in the next repetition. Thus,

$$p_{1j} + p_{2j} + p_{3j} + \cdots + p_{kj} = 1, \qquad 1 \le j \le k. \tag{3.5.1}$$

Let $p_i(n)$ denote the probability that the state s_i will occur on the nth repetition of the experiment, $1 \le i \le k$. Since one of the states s_i must occur on the nth repetition, it follows that

$$p_1(n) + p_2(n) + \cdots + p_k(n) = 1. \tag{3.5.2}$$

To derive a mathematical model of this experiment, we must define $p_i(n+1), 1 \le i \le k$, as the probability that the state s_i occurs on the $(n+1)$th repetition of the experiment. There are k ways that this can happen. The first case is where repetition n gives us s_1, and repetition $(n+1)$ produces

s_i. Since the probability of getting s_1 on the nth repetition is $p_1(n)$, and the probability of having s_i after s_1 is p_{i1}, it follows (by the multiplication principle) that the probability of the first case occurring is $p_{i1}p_1(n)$. The second case is where we get s_2 on repetition n and s_i on repetition $(n+1)$. The probability of the occurrence of the second case is $p_{i2}p_2(n)$. Repeating this for cases $3, 4, \dots, k$, and for $i = 1, 2, \dots, k$, we obtain the k-dimensional system

$$\begin{aligned} p_1(n+1) &= p_{11}p_1(n) + p_{12}p_2(n) + \cdots + p_{1k}p_k(n), \\ p_2(n+1) &= p_{21}p_1(n) + p_{22}p_2(n) + \cdots + p_{2k}p_k(n), \\ &\vdots \\ p_k(n+1) &= p_{k1}p_1(n) + p_{k2}p_2(n) + \cdots + p_{kk}p_k(n), \end{aligned}$$

or, in vector notation,

$$p(n+1) = Sp(n), \qquad n = 1, 2, 3 \dots, \tag{3.5.3}$$

where $p(n) = (p_1(n), p_2(n), \dots, p_k(n))^T$ is the *probability vector* and $S = (p_{ij})$ is a $k \times k$ *transition matrix*.

The matrix S belongs to a special class of matrices called *Markov matrices*. A matrix $A = (a_{ij})$ is said to be *nonnegative* (positive) if $a_{ij} \geq 0$ (> 0) for all entries a_{ij} of A. A nonnegative $k \times k$ matrix A is said to be *Markov* (or stochastic) if $\sum_{i=1}^{k} a_{ij} = 1$ for all $j = 1, 2, \dots, k$. It follows from Table 4.1 that $\|A\|_1 = 1$, which by inequality (4.1.3) implies that $p(A) \leq 1$. Hence $|\lambda| \leq 1$ for all the eigenvalues λ of a Markov matrix. Furthermore, $\lambda = 1$ is an eigenvalue of a Markov matrix (Exercises 3.5, Problem 3). Hence $p(A) = 1$ if A is Markov.

3.5.2 Regular Markov Chains

A regular Markov chain is one in which S^m is positive for some positive integer m. To give a complete analysis of the eigenvalues of such matrices, we need the following theorem, due to O. Perron.

Theorem 3.36 (Perron's Theorem). *Let A be a positive $k \times k$ matrix. Then $\rho(A)$ is a simple* real *eigenvalue (not repeated) of A. If λ is any other eigenvalue of A, then $|\lambda| < \rho(A)$. Moreover, an eigenvector associated with $\rho(A)$ may be assumed to be positive.*

Suppose now that S is the transition matrix of a regular Markov chain with eigenvalues $\lambda_1, \lambda_2, \dots, \lambda_k$. Then $\rho(S) = 1$. If S^m is positive, then $\rho(S^m) = 1$. As a matter of fact, the eigenvalues of S^m are $\lambda_1^m, \lambda_2^m, \dots, \lambda_k^m$. By Perron's theorem, 1 is a simple eigenvalue of S^m. Consequently, S has exactly one simple eigenvalue, say λ_1, which equals 1; all other eigenvalues satisfy $|\lambda_i| < 1, i = 2, 3, \ldots, k$. Hence, the Jordan form of S must be of the form $J = \begin{pmatrix} 1 & 0 \\ 0 & J_* \end{pmatrix}$, where the eigenvalues of J_* are $\lambda_2, \lambda_3, \ldots, \lambda_k$.

By Corollary 3.24, $J_*^n \to 0$ as $n \to \infty$, so that $J^n \to \text{diag}(1, 0, \ldots, 0)$ as $n \to \infty$. Therefore, if $S = QJQ^{-1}$, we have

$$\lim_{n\to\infty} p(n) = \lim_{n\to\infty} S^n p(0) = \lim_{n\to\infty} QJ^nQ^{-1}p(0) = (\xi_1, 0, 0, \ldots, 0)\eta = a\xi_1, \tag{3.5.4}$$

where $\xi_1 = (\xi_{11}, \xi_{21}, \ldots, \xi_{k1})^T$ is the eigenvector of S that corresponds to the eigenvalue $\lambda_1 = 1$, and a is the first component of $\eta = Q^{-1}p(0)$. Since finding the matrix Q is not a simple task, we will choose instead to devise a very easy method to find the constant a. Recall that for

$$p(n) = (p_1(n), p_2(n), \ldots, p_k(n))^T$$

we have, from formula (3.5.2), $\sum_{i=1}^k p_i(n) = 1$. Since $\lim_{n\to\infty} p(n) = a\xi_1$, it follows that

$$a\xi_{11} + a\xi_{21} + \cdots + a\xi_{k1} = 1.$$

Therefore,

$$a = \frac{1}{\xi_{11} + \xi_{21} + \cdots + \xi_{k1}}.$$

The following example illustrates a regular Markov chain.

Example 3.37. The simplest type of genetic inheritance of traits in animals occurs when a certain trait is determined by a specific pair of genes, each of which may be of two types, say G and g. An individual may have a GG combination, a Gg (which is genetically the same as gG), or a gg combination. An individual with GG genes is said to be *dominant*; a gg individual is referred to as *recessive*; a *hybrid* has Gg genes.

In the mating of two animals, the offspring inherits one gene of the pair from each parent: The basic assumption of genetics is that the selection of these genes is random.

Let us consider a process of continued matings. We begin with an individual of known genetic character (GG) and mate it with a hybrid. Assuming that there is one offspring, we mate that offspring with a hybrid, repeating this process through a number of generations. In each generation there are three possible states, $s_1 = GG, s_2 = Gg$, and $s_3 = gg$. Let $p_i(n)$ represent the probability that state s_i occurs in the nth generation and let p_{ij} be the probability that s_i occurs in the $(n+1)$th generation given that s_j occurred in the nth generation.

The difference system that models this Markov chain is denoted by

$$\begin{aligned} p_1(n+1) &= p_{11}p_1(n) + p_{12}p_2(n) + p_{13}p_3(n), \\ p_2(n+1) &= p_{21}p_1(n) + p_{22}p_2(n) + p_{23}p_3(n), \\ p_3(n+1) &= p_{31}p_1(n) + p_{32}p_2(n) + p_{33}p_3(n). \end{aligned}$$

Now, p_{11} is the probability of producing an offspring GG by mating GG and Gg. Clearly, the offspring receives a G gene from his parent GG with probability 1 and the other G from his parent Gg with probability $\frac{1}{2}$. By the multiplication principle, $p_{11} = 1 \times \frac{1}{2} = \frac{1}{2}$. The probability of creating an offspring GG from mating a Gg with a Gg is p_{12}. By similar analysis one may show that $p_{12} = \frac{1}{2} \times \frac{1}{2} = \frac{1}{4}$. Likewise, p_{13} is the probability of generating an offspring GG from mating a gg with a Gg. Obviously, $p_{13} = 0$. One may show by the same process that

$$p_{21} = \frac{1}{2}, \qquad p_{22} = \frac{1}{2}, \qquad p_{23} = \frac{1}{2}, \quad p_{31} = 0, \qquad p_{32} = \frac{1}{4}, \qquad p_{33} = \frac{1}{2}.$$

Hence, we have

$$p(n+1) = Sp(n)$$

with

$$S = \begin{pmatrix} 0.5 & 0.25 & 0 \\ 0.5 & 0.5 & 0.5 \\ 0 & 0.25 & 0.5 \end{pmatrix}.$$

Notice that all the entries for S^2 are positive, and thus this is a regular Markov chain. The eigenvalues of S are $\lambda_1 = 1, \lambda_2 = \frac{1}{2}$, and $\lambda_3 = 0$. Recall from formula (3.5.4) that

$$\lim_{n\to\infty} p(n) = a\xi_1.$$

Now,

$$\xi_1 = \begin{pmatrix} 1 \\ 2 \\ 1 \end{pmatrix}$$

and

$$a = \frac{1}{4}$$

implies that

$$\lim_{n\to\infty} p(n) = \begin{pmatrix} 0.25 \\ 0.5 \\ 0.25 \end{pmatrix}.$$

This relation dictates that as the number of repetitions approaches infinity, the probability of producing a purely dominant or a purely recessive offspring is 0.25, and the probability of creating a hybrid offspring is 0.5.

3.5.3 Absorbing Markov Chains

A state s_i in a Markov chain is said to be *absorbing* if whenever it occurs on the nth repetition of the experiment, it then occurs on every subsequent repetition. In other words, if for some i, $p_{ii} = 1$, then $p_{ij} = 0$ for $j \neq i$. A Markov chain is said to be *absorbing* if it has at least one absorbing state and if from every state it is possible to go to an absorbing state. In an absorbing Markov chain, a state that is not absorbing is called *transient.*

Example 3.38. Drunkard's Walk

A man walks along a four-block stretch. He starts at corner x. With probability $\frac{1}{2}$ he walks one block to the right, and with probability $\frac{1}{2}$ he walks one block to the left. When he comes to the next corner he again randomly chooses his direction. He continues until he reaches corner 5, which is a bar, or corner 1, which is his home. (See Figure 3.2.) If he reaches either home or the bar, he stays there. This is clearly an absorbing Markov chain.

Let us rename the states so that the absorbing states at 1 and 5 are last, and so we refer to them as s_4 and s_5. The transient states 2, 3, and 4 will be called s_1, s_2, and s_3, respectively. Accordingly, $p_1(n), p_2(n), p_3(n), p_4(n)$, and $p_5(n)$ will be, respectively, the probabilities of reaching s_1, s_2, s_3, s_4, and s_5 after n walks. The difference equation that represents this Markov chain is $p(n+1) = Sp(n)$, where the transition matrix is

$$S = \left(\begin{array}{ccc|cc} 0 & \frac{1}{2} & 0 & 0 & 0 \\ \frac{1}{2} & 0 & \frac{1}{2} & 0 & 0 \\ 0 & \frac{1}{2} & 0 & 0 & 0 \\ \cdots & \cdots & \cdots & \cdots & \cdots & \cdots \\ \frac{1}{2} & 0 & 0 & 1 & 0 \\ 0 & 0 & \frac{1}{2} & 0 & 1 \end{array}\right) = \begin{pmatrix} T & 0 \\ Q & I \end{pmatrix}.$$

Let $u(n) = (p_1(n), p_2(n), p_3(n))^T$ and $v(n) = (p_4(n), p_5(n))^T$. Then

$$\begin{pmatrix} u(n+1) \\ v(n+1) \end{pmatrix} = \begin{pmatrix} T & 0 \\ Q & I \end{pmatrix} \begin{pmatrix} u(n) \\ v(n) \end{pmatrix},$$

$$u(n+1) = Tu(n), \tag{3.5.5}$$

$$v(n+1) = v(n) + Qu(n). \tag{3.5.6}$$

Therefore,

$$u(n) = T^n u(0). \tag{3.5.7}$$

Substituting from formula (3.5.7) into formula (3.5.6) yields

$$v(n+1) = v(n) + QT^n u(0). \tag{3.5.8}$$

By formula (3.2.14), it follows that the solution of (3.5.8) is given by

$$v(n) = v(0) + \sum_{r=0}^{n-1} QT^r u(0). \tag{3.5.9}$$

The eigenvalues of T are

$$0, \quad -\sqrt{\frac{1}{2}}, \quad \sqrt{\frac{1}{2}}.$$

Hence, by Corollary 3.24, $\lim_{n\to\infty} T^n = 0$. In this case one may show that

$$\sum_{r=0}^{\infty} T^r = \lim_{n\to\infty} \sum_{r=0}^{n-1} T^r = (I-T)^{-1}$$

(Exercises 3.5, Problem 5). Using formula (3.5.9), we generate

$$\lim_{n\to\infty} v(n) = v(0) + Q(I-T)^{-1}u(0).$$

Now,

$$(I-T)^{-1} = \begin{pmatrix} \frac{3}{2} & 1 & \frac{1}{2} \\ 1 & 2 & 1 \\ \frac{1}{2} & 1 & \frac{3}{2} \end{pmatrix}.$$

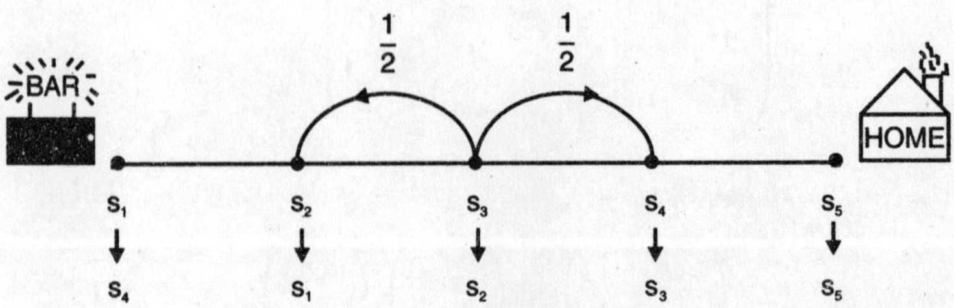

FIGURE 3.2. Drunkard's walk.

Assume that the man starts midway between home and the bar, that is, at state s_2. Then

$$u(0) = \begin{pmatrix} 0 \\ 1 \\ 0 \end{pmatrix}$$

and

$$v(0) = \begin{pmatrix} 0 \\ 0 \end{pmatrix}.$$

In this case

$$\lim_{n\to\infty} v(n) = \begin{pmatrix} \frac{1}{2} & 0 & 0 \\ 0 & 0 & \frac{1}{2} \end{pmatrix} \begin{pmatrix} \frac{3}{2} & 1 & \frac{1}{2} \\ 1 & 2 & 1 \\ \frac{1}{2} & 1 & \frac{3}{2} \end{pmatrix} \begin{pmatrix} 0 \\ 1 \\ 0 \end{pmatrix} = \begin{pmatrix} \frac{1}{2} \\ \frac{1}{2} \end{pmatrix}.$$

Thus, the probability that the man ends up at his home is 0.5. The probability that he ends up at the bar is also 0.5. Common sense could probably have told us this in the first place, but not every situation will be this simple.

3.5.4 A Trade Model

Example 3.39. Consider a model of the trade between two countries, restricted by the following assumptions:

(i) National income = consumption outlays + net investment + exports − imports.

(ii) Domestic consumption outlays = total consumption − imports.

(iii) Time is divided into periods of equal length, denoted by $n = 0, 1, 2, \ldots$.

Let, for country $j = 1, 2$,

$$\begin{aligned} y_j(n) &= \text{national income in period } n, \\ c_j(n) &= \text{total consumption in period } n, \\ i_j(n) &= \text{net investment in period } n, \\ x_j(n) &= \text{exports in period } n, \\ m_j(n) &= \text{imports in period } n, \\ d_j(n) &= \text{consumption of domestic products in period } n. \end{aligned}$$

For country 1 we then have

$$\begin{aligned} y_1(n) &= c_1(n) + i_1(n) + x_1(n) - m_1(n), \\ d_1(n) &= c_1(n) - m_1(n), \end{aligned}$$

which, combining those two equations, gives

$$y_1(n) = d_1(n) + x_1(n) + i_1(n). \tag{3.5.10}$$

Likewise, for country 2, we have

$$y_2(n) = d_2(n) + x_2(n) + i_2(n). \tag{3.5.11}$$

We now make the following reasonable assumption: The domestic consumption $d_j(n)$ and the imports $m_j(n)$ of each country at period $(n+1)$ are proportional to the country's national income $y_i(n)$ one time period earlier. Thus,

$$d_1(n+1) = a_{11}y_1(n), \qquad m_1(n+1) = a_{21}y_1(n), \tag{3.5.12}$$

$$d_2(n+1) = a_{22}y_2(n), \qquad m_2(n+1) = a_{12}y_2(n). \tag{3.5.13}$$

The constants a_{ij} are called *marginal propensities.* Furthermore, $a_{ij} > 0$, for $i, j = 1, 2$. Since we are considering a world with only two countries, the exports of one must be equal to the imports of the other, i.e.,

$$m_1(n) = x_2(n), \qquad m_2(n) = x_1(n). \tag{3.5.14}$$

Substituting from equations (3.5.12), (3.5.13), and (3.5.14) into (3.5.10) and (3.5.11) leads to

$$\begin{pmatrix} y_1(n+1) \\ y_2(n+1) \end{pmatrix} = \begin{pmatrix} a_{11} & a_{12} \\ a_{21} & a_{22} \end{pmatrix} \begin{pmatrix} y_1(n) \\ y_2(n) \end{pmatrix} + \begin{pmatrix} i_1(n+1) \\ i_2(n+1) \end{pmatrix}. \tag{3.5.15}$$

Let us further assume that the net investments $i_1(n) = i_1$ and $i_2(n) = i_2$ are constants. Then (3.5.15) becomes

$$\begin{pmatrix} y_1(n+1) \\ y_2(n+1) \end{pmatrix} = \begin{pmatrix} a_{11} & a_{12} \\ a_{21} & a_{22} \end{pmatrix} \begin{pmatrix} y_1(n) \\ y_2(n) \end{pmatrix} + \begin{pmatrix} i_1 \\ i_2 \end{pmatrix}. \tag{3.5.16}$$

By the variation of constants formula (3.2.14), we obtain

$$y(n) = A^n y(0) + \sum_{r=0}^{n-1} A^{n-r-1} I = A^n y(0) + \sum_{r=0}^{n-1} A^r I, \tag{3.5.17}$$

where $I = (i_1, i_2)^T$. To have a stable economy, common sense dictates that the sum of the domestic consumption $d_j(n+1)$ and the imports $m_j(n+1)$ in period $(n+1)$ must be less than the national income $y_j(n)$ in period n; that is,

$$d_j(n+1) + m_j(n+1) < y_j(n), \qquad j = 1, 2,$$

or

$$a_{11} + a_{21} < 1, \qquad a_{12} + a_{22} < 1. \tag{3.5.18}$$

Under conditions (3.5.18), one may show that for all the eigenvalues λ of A, $|\lambda| < 1$ (Exercises 3.5, Problem 4).

This implies from Corollary 3.24 that $A^n \to 0$ as $n \to \infty$. This fact further generates the so-called *Neumann's expansion* (Exercises 3.5, Problem 4):

$$\lim_{n\to\infty} \sum_{r=0}^{n-1} A^r = \sum_{r=0}^{\infty} A^r = (I - A)^{-1}. \tag{3.5.19}$$

It follows from formula (3.5.17) that

$$\lim_{n\to\infty} y(n) = (I - A)^{-1} i.$$

This equation says that the national incomes of countries 1 and 2 approach equilibrium values independent of the initial values of the national incomes $y_1(0), y_2(0)$.

However, as we all know, international economics involves many more factors than we can account for here. But in Exercises 3.5, Problem 11, the student will be allowed to create a model for the economic interaction among three countries.

3.5.5 *The Heat Equation*

Example 3.40. Consider the distribution of heat through a thin bar composed of a homogeneous material. Let $x_1, x_2, \ldots, x_k$ be k equidistant points on the bar. Let $T_i(n)$ be the temperature at time $t_n = (\Delta t)n$ at the point $x_i, 1 \le i \le k$. Denote the temperatures at the left and the right ends of the bar at time t_n by $T_0(n), T_{k+1}(n)$, respectively. (See Figure 3.3.)

Assume that the sides of the bar are sufficiently well insulated that no heat energy is lost through them. The only thing, then, that affects the temperature at the point x_i is the temperature of the points next to it, which are x_{i-1}, x_{i+1}. Assume that the left end of the bar is kept at b degrees Celsius and the right end of the bar at c degrees Celsius. These conditions imply that $x_0(n) = b$ and $x_{k+1}(n) = c$, for $n \ge 0$.

We assume that the temperature at a given point x_i is determined only by the temperature at the nearby points x_{i-1} and x_{i+1}. Then according to Newton's law of cooling, the change in temperature $T_i(n+1) - T_i(n)$ at a point x_i from time n to $n+1$ is directly proportional to the temperature difference between the point x_i and the nearby points x_{i-1} and x_{i+1}. In

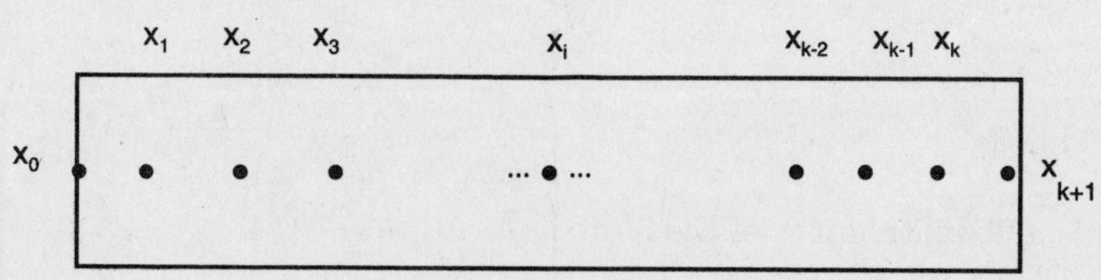

FIGURE 3.3. Heat transfer.

other words

$$\begin{aligned} T_i(n+1) - T_i(n) &= \alpha([T_{i-1}(n) - T_i(n)] + [T_{i+1}(n) - T_i(n)]) \\ &= \alpha[T_{i+1}(n) - 2T_i(n) + T_{i-1}(n)], \qquad (3.5.20) \end{aligned}$$

or

$$T_i(n+1) = \alpha T_{i-1}(n) + (1-2\alpha)T_i(n) + \alpha T_{i+1}(n), \qquad i = 2, 3, \ldots, k-1.$$

Similarly, one may also derive the following two equations:

$$\begin{aligned} T_1(n+1) &= (1-2\alpha)T_1(n) + \alpha T_2(n) + \alpha b, \\ T_k(n+1) &= \alpha T_{k-1}(n) + (1-2\alpha)T_k(n) + \alpha c. \end{aligned}$$

This correlation may be written in the compact form

$$T(n+1) = AT(n) + g,$$

where

$$A = \begin{pmatrix} (1-2\alpha) & \alpha & 0 & \cdots & 0 \\ \alpha & (1-2\alpha) & \alpha & & \vdots \\ 0 & \alpha & (1-2\alpha) & \ddots & \\ \vdots & \vdots & \vdots & \ddots & \alpha \\ 0 & 0 & 0 & \alpha & (1-2\alpha) \end{pmatrix}, \quad g = \begin{pmatrix} \alpha b \\ 0 \\ 0 \\ \vdots \\ \alpha c \end{pmatrix}.$$

This is a tridiagonal *Toeplitz* matrix.[2] Its eigenvalues may be found by the formula [111]

$$\lambda_n = (1-2\alpha) + \alpha \cos\left(\frac{n\pi}{k+1}\right), \qquad n = 1, 2, \ldots, k.$$

Hence $|\lambda| < 1$ for all eigenvalues λ of A. Corollary 3.24 then implies that

$$\lim_{n\to\infty} A^n = 0.$$

From the variation of constants formula (3.2.12), it follows that

$$T(n) = A^n T(0) + \sum_{r=0}^{n-1} A^r g.$$

---

[2] A is a Toeplitz if it is of the form $\begin{pmatrix} a_0 & a_1 & a_2 & \cdots & a_{k-1} \\ a_{-1} & a_0 & a_1 & & \vdots \\ a_{-2} & a_{-1} & a_0 & & a_2 \\ \vdots & \vdots & & & a_1 \\ a_{-k+1} & \cdots & a_{-2} & a_{-1} & a_0 \end{pmatrix}$.

Thus, $\lim_{n\to\infty} T(n) = (I - A)^{-1}g$. Finally, this equation points out that the temperature at the point $x_i, 1 \le i \le k$, approaches the ith component of the vector $(I - A)^{-1}g$, regardless of the initial temperature at the point x_i.

Consider the above problem with $k = 3, \alpha = 0.4, T_0(n) = 10°$ C , $T_4(n) =$ 20° C.

Then

$$A = \begin{pmatrix} lll0.2 & 0.4 & 0 \\ 0.4 & 0.2 & 0.4 \\ 0 & 0.4 & 0.2 \end{pmatrix}, \qquad g = \begin{pmatrix} 4 \\ 0 \\ 8 \end{pmatrix},$$

$$(I - A)^{-1} = \begin{pmatrix} 0.8 & -0.4 & 0 \\ -0.4 & 0.8 & -0.4 \\ 0 & -0.4 & 0.8 \end{pmatrix}^{-1} = \begin{pmatrix} \frac{15}{8} & \frac{5}{4} & \frac{5}{8} \\ \frac{5}{4} & \frac{5}{2} & \frac{5}{4} \\ \frac{13}{8} & \frac{5}{4} & \frac{15}{8} \end{pmatrix}.$$

Hence

$$\lim_{n\to\infty} T(n) = \begin{pmatrix} \frac{15}{8} & \frac{5}{4} & \frac{5}{8} \\ \frac{5}{4} & \frac{5}{2} & \frac{5}{4} \\ \frac{13}{8} & \frac{5}{4} & \frac{15}{8} \end{pmatrix} \begin{pmatrix} 4 \\ 0 \\ 8 \end{pmatrix} = \begin{pmatrix} \frac{25}{2} \\ 15 \\ \frac{43}{2} \end{pmatrix}.$$

Remark: Let $\Delta x = x_i - x_{i-1}$ and $\Delta t = t_{i-1} - t_i$. If we assume that the constant of proportionality α depends on both Δt and Δx, then we may write

$$\alpha = \left[\frac{\Delta t}{(\Delta x)^2}\right] \beta, \tag{3.5.21}$$

where β is a constant that depends on the material of the bar. Formula (3.5.21) simply states that the smaller the value of Δt, the smaller should be the change in the temperature at a given point. Moreover, the smaller the separation of points, the larger should be their influence on the temperature changes in nearby points. Using formula (3.5.21) in (3.5.20) yields

$$\frac{T_i(n+1) - T_i(n)}{\Delta t} = \beta \left[\frac{T_{i+1}(n) - 2T_i(n) + T_{i-1}(n)}{(\Delta x)^2}\right]. \tag{3.5.22}$$

If we let $\Delta t \to 0, \Delta x \to 0$ as $n \to \infty$ and $i \to \infty, x_i = (\Delta x)i = x$, and $t_i = (\Delta t)i = t$, then (3.5.22) gives the partial differential equation

$$\frac{\partial T(x,t)}{\partial t} = \beta \frac{\partial^2 T(x,t)}{\partial x^2}. \tag{3.5.23}$$

Equation (3.5.23) is known as the *heat equation* [137].

Exercises 3.5

1. Consider the difference system
$$P(n+1) = RP(n)$$
where
$$R = \begin{pmatrix} 0.2 & 0.1 & 0.3 \\ 0.3 & 0.5 & 0.1 \\ 0.5 & 0.4 & 0.6 \end{pmatrix}.$$
   (a) Show that R is a Markov matrix.
   (b) Find $\lim_{n\to\infty} P(n)$.
2. Consider the difference system
$$P(n+1) = RP(n)$$
where
$$R = \begin{pmatrix} 1 & 0 & 0.3 & 0.1 \\ 0 & 1 & 0.1 & 0.2 \\ 0 & 0 & 0.4 & 0.3 \\ 0 & 0 & 0.2 & 0.4 \end{pmatrix}.$$
   (a) Show that R is an absorbing Markov matrix.
   (b) Find $\lim_{n\to\infty} P(n)$.
3. Show that if A is a $k \times k$ Markov matrix, then it has an eigenvalue equal to 1.
4. Let $A = (a_{ij})$ be a $k \times k$ positive matrix such that $\sum_{j=1}^{k} a_{ij} < 1$ for $i = 1, 2, \ldots, k$. Show that $|\lambda| < 1$ for all eigenvalues λ of A.
5. Let A be a $k \times k$ matrix with $|\lambda| < 1$ for all eigenvalues λ of A. Show that:
   (i) $(I - A)$ is nonsingular.
   (ii) $\sum_{i=0}^{\infty} A^i = (I - A)^{-1}$.
6. Modify Example 3.37 by first mating a recessive individual (genes gg) with a dominant individual (genes GG). Then, continuing to mate the offspring with a dominant individual, write down the difference equation that describes the probabilities of producing individuals with genes GG, Gg, and gg. Find $\lim_{n\to\infty} p(n)$ and then interpret your results.
7. In the dark ages, Harvard, Yale, and MIT admitted only male students. Assume that at the time, 80% of the sons of Harvard men went to Harvard and the rest went to MIT, 40% of the sons of MIT men went

to MIT and the rest split evenly between Harvard and Yale; and of the sons of Yale men, 70% went to Yale, 20% to Harvard, and 10% to MIT. Find the transition matrix R of this Markov chain. Find the long-term probabilities that the descendants of Harvard men will go to Yale. (Assume that we start with N men, and each man sends one son to college.)

8. A New York governor tells person A his intention either to run or not to run in the next presidential election. Then A relays the news to B, who in turn relays the message to C, and so forth, always to some new person. Assume that there is a probability α that a person will change the answer from yes to no when transmitting it to the next person and a probability β that he will change it from no to yes. Write down the state transition matrix of this process, then find its limiting state. Note that the initial state is the governor's choice.

9. A psychologist conducts an experiment in which 20 rats are placed at random in a compartment that has been divided into rooms labeled 1, 2, and 3 as shown in Figure 3.4. Observe that there are four doors in the arrangement. There are three possible states for each rat: It can be in room 1, 2, or 3. Let us assume that the rats move from room to room. A rat in room 1 has the probabilities $p_{11} = 0, P_{21} = \frac{1}{3}$, and $p_{31} = \frac{2}{3}$ of moving to the various rooms based on the distribution of doors. Predict the distribution of the rats in the long run. What is the limiting probability that a given marked rat will be in room 2?

10. In Example 3.38 (drunkard's walk), assume that the probability of a step to the right is $\frac{2}{3}$ and that of a step to the left is $\frac{1}{3}$. Write down the transition matrix and determine $\lim_{n\to\infty} p(n)$.

11. In the trade model (Example 3.39) let $a_{11} = 0.4, a_{21} = 0.5, a_{12} = 0.3, a_{22} = 0.6, i_1 = 25$ billion dollars, and $i_2 = 20$ billion dollars. If $y_1(n)$ and $y_2(n)$ denote the national incomes of countries 1 and 2 in year n, respectively, and $y_1(0) = 500$ billion dollars and $y_2(0) = 650$ billion dollars, find $y_1(3)$ and $y_2(3)$. What are the equilibrium national incomes for nations 1 and 2?

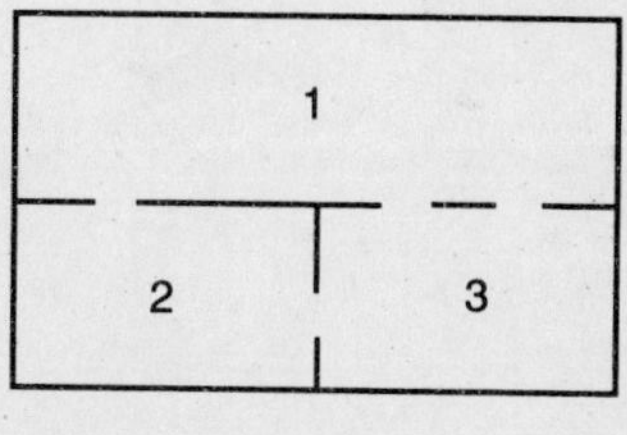

FIGURE 3.4. Diagram for Problem 9.

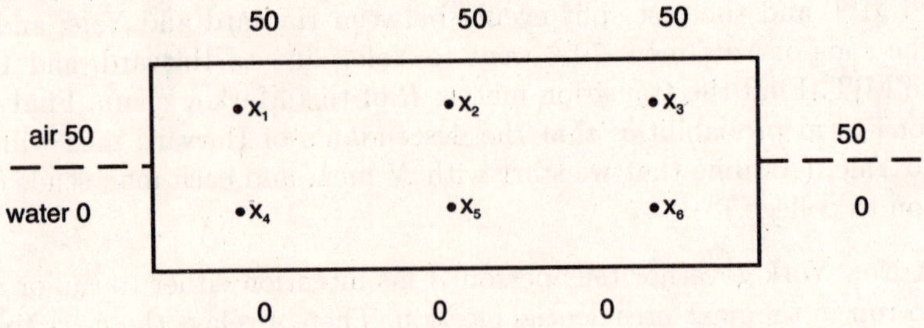

FIGURE 3.5. Heat flow diagram for Problem 14.

12. Develop a mathematical model for a foreign trade model among three countries using an argument similar to that used in Example 3.39.

13. In Example 3.40, let $k = 4, \alpha = 0.2$, and $x_0(n) = T_5(n) = 0°$ C. Compute $T_i(n), 1 \leq i \leq 4$, for $n = 1, 2, 3$; then find $\lim_{n\to\infty} T_i(n)$.

14. Suppose we have a grid of six points on a bar as shown in Figure 3.5. Part of the bar is in air that is kept at a constant temperature of 50 degrees, and part of the bar is submerged in a liquid that is kept at a constant temperature of 0 degrees. Assume that the temperature at the point $x_i, 1 \leq i \leq 6$, depends only on the temperature of the four nearest points, that is, the points above, below, to the left, and to the right.

    (i) Write a mathematical model that describes the flow of heat in this bar.

    (ii) Find the equilibrium temperature at the six points x_i.

4

Stability Theory

In Chapter 1 we studied the stability properties of first-order difference equations. In this chapter we will develop the theory for k-dimensional systems of first-order difference equations. As shown in Chapter 3, this study includes difference equations of any order. Here we are interested in the qualitative behavior of solutions without actually computing them. Realizing that most of the problems that arise in practice are nonlinear and mostly unsolvable, this investigation is of vital importance to scientists, engineers, and applied mathematicians.

In this chapter we adapt the differential methods and techniques of Liapunov [93], Perron [114], and many others, to difference equations. First, we introduce the notion of norms of vectors and matrices in Section 4.1. Next, we give definitions of various notions of stability and some simple examples to illustrate them in Section 4.2. Section 4.3 addresses the question of stability of both autonomous and nonautonomous linear systems and includes the Stable Mainfold Theorem. In Section 4.4 we study the geometrical properties of planar linear systems by means of phase space analysis. Section 4.5 introduces to the reader the basic theory of the direct method of Liapunov, by far the most advanced topic in this chapter. In Section 4.6 we present the stability of nonlinear systems by the method of linear approximation, which is widely used by scientists and engineers. And, finally, in Section 4.7 we investigate mathematical models of population dynamics and a business model. Due to the enormity of the existing literature on Liapunov theory, we have limited our exposition to autonomous equations.

(i) the l_1 norm:

$$\|x\|_1 = \sum_{i=1}^{k} |x_i|$$

(ii) the l_∞ norm:

$$\|x\|_\infty = \max_{1 \le i \le k} |x_i|$$

(iii) the Euclidean norm l_2:

$$\|x\|_2 = \left(\sum_{i=1}^{k} x_i^2 \right)^{1/2}$$

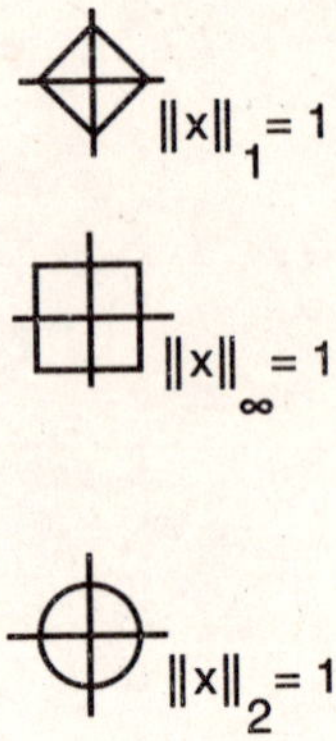

FIGURE 4.1. A circle in different norms.

4.1 A Norm of a Matrix

We start this section by introducing the notion of norms of vectors and matrices.

Definition 4.1. A real-valued function on a vector space V is called a *norm*, and is denoted by $\|\|$, if the following properties hold:

(i) $\|x\| \ge 0$ and $\|x\| = 0$ only if $x = 0$;

(ii) $\|\alpha x\| = |\alpha| \|x\|$ for all $x \in V$ and scalars α;

(iii) $\|x + y\| \le \|x\| + \|y\|$ for all $x, y \in V$.

The three most commonly used norms on $\mathbb{R}^k$ are shown in Figure 4.1.

We remark here that all norms on $\mathbb{R}^k$ are equivalent in the sense that if $\|\|$, $\|\|'$ are any two norms, then there exist constants $\alpha, \beta > 0$ such that

$$\alpha \|x\| \le \|x\|' \le \beta \|x\|.$$

Thus if $\{x_n\}$ is a sequence in $\mathbb{R}^k$, then $\|x_n\| \to 0$ as $n \to \infty$ if and only if $\|x_n\|' \to 0$ as $n \to \infty$.

Corresponding to each vector norm $\|\|$ on $\mathbb{R}^k$ one may define an operator norm $\|\|$ on a $k \times k$ matrix A as

$$\|A\| = \max_{\|x\| \ne 0} \frac{\|Ax\|}{\|x\|}. \tag{4.1.1}$$

It may be shown easily that

$$\boxed{\|A\| = \max_{\|x\| \le 1} \|Ax\| = \max_{\|x\| = 1} \|Ax\|.} \tag{4.1.2}$$

Using this definition one may easily compute $\|A\|$ relative to the above three norms as shown in Table 4.1. (For a proof see [85].)

TABLE 4.1. Vector and Matrix Norms.

Norm	$\|x\|$	$\|A\|$	
l_1	$\sum_{i=1}^{k}\|x_i\|$	$\max_{1\le j\le k}\sum_{i=1}^{k}\|a_{ij}\|$	Sum over columns
l_∞	$\max_{1\le i\le k}\|x_i\|$	$\max_{1\le i\le k}\sum_{j=1}^{k}\|a_{ij}\|$	Sum over rows
l_2	$\left(\sum_{i=1}^{k}x_i^2\right)^{\frac{1}{2}}$	$[\rho(A^TA)]^{\frac{1}{2}}$	

From (4.1.1) we may deduce that for any operator norm on A (Exercises 4.1, Problem 5),

$$\rho(A) \le \|A\|, \tag{4.1.3}$$

where $\rho(A) = \max\{|\lambda| : \lambda \text{ is an eigenvalue of } A\}$ is the spectral radius of A.

Exercises 4.1

1. Compute $\|A\|_1$, $\|A\|_\infty$, $\|A\|_2$, and $\rho(A)$ for the following matrices:

   (a) $\begin{bmatrix} 2 & 1 \\ 1 & 2 \end{bmatrix}$. (b) $\begin{bmatrix} 1 & 1 & 2 \\ 0 & 2 & -1 \\ 0 & 3 & 0 \end{bmatrix}$. (c) $\begin{bmatrix} 2 & 1 & 0 \\ 0 & 2 & 0 \\ 0 & 3 & 4 \end{bmatrix}$.

2. Give an example of a matrix A such that $\rho(A) \ne \|A\|_\infty, \|A\|_1, \|A\|_2$.

3. Let

   $$A = \begin{pmatrix} \lambda & 1 & 0 \\ 0 & \lambda & 1 \\ 0 & 0 & \lambda \end{pmatrix}.$$

   Show that for each $\varepsilon > 0$ there exists a diagonal matrix D such that $\|D^{-1}AD\| \le |\lambda| + \varepsilon$ for operator norms $\|A\|_1$ $\|A\|_\infty$.

4. Generalize Problem 3 to any $k \times k$ matrix A in the Jordan form $\operatorname{diag}(J_1, J_2, \ldots, J_r)$.

5. Prove that $\rho(A) \le \|A\|$ for any operator norm $\|\|$ on A.

6. Show that for any two norms $\|\|$, $\|\|'$ on $\mathbb{R}^k$ there are constants $\alpha, \beta > 0$ such that $\alpha\|x\| \leq \|x\|' \leq \beta\|x\|$.

7. Deduce from Problem 6 that for any sequence $\{x(n)\}$, $\|x(n)\| \to 0$ as $n \to \infty$ if and only if $\|x(n)\|' \to 0$.

4.2 Notions of Stability

Let us consider the vector difference equation

$$x(n+1) = f(n, x(n)), \qquad x(n_0) = x_0, \tag{4.2.1}$$

where $x(n) \in \mathbb{R}^k, f : \mathbb{Z}^+ \times \mathbb{R}^k \to \mathbb{R}^k$. We assume that $f(n, x)$ is continuous in x. Recall that (4.2.1) is said to be *autonomous* or time-invariant if the variable n does not appear explicitly in the right-hand side of the equation $f(n, x(n)) \equiv f(x(n))$. It is said to be *periodic* if for all $n \in \mathbb{Z}$, $f(n+N, x) = f(n, x)$ for some positive integer N.

A point x^* in $\mathbb{R}^k$ is called an *equilibrium point* of (4.2.1) if $f(n, x^*) = x^*$ for all $n \geq n_0$. In most of the literature x^* is assumed to be the origin 0 and is called the zero solution. The justification for this assumption is as follows: Let $y(n) = x(n) - x^*$. Then (4.2.1) becomes

$$y(n+1) = f(n, y(n) + x^*) - x^* = g(n, y(n)). \tag{4.2.2}$$

Notice that $y = 0$ corresponds to $x = x^*$. Since in many cases it is not convenient to make this change of coordinates, we will not assume that $x^* = 0$ unless it is more convenient to do so.

Recall that in Chapter 3 we dealt with the existence and uniqueness of solutions of linear systems, that is, the case $f(n, x(n)) = A(n)x(n)$, where $A(n)$ is a $k \times k$ matrix. The existence and uniqueness of solutions of (4.2.1) may be established in a similar fashion (Exercises 4.2, Problem 9).

We are now ready to introduce the various stability notions of the equilibrium point x^* of (4.2.1).

Definition 4.2. The equilibrium point x^* of (4.2.1) is said to be:

(i) *Stable* (S) if given $\varepsilon > 0$ and $n_0 \geq 0$ there exists $\delta = \delta(\varepsilon, n_0)$ such that $\|x_0 - x^*\| < \delta$ implies $\|x(n, n_0, x_0) - x^*\| < \varepsilon$ for all $n \geq n_0$, *uniformly stable* (US) if δ may be chosen independent of n_0, *unstable* if it is not stable.

(ii) *Attracting* (A) if there exists $\mu = \mu(n_0)$ such that $\|x_0 - x^*\| < \mu$ implies $\lim_{n\to\infty} x(n, n_0, x_0) = x^*$, *uniformly attracting* (UA) if the choice of μ is independent of n_0. The condition for *uniform attractivity* may be paraphrased by saying that there exists $\mu > 0$ such that for every ε and n_0 there exists $N = N(\varepsilon)$ independent of n_0 such that $\|x(n, n_0, x_0) - x^*\| < \varepsilon$ for all $n \geq n_0 + N$ whenever $\|x_0 - x^*\| < \mu$.

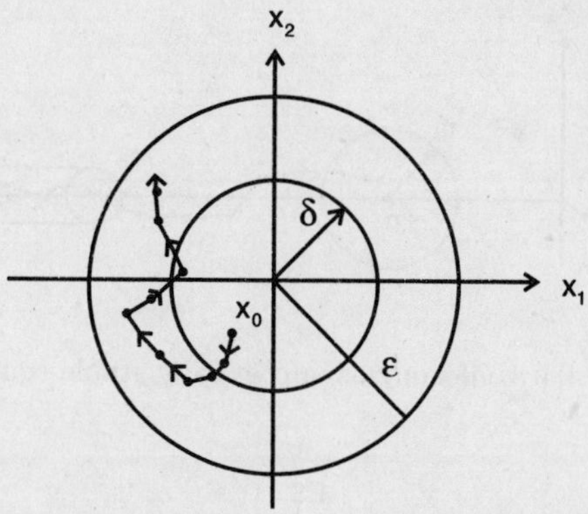

FIGURE 4.2. Stable equilibrium in phase space.

(iii) *Asymptotically stable* (AS) if it is stable and attracting, and *uniformly asymptotically stable* (UAS) if it is uniformly stable and uniformly attracting.

(iv) *Exponentially stable* (ES) if there exist $\delta > 0$, $M > 0$, and $\eta \in (0,1)$ such that $\|x(n, n_0, x_0) - x^*\| \leq M\|x_0 - x^*\|\eta^{n-n_0}$, whenever $\|x_0 - x^*\| < \delta$.

(v) A solution $x(n, n_0, x_0)$ is bounded if for some positive constant M, $\|x(n, n_0, x_0)\| \leq M$ for all $n \geq n_0$, where M may depend on each solution.

If in parts (ii), (iii) $\mu = \infty$ or in part (iv) $\delta = \infty$, the corresponding stability property is said to be *global*. In Figure 4.2, we suppress the (time) n and show only the movement of a solution that starts inside a ball of radius δ. The figure illustrates that all future states $x(n, n_0, x_0), n \geq n_0$, will stay

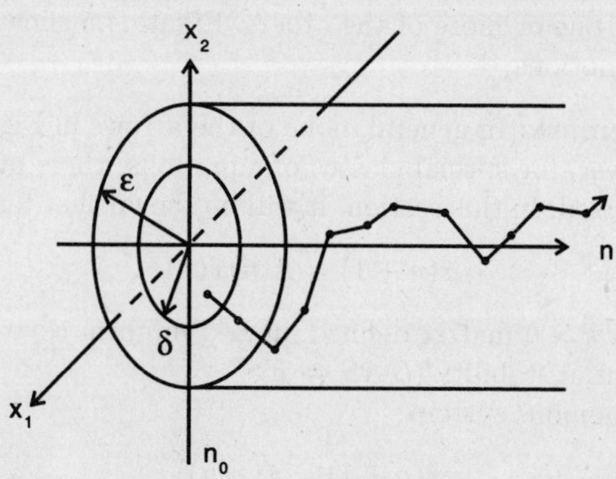

FIGURE 4.3. Stable equilibrium.

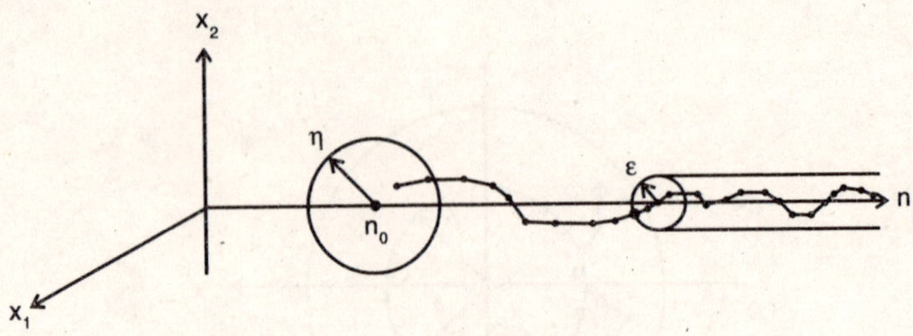

FIGURE 4.4. Uniformly asymptotically stable equilibrium.

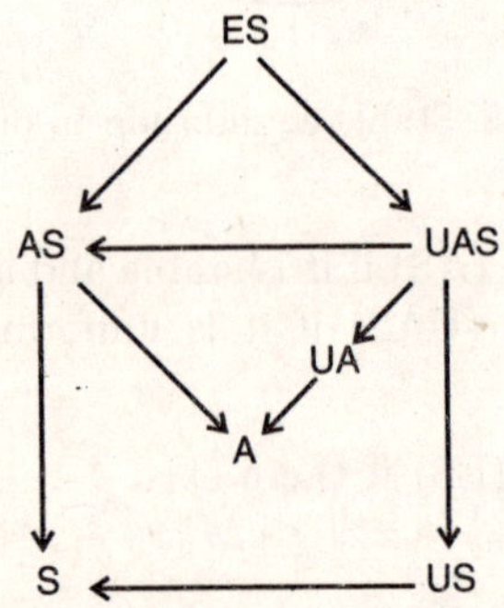

FIGURE 4.5. Hierarchy of stability notions.

within the ε ball. This diagram is called a *phase space* portrait and will be used extensively in later sections. In Figure 4.3 the time n is considered part of a three-dimensional coordinate system that provides another perspective on stability. Figure 4.4 depicts the uniform asymptotic stability of the zero solution.

Note that in the above definitions, some of the stability properties automatically imply one or more of the others. Figure 4.5 shows the hierarchy of the stability notions.

Important Remark: In general, none of the arrows in Figure 4.5 may be reversed. However, for special classes of equations, these arrows in Figure 4.5 may be reversed. In this section, it will be shown that for linear systems

$$x(n+1) = A(n)x(n) \tag{4.2.3}$$

where $A(n)$ is a $k \times k$ matrix defined on $\mathbb{Z}^+$, uniform asymptotic stability implies exponential stability ($UAS \Leftrightarrow ES$).

For the autonomous system

$$x(n+1) = f(x(n)) \tag{4.2.4}$$

we have the following result.

Theorem 4.3. *For the autonomous system (4.2.4), the following statements hold for the equilibrium point x^*:*

(i) $S \leftrightarrow US$.

(ii) $AS \leftrightarrow UAS$.

(iii) $A \leftrightarrow UA$.

PROOF.

(i) Let $x(n, n_0, x_0)$ and $y(n, m_0, x_0)$ be two solutions of (4.2.4), with $m_0 = n_0 + r_0$, $r_0 \geq 0$. Notice that $x(n-r_0, n_0, x_0)$ intersects with $y(n, m_0, x_0)$ at $n = m_0$. By uniqueness of solutions, it follows that $y(n, m_0, x_0) = x(n - r_0, n_0, x_0)$. This implies that the δ in the definition of stability is independent of the initial time n_0 which establishes our result.

The proofs of (ii) and (iii) are similar to the proof of (i). □

The following examples serve to illustrate the definitions.

1. The solution of the scalar equation $x(n+1) = x(n)$ is given by $x(n, n_0, x_0) = x_0$; hence the zero solution is uniformly stable but not asymptotically stable.

2. The solutions of the scalar equation $x(n+1) = a(n)x(n)$ are

$$x(n, n_0, x_0) = \left[\prod_{i=n_0}^{n-1} a(i)\right] x_0. \tag{4.2.5}$$

Hence one may conclude the following:

(i) The zero solution is stable if and only if

$$\left|\prod_{i=n_0}^{n-1} a(i)\right| \leq M(n_0) \equiv M, \tag{4.2.6}$$

where M is a positive constant that depends on n_0 (Exercises 4.2, Problem 2). This condition holds if $a(i) = (1 + \eta^i)$, where $0 < \eta < 1$.

To show this we write the solution as $x(n, n_0, x_0) = \Phi(n)x_0$, where $\Phi(n) = \prod_{i=n_0}^{n-1}(1 + \eta^i)$. Since $1 + \eta^i < \exp(\eta^i)$, it follows that

$$\Phi(n) \leq \exp\left(\sum_{i=n_0}^{n-1} \eta^i\right) \leq \exp\left(\sum_{i=n_0}^{\infty} \eta^i\right) \leq \exp\left(\frac{\eta^{n_0}}{1-\eta}\right)$$
$$= M(n_0) = M.$$

Given $\varepsilon > 0$ and $n_0 \geq 0$, if we let $\delta = \varepsilon/(2M)$, then $|x_0| < \delta$ implies $|x(n, n_0, x_0)| = \Phi(n)x_0 < \varepsilon$.

(ii) The zero solution is uniformly stable if and only if

$$\left|\prod_{i=n_0}^{n-1} a(i)\right| \leq M, \tag{4.2.7}$$

where M is a positive constant independent of n_0 (Exercises 4.2, Problem 5). This condition holds if $a(i) = \sin(i+1)$.

(iii) The zero solution is asymptotically stable if and only if

$$\lim_{n\to\infty}\left|\prod_{i=n_0}^{n-1} a(i)\right| = 0 \tag{4.2.8}$$

(Exercises 4.2, Problem 5). This condition clearly holds if $a(i) = \frac{i+1}{i+2}$. The solution is given by $x(n, n_0, x_0) = (n_0+1)/(n+1)x_0$. Thus, the zero solution is uniformly stable and asymptotically stable (globally), but not uniformly asymptotically stable. (Why?) (See Exercises 4.2, Problem 6.)

(iv) The zero solution is uniformly asymptotically stable (and thus exponentially stable) if and only if

$$\left|\prod_{i=n_0}^{n-1} a(i)\right| \leq M\eta^{n-n_0}, \tag{4.2.9}$$

for some $M > 0, 0 < \eta < 1$. This may be satisfied if $a(i) = 1/i$ (Exercises 4.2, Problem 8).

Now we give two important examples. In the first example we show that the zero solution is stable but not uniformly stable. In the second example the zero solution is attracting but not stable (personal communication by Professor Bernd Aulbach).

Example 4.4. The solution of the equation $x(n+1) = \left(\frac{n+1}{2}\right)[x(n)]^2$ is given by

$$x(n, n_0, x_0) = \left(\frac{n}{2}\right)\left(\frac{n-1}{2}\right)^2\left(\frac{n-2}{2}\right)^4 \cdots \left(\frac{n_0+1}{2}\right)^{2^{n-n_0-1}} (x_0)^{2^{n-n_0}},$$

$$x(n_0) = x_0.$$

If $|x_0|$ is sufficiently small, then $\lim_{n\to\infty} x(n) = 0$. Thus, the zero solution is attracting. However, it is not uniformly attracting. For if $\delta > 0$ is given and n_0 is chosen such that $(n_0+1)\delta^2 \geq 2$, then, for $|x_0| = \delta$,

$$|x(n_0+1, n_0, x_0)| = \left(\frac{n_0+1}{2}\right)|x_0|^2 \geq 1.$$

Let us now check the stability of the zero solution. Given $\varepsilon > 0$ and $n_0 \geq 0$, let $\delta = \varepsilon/(n_0+1)$. If $|x_0| < \delta$, then $|x(n, n_0, x_0)| < \varepsilon$ for all $n \geq n_0$. Since δ

depends on the choice of n_0, the zero solution is stable but not uniformly stable.

Example 4.5. Consider the difference equation (in polar coordinates)

$$r(n+1) = \sqrt{r(n)}, \qquad r > 0,$$
$$\theta(n+1) = \sqrt{2\pi\theta(n)}, \qquad 0 \le \theta \le 2\pi.$$

We claim that the equilibrium point $(1, 0)$ is attracting but not stable. To show this, observe that

$$r(n) = r_0^{2^{-n}}, \qquad r_0 = r(0),$$
$$\theta(n) = (2\pi)^{\left(1-2^{-n}\right)} \cdot (\theta_0)^{2^{-n}}, \qquad \theta_0 = \theta(0).$$

Clearly, $\lim_{n\to\infty} r(n) = 1$ and $\lim_{n\to\infty} \theta(n) = 2\pi$. Now, if $r_0 \neq 0$, $\theta_0 = 0$, then $(r(n), \theta(n)) = ((r_0)^{2^{-n}}, 0)$, which converges to the equilibrium point $(1, 0)$. However, if $\theta_0 = \delta\pi, 0 < \delta < 1$, then the orbit of (r_0, θ_0) will spiral around the circle counterclockwise to converge to the equilibrium point $(1, 0)$. Hence the equilibrium point $(1, 0)$ is attracting but not stable. (See Figure 4.6.)

Remark: The situation in Example 4.5 is a higher dimension phenomenon. In 1997, Sedaghat [132] showed that a continuous map on the real line cannot have an attracting unstable fixed point.

To demonstrate this phenomenon, let us contemplate the following example:

Example 4.6. Consider the map

$$G_\mu(x) = \begin{cases} -2x & \text{if } x < \mu, \\ 0 & \text{if } x \ge \mu, \end{cases}$$

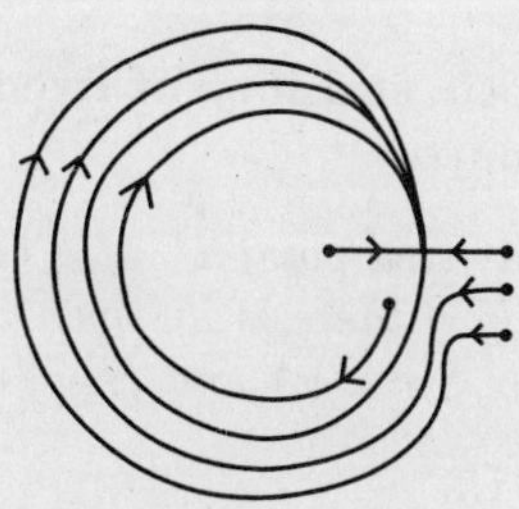

FIGURE 4.6. Attracting but not stable equilibrium.

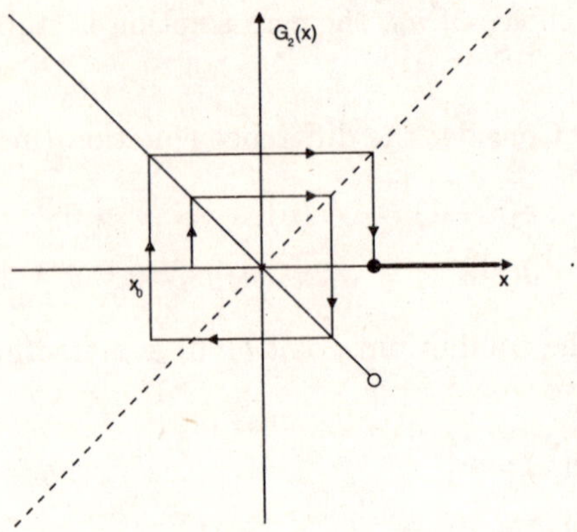

FIGURE 4.7. $G_2(x)$.

where $\mu \in \mathbb{R}^+$. Equivalently, we have the difference equation $x(n+1) = G_\mu(x(n))$ whose solution is given by

$$x(n) = G_\mu^n(x_0) = \begin{cases} (-2)^n x_0 & \text{if } (-2)^{n-1}x_0 < \mu, \\ 0 & \text{if } (-2)^{n-1}x_0 \geq \mu, \end{cases}$$

where $x(0) = x_0$.

Now, if $x_0 \geq \mu$, then $G_\mu^n(x_0) = 0$ for all $n \geq 1$. On the other hand, if $x_0 < \mu$, then for some $k \in \mathbb{Z}^+$, $G_\mu^k(x_0) \geq \mu$. Hence, $G_\mu^n(x_0) = 0$ for all $n \geq k$.

Hence the fixed point $x^* = 0$ is *globally attracting*. However, $x^* = 0$ is unstable, for points x_0 that are close to 0 are mapped to points further away from 0 until they exceed μ (see Figure 4.7 for G_2).

Theorem 4.7 [132]. *A continuous map f on the real line cannot have an attracting unstable fixed point.*

To facilitate the proof of the theorem, we first establish a stability result that is of independent interest, since it does not require differentiability of f.

Criterion for asymptotic stability of fixed points of nondifferentiable maps.

Theorem 4.8 [135]. *A fixed point x^* of a continuous map f is asymptotically stable if and only if there is an open interval (a,b) containing x^* such that $f^2(x) > x$ for $a < x < x^*$ and $f^2(x) < x$ for $x^* < x < b$.*

PROOF. See Appendix C. □

PROOF OF THEOREM 4.7. Let f be a continuous map on $\mathbb{R}$ that has an unstable globally attracting fixed point x^*. This implies that the equation

$f^2(x) = x$ has only one solution $x = x^*$. Hence, there are only two possible cases:

(a) $f^2(x) > x$ for $x < x^*$ and $f^2(x) < x$ for $x > x^*$;

(b) $f^2(x) < x$ for $x < x^*$ and $f^2(x) > x$ for $x > x^*$.

Notice that by Theorem 4.8, case (a) implies that x^* is asymptotically stable and must be discarded. It remains to rule out case (b). So assume that $f^2(x) < x$ for $x < x^*$. Now let $x_0 < x^*$. Then by iteration, we have $\cdots < f^4(x_0) < f^2(x_0) < x_0 < x^*$.

Thus, $f^{2n}(x_0)$ does not converge to x^*, which contradicts the global attractivity of x^*. The case $f^2(x) > x$ for $x > x^*$ is similar and will lead to a contradiction. Hence, our assumption is false, which proves the assertion of the theorem. □

Exercises 4.2

1. Meditate upon the scalar equation $x(n+1) = ax(n)$. Prove that:
   (i) If $|a| < 1$, the zero solution is uniformly asymptotically stable.
   (ii) If $|a| = 1$, the zero solution is uniformly stable.
   (iii) If $|a| > 1$, the zero solution is not stable.
2. (a) Prove that the zero solution of the scalar equation $x(n+1) = a(n)x(n)$ is stable if and only if $\left|\prod_{i=n_0}^{n-1} a(i)\right| \le M(n_0)$, where M depends on n_0.
   (b) Show that the zero solution of the equation $x(n+1) = (1 + \eta^n)x(n), 0 < \eta < 1$, is stable.
3. (a) Prove that the zero solution of the equation $x(n+1) = a(n)x(n)$ is uniformly stable if and only if $\left|\prod_{i=n_0}^{n-1} a(i)\right| \le M$, where M is a positive constant independent of n_0.
   (b) Show that the zero solution of the equation $x(n+1) = \sin(n+1)x(n)$ is uniformly stable.
4. Show that the zero solution of the equation $x(n+1) = \frac{n+1}{n+2}x(n)$ is asymptotically stable.
5. Prove that the zero solution of the equation $x(n+1) = a(n)x(n)$ is asymptotically stable if and only if $\lim_{n\to\infty} \left|\prod_{i=n_0}^{n-1} a(i)\right| = 0$.
6. Show that the zero solution of the equation in Problem 4 is not uniformly asymptotically stable.
7. Prove that the zero solution of the equation $x(n) = a(n)x(n)$ is uniformly asymptotically stable if and only if $\left|\prod_{i=n_0}^{n-1} a(i)\right| \le M\eta^{n-n_0}$, for some $M > 0, 0 < \eta < 1$.

8. Show that the zero solution of the scalar equation $x(n+1) = (1/n)x(n), n \geq 1$, is uniformly asymptotically stable.

9. Establish the existence and uniqueness of solutions of (4.2.1).

10. Consider the system

$$x(n+1) = x(n) + \frac{x^2(n)(y(n)-x(n)) + y^5(n)}{[x^2(n)+y^2(n)] + [x^2(n)+y^2(n)]^3},$$
$$y(n+1) = y(n) + \frac{y^2(n)(y(n)-2x(n))}{[x^2(n)+y^2(n)] + [x^2(n)+y^2(n)]^3},$$

which can be written as

$$x(n+1) = x(n) + g_1(x(n), y(n)),$$
$$y(n+1) = y(n) + g_2(x(n), y(n)).$$

Show that the zero solution is globally attracting but unstable.

11. Define the difference equation on the unit circle as

$$r(n+1) = 1,$$
$$\theta(n+1) = \sqrt{2\pi\theta}, \quad 0 \leq \theta < 2\pi.$$

Show that the fixed point $(1, 0)$ is globally attracting but unstable.

4.3 Stability of Linear Systems

4.3.1 Nonautonomous Linear Systems

In this subsection we investigate the stability of the linear nonautonomous (time-variant) system given by

$$x(n+1) = A(n)x(n), \quad n \geq n_0 \geq 0. \tag{4.3.1}$$

It is always assumed that $A(n)$ is nonsingular for all $n \geq n_0$.

If $\Phi(n)$ is any fundamental matrix of system (4.3.1) or (4.3.6), then recall that $\Phi(n,m) = \Phi(n)\Phi^{-1}(m)$ is the state transition matrix. In the following result we express the conditions for stability in terms of a fundamental matrix $\Phi(n)$ of system (4.3.1).

Theorem 4.9. *Consider system* (4.3.1). *Then its zero solution is*

(i) *stable if and only if there exists a positive constant M such that*

$$\|\Phi(n)\| \leq M \quad \text{for } n \geq n_0 \geq 0; \tag{4.3.2}$$

(ii) *uniformly stable if and only if there exists a positive constant M such that*

$$\|\Phi(n,m)\| \leq M \quad \text{for } n_0 \leq m \leq n < \infty; \tag{4.3.3}$$

(iii) *asymptotically stable if and only if*

$$\lim_{n\to\infty} \|\Phi(n)\| = 0; \tag{4.3.4}$$

(iv) *uniformly asymptotically stable if and only if there exist positive constants M and $\eta \in (0,1)$ such that:*

$$\|\Phi(n,m)\| \le M\eta^{n-m} \quad \text{for } n_0 \le m \le n < \infty. \tag{4.3.5}$$

PROOF. Without loss of generality we may assume that $\Phi(n_0) = I$, since conditions (4.3.2) through (4.3.5) hold true for every fundamental matrix if they hold for one. Thus $x(n, n_0, x_0) = \Phi(n)x_0$.

(i) Suppose that inequality (4.3.2) holds. Then $\|x(n, n_0, x_0)\| \le M\|x_0\|$. So for $\varepsilon > 0$, let $\delta < \varepsilon/M$. Then $\|x_0\| < \delta$ implies $\|x(n, n_0, x_0)\| < \varepsilon$ and, consequently, the zero solution is stable. Conversely, suppose that $\|x(n, n_0, x_0)\| = \|\Phi(n)x_0\| < \varepsilon$ whenever $\|x_0\| \le \delta$. Observe that $\|x_0\| \le \delta$ if and only if $\frac{1}{\delta}\|x_0\| \le 1$. Hence

$$\|\Phi(n)\| = \sup_{\|\xi\|\le 1} \|\Phi(n)\xi\| = \frac{1}{\delta} \sup_{\|x_0\|\le\delta} \|\Phi(n)x_0\| \le \frac{\varepsilon}{\delta} = M.$$

Parts (ii) and (iii) remain as Exercises 4.3, Problems 9 and 10.

(iv) Suppose finally that inequality (4.3.5) holds. The zero solution of system (4.3.1) would then be uniformly stable by part (ii). Furthermore, for $\varepsilon > 0, 0 < \varepsilon < M$, take $\mu = 1$ and N such $\eta^N < \varepsilon/M$. Hence, if $\|x_0\| < 1$, then $\|x(n, n_0, x_0)\| = \|\Phi(n, n_0)x_0\| \le M\eta^{n-n_0} < \varepsilon$ for $n \ge n_0 + N$. The zero solution would be uniformly asymptotically stable. Conversely, suppose that the zero solution is uniformly asymptotically stable. It is also then uniformly stable, and thus by Theorem 5.1(ii), $\|\Phi(n,m)\| \le M$ for $0 \le n_0 \le m \le n < \infty$. From uniform attractivity, there exists $\mu > 0$ such that for ε with $0 < \varepsilon < 1$ there exists N such that $\|\Phi(n, n_0)\| < \varepsilon$ for $n \ge n_0 + N$, whenever $\|x_0\| < \mu$. This implies that $M\Phi(n, n_0)\| \le \varepsilon$ for $n \ge n_0 + N$. Then for $n\varepsilon[n_0 + mN, n_0 + (m+1)N], m > 0$, we have

$$\begin{aligned}
\|\Phi(n, n_0)\| &\le \|\Phi(n, n_0 + mN)\|\|\Phi(n_0 + mN, n_0 + (m-1)N)\| \times \\
&\quad \cdots \times \|\Phi(n_0 + N, n_0)\| \\
&\le M\varepsilon^m \le \frac{M}{\varepsilon}\left(\varepsilon^{\frac{1}{N}}\right)^{(m+1)N} = \tilde{M}\eta^{(m+1)N}, \\
&\le \tilde{M}\eta^{(n-n_0)},
\end{aligned}$$

for $mN \le n - n_0 \le (m+1)N$ where $\tilde{M} = \frac{M}{\varepsilon}, \eta = \varepsilon^{\frac{1}{N}}$. This concludes the proof of the theorem. □

The following result arises as an immediate consequence of the above theorem. [See the Important Remark, part (i).]

Corollary 4.10. *For the linear system* (4.3.1) *the following statements hold:*

(i) *The zero solution is stable if and only if all solutions are bounded.*

(ii) *The zero solution is exponentially stable if and only if it is uniformly asymptotically stable.*

PROOF. Statements (i) and (ii) follow immediately from conditions (4.3.3) and (4.3.5), respectively (Exercises 4.3, Problem 6). □

The following is another *important consequence* of Theorem 4.9:

Corollary 4.11. *For system* (4.3.1), *every local stability property of the zero solution implies the corresponding global stability property.*

PROOF. Use Theorem 4.9 (Exercises 4.3, Problem 7). □

We now give a simple but powerful criterion for uniform stability and uniform asymptotic stability.

Theorem 4.12 [17].

(i) *If* $\sum_{i=1}^{k} |a_{ij}(n)| \leq 1, 1 \leq j \leq k, n \geq n_0$, *then the zero solution of system* (3.2.15) *is uniformly stable.*

(ii) *If* $\sum_{i=1}^{k} |a_{ij}(n)| \leq 1 - \nu$ *for some* $\nu > 0, 1 \leq j \leq k, n \geq n_0$, *then the zero solution is uniformly asymptotically stable.*

PROOF.

(i) From condition (i) in Theorem 4.12, $\|A(n)\|_1 \leq 1$ for all $n \geq n_0$. Thus,

$$\|\Phi(n,m)\|_1 = \left\| \prod_{i=m}^{n-1} A(i) \right\|_1 \leq \|A(n-1)\|_1 \|A(n-2)\|_1 \cdots \|A(m)\|_1 \leq 1.$$

This now implies uniform stability by Theorem 4.9, part (ii).

(ii) The proof of statement (ii) is so similar to the proof of statement (i) that we will omit it here. □

4.3.2 Autonomous Linear Systems

In this subsection we specialize the results of the previous section to autonomous (time-invariant) systems of the form

$$x(n+1) = Ax(n). \tag{4.3.6}$$

In the next theorem we summarize the main stability results for the linear autonomous systems (4.3.6).

Theorem 4.13. *The following statements hold:*

(i) *The zero solution of* (4.3.6) *is stable if and only if* $\rho(A) \leq 1$ *and the eigenvalues of unit modulus are semisimple.*[1]

(ii) *The zero solution of* (4.3.6) *is asymptotically stable if and only if* $\rho(A) < 1$.

PROOF.

(i) Let $A = PJP^{-1}$, where $J = \text{diag}(J_1, J_2, \ldots, J_r)$ is the Jordan form of A and

$$J_i = \begin{pmatrix} \lambda_i & 1 & & 0 \\ & \lambda_i & & \\ & & \ddots & \ddots & \\ & & & 1 \\ 0 & & & \lambda_i \end{pmatrix}.$$

From Theorem 4.9 the zero solution of (4.3.6) is stable if and only if $\|A^n\| = \|PJ^nP^{-1}\| \leq M$ or $\|J^n\| \leq \tilde{M}$, where $\tilde{M} = M/(\|P\|\|P^{-1}\|)$. Now, $J^n = \text{diag}(J_1^n, J_2^n, \ldots, J_r^n)$, where

$$J_i^n = \begin{pmatrix} \lambda_i^n & \binom{n}{1}\lambda_i^{n-1} & \cdots & \binom{n}{s_i - 1}\lambda_i^{n-s_i+1} \\ 0 & \lambda_i^n & \cdots & \vdots \\ & & & \binom{n}{1}\lambda_i^{n-1} \\ 0 & 0 & \cdots & \lambda_i^n \end{pmatrix}.$$

Obviously, J_i^n becomes unbounded if $|\lambda_i| > 1$ or if $|\lambda_i| = 1$ and J_i is not 1×1. If $|\lambda_i| < 1$, then $J_i^n \to 0$ as $n \to \infty$. To prove this conclusion it suffices to show that $|\lambda_i|^n\, n^\ell \to 0$, as $n \to \infty$ for any positive integer ℓ. This conclusion follows from L'Hôpital's rule, since $|\lambda_i|^n\, n^\ell = n^\ell e^{(\ln|\lambda_i|)n}$ (Exercises 4.3, Problem 8).

(ii) The proof of statement (ii) has already been established by the above argument. This completes the proof of the theorem. □

Explicit Criteria for Stability of Two-Dimensional Systems

In many applications one needs explicit criteria on the entries of the matrix for the eigenvalues to lie inside the unit disk. So consider the matrix

$$A = \begin{pmatrix} a_{11} & a_{12} \\ a_{21} & a_{22} \end{pmatrix}$$

[1]An eigenvalue is said to be semisimple if the corresponding Jordan block is diagonal.

whose characteristic equation is given by

$$\lambda^2 - (a_{11} + a_{22})\lambda + (a_{11}a_{22} - a_{12}a_{21}) = 0$$

or

$$\lambda^2 - (\operatorname{tr} A)\lambda + \det A = 0. \tag{4.3.7}$$

Comparing (4.3.7) with the equation

$$\lambda^2 + p_1\lambda + p_2 = 0,$$

where $p_1 = -\operatorname{tr} A$, $p_2 = \det A$, we conclude from Theorem 2.37 that the eigenvalues of A lie inside the unit disk if and only if

$$1 + \operatorname{tr} A + \det A > 0, \quad 1 - \operatorname{tr} A + \det A > 0, \quad 1 - \det A > 0 \tag{4.3.8}$$

or, equivalently,

$$\boxed{|\operatorname{tr} A| < 1 + \det A < 2.} \tag{4.3.9}$$

It follows that under condition (4.3.9), the zero solution of the equation

$$x(n+1) = Ax(n)$$

is asymptotically stable.

We now describe the situation when some eigenvalues of A in (4.3.6) are inside the unit disk and some eigenvalues are outside the unit disk. The result below is called the Stable Subspace (Manifold) Theorem. The result does not require that A is invertible.

Let λ be an eigenvalue of A of multiplicity m and let $\xi_1, \xi_2, \ldots, \xi_m$ be the generalized eigenvectors corresponding to λ. Then for each i, $1 \le i \le m$, either

$$A\xi_i = \lambda\xi_i \quad (\xi_i \text{ is an eigenvector of } A), \quad \text{or}$$
$$A\xi_i = \lambda\xi_i + \xi_{i-1}.$$

It follows that the generalized eigenvectors corresponding to λ are the solutions of the equation

$$(A - \lambda J)^m \xi = 0. \tag{4.3.10}$$

The set of all linear combinations, or the span of the generalized eigenvectors corresponding to λ is invariant under A and is called the generalized eigenspace E_λ of the eigenvalue of A. Clearly, if $\lambda_1 \neq \lambda_2$, then $E_{\lambda_1} \cap E_{\lambda_2} = \{0\}$. Notice that each eigenspace E_λ includes the zero vector.

Assume that A is hyperbolic, that is, none of the eigenvalues of A lie on the unit circle. Arrange the eigenvalues of A such that $\Delta_s = \{\lambda_1, \lambda_2, \ldots, \lambda_r\}$ are all the eigenvalues of A with $|\lambda_i| < 1$, $1 \le i \le r$ and $\Delta_u = \{\lambda_{r+1}, \lambda_{r+s}, \ldots, \lambda_k\}$ are all the eigenvalues of A with $|\lambda_i| > 1$, $r+1 \le i \le k$. The eigenspace spanned by the eigenvalues in Δ_s is denoted by W^s, where $W^s = \bigcup_{i=1}^{r} \lambda_i$ and the eigenspace spanned by the eigenvalues in Δ_u is denoted by W^u, where $W^u = \bigcup_{i=r+1}^{k} \lambda_i$.

Theorem 4.14 (The Stable Subspace (Manifold) Theorem). *If A is hyperbolic, then the following statements hold true:*

(i) *If $x(n)$ is a solution of* (4.3.6) *with $x(0) \in W^s$, then for each n, $x(n) \in W^s$. Furthermore,*

$$\lim_{n \to \infty} x(n) = 0.$$

(ii) *If $x(n)$ is a solution of* (4.3.6) *with $x(0) \in W^u$, then $x(n) \in W^u$ for each n. Moreover,*

$$\lim_{n \to -\infty} x(n) = 0.$$

PROOF.

(i) Let $x(n)$ be a solution of (4.3.6) with $x(0) \in W^s$. Since $AE_\lambda = E_\lambda$, it follows that $AW^s = W^s$. Hence $x(n) \in W^s$ for all $n \in \mathbb{Z}^+$. To prove the second statement, observe that $x(0) = \sum_{i=1}^{r} c_i \xi_i$, where $1 \leq \xi_i \leq r$ are the generalized eigenvectors corresponding to elements in Δ_s. Let $J = P^{-1}AP$ be the Jordan form of A. Then J may be written in the form

$$J = \begin{pmatrix} J_s & 0 \\ 0 & J_u \end{pmatrix}$$

where J_s has the eigenvalues in Δ_s and J_u has the eigenvalues in $\Delta_{u'}$. By Lemma 3.27 in Chapter 3, the corresponding generalized eigenvectors $\tilde{\xi}_i$, $1 \leq i \leq r$, of J_s are of the form $\tilde{\xi}_i = P^{-1}\xi_i = (a_{i1}, a_{i2}, \ldots, a_{ir}, 0, 0, \ldots, 0)^T$. Now

$$\begin{aligned} x(n) &= A^n x(0) \\ &= PJ^nP^{-1} \sum_{i=1}^{r} c_i \xi_i \\ &= PJ^n \sum_{i=1}^{r} c_i \tilde{\xi}_i \\ &= P \sum_{i=1}^{r} c_i \begin{pmatrix} J_s^n & \tilde{\xi}_i \\ 0 & 0 \end{pmatrix}. \end{aligned}$$

Thus $\lim_{n \to \infty} x(n) = 0$ since $J_s^n \to 0$ as $n \to \infty$.

(ii) The proof of (ii) is analogous to (i) and will be left to the reader as Problem 11. □

Remark:

(i) Part (i) may be obtained without the condition of hyperbolicity of A, and similarly for part (ii).

(ii) The General Stable Manifold Theorem for Nonlinear Maps will be given in Appendix D.

We now use the above result to investigate the stability of the periodic system

$$x(n+1) = A(n)x(n), \qquad A(n+N) = A(n). \tag{4.3.11}$$

Recall from Chapter 3 that if $\Phi(n, n_0)$ is a fundamental matrix of (4.3.11), then there exist a constant matrix B whose eigenvalues are called the Floquet exponents and a periodic matrix $P(n, n_0)$ such that $\Phi(n, n_0) = P(n, n_0)B^{n-n_0}$, where $P(n+N, n_0) = P(n, n_0)$. Thus if B^n is bounded, then so is $\Phi(n, n_0)$, and if $B^n \to 0$ as $n \to \infty$, then it follows that $\Phi(n, n_0) \to 0$ as $n \to \infty$. This proves the following result.

Theorem 4.15. *The zero solution of* (4.3.11) *is:*

(i) *stable if and only if the Floquet exponents have modulus less than or equal to* 1*; those of modulus of* 1 *are semisimple;*

(ii) *asymptotically stable if and only if all the Floquet exponents lie inside the unit disk.*

For practical purposes, the following corollary is of paramount importance.

Corollary 4.16. *The zero solution of* (4.3.11) *is:*

(i) *stable if and only if each eigenvalue of the matrix* $C = A(N-1)A(N-2) \cdots A(0)$ *has modulus less than or equal to* 1*; those solutions with modulus of value* 1 *are semisimple;*

(ii) *asymptotically stable if and only if each eigenvalue of* $C = A(N-1)A(N-2)\cdots A(0)$ *has modulus less than* 1.

Let us summarize what we have learned thus far. First, for the autonomous (time-invariant) linear system $x(n+1) = Ax(n)$, the eigenvalues of A determine the stability properties of the system (Theorem 4.13). But for a *periodic system* $x(n+1) = A(n)x(n)$, the eigenvalues of $A(n)$ do not play any role in the determination of the stability properties of the system. Instead, the *Floquet multipliers* of $A(n)$ determine those properties. The following example should dispel any wrong ideas concerning the role of eigenvalues in a nonautonomous system.

Example 4.17. Let us again consider the periodic system in Example 3.35 where

$$A(n) = \begin{pmatrix} 0 & \dfrac{2+(-1)^n}{2} \\ \dfrac{2-(-1)^n}{2} & 0 \end{pmatrix}.$$

Here the eigenvalues of A are $\pm\sqrt{3}/2$, and thus $\rho[A(n)] < 1$. By applying Corollary 4.16, one may quickly check the stability of this system. We have

$$C = A(1)A(0) = \begin{pmatrix} 0 & \frac{3}{2} \\ \frac{1}{2} & 0 \end{pmatrix} \begin{pmatrix} 0 & \frac{1}{2} \\ \frac{3}{2} & 0 \end{pmatrix} = \begin{pmatrix} \frac{9}{4} & 0 \\ 0 & \frac{1}{4} \end{pmatrix}.$$

Hence, by Corollary 4.16, the zero solution is unstable, since C has an eigenvalue $9/4$ which is greater than 1.

For the eager reader, perpetually searching for a challenge, we might determine the stability by explicitly transcribing the fundamental matrix as follows:

$$\Phi(n) = \begin{pmatrix} \dfrac{2^{1-n} - (-2)^{1-n}}{2} & \dfrac{\left(\frac{3}{2}\right)^n - \left(-\frac{3}{2}\right)^n}{2} \\ \dfrac{2^{-n} - (-2)^{-n}}{2} & \dfrac{\left(\frac{3}{2}\right)^n - \left(-\frac{3}{2}\right)^n}{2} \end{pmatrix}.$$

Hence, these are unbounded solutions. Consequently, the zero solution is unstable. This example demonstrates without any doubt that eigenvalues do not generally provide any information about the stability of nonautonomous difference systems.

Exercises 4.3

1. Determine whether the zero solution of the system $x(n+1) = Ax(n)$ is stable, asymptotically stable, or unstable if the matrix A is:

(a) $\begin{pmatrix} 1 & 0 \\ -2 & 1 \end{pmatrix}$. (b) $\begin{pmatrix} \frac{5}{12} & 0 & \frac{1}{2} \\ -1 & -\frac{1}{2} & \frac{5}{4} \\ \frac{1}{3} & 0 & 0 \end{pmatrix}$.

(c) $\begin{pmatrix} -1 & 5 \\ -0.5 & 2 \end{pmatrix}$. (d) $\begin{pmatrix} 1.5 & 1 & -1 \\ -1.5 & -0.5 & 1.5 \\ 0.5 & 1 & 0 \end{pmatrix}$.

2. Give another example (see Example 4.17) of a matrix $A(n)$ such that $\rho[A(n)] < 1$ and the zero solution of $x(n+1) = A(n)x(n)$ is unstable.

3. Give an example of a stable matrix A (i.e., $\rho(A) < 1$) with $\|A\| > 1$, for some matrix norm $\|\|$.

4. Consider the autonomous (time-invariant) system (4.3.6). Prove the following statements:

   (i) The zero solution is stable if and only if it is uniformly stable.

(ii) The zero solution is asymptotically stable if and only if it is uniformly asymptotically stable.

5. Use Corollary 4.16 to determine whether or not the zero solution of $x(n+1) = A(n)x(n)$ is uniformly stable or uniformly asymptotically stable, where $A(n)$ is the matrix:

(a) $\begin{pmatrix} -1 & \frac{1}{4}\cos(n) \\ 0 & \frac{1}{2}\sin(n) \end{pmatrix}$. (b) $\begin{pmatrix} \frac{n}{n+1} & 0 \\ -1 & 1 \end{pmatrix}$.

(c) $\begin{pmatrix} \frac{1}{n+1} & 0 & \frac{1}{2}\sin(n) \\ \frac{1}{4} & \frac{1}{2}\sin(n) & \frac{1}{4}\cos(n) \\ \frac{1}{5} & 0 & 0 \end{pmatrix}$. (d) $\begin{pmatrix} \frac{n+2}{n+1} & 0 & 0 \\ 0 & 1 & 0 \\ \frac{1}{n+1} & 0 & 1 \end{pmatrix}$.

6. Prove Corollary 4.10.

7. Prove Corollary 4.11.

8. Show that if $|\lambda| < 1$, then $\lim_{n\to\infty} |\lambda|^n n^s = 0$ for any given positive integer s.

9. Prove that the zero solution of system (4.3.1) is uniformly stable if and only if there exists $M > 0$ such that $\|\Phi(n, m)\| \leq M$, for $n_0 \leq m \leq n < \infty$.

10. Prove that the zero solution of system (4.3.1) is asymptotically stable if and only if $\lim_{n\to\infty} \|\Phi(n)\| = 0$.

11. Prove Theorem 4.14, part (ii).

Iterative Methods

Consider the system of linear algebraic equations

$$Ax = b, \tag{4.3.12}$$

where $A = (a_{ij})$ is a $k \times k$ matrix.

Iterative methods are used widely to solve (4.3.12) numerically. We generate a sequence $x(n)$ using the difference equation

$$x(n+1) = Bx(n) + d, \tag{4.3.13}$$

where the choice of B and d depends on the iterative method used. The iterative method (4.3.13) is *consistent* with (4.3.12) if a solution x^* of (4.3.12) is an equilibrium point of (4.3.13), i.e., if

$$Bx^* + d = x^*. \tag{4.3.14}$$

We now describe one such consistent method, the *Jacobi iterative method.* Assuming that the diagonal elements a_{ii} of A are nonzero, then $D =$

$\text{diag}(a_{11}, a_{22}, \ldots, a_{kk})$ is nonsingular. In (4.3.13) define

$$B = I - D^{-1}A, \qquad d = D^{-1}b. \tag{4.3.15}$$

This method is consistent (Exercises 4.3, Problem 12). If A is nonsingular, then x^* is unique. The associated error equation may be derived by letting $e(n) = x(n) - x^*$. Equations (4.3.13) and (4.3.14) then yield the equation

$$e(n+1) = Be(n). \tag{4.3.16}$$

The quantity $e(n)$ represents the error in approximating the solution x^* by the nth iterate $x(n)$ of (4.3.13).

12. (i) Prove that $x(n) \to x^*$ as $n \to \infty$ if and only if $e(n) \to 0$ as $n \to \infty$. In other words, the iterative method ((4.3.13) converges to the solution x^* of (4.3.12) if and only if the zero solution of (4.3.16) is asymptotically stable).

    (ii) Use Theorem 4.9 to show that the iterative method converges if and only if $\rho(B) < 1$.

13. Show that the Jacobi iterative method is consistent.

14. Consider (4.3.12) and (4.3.13) with the assumption that the diagonal elements of A are nonzero. Let L be the lower triangular part of A and let U be the strictly upper triangular part of A (i.e., the main diagonal of U is zero). Then $A = L + U$. The Gauss–Seidel iterative method defines $B = -L^{-1}U$ and $d = L^{-1}b$. Show that this method is consistent.

In Problem 15 we consider the k-dimensional system $x(n+1) = A(n)x(n)$.

*15. (a) Define $H(n) = A^T(n)A(n)$. Prove the *Lagrange identity*

$$\|x(n+1)\|_2^2 = x^T(n+1)x(n+1) = x^T(n)H(n)x(n). \tag{4.3.17}$$

(b) Show that all eigenvalues of $H(n)$ as defined in part (a) are real and nonnegative.

(c) Let the eigenvalues of $H(n)$ be ordered as $\lambda_1(n) \le \lambda_2(n) \le \cdots \le \lambda_k(n)$. Show that, for all $x \in R^n$,

$$\lambda_1(n)x^T x \le x^T H(n)x \le \lambda_k(n)x^T x. \tag{4.3.18}$$

(d) Use the Lagrange identity (4.3.17) in formula (4.3.18) to show that

$$\left(\prod_{i=n_0}^{n-1} \lambda_1(i)\right) x^T(n_0)x(n_0) \le x^T(n)x(n)$$

$$\le \left(\prod_{i=n_0}^{n-1} \lambda_k(i)\right) x^T(n_0)x(n_0).$$

(e) Show that

$$\prod_{i=n_0}^{n-1} \lambda_1(i) \leq \|\Phi(n, n_0)\|^2 \leq \prod_{i=n_0}^{n-1} \lambda_k(i). \tag{4.3.19}$$

4.4 Phase Space Analysis

In this section we will study the stability properties of the second-order linear autonomous (time-invariant) systems

$$\begin{aligned} x_1(n+1) &= a_{11}x_1(n) + a_{12}x_2(n), \\ x_2(n+1) &= a_{21}x_1(n) + a_{22}x_2(n), \end{aligned}$$

or

$$x(n+1) = Ax(n), \tag{4.4.1}$$

where

$$A = \begin{pmatrix} a_{11} & a_{12} \\ a_{21} & a_{22} \end{pmatrix}.$$

Recall that x^* is an equilibrium point of system (4.4.1) if $Ax^* = x^*$ or $(A - I)x^* = 0$. So if $(A - I)$ is nonsingular, then $x^* = 0$ is the only equilibrium point of system (4.4.1). On the other hand, if $(A-I)$ is singular, then there is a family of equilibrium points, as illustrated in Figure 4.8. In the latter case we let $y(n) = x(n) - x^*$ in (4.4.1) to obtain the system $y(n+1) = Ay(n)$, which is identical to system (4.4.1). Thus the stability properties of any equilibrium point $x^* \neq 0$ are the same as those of the equilibrium point $x^* = 0$. Henceforth, we will assume that $x^* = 0$ is the only equilibrium point of system (4.4.1).

Let $J = P^{-1}AP$ be the Jordan form of A. Then J may have one of the following canonical forms:

$$\begin{pmatrix} \lambda_1 & 0 \\ 0 & \lambda_2 \end{pmatrix}, \qquad \begin{pmatrix} \lambda & 1 \\ 0 & \lambda \end{pmatrix}, \qquad \begin{pmatrix} \alpha & \beta \\ -\beta & \alpha \end{pmatrix}. \tag{4.4.2}$$

(a)	(b)	(c)
distinct real eigenvalues λ_1, λ_2	repeated real eigenvalue λ	complex conjugate eigenvalues $\lambda = \alpha \pm i\beta$

If we let

$$y(n) = P^{-1}x(n),$$

or

$$x(n) = Py(n), \tag{4.4.3}$$

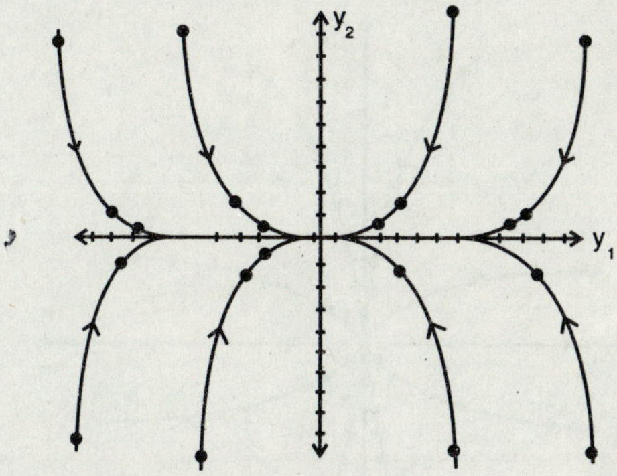

FIGURE 4.8. $\lambda_1 < \lambda_2 < 1$, asymptotically stable node.

then system (4.4.1) becomes

$$y(n+1) = Jy(n). \tag{4.4.4}$$

If $x(0) = x_0$ is an initial condition for system (4.4.1), then $y(0) = y_0 = P^{-1}x_0$ will be the corresponding initial condition for system (4.4.4). Notice that the qualitative properties of the equilibrium points of systems (4.4.1) and (4.4.4) are identical.

Our program is to sketch the phase space of system (4.4.4) in cases (a), (b), and (c). Starting with an initial value

$$y_0 = \begin{pmatrix} y_{10} \\ y_{20} \end{pmatrix}$$

in the y_1y_2-plane, we trace the movement of the points $y(1), y(2), y(3), \ldots$. Essentially, we draw the orbit $\{y(n, 0, y_0) | n \geq 0\}$. An arrow on the orbit indicates the direction of motion as time increases.

Case (a). In this case the system becomes

$$y_1(n+1) = \lambda_1 y_1(n),$$
$$y_2(n+1) = \lambda_2 y_2(n).$$

Hence

$$\begin{pmatrix} y_1(n) \\ y_2(n) \end{pmatrix} = \begin{pmatrix} \lambda_1^n y_{10} \\ \lambda_2^n y_{20} \end{pmatrix},$$

and thus

$$\frac{y_2(n)}{y_1(n)} = \left(\frac{\lambda_2}{\lambda_1}\right)^n \left(\frac{y_{20}}{y_{10}}\right).$$

If $|\lambda_1| > |\lambda_2|$, then $\lim_{n\to\infty} y_2(n)/y_1(n) = 0$, and if $|\lambda_1| < |\lambda_2|$, then $\lim_{n\to\infty} \frac{|y_2(n)|}{|y_1(n)|} = \infty$ (see Figures 4.8, 4.9, 4.10, 4.11, 4.12).

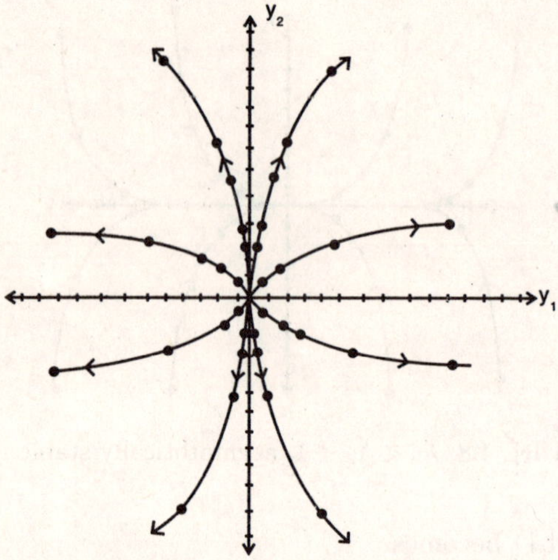

FIGURE 4.9. $\lambda_1 > \lambda_2 > 1$, unstable node.

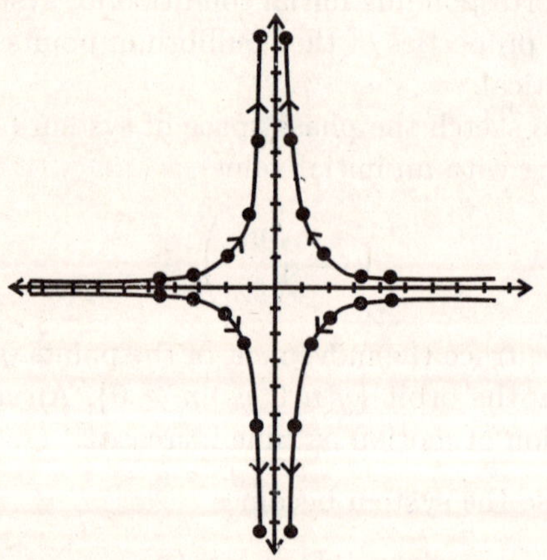

FIGURE 4.10. $0 < \lambda_1 < 1, \lambda_2 > 1$, saddle (unstable).

Case (b). In this case,

$$\begin{pmatrix} y_1(n) \\ y_2(n) \end{pmatrix} = J^n \left(\begin{pmatrix} y_{10} \\ y_{20} \end{pmatrix} \right) = \begin{pmatrix} \lambda^n & n\lambda^{n-1} \\ 0 & \lambda^n \end{pmatrix} \left(\begin{pmatrix} y_{10} \\ y_{20} \end{pmatrix} \right),$$

or

$$\begin{aligned} y_1(n) &= \lambda^n y_{10} + n\lambda^{n-1} y_{20}, \\ y_2(n) &= \lambda^n y_{20}. \end{aligned}$$

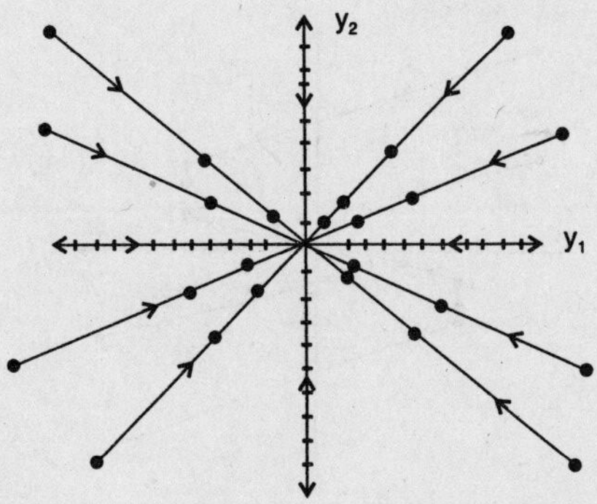

FIGURE 4.11. $0 < \lambda_1 = \lambda_2 < 1$, asymptotically stable node.

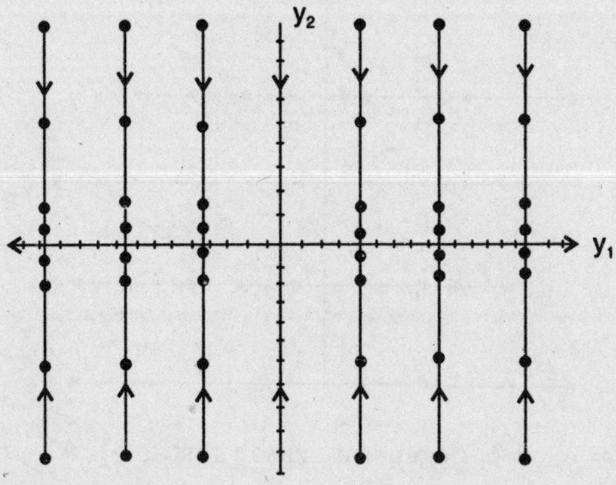

FIGURE 4.12. $\lambda_1 = 1, \lambda_2 < \lambda_1$, degenerate node.

Thus

$$\lim_{n\to\infty} \frac{y_2(n)}{y_1(n)} = 0.$$

(See Figures 4.13, 4.14.)

Case (c). In this case, the matrix A has two complex conjugate eigenvalues,

$$\lambda_1 = \alpha + i\beta \quad \text{and} \quad \lambda_2 = \alpha - i\beta, \qquad \beta \neq 0.$$

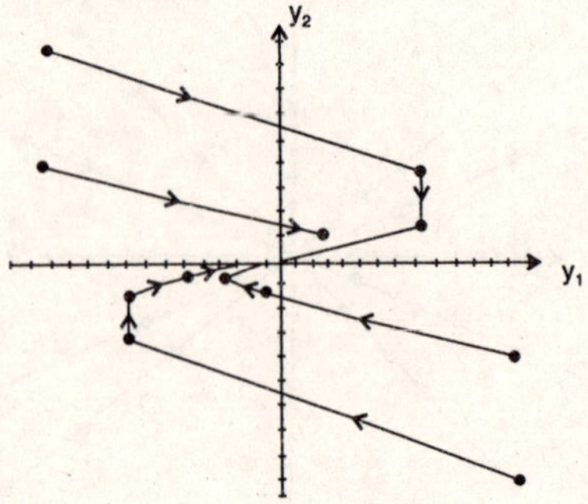

FIGURE 4.13. $\lambda_1 = \lambda_2 < 1$, asymptotically stable.

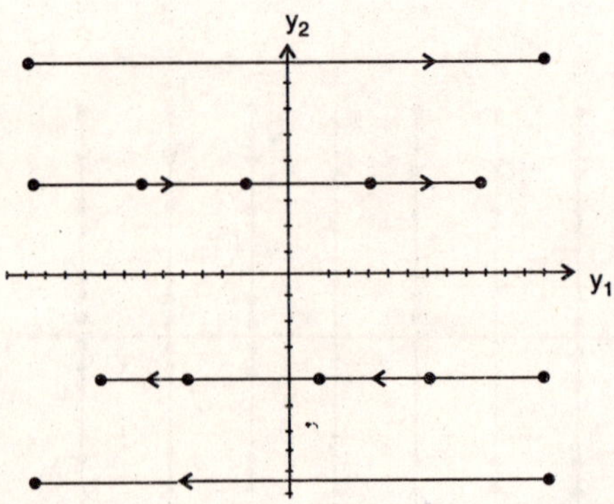

FIGURE 4.14. $\lambda_1 = \lambda_2 = 1$, degenerate case (unstable). All points on the y_1-axis are equilibrium points.

The eigenvector corresponding to $\lambda_1 = \alpha + i\beta$ is given by $\xi_1 = \begin{pmatrix} 1 \\ i \end{pmatrix}$, and the solution may be given by

$$\begin{aligned} \begin{pmatrix} 1 \\ i \end{pmatrix} (\alpha + i\beta)^n &= \begin{pmatrix} 1 \\ i \end{pmatrix} |\lambda_1|^n (\cos n\omega + i \sin n\omega), \\ &= |\lambda_1|^n \begin{pmatrix} \cos n\omega \\ -\sin n\omega \end{pmatrix} + i|\lambda_1|^n \begin{pmatrix} \sin n\omega \\ \cos n\omega \end{pmatrix}, \end{aligned}$$

where $\omega = \tan^{-1}(\beta/\alpha)$.

A general solution may then be given by

$$\begin{pmatrix} y_1(n) \\ y_2(n) \end{pmatrix} = |\lambda_1|^n \begin{pmatrix} c_1 \cos n\omega + c_2 \sin n\omega \\ -c_1 \sin n\omega + c_2 \cos n\omega \end{pmatrix}.$$

Given the initial values $y_1(0) = y_{10}$ and $y_2(0) = y_{20}$, one may obtain $c_1 = y_{10}$ and $c_2 = y_{20}$. The solution is denoted by

$$y_1(n) = |\lambda_1|^n (y_{10} \cos n\omega + y_{20} \sin n\omega),$$
$$y_2(n) = |\lambda_1|^n (-y_{10} \sin n\omega + y_{20} \cos n\omega).$$

If we let $\cos\gamma = y_{10}/r_0$ and $\sin\gamma = y_{20}/r_0$, where $r_0 = \left(y_{10}^2 + y_{20}^2\right)$, we have $y_1(n) = |\lambda_1|^n r_0 \cos(n\omega - \gamma)$ and $y_2(n) = -|\lambda_1|^n r_0 \sin(n\omega - \gamma)$. Using polar coordinates we may now write the solution as

$$r(n) = r_0 |\lambda_1|^n, \qquad \theta(n) = -(n\omega - \gamma).$$

If $|\lambda_1| < 1$, we have an asymptotically stable *focus*, as illustrated in Figure 4.15. If $|\lambda_1| > 1$, we find an unstable *focus*, as shown in Figure 4.16. When $|\lambda_1| = 1$, we obtain a center where orbits are circles with radii (Figure 4.17)

$$r_0 = \sqrt{y_{10}^2 + y_{20}^2}.$$

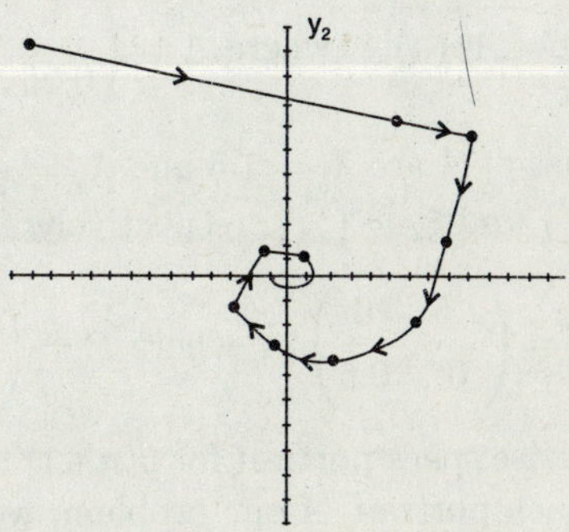

FIGURE 4.15. $|\lambda| < 1$, asymptotically stable focus.

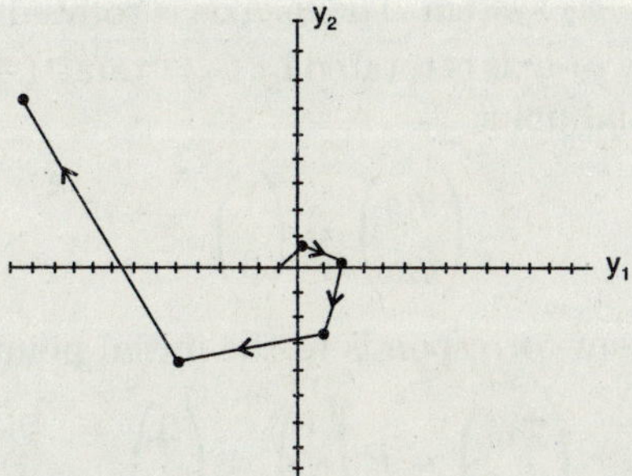

FIGURE 4.16. $|\lambda| > 1$, unstable focus.

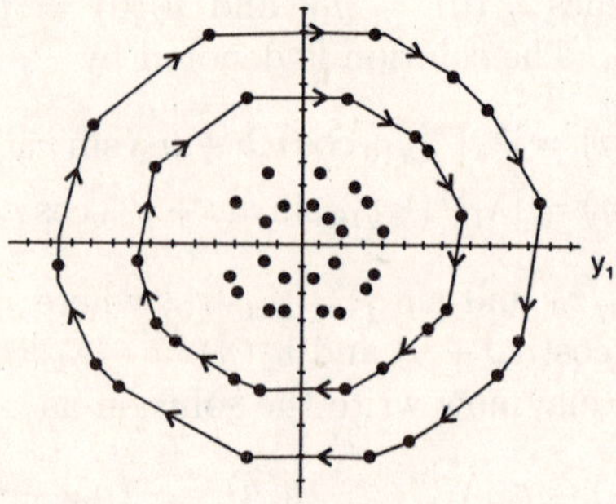

FIGURE 4.17. $|\lambda| = 1$, center (stable).

Using (4.4.3) one may sketch the corresponding phase space portraits in the x_1x_2-plane for the system of equations (4.4.1). The following example illustrates this method.

Example 4.18. Sketch the phase space portrait of the system

$$x(n+1) = Ax(n), \quad \text{where } A = \begin{pmatrix} 1 & 1 \\ 0.25 & 1 \end{pmatrix}.$$

Solution The eigenvalues of A are $\lambda_1 = 1.5$ and $\lambda_2 = \frac{1}{2}$; the corresponding eigenvectors are $\xi_1 = \binom{2}{1}$ and $\xi_2 = \binom{2}{-1}$, respectively. Thus

$$P^{-1}AP = J = \begin{pmatrix} 1.5 & 0 \\ 0 & 0.5 \end{pmatrix}, \quad \text{where } P = \begin{pmatrix} 2 & 2 \\ 1 & -1 \end{pmatrix}.$$

Figure 4.18 shows the phase space portrait for $y(n+1) = Jy(n)$. To find the corresponding phase space portrait of our problem, we let $x(n) = Py(n)$. We define the relationship between the y_1–y_2 system and the x_1–x_2 system by noticing that $\binom{1}{0}$ in the y_1–y_2 system corresponds to $P\binom{1}{0} = \binom{2}{1}$ in the x_1–x_2 system, and $\binom{0}{1}$ in the y_1-y_2 system corresponds to the point $P\binom{0}{1} = \binom{2}{-1}$ in the x_1–x_2 system. The y_1-axis is rotated by $\theta_1 = \tan^{-1}(0.5)$ to the x_1-axis, and the y_2-axis is rotated by $\theta_2 = \tan^{-1}(-0.5)$ to the x_2-axis. Furthermore, the initial point

$$\begin{pmatrix} y_{10} \\ y_{20} \end{pmatrix} = \begin{pmatrix} 1 \\ 0 \end{pmatrix}$$

for the canonical system corresponds to the initial point

$$\begin{pmatrix} x_{10} \\ x_{20} \end{pmatrix} = P\begin{pmatrix} 1 \\ 0 \end{pmatrix} = \begin{pmatrix} 2 \\ 1 \end{pmatrix}.$$

The phase space portrait of our system is shown in Figure 4.19. Basically, the axis x_1 is $c\xi_1 = \binom{2c}{c}$, $c \in \mathbb{R}$, and the axis x_2 is $c\xi_2 = \binom{2c}{-c}, c \in \mathbb{R}$.

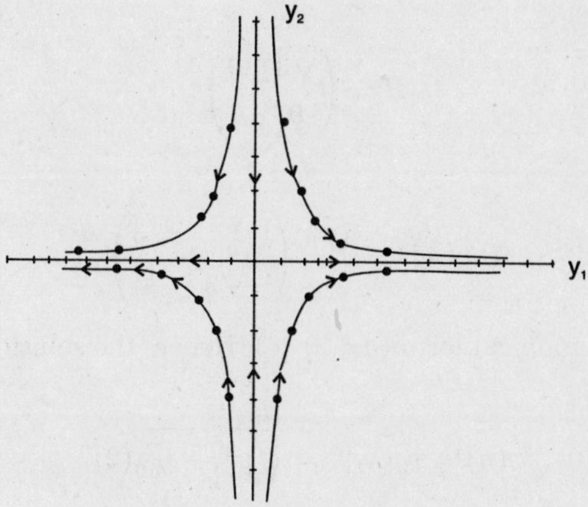

FIGURE 4.18. Canonical saddle.

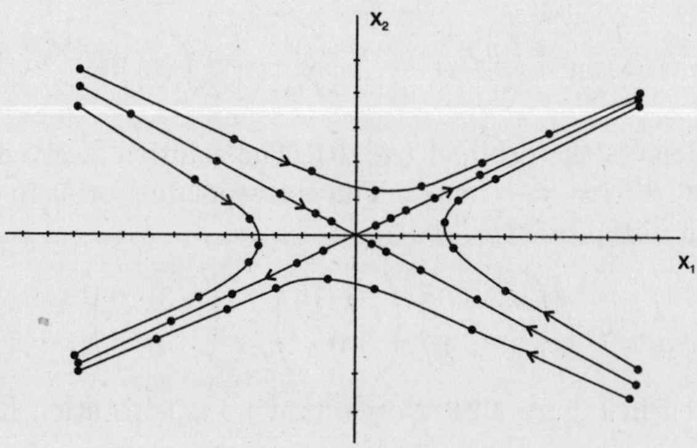

FIGURE 4.19. Actual saddle.

Example 4.19. Sketch the phase space portrait of the system $x(n+1) = Ax(n)$ with

$$A = \begin{pmatrix} 1 & 3 \\ -1 & 1 \end{pmatrix}.$$

Solution The eigenvalues of A are $\lambda_1 = 1 + \sqrt{3}\,i$ and $\lambda_2 = 1 - \sqrt{3}\,i$. The eigenvector corresponding to λ_1 is

$$\xi_1 = \begin{pmatrix} \sqrt{3} \\ i \end{pmatrix} = \begin{pmatrix} \sqrt{3} \\ 0 \end{pmatrix} + i \begin{pmatrix} 0 \\ 1 \end{pmatrix}.$$

If we let

$$P = \begin{pmatrix} \sqrt{3} & 0 \\ 0 & 1 \end{pmatrix},$$

then

$$P^{-1}AP = J = \begin{pmatrix} 1 & \sqrt{3} \\ -\sqrt{3} & 1 \end{pmatrix},$$

which is in the canonical form (4.4.2) (c). Hence, the solution of $y(n+1) = Jy(n)$ is

$$r(n) = r_0|\lambda_1|^n = \sqrt{y_{10}^2 + y_{10}^2}(2)^n$$

and

$$\theta(n) = \alpha - n\omega,$$

where

$$\alpha = \tan^{-1}\left(\frac{y_{20}}{y_{10}}\right), \qquad \omega = \tan^{-1}\left(\sqrt{3}\right) = \frac{\pi}{3}.$$

Figure 4.20 depicts the orbit of $\left(\frac{-1}{16}, 0\right)$. The solution is given by $r(n) = \frac{1}{16}(2)^n = 2^{n-4}, \theta(n) = \pi - (n\pi)/3$. The corresponding orbit in the original system has an initial point

$$x_0 = \begin{pmatrix} \sqrt{3} & 0 \\ 0 & 1 \end{pmatrix} \begin{pmatrix} -1/16 \\ 0 \end{pmatrix} = \begin{pmatrix} \sqrt{3}/16 \\ 0 \end{pmatrix}$$

and is depicted in Figure 4.21. Notice that no axis rotation has occurred here.

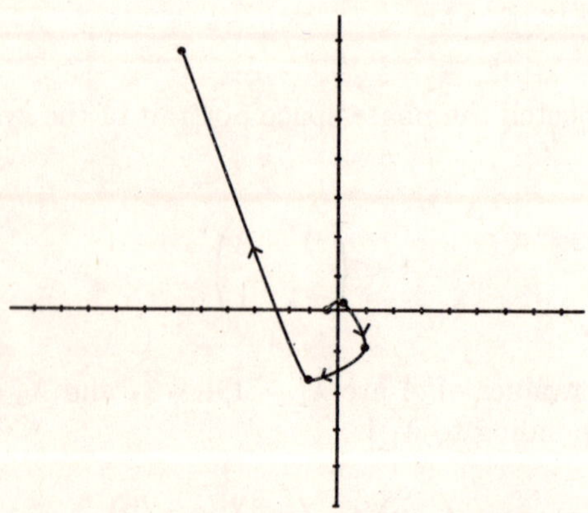

FIGURE 4.20. Canonical unstable focus.

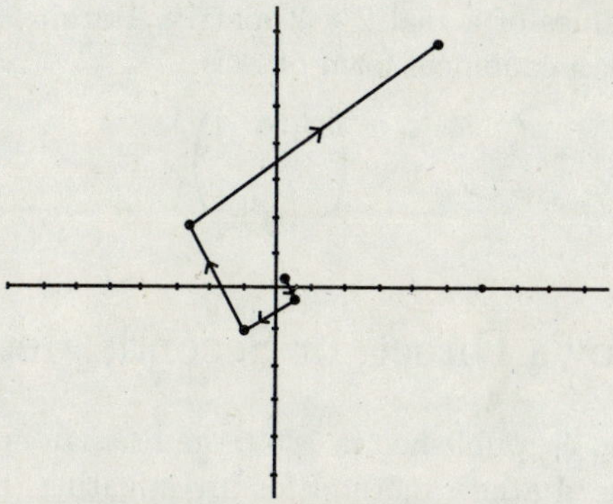

FIGURE 4.21. Actual unstable focus.

Exercises 4.4

1. Sketch the phase space diagram and determine the stability of the equation $x(n+1) = Ax(n)$, where A is given by

   (a) $\begin{pmatrix} 0.5 & 0 \\ 0 & 0.5 \end{pmatrix}$. (b) $\begin{pmatrix} 0.5 & 0 \\ 0 & 2 \end{pmatrix}$.

   (c) $\begin{pmatrix} 2 & 1 \\ 0 & 2 \end{pmatrix}$. (d) $\begin{pmatrix} -0.5 & 1 \\ 0 & -0.5 \end{pmatrix}$.

2. Sketch the phase space diagram and determine the stability of the system $x(n+1) = Ax(n)$, where A is given by

   (a) $\begin{pmatrix} 0 & 2 \\ -2 & 0 \end{pmatrix}$. (b) $\begin{pmatrix} 0.6 & -0.5 \\ 0.5 & 0.6 \end{pmatrix}$.

   (c) $\begin{pmatrix} 1 & 0.5 \\ -0.5 & 1 \end{pmatrix}$. (d) $\begin{pmatrix} 0.6 & 0.8 \\ -0.8 & 0.6 \end{pmatrix}$.

In Problems 3 through 6, sketch the phase space diagram and determine the stability of the system $x(n+1) = Ax(n)$.

3. $A = \begin{pmatrix} 1 & 1 \\ -1 & 3 \end{pmatrix}$.

4. $A = \begin{pmatrix} -2 & 1 \\ -1 & 3 \end{pmatrix}$.

5. $A = \begin{pmatrix} -2 & 1 \\ -7 & 3 \end{pmatrix}$.

6. $A = \begin{pmatrix} 1 & 2 \\ -1 & -1 \end{pmatrix}$.

7. If the eigenvalues of a real 2×2 matrix A are $\alpha + i\beta$, $\alpha - i\beta$, show that the Jordan canonical form of A is

$$\begin{pmatrix} \alpha & \beta \\ -\beta & \alpha \end{pmatrix}.$$

4.5 Liapunov's Direct, or Second, Method

In his famous memoir, published in 1892, the Russian mathematician A.M. Liapunov introduced a new method for investigating the stability of nonlinear differential equations. This method, known as Liapunov's direct method, allows one to investigate the qualitative nature of solutions without actually determining the solutions themselves. Therefore, we regard it as one of the major tools in stability theory. The method hinges upon finding certain real-valued functions, which are named after Liapunov. The major drawback in the direct method, however, lies in determining the appropriate Liapunov function for a given equation.

In this section we adapt Liapunov's direct method to difference equations. We begin our study with the autonomous difference equation

$$x(n+1) = f(x(n)), \tag{4.5.1}$$

where $f : G \to \mathbb{R}^k, G \subset \mathbb{R}^k$, is continuous. We assume that x^* is an equilibrium point of (4.5.1), that is, $f(x^*) = x^*$.

Let $V : \mathbb{R}^k \to \mathbb{R}$ be defined as a real-valued function. The variation of V relative to (4.5.1) would then be defined as

$$\Delta V(x) = V(f(x)) - V(x)$$

and

$$\Delta V(x(n)) = V(f(x(n))) - V(x(n)) = V(x(n+1)) - V(x(n)).$$

Notice that if $\Delta V(x) \leq 0$, then V is nonincreasing along solutions of (4.5.1). The function V is said to be a *Liapunov function* on a subset H of $\mathbb{R}^k$ if:

(i) V is continuous on H, and

(ii) $\Delta V(x) \leq 0$, whenever x and $f(x)$ belong to H.

Let $B(x, \gamma)$ denote the open ball in $\mathbb{R}^k$ of radius γ and center x defined by $B(x, \gamma) = \{y \in \mathbb{R}^k | \|y - x\| < \gamma\}$. For the sake of brevity, $B(0, \gamma)$ will henceforth be denoted by $B(\gamma)$. We say that the real-valued function V is *positive definite* at x^* if:

(i) $V(x^*) = 0$, and

(ii) $V(x) > 0$ for all $x \in Bx^*, \gamma)$, $x \neq x^*$, for some $\gamma > 0$.

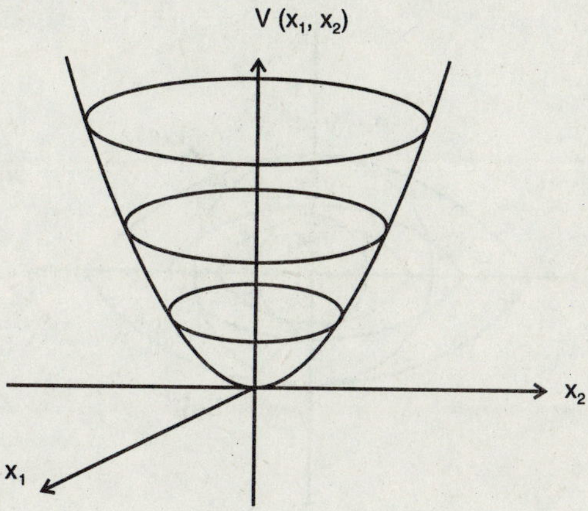

FIGURE 4.22. A quadratic Liapunov function.

We now present to the reader an informal geometric discussion of the first Liapunov stability theorem. For simplicity, we will assume that our system is planar with $x^* = 0$ as the equilibrium point. Suppose that (4.5.1) has a positive definite Liapunov function V defined on $B(\eta)$. Figure 4.22 then illustrates the graph of V in a three-dimensional coordinate system, while Figure 4.23 gives the level curves $V(x_1, x_2) = c$ of V in the plane. If we now let $\varepsilon > 0, B(\varepsilon)$ then contains one of the level curves of V, say $V(x) = \tilde{c}_2$. The level curve $V(x) = \tilde{c}_2$ contains the ball $B(\delta)$ for some δ with $0 < \delta \leq \varepsilon$. If a solution $x(n, 0, x_0)$ starts at $x_0 \in B(\delta)$, then $V(x_0) \leq \tilde{c}_2$. Since $\Delta V \leq 0$, V is a monotonic nonincreasing function along solutions of (4.5.1). Hence, $V(x(n)) \leq V(x_0) \leq \tilde{c}_2$ for all $n \geq 0$. Thus, the solution $x(n, 0, x_0)$ will stay forever in the ball $B(\varepsilon)$. Consequently, the zero solution is stable. The above argument contains the essence of the proof of the first Liapunov stability theorem.

Theorem 4.20 (Liapunov Stability Theorem). *If V is a Liapunov function for* (4.5.1) *in a neighborhood H of the equilibrium point x^*, and V is positive definite with respect to x^*, then x^* is stable. If, in addition, $\Delta V(x) < 0$ whenever $x, f(x) \in H$ and $x \neq x^*$, then x^* is asymptotically stable. Moreover, if $G = H = \mathbb{R}^k$ and*

$$V(x) \to \infty \quad \text{as} \quad \|x\| \to \infty, \tag{4.5.2}$$

then x^* is globally asymptotically stable.

PROOF. Choose $\alpha_1 > 0$ such that $B(x^*, \alpha_1) \subset G \cap H$. Since f is continuous, there is $\alpha_2 > 0$ such that if $x \in Bx^*, \alpha_2)$, then $f(x) \in B(x^*, \alpha_1)$. Let $0 < \varepsilon \leq \alpha_2$ be given. Define $\psi(\varepsilon) = \min\{V(x) | \varepsilon \leq \|x - x^*\| \leq \alpha_1\}$. By the Intermediate Value Theorem, there exists $0 < \delta < \varepsilon$ such that $V(x) < \psi(\varepsilon)$ whenever $\|x - x^*\| < \delta$.

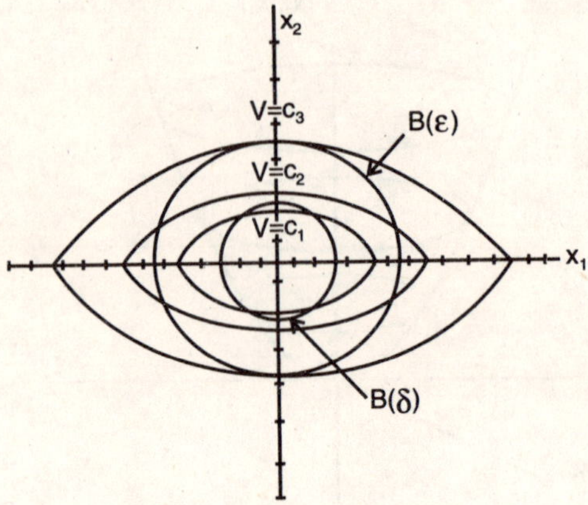

FIGURE 4.23. Level curves.

Realize now that if $x_0 \in B(x^*, \delta)$, then $x(n) \in B(x^*, \varepsilon)$ for all $n \geq 0$. This claim is true because, if not, there exist $x_0 \in B(x^*, \delta)$ and a positive integer m such that $x(r) \in B(x^*, \varepsilon)$ for $1 \leq r \leq m$ and $x(m+1) \notin B(x^*, \varepsilon)$. Since $x(m) \in B(x^*, \varepsilon) \subset B(x^*, \alpha_2)$, it follows that $x(m+1) \in B(x^*, \alpha_1)$. Consequently, $V(x(m+1)) \geq \psi(\varepsilon)$. However, $V(x(m+1)) \leq \cdots \leq V(x_0) < \psi(\varepsilon)$, and we thus have a contradiction. This establishes stability.

To prove asymptotic stability, assume that $x_0 \in B(x^*, \delta)$. Then $x(n) \in B(x^*, \varepsilon)$ holds true for all $n \geq 0$. If $\{x(n)\}$ does not converge to x^*, then it has a subsequence $\{x(n_i)\}$ that converges to $y \in R^k$. Let $E \subset B(x^*, \alpha_1)$ be an open neighborhood of y with $x^* \notin E$. Having already defined on E the function $h(x) = V(f(x))/V(x)$, we may consider h as well-defined and continuous, and $h(x) < 1$ for all $x \in E$. Now, if $\eta \in (h(y), 1)$, then there exists $\alpha > 0$ such that $x \in B(y, \alpha)$ implies $h(x) \leq \eta$. Thus, for sufficiently large n_i,

$$V\left(f(x(n_i))\right) \leq \eta V\left(x(n_i - 1)\right) \leq \eta^2 V\left(x(n_i - 2)\right) \leq \cdots \leq \eta^{n_i} V(x_0).$$

Hence,

$$\lim_{n_i \to \infty} V\left(x(n_i)\right) = 0.$$

But since $\lim_{n_i \to \infty} V\left(x(n_i)\right) = V(y)$, this statement implies that $V(y) = 0$ and, consequently, $y = x^*$.

To prove the global asymptotic stability, it suffices to show that all solutions are bounded and then repeat the above argument. Begin by assuming that there exists an unbounded solution $x(n)$, and then some subsequence $\{x(n_i)\} \to \infty$ as $n_i \to \infty$. By condition (4.5.2), this assumption implies that $V\left(x(n_i)\right) \to \infty$ as $n_i \to \infty$, which is a contradiction, since $V(x_0) > V\left(x(n_i)\right)$ for all i. This concludes the proof. □

The result on boundedness has its own independent importance, so we give it its due respect by stating it here as a separate theorem.

Theorem 4.21. *If V is a Liapunov function on the set $\{x \in \mathbb{R}^k | \|x\| > \alpha\}$ for some $\alpha > 0$, and if condition* (4.5.2) *holds, then all solutions of* (4.5.1) *are bounded.*

PROOF. (Exercises 4.5, Problem 7.) □

Example 4.22. Consider the following second-order difference equation:

$$x(n+1) = \frac{\alpha x(n-1)}{1+\beta x^2(n)}, \quad \beta > 0.$$

This equation is often called an *equation with delay.* There are three equilibrium points, namely, $x^* = 0$ and

$$x^* = \pm\sqrt{\frac{(\alpha-1)}{\beta}}$$

if $\alpha > 1$. Let us first change the equation into a system by letting $y_1(n) = x(n-1)$ and $y_2(n) = x(n)$. Then we obtain the system

$$y_1(n+1) = y_2(n),$$

$$y_2(n+1) = \frac{\alpha y_1(n)}{1+\beta y_2^2(n)}.$$

Consider the stability of the equilibrium point $(0,0)$. Our first choice of a Liapunov function will be $V(y_1, y_2) = y_1^2 + y_2^2$. This is clearly continuous and positive definite on $\mathbb{R}^2$:

$$\Delta V\left(y_1(n), y_2(n)\right) = y_1^2(n+1) + y_2^2(n+1) - y_1^2(n) - y_2^2(n).$$

Thus,

$$\Delta V\left(y_1(n), y_2(n)\right) = \left(\frac{\alpha^2}{[1+\beta y_2^2(n)]^2} - 1\right) y_1^2(n) \leq (\alpha^2 - 1) y_1^2(n). \quad (4.5.3)$$

If $\alpha^2 \leq 1$, then $\Delta V \leq 0$. In this case $x^* = 0$ would be the only equilibrium point, and by Theorem 4.20, the origin is stable (Figure 4.24). Since $\lim_{\|x\|\to\infty} V(x) = \infty$, Theorem 4.21 implies that all solutions are bounded. Since $\Delta V = 0$ for all points on the y_2-axis, Theorem 4.20 fails to determine asymptotic stability for this equation.

This situation is typical in most of the problems encountered in applications in science and engineering. Therefore, a finer and more precise analysis is required. This need leads us to *LaSalle's invariance principle*, which will be presented shortly.

To prepare for the introduction of our major theorem, we ought to familiarize ourselves with some vital terminology:

(i) For a subset $G \subset \mathbb{R}^k$, x is a *limit point* of G if there exists a sequence $\{x_i\}$ in G with $x_i \to x$ as $i \to \infty$.

(ii) The *closure* $\overline{G}$ of G is defined to be the union of G and all of its limit points.

(iii) After considering (4.5.1), the positive orbit $O^+(x_0)$ is defined as $O^+(x_0) = \{x(n, 0, x_0) | n \in \mathbb{Z}^+\}$. Since we will only deal with positive orbits, $O^+(x)$ will be denoted by $O(x)$. We will denote $O^+(x_0)$ by $O(x_0)$.

(iv) *The limit set* $\Omega(x_0)$, also referred to as the *positive limit set*, of x_0 is the set of all positive limit points of x_0. Thus, $\Omega(x_0) = \{y \in R^k | x(n_i) \to y$ as $n_i \to \infty$ for some subsequence $\{n_i\}$ of $\mathbb{Z}^+\}$.

(v) A set A is positively *invariant* if $O(x_0) \subset A$ for every $x_0 \in A$. One may easily show that both $O(x_0)$ and $\Omega(x_0)$ are (positively) invariant.

The nagging question still persists as to whether or not $\Omega(x_0)$ is nonempty for a given $x_0 \in R^k$. The next lemma satisfies that question.

Theorem 4.23. *Let $x_0 \in \mathbb{R}^k$ and let $\Omega(x_0)$ be its limit set in* (4.5.1). *Then the following statements hold true:*

(i) $\Omega(x_0) = \bigcap_{i=0}^{\infty} \overline{\bigcup_{n=i}^{\infty} \{f^n(x_0)\}} = \bigcap_{i=0}^{\infty} \overline{\bigcup_{n=i}^{\infty} \{x(n)\}}$.

(ii) *If $f^j(x_0) = y_0$, $j \in \mathbb{Z}^+$, then $\Omega(y_0) = \Omega(x_0)$* .

(iii) *$\Omega(x_0)$ is closed and invariant.*

(iv) *If the orbit $O(x_0)$ is bounded, then $\Omega(x_0)$ is nonempty and bounded.*

PROOF.

(i) Let $y \in \Omega(x_0)$. Then $f^{n_i}(x_0) \to y$ as $n_i \to \infty$. Now for each i, there exists a positive integer N_i such that $f^{n_j}(x_0) \in \bigcup_{i=0}^{\infty} \{f^n(x_0)\}$ for all $n_j \geq N_i$. Thus $y \in \overline{\bigcup_{n=i}^{\infty} \{f^n(x_0)\}}$ for every N and, consequently, $y \in \bigcap_{i=0}^{\infty} \overline{\bigcup_{n=i}^{\infty} \{f^n(x_0)\}}$. This proves one inclusion and, conversely, let $y \in \bigcap_{i=0}^{\infty} \overline{\bigcup_{n=i}^{\infty} \{f^n(x_0)\}}$. Then for each i, $y \in \overline{\bigcup_{n=i}^{\infty} \{f^n(x_0)\}}$. Thus for each i there exists $f^{n_i}(x_0) \in B_y(x_0)$, with $n_1 < n_2 < n_3 < \cdots$ and $n_i \to \infty$ as $i \to \infty$. Clearly, $f^{n_i}(x_0) \to y$ as $n_N \to \infty$ and hence $y \in \Omega(x_0)$.

(ii) The proof of (ii) is left to the reader as Problem 5.

(iii) Since the closure of a set is closed, $\overline{\bigcup_{n=i}^{\infty} \{f^n(x_0)\}}$ is closed. Now that $\Omega(x_0)$ is closed follows from the fact that the intersection of closed sets is closed.

To show that $\Omega(x_0)$ is invariant, let $y \in \Omega(x_0)$. Then $f^{n_i}(x_0) \to y$ as $n_i \to \infty$. Since f is continuous, it follows that $f^{n_i+1}(x_0) = f(f^{n_i}(x_0) \to f(y))$. Hence $f(y) \in \Omega(x_0)$ and $\Omega(x_0)$ is thus invariant.

(iv) This is left to the reader as Problem 6. □

Let V be a positive Liapunov function on a subset G of $\mathbb{R}^k$. Define

$$E = \{x \in \overline{G} | \Delta V(x) = 0\}.$$

Let M be the maximal invariant subset of E, that is, define M as the union of all invariant subsets of E.

Theorem 4.24 (LaSalle's Invariance Principle) [88]. *Suppose that V is a positive definite Liapunov function for (4.5.1) in $G \subset \mathbb{R}^k$. Then for each bounded solution $x(n)$ of (4.5.1) that remains in G for all $n \in \mathbb{Z}^+$, there exists a number c such that $x(n) \to M \cap V^{-1}(c)$ as $n \to \infty$.*

Proof. Let $x(n)$ be a bounded solution of (4.5.1) with $x(0) = x_0$ and such that $x(n)$ is bounded and remains in G. Then, by Theorem 4.23, $\phi \neq \Omega(x_0) \subset \overline{G}$. Thus, if $y \in \Omega(x_0)$, *then* $x(n_i) \to y$ as $n_i \to \infty$ for some subsequence $n_i \in \mathbb{Z}^+$. Since $V(x(n))$ is nonincreasing and bounded below, $\lim_{n\to\infty} V(x(n)) = c$ for some number c. By the continuity of V, it follows that $V(x(n_i)) \to V(y)$ as $n_i \to \infty$, and thus $V(y) = c$. This implies that $V(\Omega(x_0)) = c$ and, consequently, $\Omega(x_0) \subset V^{-1}(c)$. Moreover, $\Delta V(y) = 0$ for every $y \in \Omega(x_0)$. This implies that $\Omega(x_0) \subset E$. But, since $\Omega(x_0)$ is (positively) invariant, $\Omega(x_0) \subset M$. Therefore, $x(n) \to \Omega(x_0) \subset M \cap V^{-1}(c)$ as $n \to \infty$. □

Example 4.22 revisited. Let us reexamine Example 4.22 in light of LaSalle's invariance principle. We will consider three cases:

Case 1. $\alpha^2 = 1$. The set E consists of all the points on the x- and y-axes. We have two subcases to consider. *Subcase* (i): $\alpha = 1$. If $y_1(0) = a$ and $y_2(0) = 0$, then $y_1(1) = 0$ and $y_2(1) = a$, and $y_1(2) = a$, $y_2(2) = 0$. Therefore, any solution starting on either axis is of period 2, and $M = E$. *Subcase* (ii): $\alpha = -1$. Then $0^+(a,0) = \{(a,0),(0,-a),(-a,0),(0,a)\}$. Thus any solution starting on either axis is of period 4, and $M = E$ again. Hence all solutions converge to $(a,0)$, $(-a,0)$, $(0,a)$, or $(0,-a)$. Clearly, the zero solution is not asymptotically stable.

Case 2. Because $\alpha^2 < 1$, E is equal to the y-axis and $M = \{(0,0)\}$. Thus, all solutions converge to the origin. Hence the origin is globally asymptotically stable. Figure 4.24 depicts the phase portrait for $\alpha = 0.5$. Notice the difference in the way solutions in quadrants I and III begin, compared to the way the solutions in quadrants II and IV commence.

Case 3. $\alpha^2 > 1$. In this case, LaSalle's invariance principle does not aid us in determining the stability of the solution. In other words, the stability is indeterminable.

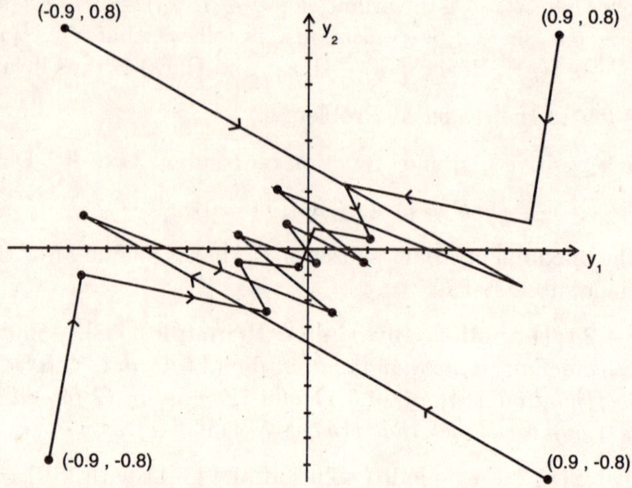

FIGURE 4.24. A globally asymptotically stable equilibrium.

Sometimes, we may simplify the difference equation by applying a simple basic transformation to the system. For instance, one might translate the system into polar coordinates (r, θ), where $x_1 = r\cos\theta$, $x_2 = r\sin\theta$. The following example demonstrates the effectiveness of this method.

Example 4.25. Consider the difference system

$$x_1(n+1) = x_1^2(n) - x_2^2(n),$$
$$x_2(n+1) = 2x_1(n)x_2(n).$$

Let $x_1(n) = r(n)\cos\theta(n)$ and $x_2(n) = r(n)\sin\theta(n)$.

Then

$$r(n+1)\cos\theta(n+1) = r^2(n)\cos^2\theta(n) - r^2(n)\sin^2\theta(n)$$
$$= r^2(n)\cos 2\theta(n), \tag{4.5.4}$$

and

$$r(n+1)\sin\theta(n+1) = 2r^2\sin\theta(n)\cos\theta(n)$$
$$= r^2(n)\sin 2\theta(n). \tag{4.5.5}$$

Dividing (4.5.4) by (4.5.5), we get

$$\theta(n+1) = 2\theta(n).$$

Substituting this into (4.5.4), we obtain

$$r(n+1) = r^2(n).$$

We may write this solution as $r(n) = [r(0)]^{2^n}$ and $\theta(n) = 2^n\theta(0)$.

The equilibrium points are $(0, 0)$ and $(1, 0)$.

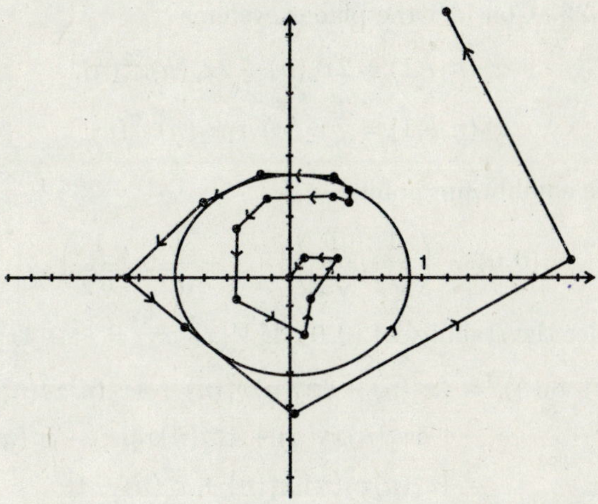

FIGURE 4.25. Unstable limit cycle. Three initial values (0.6, 0.8), (0.6, 0.81), (0.6, 0.79).

For $r(0) < 1, \lim_{n\to\infty} r(n) = 0$. Thus solutions starting inside the unit disk spiral toward the origin. Consequently, the origin is asymptotically stable (not globally), as shown in Figure 4.25.

For $r(0) > 1$, we have $\lim_{n\to\infty} r(n) = \infty$, and hence solutions that start outside the unit disk spiral away from the unit circle to ∞. This occurrence makes the equilibrium point $(1, 0)$ unstable.

For $r(0) = 1, r(n) = 1$, for all $n \geq 0$. Therefore, the circle is an invariant set, with very complicated dynamics. For instance, the solution starting at $\left(1, \frac{\pi}{4}\right)$ will reach the equilibrium point $(1, 0)$ in three iterations: $\left(1, \frac{\pi}{4}\right), \left(1, \frac{\pi}{2}\right) (1, \pi), (1, 0)$. However, the solution that starts at $\left(1, \frac{2\pi}{3}\right)$ is a 2-cycle. In general, $(1, \theta)$ is periodic, with period m, if and only if $2^m\theta = \theta + 2k\pi$ for some integer k, i.e., if and only if $\theta = (2k\pi)/2^m - 1, k = 0, 1, 2, \ldots, 2^m$. For $m = 3, \theta = \frac{2\pi}{7}, \frac{4\pi}{7}, \frac{6\pi}{7}, \frac{8\pi}{7}, \frac{10\pi}{7}, \frac{12\pi}{7}$. For $m = 4, \theta = \frac{2\pi}{15}, \frac{4\pi}{15}, \frac{2\pi}{5}, \frac{8\pi}{15}, \frac{2\pi}{3}, \frac{4\pi}{5}, \frac{14\pi}{15}, \frac{16\pi}{15}, \ldots$.

Notice here that θ is essentially the $(2^m - 1)$th root of 1. Hence, the set of periodic points $(1, \theta)$ densely fills the unit circle (Exercises 4.5, Problem 8). Furthermore, for every $m = 1, 2, \ldots$, there is a periodic point on the unit circle of that period m.

Now, if $\theta = \alpha\pi$, α irrational, then obviously, $\theta \neq (2k\pi)/2^m - 1$ for any m, and thus any solution starting at $(1, \alpha\pi)$ cannot be periodic. However, its orbit is dense within the unit circle, that is, $\overline{O(x)}$ is the unit circle (Exercises 4.5, Problem 8).

Sometimes, some simple intuitive observations make it much easier to show that an equilibrium point is not asymptotically stable. The following example illustrates this remark.

Example 4.26. Consider the planar systems

$$x_1(n+1) = 2x_2(n) - 2x_2(n)x_1^2(n),$$
$$x_2(n+1) = \frac{1}{2}x_1(n) + x_1(n)x_2^2(n).$$

We find three equilibrium points:

$$(0,0), \quad \left(\frac{1}{\sqrt{2}}, \frac{1}{\sqrt{2}}\right), \quad \left(-\frac{1}{\sqrt{2}}, -\frac{1}{\sqrt{2}}\right).$$

Let us consider the stability of $(0,0)$. If $V(x_1, x_2) = x_1^2 + 4x_2^2$, then

$$\begin{aligned}\Delta V\left(x_1(n), x_2(n)\right) &= 4x_2^2(n) - 8x_2^2(n)x_1^2(n) + 4x_2^2(n)x_1^4(n) + x_1^2(n)\\ &\quad + 4x_1^2(n)x_2^2(n) + 4x_1^2(n)x_2^4(n) - x_1^2(n) - 4x_2^2(n)\\ &= 4x_1^2(n)x_2^2(n)[x_1^2(n) + x_2^2(n) - 1].\end{aligned}$$

If $x_1^2 + x_2^2 \leq 1$, then $\Delta V(x_1, x_2) \leq 0$.

For any real number a, the solution with an initial value of $x_0 = \binom{a}{0}$ is periodic with period 2 and with orbit $\left\{\binom{a}{0}, \binom{0}{a/2}\right\}$, and a solution with an initial value of $x_0 = \binom{0}{a}$ is also periodic with period 2. Hence, the zero solution cannot be asymptotically stable. However, it is stable according to Theorem 4.20. (Figure 4.26 depicts the phase space portrait near the origin.)

We now turn our attention to the question of instability. We are interested in finding conditions on Liapunov functions under which the zero solution is unstable. Here is a widely used theorem in this area.

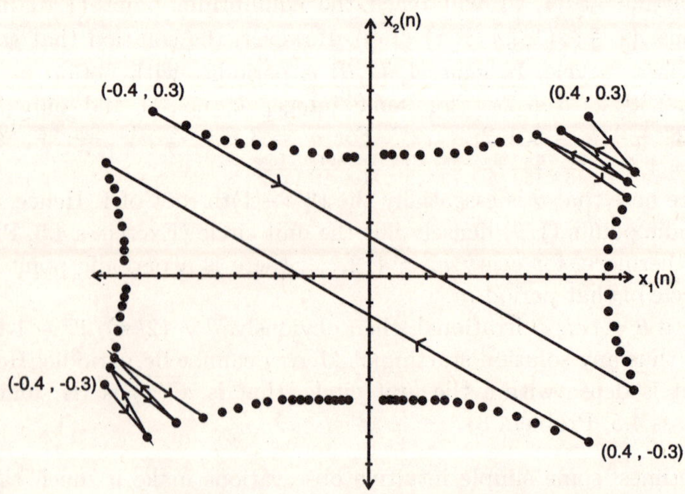

FIGURE 4.26. Stable equilibrium.

Theorem 4.27. *If ΔV is positive definite in a neighborhood of the origin and there exists a sequence $a_i \to 0$ with $V(a_i) > 0$, then the zero solution of (4.5.1) is unstable.*

PROOF. Let $\Delta V(x) > 0$ for $x \in B(\eta), x \neq 0, V(0) = 0$. We will prove Theorem 4.27 by contradiction, first assuming that the zero solution is stable, in which case, for $\varepsilon < \eta$, there would exist $\delta < \varepsilon$ such that $\|x_0\| < \delta$ implies $\|x(n, 0, x_0)\| < \varepsilon, n \in \mathbb{Z}^+$.

Since $a_i \to 0$, pick $x_0 = a_j$ for some j with $\Delta V(x_0) > 0$ and $\|x_0\| < \delta$. Hence $\overline{0(x_0)} \subset \overline{B(\varepsilon)} \subset B(\eta)$ is closed and bounded (compact). Since its domain is compact, $V(x(n))$ is also compact, and therefore bounded above. Since $V(x(n))$ is also increasing, it follows that $V(x(n)) \to c$. Following the proof of LaSalle's invariance principle, it is easy to see that $\lim_{n\to\infty} x(n) = 0$. Therefore, we would be led to believe that $0 < V(x_0) < \lim_{n\to\infty} V(x(n)) = 0$. This statement is infeasible—so the zero solution cannot be stable, as we first assumed. The zero solution of (4.5.1) is thus unstable.

The conclusion of the theorem also holds if ΔV is negative definite and $V(a_i) < 0$. □

Example 4.28. Consider the systems

$$x_1(n+1) = 4x_2(n) - 2x_2(n)x_1^2(n),$$
$$x_2(n+1) = \frac{1}{2}x_1(n) + x_1(n)x_2^2(n).$$

Let $V(x_1, x_2) = x_1^2 + 16x_2^2$. Then

$$\Delta V(x_1(n), x_2(n)) = 3x_1^2(n) + 16x_1^2(n)x_2^4(n) + 4x_1^4x_2^2 > 0 \quad \text{if } x_1(n) \neq 0.$$

Hence, by Theorem 4.27 the zero solution is unstable.

Example 4.29. First, contemplate the systems

$$x_1(n+1) = x_1(n) + x_2^2(n) + x_1^2(n), \tag{4.5.6}$$
$$x_2(n+1) = x_2(n). \tag{4.5.7}$$

Notice that $(0,0)$ is an equilibrium of the systems. Its linear component is denoted by $x(n+1) = Ax(n)$, where

$$A = \begin{pmatrix} 1 & 0 \\ 0 & 1 \end{pmatrix},$$

and thus $\rho(A) = 1$. Let $V(x) = x_1 + x_2$ be a Liapunov function. Then

$$\Delta V[x(n)] = x_1^2 + x_2^2 > 0, \quad \text{if } (x_1, x_2) \neq (0,0).$$

Theorem 4.27 implies that the zero solution of system (4.5.6) is unstable.

Let us now ponder system (4.5.7), with the same linear component as system (4.5.6):

$$x_1(n+1) = x_1(n) - x_1^3(n)x_2^2(n),$$
$$x_2(n+1) = x_2(n). \tag{4.5.8}$$

We let $V(x) = x_1^2 + x_2^2$ be a Liapunov function for system (4.5.7). Then

$$\Delta V[x(n)] = x_1^4(n)x_2^2(n)\left[-2 + x_1^2(n)x_2^2(n)\right].$$

Hence, $\Delta V(x) \leq 0$ if $x_1^2 x_2^2 \leq 2$. It follows from Theorem 4.27 that the zero solution of system (4.5.7) is stable.

We conclude from this discussion that if $\rho(A) = 1$, then the zero solution of the nonlinear equation may be either stable or unstable, thus proving part (i) of Theorem 4.38.

We conclude this section with a brief discussion of Liapunov functions for linear autonomous systems. In Section 4.3, we noticed that the condition for asymptotic stability of the difference equation (4.3.6) is that $\rho(A) < 1$. This condition requires the computation of the eigenvalues of A. Using the second method of Liapunov, such computation is unnecessary. Before introducing Liapunov's method, however, we need to recall the definition of a positive definite matrix. Consider the quadratic form $V(x)$ for a $k \times k$ real symmetric matrix $B = (b_{ij})$:

$$V(x) = x^T Bx = \sum_{i=1}^{k}\sum_{j=1}^{k} b_{ij}x_i x_j.$$

A matrix B is said to be positive definite if $V(x)$ is positive definite. Sylvester's criterion is the simplest test for positive definiteness of a matrix. It merely notes that a real symmetric matrix B is positive definite if and only if the determinants of its leading principal minors are positive, i.e., if and only if

$$b_{11} > 0, \quad \begin{vmatrix} b_{11} & b_{12} \\ b_{12} & b_{22} \end{vmatrix} > 0, \quad \begin{vmatrix} b_{11} & b_{12} & b_{13} \\ b_{21} & b_{22} & b_{23} \\ b_{31} & b_{32} & b_{33} \end{vmatrix} > 0, \ldots, \det B > 0.$$

The leading principal minors of matrix B are B itself and the minors obtained by removing successively the last row and the last column. For instance, the leading principal minors of

$$B = \begin{pmatrix} 3 & 2 & 0 \\ 2 & 5 & -1 \\ 0 & -1 & 1 \end{pmatrix}$$

are

$$(3), \quad \begin{pmatrix} 3 & 2 \\ 2 & 5 \end{pmatrix}, \quad B,$$

all of which have positive determinants. Hence, B is positive definite. Notice that, for $x = (x_1, x_2, x_3)^T$,

$$V(x) = x^T Bx = 3x_1^2 + 5x_2^2 + x_3^2 + 4x_1x_2 - 2x_2x_3 > 0$$

for all $x \neq 0$, and $V(0) = 0$.

On the other hand, given

$$V(x) = ax_1^2 + bx_2^2 + cx_3^2 + dx_1x_2 + ex_1x_3 + fx_2x_3,$$

one may write

$$V(x) = x^T Bx,$$

where

$$B = \begin{pmatrix} a & d/2 & e/2 \\ d/2 & b & f/2 \\ e/2 & f/2 & c \end{pmatrix}.$$

Hence V is positive definite if and only if B is. We now make a useful observation. Note that if B is a positive definite symmetric matrix, then all eigenvalues of B are positive (Exercises 4.5, Problem 14). Furthermore, if $\lambda_1, \lambda_2, \ldots, \lambda_k$ are the eigenvalues of B with

$$\begin{aligned} \lambda_{\min} &= \min\{|\lambda_i||1 \le i \le k\}, \\ \lambda_{\max} &= \rho(A) = \max\{|\lambda_i||1 \le i \le k\}, \end{aligned}$$

then

$$\lambda_{\min}\|x\|^2 \le V(x) \le \lambda_{\max}\|x\|^2, \tag{4.5.9}$$

for all $x \in \mathbb{R}^k$, where $V(x) = x^T Bx$, and $\|.\|$ is the Euclidean norm (Exercises 4.5, Problem 15).

If B is a positive definite matrix, we let $V(x) = x^T Bx$ be a Liapunov function of (4.3.6). Then, relative to (4.3.6),

$$\begin{aligned} \Delta V(x(n)) &= x^T(n)A^T BAx(n) - x^T(n)B(n) \\ &= x^T(A^T BA - B)x. \end{aligned} \tag{4.5.10}$$

Thus $\Delta V < 0$ if and only if

$$A^T BA - B = -C \tag{4.5.11}$$

for some positive definite matrix C. Equation (4.5.11) is labeled the Liapunov equation of the system of equations (4.3.6). The above argument establishes a sufficient condition for the asymptotic stability of the zero solution of (4.3.6). It is also a necessary and vital condition, as may be seen by the following result.

Theorem 4.30. *The zero solution of* (4.3.6) *is asymptotically stable if and only if for every positive definite symmetric matrix* C, (4.5.11) *has a unique solution* B *that is also symmetric and positive definite.*

PROOF. Assume that the zero solution of (4.3.6) is asymptotically stable. Let C be a positive definite symmetric matrix. We will show that the Liapunov equation (4.5.11) has a unique solution B. Multiply (4.5.11) from the left by $(A^T)^r$ and from the right by A^r to obtain

$$(A^T)^{r+1}BA^{r+1} - (A^T)^r BA^r = -(A^T)^r CA^r.$$

Hence

$$\lim_{n\to\infty}\sum_{r=0}^{n}\left[(A^T)^{r+1}BA^{r+1} - (A^T)^r BA^r\right] = -\lim_{n\to\infty}\sum_{r=0}^{n}(A^T)^r CA^r$$

and

$$\lim_{n\to\infty}\left[B-(A^T)^{n+1}BA^{n+1}\right] = \sum_{r=0}^{\infty}(A^T)^r CA^r. \tag{4.5.12}$$

Using Theorem 4.13, part (ii), we conclude that $\rho(A) < 1$ and, consequently, $\rho(A^T) < 1$. This implies that $\lim_{n\to\infty}(A^T)^{n+1}BA^{n+1} = 0$. Thus formula (4.5.12) yields

$$B = \sum_{r=0}^{\infty}(A^T)^r CA^r. \tag{4.5.13}$$

It is straightforward to prove that formula (4.5.13) gives a solution of (4.5.11) (Exercises 4.5, Problem 16). But since there is a norm such that $\|A^T\| < 1$ and $\|A\| < 1$, it may be shown that the series in formula (4.5.13) converges (Exercises 4.5, Problem 16). It is easy to verify that B is symmetric and positive definite (Exercises 4.5, Problem 16). □

Remark: Note that from the proof preceding the statement of Theorem 4.30, the zero solution of (4.3.6) is asymptotically stable if (4.5.11) has a unique, symmetric, and positive definite matrix B for some (not all) positive definite matrices C. Indeed, one may allow C to be the identity matrix I. In this case a solution of (4.5.11) is given by

$$B = \sum_{r=0}^{\infty}(A^T)^r A^r. \tag{4.5.14}$$

Corollary 4.31. *If* $\rho(A) > 1$, *then there exists a real symmetric matrix* B *that is not positive semidefinite such that* (4.5.11) *holds for some positive definite matrix* C.

PROOF. This follows from Theorem 4.30 and is left to the reader as Problem 17 of Exercises 4.5. □

Exercises 4.5

1. Consider the planar system

$$x_1(n+1) = x_2(n)/[1 + x_1^2(n)], \qquad x_2(n+1) = x_1(n)/[1 + x_2^2(n)].$$

Find the equilibrium points and determine their stability.

2. Consider the planar system

$$x_1(n+1) = g_1(x_1(n), x_2(n)),$$
$$x_2(n+1) = g_2(x_1(n), x_2(n)),$$

with $g_1(0,0) = g_2(0,0) = 0$ and $g_1(x_1, x_2)g_2(x_1, x_2) > x_1x_2$, for every point $x = (x_1, x_2)$ in a neighborhood of the origin. Show that the origin is unstable.

*3. Consider the system

$$x_1(n+1) = ax_2(n)/[1 + x_1^2(n)], \qquad x_2(n+1) = bx_1(n)/[1 + x_2^2(n)].$$

(i) Find conditions on a and b under which:

(a) the zero solution is stable, and

(b) the zero solution is asymptotically stable.

(ii) Find the attractor when $a^2 = b^2 = 1$.

4. Prove that the zero solution of

$$x_1(n+1) = x_2(n) - x_2(n)[x_1^2(n) + x_2^2(n)],$$
$$x_2(n+1) = x_1(n) - x_1(n)[x_1^2(n) + x_2^2(n)],$$

is asymptotically stable.

5. Prove Theorem 4.23, part (ii).

6. Prove Theorem 4.23, part (iv).

7. Prove Theorem 4.21.

8. In Example 4.25:

(a) Show that the orbit starting at the point $(1, \alpha\pi)$, where α is any irrational number, is dense on the unit circle.

(b) Show that the set of periodic points $(1, \theta)$ is dense on the unit circle.

*9. Suppose that:

(i) V is a Liapunov function of system equation (4.5.1) on $\mathbb{R}^k$,

(ii) $G_\lambda = \{x|V(x) < \lambda\}$ is bounded for each λ, and

(iii) M is closed and bounded (where M is the maximal invariant set in E).

(a) Prove that M is a global attractor, i.e., $\Omega(x_0) \subset M$ for all $x_0 \in \mathbb{R}^k$.

(b) Suppose that $M = \{0\}$. Verify that the origin is globally asymptotically stable.

10. Show that the sets G_λ defined in the preceding problem are bounded if $V(x) \to \infty$ as $\|x\| \to \infty$.

*11. (Project). Suppose that $V : \mathbb{R}^k \to \mathbb{R}$ is a continuous function with $\Delta^2 V(x(n)) > 0$ for $x(n) \neq 0$, where $x(n)$ is a solution of (4.5.1). Prove that for any $x_0 \in \mathbb{R}^k$, either $x(n, x_0)$ is unbounded or $\lim_{n\to\infty} x(n, x_0) = 0$.

*12. (Project). Wade through Problem 11 again, after replacing the condition $\Delta^2 V(x(n)) > 0$ by $\Delta^2 V(x(n)) < 0$.

13. Contemplate the planar system

$$\begin{aligned} x(n+1) &= y(n), \\ y(n+1) &= x(n) + f(x(n)). \end{aligned}$$

If $\Delta[y(n) f(x(n))] > 0$ for all $n \in \mathbb{Z}^+$, prove that the solutions are either unbounded or tend to the origin.

14. Prove that if B is a positive definite symmetric matrix, then all its eigenvalues are positive.

15. Let B be a positive definite symmetric matrix with eigenvalues $\lambda_1 \leq \lambda_2 \leq \cdots \leq \lambda_k$. For $V(x) = x^T B x$, show that $\lambda_1 \|x\|_2^2 \leq V(x) \leq \lambda_2 \|x\|_2^2$ for all $x \in \mathbb{R}^k$.

16. (a) Show that the matrix $B = \sum_{r=0}^{\infty} (A^T)^r C A^r$ is symmetric and positive definite if $\|A\| < 1$ and C is a positive definite symmetric matrix.

(b) Show that the matrix B in formula (4.5.13) is a solution of (4.5.11).

17. Prove Corollary 4.31.

4.6 Stability by Linear Approximation

The linearization method is the oldest method in stability theory. Scientists and engineers frequently use this method in the design and analysis of control systems and feedback devices. The mathematicians Liapunov and Perron originated the linearization method, each with his own unique approach, in their work with the stability theory of differential equations. In this section we adapt Perron's approach to our study of the nonlinear systems of difference equations

$$y(n+1) = A(n)y(n) + g(n, y(n)) \tag{4.6.1}$$

using their linear component

$$z(n+1) = A(n)z(n), \tag{4.6.2}$$

where $A(n)$ is a $k \times k$ matrix for all $n \in \mathbb{Z}^+$ and $g : \mathbb{Z}^+ \times G \to \mathbb{R}^k, G \subset \mathbb{R}^k$, is a continuous function. One may perceive system (4.6.1) as a perturbation of system (4.6.2). The function $g(n, y(n))$ represents the perturbation due to noise, inaccuracy in measurements, or other outside disturbances. System (4.6.1) may arise from the linearization of nonlinear equations of the form

$$x(n+1) = f(n, x(n)), \tag{4.6.3}$$

where $f : \mathbb{Z}^+ \times G \to \mathbb{R}^k, G \subset \mathbb{R}^k$, is continuously differentiable at an equilibrium point y^* (i.e., $\frac{\partial f}{\partial y_i}|y^*$ exists and is continuous on an open neighborhood of y^* for $1 \leq i \leq k$). We now describe the linearization method applied to system (4.6.3). Let us write $f = (f_1, f_2, \ldots, f_k)^T$. The

$$\left.\frac{\partial f(n,y)}{\partial y}\right|_{y=0} = \frac{\partial f(n,0)}{\partial y} = \begin{pmatrix} \frac{\partial f_1(n,0)}{\partial y_1} & \frac{\partial f_1(n,0)}{\partial y_2} & \cdots & \frac{\partial f_1(n,0)}{\partial y_k} \\ \frac{\partial f_2(n,0)}{\partial y_1} & \frac{\partial f_2(n,0)}{\partial y_2} & \cdots & \frac{\partial f_2(n,0)}{\partial y_k} \\ \vdots & \vdots & & \vdots \\ \frac{\partial f_k(n,0)}{\partial y_1} & \frac{\partial f_n(n,0)}{\partial y_2} & \cdots & \frac{\partial f_k(n,0)}{\partial y_k} \end{pmatrix}.$$

For simplicity, $\frac{\partial f(n,x^*)}{\partial x}$ is denoted by $Df(n, x^*)$. Letting

$$y(n) = x(n) - x^* \tag{4.6.4}$$

in (4.6.3) yields

$$\begin{aligned} y(n+1) &= f(n, y(n) + x^*) - x^* \\ &= \frac{\partial f}{\partial x}(n, x^*)y(n) + g(n, y(n)) \end{aligned}$$

where $g(n, y(n)) = f(n, y(n) + x^*) - x^* - \frac{\partial f}{\partial x}(n, x^*)y(n)$.

If we let $A(n) = \frac{\partial f}{\partial x}(n, x^*)$, then we obtain (4.6.1). From the assumptions on f, we conclude that $g(n, y) = o(\|y\|)$ as $\|y\| \to 0$. This means, given $\varepsilon > 0$, there exists $\delta > 0$ such that $\|g(n, y)\| \leq \varepsilon\|y\|$ whenever $\|y\| < \delta$, for all $n \in \mathbb{Z}^+$.

Notice that when $x^* = 0$, we have

$$\begin{aligned} g(n, y(n)) &= f(n, y(n)) - Df(n, 0)y(n) \\ &= f(n, y(n)) - A(n)y(n). \end{aligned}$$

An important special case of system (4.6.3) is the autonomous system

$$y(n+1) = f(y(n)), \tag{4.6.5}$$

which may be written as

$$y(n+1) = Ay(n) + g(y(n)), \tag{4.6.6}$$

where $A = f'(0)$ is the Jacobian matrix of f at 0, and $g(y) = f(y) - Ay$. Since f is differentiable at 0, it follows that $g(y) = o(y)$ as $\|y\| \to 0$. Equivalently,

$$\lim_{\|y\| \to 0} \frac{\|g(y)\|}{\|y\|} = 0.$$

Remarks:

(a) Observe that whether the linearization is about a nontrivial equilibrium point $x^* \neq 0$ or a trivial equilibrium $x^* = 0$, $g(n, 0) = 0$ ($g(0) = 0$) for all $n \in \mathbb{Z}^+$. Hence the zero solution of (4.6.1) corresponds to the equilibrium point x^* that we linearize about.

(b) If one wishes to study a nontrivial equilibrium point $x^* \neq 0$, then by virtue of (a), we have two options. The first option is to linearize about x^*. The second option is to make the change of variable $y(n) = x(n) - x^*$ as in (4.6.4). In the new system, $y^* = 0$ corresponds to x^*. Then we linearize the new system about $y^* = 0$. The latter option is simple in computation as it is usually used if x^* is given explicitly. The former option is used normally if x^* is given implicitly or we have multiequilibria.

Before commencing our stability analysis we must consider a simple but important lemma. This lemma is the discrete analogue of the so-called Gronwall inequality, which is used, along with its variations, extensively in differential equations.

Lemma 4.32 (Discrete Gronwall Inequality). *Let $z(n)$ and $h(n)$ be two sequences of real numbers, $n \geq n_0 \geq 0$ and $h(n) \geq 0$. If*

$$z(n) \leq M\left[z(n_0) + \sum_{j=n_0}^{n-1} h(j)z(j)\right]$$

for some $M > 0$, then

$$z(n) \le z(n_0) \prod_{j=n_0}^{n-1} [(1 + Mh(j)], \quad n \ge n_0, \tag{4.6.7}$$

$$z(n) \le z(n_0) \exp\left[\sum_{j=n_0}^{n-1} Mh(j)\right], \quad n \ge n_0. \tag{4.6.8}$$

PROOF. Let

$$u(n) = M\left[u(n_0) + \sum_{j=n_0}^{n-1} h(j)u(j)\right], \qquad u(n_0) = z(n_0). \tag{4.6.9}$$

Since $h(j) \ge 0$ for all $j \ge n_0$, it follows that $z(n) \le u(n)$ for all $n \ge n_0$. From (4.6.9) we have $u(n+1) - u(n) = Mh(n)u(n)$, or $u(n+1) = [1 + Mh(n)]u(n)$. By formula (1.2.3) we obtain

$$u(n) = \prod_{j=n_0}^{n-1} [1 + Mh(j)]u(n_0).$$

This proves formula (4.6.7). The conclusion of formula (4.6.8) follows by noting that $1 + Mh(j) \le \exp(Mh(j))$. □

Theorem 4.33. *Assume that $g(n,y) = o(\|y\|)$ uniformly as $\|y\| \to 0$. If the zero solution of the linear system (4.6.2) is uniformly asymptotically stable, then the zero solution of the nonlinear system (4.6.1) is exponentially stable.*

PROOF. From (4.3.5) it follows that $\|\Phi(n,m)\| \le M\eta^{(n-m)}, n \ge m \ge n_0$, for some $M \ge 1$ and $\eta \in (0,1)$. By the variation of constants formula (3.2.12), the solution of (4.6.6) is given by

$$y(n, n_0, y_0) = \Phi(n, n_0)y_0 + \sum_{j=n_0}^{n-1} \Phi(n, j+1)g(j, y(j)).$$

Thus

$$\|y(n)\| \le M\eta^{(n-n_0)}\|y_0\| + M\eta^{-1} \sum_{j=n_0}^{n-1} \eta^{(n-j)}\|g(j, y(j))\|. \tag{4.6.10}$$

For a given $\varepsilon > 0$ there is $\delta > 0$ such that $\|g(j,y)\| < \varepsilon\|y\|$ whenever $\|y\| < \delta$. So as long as $\|y(j)\| < \delta$, (4.6.10) becomes

$$\eta^{-n}\|y(n)\| \le M\left[\eta^{-n_0}\|y_0\| + \sum_{j=n_0}^{n-1} \varepsilon\eta^{-j-1}\|y(j)\|\right]. \tag{4.6.11}$$

Letting $z(n) = \eta^{-n}\|y(n)\|$ and then applying the Gronwall inequality (4.6.7), one obtains

$$\eta^{-n}\|y(n)\| \leq \eta^{-n_0}\|y_0\| \prod_{j=n_0}^{n-1} [1 + \varepsilon\eta^{-1}M].$$

Thus,

$$\|y(n)\| \leq \|y_0\|(\eta + \varepsilon M)^{(n-n_0)}. \tag{4.6.12}$$

Choose $\varepsilon < (1-\eta)/M$. Then $\eta + \varepsilon M < 1$. Thus $\|y(n)\| \leq \|y_0\| < \delta$ for all $n \geq n_0 \geq 0$. Therefore, formula (4.6.11) holds and, consequently, by virtue of formula (4.6.12), we obtain exponential stability. □

Corollary 4.34. *If* $\rho(A) < 1$, *then the zero solution of* (4.6.6) *is exponentially stable.*

PROOF. Using Theorem 4.13, the corollary follows immediately from Theorem 4.33. □

Corollary 4.35. *If* $\|f'(0)\| < 1$, *then the zero solution of* (4.6.5) *is exponentially stable.*

PROOF. Since $\rho(f'(0)) \leq \|f'(0)\|$, the proof follows from Corollary 4.34. □

A Remark about Corollaries 4.34 and 4.35

It is possible that $\|A\| \geq 1$ but $\rho(A) < 1$. For example,

$$A = \begin{pmatrix} 0.5 & 1 \\ 0 & 0.5 \end{pmatrix}, \qquad \|A\|_2 = \sqrt{\rho(A^TA)} = \sqrt{0.75 + (\sqrt{2}/2)} > 1,$$

$$\|A\|_\infty = \frac{3}{2}, \qquad \|A\|_1 = \frac{3}{2}.$$

However, $\rho(A) = \frac{1}{2}$. With the above matrix A, the zero solution of the system $x(n+1) = Ax(n) + g(x(n))$ is exponentially stable, provided that $g(x) = o(x)$ as $\|x\| \to 0$. Obviously, Corollary 4.35 fails to help us in determining the stability of the system. However, even with all its shortcomings, Corollary 4.35 is surprisingly popular among scientists and engineers.

It is also worthwhile to mention that if $\rho(A) < 1$, there exists a nonsingular matrix Q such that $\|Q^{-1}AQ\| < 1$ [85]. One may define a new norm on A, $\|A\| = \|Q^{-1}AQ\|$, and then apply Corollary 4.35 in a more useful way.

Let us return to our example where

$$A = \begin{pmatrix} 0.5 & 1 \\ 0 & 0.5 \end{pmatrix}.$$

Let

$$Q = \begin{pmatrix} 1 & 0 \\ 0 & \alpha \end{pmatrix}.$$

Then

$$Q^{-1} = \begin{pmatrix} 1 & 0 \\ 0 & 1/\alpha \end{pmatrix}$$

and

$$Q^{-1}AQ = \begin{pmatrix} 0.5 & \alpha \\ 0 & 0.5 \end{pmatrix}.$$

We have $\|Q^{-1}AQ\|_1 = \alpha + 0.5$. If we choose $\alpha < 0.5$, then $\|Q^{-1}AQ\|_1 < 1$.

The above procedure may be generalized to any Jordan block

$$A = \begin{pmatrix} \lambda & 1 & \cdots & 0 \\ \vdots & \lambda & & \vdots \\ \vdots & \vdots & & 1 \\ 0 & 0 & & \lambda \end{pmatrix}.$$

In this case, we let $Q = \text{diag}(1, \alpha, \alpha^2, \ldots, \alpha^{k-1})$, where k is the order of A.

Hence,

$$\|Q^{-1}AQ\| = \begin{pmatrix} \lambda & \alpha & \cdots & 0 \\ \vdots & \lambda & & \vdots \\ \vdots & \vdots & & \alpha \\ 0 & 0 & & \lambda \end{pmatrix},$$

and $\|Q^{-1}AQ\|_1 = |\lambda| + |\alpha|$ (see Exercises 4.1, Problems 3 and 4). Consequently, if $|\lambda| < 1$, one may choose an α such that $|\lambda| + |\alpha| < 1$, so that under the matrix norm $\|A\| = \|Q^{-1}AQ\|_1$, $\|A\| < 1$. We now give two examples to illustrate the preceding conclusions.

Example 4.36. Investigate the stability of the zero solution of the planar system

$$\begin{aligned} y_1(n+1) &= ay_2(n)/[1 + y_1^2(n)], \\ y_2(n+1) &= by_1(n)/[1 + y_2^2(n)]. \end{aligned} \tag{4.6.13}$$

Solution Let $f = (f_1, f_2)^T$, where $f_1 = ay_2(n)/[1 + y_1^2(n)]$ and $f_2 = by_1(n)/[1 + y_2^2(n)]$. Then the Jacobian matrix is given by

$$\left.\frac{\partial f}{\partial y}\right|(0,0) = \begin{pmatrix} \dfrac{\partial f_1(0,0)}{\partial y_1} & \dfrac{\partial f_1(0,0)}{\partial y_2} \\ \dfrac{\partial f_2(0,0)}{\partial y_1} & \dfrac{\partial f_2(0,0)}{\partial y_2} \end{pmatrix} = \begin{pmatrix} 0 & a \\ b & 0 \end{pmatrix}.$$

Hence system (4.6.13) may be written as

$$\begin{pmatrix} y_1(n+1) \\ y_2(n+1) \end{pmatrix} = \begin{pmatrix} 0 & a \\ b & 0 \end{pmatrix} \begin{pmatrix} y_1(n) \\ y_2(n) \end{pmatrix} + \begin{pmatrix} -ay_2(n)y_1^2(n)/[1 + y_1^2(n)] \\ -by_2^2(n)y_1(n)/[1 + y_2^2(n)] \end{pmatrix},$$

or as

$$y(n+1) = Ay(n) + g(y(n)).$$

The eigenvalues of A are $\lambda_1 = \sqrt{ab}, \lambda_2 = -\sqrt{ab}$. Hence, if $|ab| < 1$, the zero solution of the linear part $x(n+1) = Ax(n)$ is asymptotically stable. Since $g(y)$ is continuously differentiable at $(0,0), g(y)$ is $o(y)$. Corollary 4.34 then implies that the zero solution of (4.6.13) is exponentially stable.

Example 4.37. Pielou Logistic Delay Equation [119]

In Example 2.39 we investigated the Pielou logistic equation

$$x(n+1) = \frac{\alpha x(n)}{1 + \beta x(n)}.$$

If we now assume that there is a delay of time period 1 in the response of the growth rate per individual to density change, then we obtain the difference delay equation

$$y(n+1) = \frac{\alpha y(n)}{1 + \beta y(n-1)}, \qquad \alpha > 1, \quad \beta > 0. \tag{4.6.14}$$

An example of a population that can be modeled by (4.6.14) is the blowfly (*Lucilia cuprina*) (see [107]). Find conditions on α, β for which the positive equilibrium point $y^* = \frac{\alpha-1}{\beta}$ is asymptotically stable.

Solution

Method (1): Let $\overline{y}(n) = y(n) - (\alpha - 1)/\beta$. Then (4.6.14) becomes

$$\overline{y}(n+1) = \frac{\alpha\overline{y}(n) - (\alpha - 1)\overline{y}(n-1)}{\alpha + \beta\overline{y}(n-1)}. \tag{4.6.15}$$

The equilibrium point $\overline{y}^*(n) = 0$ of (4.6.15) corresponds to $y^* = (\alpha - 1)/\beta$. To change (4.6.15) to a planar system, we let

$$x_1(n) = \overline{y}(n-1) \quad \text{and} \quad x_2(n) = \overline{y}(n).$$

Then

$$\begin{pmatrix} x_1(n+1) \\ x_2(n+1) \end{pmatrix} = \begin{pmatrix} x_2(n) \\ \dfrac{\alpha x_2(n) - (\alpha-1)x_1(n)}{\alpha + \beta x_1(n)} \end{pmatrix}. \tag{4.6.16}$$

By linearizing (4.6.16) around (0, 0) we give it the new form

$$x(n+1) = Ax(n) + g(x(n)),$$

where

$$A = \begin{pmatrix} 0 & 1 \\ \dfrac{1-\alpha}{\alpha} & 1 \end{pmatrix}$$

and

$$g(x) = \begin{pmatrix} 0 \\ \dfrac{\beta(\alpha-1)x_1^2 - \alpha\beta x_1 x_2}{\alpha(\alpha+\beta x_1)} \end{pmatrix}.$$

The characteristic equation of A is $\lambda^2 - \lambda + \frac{\alpha-1}{\alpha} = 0$. Thus by condition (4.3.3) the eigenvalues of A are inside the unit disk if and only if $1 < \frac{\alpha-1}{\alpha} + 1 < 2$, or $0 < \frac{\alpha-1}{\alpha} < 1$, which is always valid, since $\alpha > 1$.

Therefore, $\rho(A) < 1$ for all $\alpha > 1$. Since $g(x)$ is continuously differentiable at $(0,0)$, the zero solution of (4.6.16) is uniformly asymptotically stable. Consequently, the equilibrium point $x^* = (\alpha - 1)/\beta$ of (4.6.14) is asymptotically stable.

Method (2): Letting $y(n) = (\alpha-1)/\beta \exp(x(n))$ in (4.6.14), we obtain the new equation

$$\exp(x(n+1)) = \frac{\exp(x(n))}{\{1 + (\alpha-1)\exp(x(n-1))\}/\alpha}.$$

Taking the logarithm of both sides, we get

$$x(n+1) - x(n) + \frac{\alpha-1}{\alpha} f[x(n-1)] = 0,$$

or

$$x(n+2) - x(n+1) + \frac{\alpha-1}{\alpha} f[x(n)] = 0, \tag{4.6.17}$$

where

$$f(x) = \frac{\alpha}{\alpha-1} \ln\left[\frac{(\alpha-1)e^x + 1}{\alpha}\right].$$

The Taylor expansion of f around 0 is given by $f(x) = x + g(x)$, where $g(x)$ is a polynomial in x that contains terms of degree higher than or equal to 2. Thus $g(x) = o(x)$. The linearized equation of (4.6.17) is denoted by

$$x(n+2) - x(n+1) + \frac{\alpha-1}{\alpha} x(n) = 0. \tag{4.6.18}$$

Since the characteristic roots of (4.6.18) are the same as the eigenvalues of A, it follows that the zero solution (4.6.18) is asymptotically stable. Corollary 4.34 then implies that the zero solution of (4.6.17) is asymptotically stable. Since the equilibrium point $y^* = (\alpha - 1)/\beta$ corresponds to the zero solution of (4.6.17), it then follows that $y^* = (\alpha - 1)/\beta$ is an asymptotically stable equilibrium point of (4.6.15).

Our final result deals with the cases $\rho(A) = 1$ and $\rho(A) > 1$.

Theorem 4.38. *The following statements hold:*

(i) *If $\rho(A) = 1$, then the zero solution of* (4.6.6) *may be stable or unstable.*

(ii) *If $\rho(A) > 1$ and $g(x)$ is $o(x)$ as $\|x\| \to 0$, then the zero solution of* (4.6.6) *is unstable.*

PROOF.

(i) See Example 4.29.

(ii) Assume that $\rho(A) > 1$. Then by Corollary 4.31, there exists a real symmetric matrix B that is not positive semidefinite for which $B^T AB - B = -C$ is negative definite. Thus the Liapunov function $V(x) = x^T Bx$ is negative at points arbitrarily close to the origin. Furthermore, $\Delta V(x) = -x^T Cx + 2x^T A^T B\, g(x) + V(g(x))$. Now, (4.5.6) allows us to pick $\gamma > 0$ such that $x^T Cx \geq 4\gamma\|x\|^2$ for all $x \in \mathbb{R}^k$. There exists $\delta > 0$ such that if $\|x\| < \delta$, then $\|Bg(x)\| \leq \gamma\|x\|$ and $V(g(x)) \leq \gamma\|x\|$. Hence $\Delta V(x(n)) \leq -\gamma\|x(n)\|^2$. Hence by Theorem 4.27,. the zero solution is unstable. □

Example 4.39. Let $S(n)$ and $I(n)$ denote the number of susceptibles and infectives, respectively, of a population at time n. Let $d > 0$ be the per capita natural death rate of the population and $\alpha \geq 0$ be the disease related death rate. In the following model, suggested by Elaydi and Jang [47], a simple mass action βSI is used to model disease transmission, where $\beta > 0$ and a fraction $\gamma \geq 0$ of these infectives recover. Hence we have the following system

$$
\begin{aligned}
S(n+1) &= \frac{S(n) + A + \gamma I(n)}{1 + \beta hI(n) + dh}, \\
I(n+1) &= \frac{I(n) + \beta S(n)I(n)}{1 + (d + \gamma + \alpha)}, \qquad (4.6.19) \\
& S(0),\ I(0) \geq 0.
\end{aligned}
$$

We make the assumption that

$$\omega = \beta A - d(d + \gamma + \alpha) > 0 \qquad (4.6.20)$$

under assumption (4.6.20) equation (4.6.19) has two equilibria

$$X_1^* = \left(\frac{A}{d}, 0\right) \quad \text{and} \quad X_2^* = \left(\frac{d+\gamma+\alpha}{\beta}, \frac{\beta A - d(d+\gamma+\alpha)}{(d+\alpha)\beta}\right).$$

The linearization of (4.6.19) about $X_2^* = (S^*, J^*)$ yields the Jacobian matrix

$$J = \begin{pmatrix} \dfrac{1}{1+\beta I^* + d} & \dfrac{\gamma + d\gamma - S^*\beta - AB}{(1+\beta I^* + d)^2} \\ \dfrac{\beta I^*}{1+d+\gamma+\alpha} & 1 \end{pmatrix}.$$

Notice that

$$\det J = \frac{1}{1+\beta I^* + d} - \frac{\beta I^*(\gamma + d\gamma - S^*\beta - A\beta)}{(1+d+\gamma+\alpha)(1+\beta I^* + d)^2} > 0$$

and

$$\operatorname{tr} J = 1 + \frac{1}{1+\beta I^* + d} > 0.$$

One may show that

$$\operatorname{tr} J < 1 + \det J < 2.$$

Hence by Theorem 4.33 and equation (4.3.9), the equilibrium point X_2^* is asymptotically stable.

We now turn our attention to the equilibrium point $X_1^* = \left(\frac{A}{d}, 0\right)$. The linearization of (4.6.19) about X_1^* yields the Jacobian matrix

$$J = \begin{pmatrix} \dfrac{1}{1+d} & \dfrac{\gamma + d\gamma - \frac{A}{d}\beta - AB}{(1+d)^2} \\ 0 & \dfrac{1+\beta\frac{A}{d}}{1+d+\gamma+\alpha} \end{pmatrix}.$$

The eigenvalues of J are given by

$$\lambda_1 = \frac{1}{1+d} \quad \text{and} \quad \lambda_2 = \frac{1+\beta\frac{A}{d}}{1+d+\gamma+\alpha}.$$

By virtue of assumption (4.6.20), $\lambda_2 > 1$ and hence by Theorem 4.38(ii), the equilibrium point X_1^* is unstable.

Exercises 4.6

1. Determine the stability of the zero solution of the equation

$$x(n+2) - \frac{1}{2}x(n+1) + 2x(n+1)x(n) + \frac{13}{16}x(n) = 0.$$

2. Judge the stability of the zero solution of the equation

$$x(n+3) - x(n+1) + 2x^2(n) + 3x(n) = 0.$$

3. Consider Example 4.25. Determine the stability and asymptotic stability for the equilibrium points $\left(\frac{1}{\sqrt{2}}, \frac{1}{\sqrt{2}}\right), \left(-\frac{1}{\sqrt{2}}, -\frac{1}{\sqrt{2}}\right)$.

4. (a) Hunt down the equilibrium points of the system:
$$x_1(n+1) = x_1(n) - x_2(n)(1 - x_2(n)),$$
$$x_2(n+1) = x_1(n),$$
$$x_3(n+1) = \frac{1}{2}x_3(n).$$

   (b) Determine the stability of all the equilibrium points in part (a).

5. Investigate the stability of the zero solution of the system:
$$x_1(n+1) = \frac{1}{2}x_1(n) - x_2^2(n) + x_3(n),$$
$$x_2(n+1) = x_1(n) - x_2(n) + x_3(n),$$
$$x_3(n+1) = x_1(n) - x_2(n) + \frac{1}{2}x_3(n).$$

6. Linearize the equation
$$x_1(n+1) = \sin(x_2) - 0.5x_1(n),$$
$$x_2(n+1) = x_2/(0.6 + x_1(n)),$$
around the origin and then determine whether the zero solution is stable.

7. (a) Find the equilibrium points of the system:
$$x_1(n+1) = \cos x_1(n) - x_2(n),$$
$$x_2(n+1) = -x_1(n).$$

   (b) Is the point $(\pi/2, -\pi/2)$ asymptotically stable?

8. Determine conditions for the asymptotic stability of the zero solution of the system
$$x_1(n+1) = ax_1(n)/[1 + x_2(n)],$$
$$x_2(n+1) = [bx_2(n) - x_1(n)][1 + x_1(n)].$$

9. The following model of combat was proposed by Epstein [52], Sedaghat [133].
$$u(n+1) = u(n) + \frac{1}{a}(a - u(n))[a - u(n)(1 - v(n))],$$
$$v(n+1) = v(n) + \frac{1 - v(n)}{1 - d}[u(n)(1 - v(n)) - d)],$$
where $d < a$, $a > 0$.

Investigate the stability of the positive equilibrium point.

10. The following system represents a discrete epidemic model. The population is divided into three groups: susceptibles $S(n)$, infectives $I(n)$, and immune or removed individuals $R(n)$, $n \in \mathbb{Z}^+$. If we assume the total population size equals N for all time, then $S(n)+I(n)+R(n)=N$ and we then can eliminate one of the variables, say $R(n)$. The model is given by

$$S(n+1) = S(n) - \frac{\alpha}{N} I(n)S(n) + \beta(N - S(n)),$$
$$I(n+1) = I(n)(1-\gamma-\beta) + \frac{\alpha}{N} I(n)S(n),$$

with $0 < \beta + \gamma < 1$ and $0 < \alpha < 1$. This model is called an SIR epidemic model.

(a) Find all the equilibrium points.

(b) Determine the stability of the equilibrium points.

11. Consider system (4.6.19) under the assumption that

$$\sigma = \beta A - d(d+\gamma+\alpha) < 0.$$

(i) Show that there is only one equilibrium point $X_1^* = \left(\frac{A}{d}, 0\right)$.

(ii) Show that X_1^* is asymptotically stable.

(iii)*Show that X_1^* is globally asymptotically stable.

12. Show that if the zero solution of (4.6.2) is uniformly stable (uniformly asymptotically stable), then the zero solution of (4.6.1) is also uniformly stable (uniformly asymptotically stable), provided that

$$\|g(n, y(n))\| \le a_n \|y(n)\|, \quad \text{where } a_n > 0 \text{ and } \sum_{n=0}^{\infty} a_n < \infty.$$

13. Suppose that the zero solution of $x(n+1) = Ax(n)$ is asymptotically stable. Prove that the zero solution of $y(n+1) = [A + B(n)]y(n)$ is asymptotically stable if $\sum_{n=0}^{\infty} \|B(n)\| < \infty$.

4.7 Applications

4.7.1 One Species with Two Age Classes

Consider a single-species, two-age-class system, with $X(n)$ being the number of young and $Y(n)$ that of adults, in the nth time interval:

$$X(n+1) = bY(n),$$
$$Y(n+1) = cX(n) + s\,Y(n) - D\,Y^2(n). \tag{4.7.1}$$

A proportion c of the young become adult, and the rest will die before reaching adulthood. The adults have a fecundity rate b and a density-dependent survival rate $sY(n) - DY^2(n)$. Equation (4.7.1) may be written in a more convenient form by letting $\tilde{X}(n) = DX(n)/b$ and $\tilde{Y}(n) = DX(n)$. Hence we have

$$\begin{aligned} \tilde{X}(n+1) &= \tilde{Y}(n+1), \\ \tilde{Y}(n+1) &= a\tilde{X}(n) + s\,Y(n) - Y^2(n), \end{aligned} \tag{4.7.2}$$

with $a = cb > 0$.

The nontrivial fixed point is $(\tilde{X}^*, \tilde{Y}^*)$, with $\tilde{X}^* = \tilde{Y}^*$ and $\tilde{Y}^* = a+s-1$. Note that the equilibria $\tilde{X}^*$ and $\tilde{Y}^*$ must be positive in order for the model to make sense biologically. This implies that $a+s-1 > 0$. Since it is easier to do stability analysis on the zero equilibrium point, we let $x(n) = \tilde{X}(n) - \tilde{X}^*$ and $y(n) = \tilde{Y}(n) - \tilde{Y}^*$. This yields the system

$$\begin{aligned} x(n+1) &= y(n), \\ y(n+1) &= ax(n) + ry(n) - y^2(n), r = 2 - 2a - s. \end{aligned} \tag{4.7.3}$$

The fixed point $(0,0)$ corresponds to the fixed point $(\tilde{X}^*, \tilde{Y}^*)$. Local stability can now be obtained by examining the linearized system

$$\begin{aligned} x(n+1) &= y(n), \\ y(n+1) &= ax(n) + ry(n), \end{aligned}$$

whose eigenvalues are the roots of the characteristic equation

$$\lambda^2 - r\lambda - a = 0.$$

By criteria (4.3.9), the trivial solution is asymptotically stable if and only if:

(i) $1 - r - a > 0$ or $a + s > 1$, and

(ii) $1 + r - a > 0$ or $3a + s < 3$.

Hence the range of values of a and s for which the trivial solution is asymptotically stable is bounded by the region $a = 1$, $s = 1$, $a + s = 1$, and $3a + s = 3$, as shown in Figure 4.27.

The shaded region represents the range of parameters a, s for which the trivial solution is asymptotically stable.

To find the region of stability (or the basin of attraction) of the trivial solution we resort to the methods of Liapunov functions. Let

$$V(x,y) = a^2x^2 + \frac{2arxy}{1-a} + y^2.$$

Recall from calculus that $Ax^2 + 2Bxy + Cy^2 = D$ is an ellipse if $AC - B^2 > 0$, or $a^2 - \frac{a^2r^2}{(1-a)^2} > 0$, or $a - 1 < r < 1 - a$. This reduces to $s + a > 1$ and $s < 3 - 3a$, which is the shaded region in Figure 4.27. By rotating the

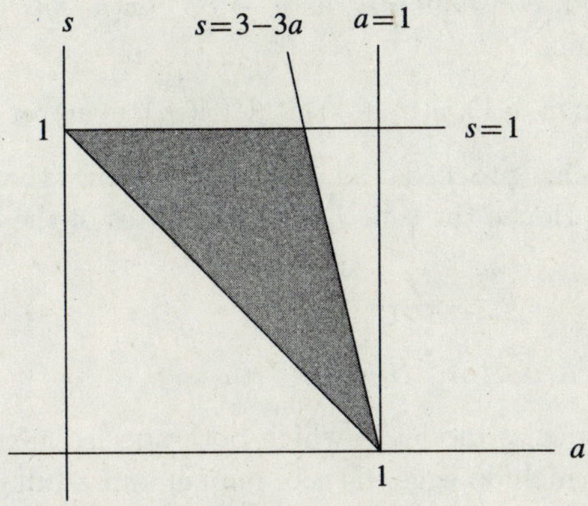

FIGURE 4.27.

axes, one may eliminate the mixed term xy to obtain $A'x^2 + C'y^2 = D$, with $A' + C' = a^2 + 1 > 0$. Moreover, $A'C' > 0$. Hence both A' and C' are positive and, consequently, D is positive. Thus in the shaded region in Figure 4.27, $V(x, y)$ is positive definite.

After some computation we obtain

$$\Delta V(x, y) = y^2 W(x, y),$$

where

$$W(x, y) = (y - r)^2 - 2ax - \frac{2ar(r - y)}{1 - a} + a^2 - 1.$$

Hence $\Delta V(x, y) \leq 0$ if $W(x, y) < 0$, that is, if (x, y) is in the region

$$G = \left\{ (x, y) : (y - r)^2 - 2ax - \frac{2ar(r - y)}{1 - a} + a^2 - 1 < 0 \right\}.$$

The region G is bounded by the parabola $W(x, y) = 0$. Now, in the region $G, \Delta V(x, y) = 0$ on the x-axis $y = 0$. Hence E is the x-axis. But since $(c, 0)$ is mapped to $(0, ac)$, the largest invariant set M in E is the origin. Hence by Theorem 4.24 every bounded solution that remains in G will converge to the origin.

We now give a crude estimate of the basin of attraction, that is, the set of all points in G that converges to the origin. Define

$$V_{\min} = \min\{V(x_0, y_0) : (x_0, y_0) \in \partial G\},$$
$$J_m = \{\tilde{X}, \tilde{Y}\} : \tilde{X} = x_0 + \tilde{X}^*, \tilde{Y} = y_0 + \tilde{Y}^*,$$
$$V(x(m), y(m)) < V_{\min}, \quad m = 0, 1, 2, \ldots.$$

Now, if $(x_0, y_0) \in J_0$, then $V(x(1), y(1)) \leq V(x_0, y_0) < V_{\min}$, and hence $(x(1), y(1)) \in J_0$. Similarly, one may show that $(x(n), y(n)) \in J_0$ for

$n = 1, 2, 3, \ldots$ and, consequently, as $n \to \infty$, $(x(n), y(n)) \to (0, 0)$. Now, if $(x_0, y_0) \in J_m$, then

$$V(x(m+1), y(m+1)) \leq V(x(m), y(m)) < V_{\min},$$

and the argument proceeds as before to show that $(x(n), y(n)) \to (0, 0)$ as $n \to \infty$. Hence the sets J_m are estimates of the basin of attraction of $(\tilde{X}, \tilde{Y})$.

4.7.2 Host–Parasitoid Systems

Consider a two-species model in which both species have a number of life-cycle stages that include eggs, larvae, pupae, and adults. Let

$H(n) =$ the density of host species in generation n,

$P(n) =$ the density of parasitoid in generation n,

$f(H(n), P(n)) =$ fraction of hosts not parasitized,

$\lambda =$ host reproductive rate,

$c =$ average number of viable eggs laid by a parasitoid on a single host.

An adult female parasitoid finds a host on which to deposit its eggs. The larval parasitoids consume and eventually kill their host. The life–cycle of both species is depicted in Figure 4.28. Then

$H(n+1) =$ number of hosts in generation $n \times$ fraction not parasitized $\times$ reproductive rate λ,

$P(n+1) =$ number of hosts parasitized in generation $n \times$ fecundity of parasitoids c.

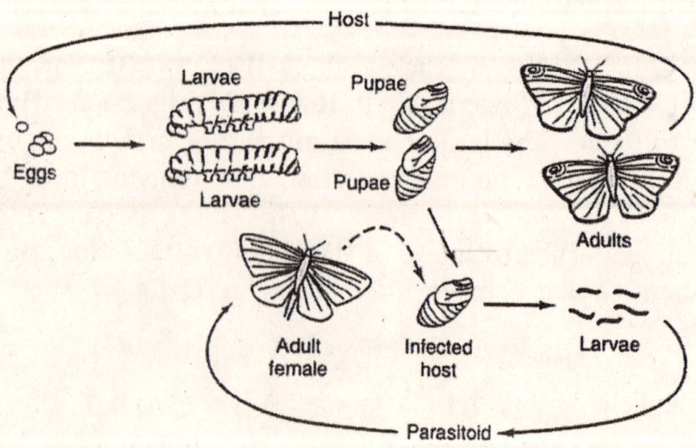

FIGURE 4.28. Host–Parasitoid. Schematic representation of host–parasite system.

Thus we have

$$H(n+1) = \lambda H(n) f(H(n), P(n)), \tag{4.7.4}$$
$$P(n+1) = cH(n)[1 - f(H(n), P(n))]. \tag{4.7.5}$$

4.7.3 A Business Cycle Model

One of the first formal mathematical models for business cycles was due to Paul Samuelson (1939). The model was later modified by Sir John Hicks (1950). Let $I(n)$ denote the investment at time period n and $Y(n)$ is the income at time period n. In the Samuelson–Hicks model, it is assumed that investment is proportional to income change, i.e.,

$$I(n) = v(Y(n-1) - Y(n-2)). \tag{4.7.6}$$

Likewise, consumption $C(n)$ is proportional to income $Y(n-1)$ in the previous period, i.e.,

$$C(n) = (1-s)Y(n) \tag{4.7.7}$$

where $0 \leq s \leq 1$ is the "complementary" proportion used. Introducing the accounting identity for a closed economy:

$$Y(n) = C(n) + I(n) \tag{4.7.8}$$

we derive a simple second–order difference equation

$$Y(n) = (1 + v - s)Y(n-1) - vY(n-2). \tag{4.7.9}$$

The linear model (4.7.9) does not adequately represent a business since it does not produce oscillatory solutions (or periodic cycles) except for special cases (such as $v = 1$).

A nonlinear cubic model

$$I(n) = v(Y(n-1) - Y(n-2)) - v(Y(n-1) - Y(n-2))^3, \tag{4.7.10}$$
$$C(n) = (1-s)Y(n-1) + \varepsilon s Y(n-2), \tag{4.7.11}$$

was proposed in Puu [124] and [125]. A fraction $0 \leq \varepsilon \leq 1$ of savings was assumed to be spent after being saved for one period. So for $\varepsilon = 0$, the original Hicks model (4.7.7) is recovered.

Let us introduce a new variable

$$\tilde{Z}(n-1) = \frac{I(n)}{v} = Y(n-1) - Y(n-2). \tag{4.7.12}$$

Adding (4.7.10) and (4.7.11) and using (4.7.8) yields

$$\begin{aligned} Y(n+1) &= I(n+1) + C(n+1) \\ &= v(Y(n) - Y(n-1)) + (1-s)Y(n) \\ &\quad + \varepsilon s Y(n-1) - v(Y(n) - Y(n-1))^3. \end{aligned}$$

Subtracting $Y(n)$ from both sides yields

$$\tilde{Z}(n) = (v - \varepsilon s)\tilde{Z}(n-1) - v\tilde{Z}^3(n) + (\varepsilon - 1)sY(n-1).$$

Let

$$\tilde{Z}(n) = \sqrt{\frac{1 + v - \varepsilon s}{v}} Z(n).$$

Then

$$Z(n) = (v - \varepsilon s)Z(n-1) - (1 + v - \varepsilon s)Z^3(n-1) + (\varepsilon - 1)sY(n-1).$$

Let $a = (v - \varepsilon s)$, $(\varepsilon - 1)s = b$. We get

$$Z(n+1) = aZ(n) - (1+a)Z^3(n) + bY(n) \tag{4.7.13}$$

where $b = (1 - \varepsilon)s$ represents a sort of eternal rate of saving. Using (4.7.12) and (4.7.13) we now have the two-dimensional system

$$\begin{aligned} Y(n+1) &= Y(n) + Z(n), \\ Z(n+1) &= aZ(n) - (a+1)Z^3(n) - bY(n). \end{aligned} \tag{4.7.14}$$

System (4.7.14) has a single equilibrium point $X^* = (Y^*, Z^*) = (0,0)$. Local stability can now be obtained by examining the linearized system

$$\begin{pmatrix} Y(n+1) \\ Z(n+1) \end{pmatrix} = \begin{pmatrix} 1 & 1 \\ -b & a \end{pmatrix} \begin{pmatrix} Y(n) \\ Z(n) \end{pmatrix}. \tag{4.7.15}$$

The eigenvalues are given by

$$\lambda_{1,2} = \frac{a + 1 \pm \sqrt{(a-1)^2 - 4b}}{2}. \tag{4.7.16}$$

By criteria (4.3.9), the trivial solution is asymptotically stable if and only if:

(i) $2 + 2a + b > 0$,

(ii) $b > 0$,

(iii) $1 - a - b > 0$.

Taking into account that $a > 0$ and $0 < b < 1$, the region of stability S is given by

$$S = \{(b,a) \mid 0 < b < 1,\ 0 < a < 1 - b\}.$$

We conclude that if $(b,a) \in S$, then equilibrium $X^* = (0,0)$ is asymptotically stable (Figure 4.29).

Notice that the eigenvalues λ_1, λ_2 are complex numbers if $1 - 2\sqrt{b} < a < 1 + 2\sqrt{b}$. But this is fulfilled if $(b,a) \in S$. Thus if $(b,a) \in S$, the equilibrium point $X^* = (0,0)$ is a stable focus.

At $a = 1 - b$, the equilibrium point $X^* = (0,0)$ loses it stability and possible appearance of cycles. For example, for $a = 0$, $b = 1$, an attracting

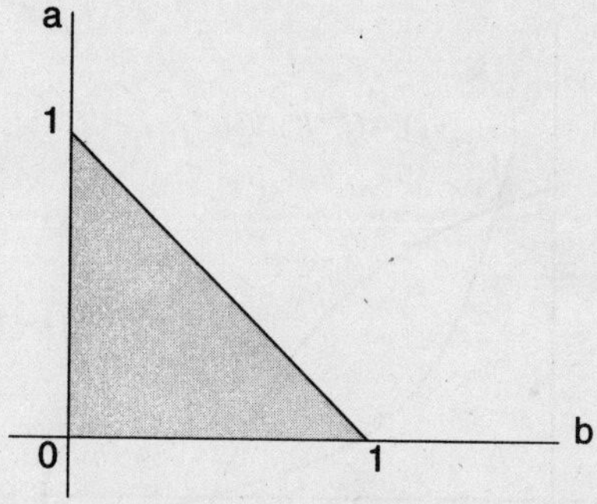

FIGURE 4.29. The region of stability S is the shaded area.

cycle of period 6 and a saddle cycle of period 7 appear. At $a = \sqrt{2} - 1$, $b = 2 - \sqrt{2}$ and attracting and saddle cycles of period 8 appear and so on [126].

4.7.4 The Nicholson–Bailey Model

In [107] the function f was specified under two assumptions:

1. The number of encounters H_e of hosts by parasitoids is proportional to the product of their densities, that is,
$$H_e = aH(n)P(n). \tag{4.7.17}$$
2. The first encounter between a host and a parasitoid is the only significant encounter. Since the encounters are assumed to be random, it is appropriate to use a Poisson probability distribution to describe these encounters.

If μ is the average number of events in a given time interval, then the probability of r events (such as encounters between host and its parasitoid) is

$$p(r) = \frac{\bar{e}^{\mu}\mu^r}{r!},$$

with $\mu = \frac{H_e}{H(n)}$. It follows from equation (4.7.17) that

$$\mu = aP(n). \tag{4.7.18}$$

Since the likelihood of escaping parasitism is the same as the probability of no encounters during the host lifetime, $f(H(n), P(n)) = \bar{e}^{aP(n)}$. Equations (4.7.4) and (4.7.5) now become

$$H(n+1) = \lambda H(n)\bar{e}^{aP(n)}, \tag{4.7.19}$$
$$P(n+1) = cH(n)(1 - \bar{e}^{aP(n)}). \tag{4.7.20}$$

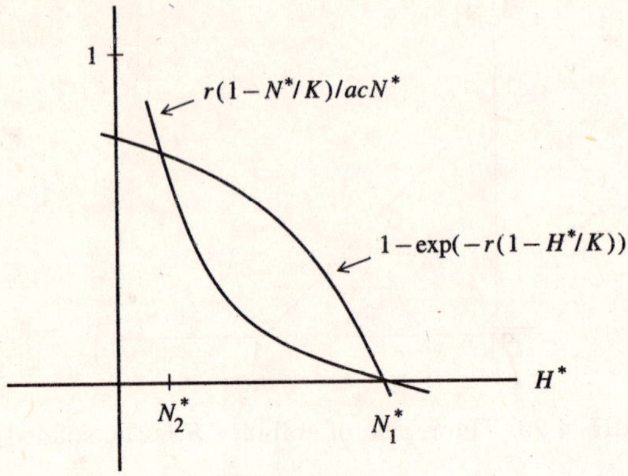

FIGURE 4.30.

The nontrivial equilibrium points are given by

$$H^* = \frac{\lambda^{\ln\lambda}}{(\lambda-1)\,ac}, \quad P^* = \frac{1}{a}\ln\lambda.$$

It can be shown by linearization that (H^*, P^*) is unstable. Thus this model is too simple for any practical applications except possibly under contrived laboratory conditions.

It is reasonable to modify the $H(n)$ equation (4.7.4) to incorporate some saturation of the prey population, or, in terms of predator encounters, a prey-limiting model. Hence a more realistic model is given by

$$\begin{aligned} H(n+1) &= H(n)\exp\left[r\left(1-\frac{H(n)}{k}\right) - aP(n)\right], \quad r>0, \\ P(n+1) &= cH(n)(1-\exp(-aP(n))). \end{aligned} \tag{4.7.21}$$

The equilibrium points are solutions of

$$1 = \exp\left[r\left(1-\frac{H^*}{K}\right) - aP^*\right], \quad P^* = cH^*(1-\exp(-aP^*)).$$

Hence

$$P^* = \frac{r}{a}\left[1-\frac{H^*}{K}\right] = \frac{r}{a}(1-q), \quad H^* = \frac{P^*}{(1-\bar{e}^{ap^*})}. \tag{4.7.22}$$

Thus

$$\frac{r(1-\frac{H^*}{K})}{acH^*} = 1 - \exp\left[-r\left(1-\frac{H^*}{K}\right)\right]. \tag{4.7.23}$$

Clearly, $H_1^* = K$, $P_1^* = 0$ is an equilibrium state. The other equilibrium point may be obtained by plotting the left- and right-hand sides of (4.7.19) against H^*. From Figure 4.30 we see that there is another equilibrium point with $0 < H_2^* < K$. Then we may find P_2^* from (4.7.18).

To perform the stability analysis of the equilibrium point (H_2^*, P_2^*), we put $H(n) = x(n) + H_2^*$, $P(n) = y(n) + P_2^*$. Hence we obtain

$$x(n+1) = -H_2^* + (x(n) + H_2^*) \exp\left[r\left(1 - \frac{x(n) + H_2^*}{K}\right) - a(y(n) + P_2^*)\right],$$

$$y(n+1) = -P_2^* + c(x(n) + H_2^*)[1 - \exp(-a(y(n) + P_2^*))]. \tag{4.7.24}$$

By linearizing around $(0,0)$ we obtain the linear system

$$\begin{pmatrix} x(n+1) \\ y(n+1) \end{pmatrix} = A \begin{pmatrix} x(n) \\ y(n) \end{pmatrix} \tag{4.7.25}$$

with

$$A = \begin{pmatrix} 1 - rq & -arq \\ c(1 - \exp(-r(1-q)) & \varphi - r(1-q) \end{pmatrix}, \tag{4.7.26}$$

where $q = \frac{H_2^*}{K}$ and

$$\varphi = \frac{r(1-q)}{1 - \exp(-r(1-q))}.$$

The details of obtaining the matrix A will be left to the reader. Observe that the value of $q = \frac{H_2^*}{K}$ is a measure of the extent to which the predator can depress the prey below the carrying capacity.

The characteristic equation of A is given by

$$\lambda^2 - \lambda(1 - r + \varphi) + (1 - rq)\varphi + r^2 q(1-q) = 0. \tag{4.7.27}$$

By criterion (4.3.9), the eigenvalues of A lie inside the unit disk if and only if

$$|1 - r + \varphi| < 1 + (1 - rq)\varphi + r^2 q(1-q) < 2.$$

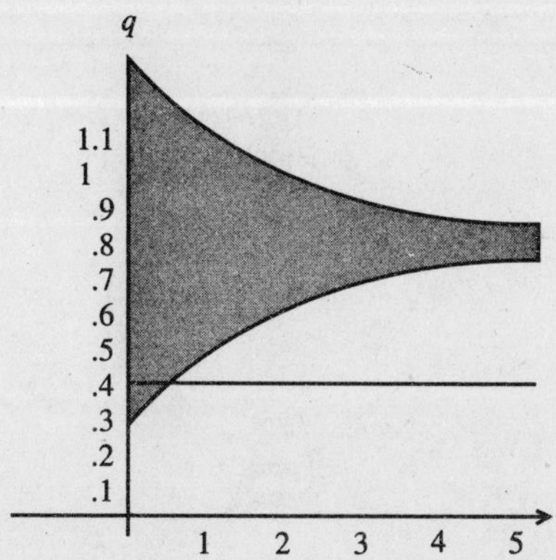

FIGURE 4.31. The origin is asymptotically stable within the shaded area.

Hence

$$(1 - rq)\varphi + r^2 q(1 - q) < 1, \tag{4.7.28}$$

$$1 + (1 - rq)\varphi + r^2 q(1 - q) > |1 - r + \varphi|. \tag{4.7.29}$$

Plotting (4.7.24) and (4.7.25) gives the region of asymptotic stability indicated by the shaded area in Figure 4.31.

The origin is asymptotically stable within the shaded area. Note that the area of stability narrows as r increases.

Beddington et al. [7] conducted a numerical simulation for the specific value $q = 0.4$. As r grows past a certain value, the equilibrium point becomes unstable and a hierarchy of stable limit cycles of increasing, non-integral period, ultimately breaking down to cycles of period 5, appears. These are followed by cycles of period $2 \times 5,\ 2^2 \times 5, \ldots, 2^n \times 5, \ldots$.

4.7.5 The Flour Beetle Case Study

The team of R.F. Costantino, J.M. Cushing, B. Dennis, R.A. Desharnais, and S.M. Henson [27] have studied the flour beetles extensively. They conducted both theoretical studies as well as experimental studies in the laboratory. To describe their model, we will give a brief background of the life-cycle of the flour beetles. The life-cycle consists of larval and pupal stages each lasting approximately two weeks, followed by an adult stage (see Figure 4.32).

As is shown in Figure 4.32, cannibalism occurs among the various groups. Adults eat pupae and eggs, larvae eat eggs. Neither larvae nor adults eat mature adults. Moreover, larvae do not feed on larvae. The cannibalism of larvae by adults and of pupae by larvae is assumed negligible since it typically occurs at much reduced rates.

Let $L(n)$ be the larvae population at time period n, let $P(n)$ be the pupal population at time period n, and let $A(n)$ be the adult population at time

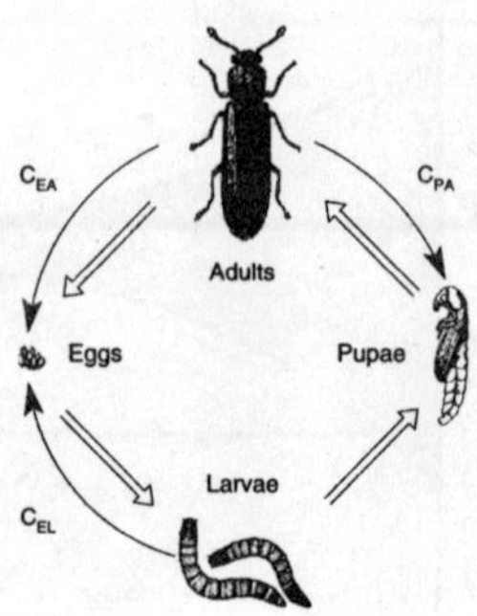

FIGURE 4.32. The arrows show the cannibalistic interaction between difference life-cycle stages.

period n. Then the larval–pupal–adult (LPA) model is given by

$$\begin{aligned} L(n+1) &= bA(n)\exp(-c_{EA}A(n) - c_{EL}L(n)), \\ P(n+1) &= (1-\mu_L)L(n), \\ A(n+1) &= P(n)\exp(-c_{PA}A(n)) + (1-\mu_A)A(n), \end{aligned} \tag{4.7.30}$$

where $L(0) \geq 0$, $P(0) \geq 0$, $A(0) \geq 0$.

The constants μ_L, μ_A are the larval and adult probability of dying from causes other than cannibalism, respectively. Thus $0 \leq \mu_L \leq 1$ and $0 \leq \mu_A \leq 1$. The term $\exp(-c_{EA}A(n))$ represents the probability that an egg is not eaten in the presence of $A(n)$ adults, $\exp(-c_{EL}L(n))$ represents the probability that an egg is not eaten in the presence of $L(n)$ larvae, and $\exp(-c_{PA}A(n))$ is the survival probability of a pupa in the presence of $A(n)$ adults. The constants $c_{EA} \geq 0$, $c_{EL} \geq 0$, $c_{PA} \geq 0$ are called the cannibalism coefficients. We assume that adult cannibalism is the only significant cause of pupal mortality.

There are two equilibrium points $(0,0,0)^T$ and $(L^*, P^*, A^*) \in \mathbb{R}^3_+$, $L^* > 0$, $P^* > 0$, $A^* > 0$. The positive equilibrium point may be obtained by solving the three equations

$$\begin{aligned} L\exp(c_{EL}L) &= bA\exp(-c_{EA}A), \\ P &= (1-\mu_L)L, \\ \mu_A\exp(c_{PA}A) &= P. \end{aligned} \tag{4.7.31}$$

Eliminating P yields

$$\begin{aligned} (1-\mu_L)L &= \mu_A A\exp(c_{PA}A), \\ L\exp(c_{EL}L) &= bA\exp(-c_{EA}A). \end{aligned}$$

Dividing the second equation by the first yields

$$\exp(c_{EL}L) = \frac{b(1-\mu_L)}{\mu_A}\exp[(-c_{EA} - c_{PA})A]. \tag{4.7.32}$$

The number

$$N = \frac{b(1-\mu_L)}{\mu_A}$$

is called the *inherent net reproductive number.* This number will play a significant role in our stability analysis. Observe that if $N < 1$, equation (4.7.32) has no solution and we have no positive equilibrium point. However, if $N > 1$, then equation (4.7.32) has a solution which is the intersection of the curve $(1-\mu_L)L = \mu_A\exp(c_{PA}A)$ and the straight line from $(0, \ln N/c_{EL})$ to $(\ln N/(c_{EA} + c_{PA}), 0)$ in the (A, L)-plane represented by equation (4.7.32). To investigate the local stability of the equilibrium point

(L, P, A) of equation (4.7.30), we compute the Jacobian J,

$$J = \begin{pmatrix} -c_{EL}bAe^{(-c_{EA}A-c_{EL}L)} & 0 & be^{(-c_{EL}L-c_{EA}A)(1-c_{EA}A)} \\ 1-\mu_L & 0 & 0 \\ 0 & e^{(-c_{PA}A)} & 1-\mu_A-c_{PA}Pe^{(-c_{PA}A)} \end{pmatrix}. \tag{4.7.33}$$

At the equilibrium point $(0, 0, 0)^T$ we have

$$J_1 = J|_{(0,0,0)^T} = \begin{pmatrix} 0 & 0 & b \\ 1-\mu_L & 0 & 0 \\ 0 & 1 & 1-\mu_A \end{pmatrix}.$$

The characteristic polynomial of J_1 is given by

$$P(\lambda) = \lambda^3 - (1-\mu_A)\lambda^2 - b(1-\mu_L) = 0 \tag{4.7.34}$$

which is of the form

$$P(\lambda) = \lambda^3 + p_1\lambda^2 + p_2\lambda + p_3 = 0,$$

with $p_1 = -(1-\mu_A)$, $p_2 = 0$, $p_3 = -b(1-\mu_L)$.

According to (5.1.17), the eigenvalues of J_1 are inside the unit circle if and only if

$$|p_3 + p_1| < 1 + p_2 \quad \text{and} \quad |p_2 - p_3p_1| < 1 - p_3^2.$$

Applying the first condition yields

$$|-b(1-\mu_L) - (1-\mu_A)| < 1,$$
$$b(1-\mu_L) + (1-\mu_A) < 1,$$

or

$$N = \frac{b(1-\mu_L)}{\mu_A} < 1. \tag{4.7.35}$$

The second condition gives

$$|-(1-\mu_A)(1-\mu_L)b| < 1 - b^2(1-\mu_L)^2,$$
$$b^2(1-\mu_L)^2 + (1-\mu_A)(1-\mu_L)b < 1.$$

But this inequality is satisfied if we assume (4.7.35). For if $N < 1$ we have

$$b^2(1-\mu_L)^2 + (1-\mu_A)(1-\mu_L)b < \mu_A^2 + \mu_A(1-\mu_A) = \mu_A \leq 1.$$

We conclude that the trivial equilibrium is asymptotically stable if and only if $N < 1$, and thus attracts all orbits in the nonnegative cone. As N increases past 1, a "bifurcation" occurs which results in the instability of the trivial equilibrium and the creation of the positive equilibrium. In fact, for $N > 1$ there exists one and only one positive equilibrium. The Jacobian at the positive equilibrium (L^*, P^*, A^*) satisfying (4.7.31) is given by

$$J_2 = J|_{(L^*,P^*,A^*)} = \begin{pmatrix} -c_{EL}L^* & 0 & \dfrac{L^*}{A^*} - c_{EA}L^* \\ 1-\mu_L & 0 & 0 \\ 0 & A^*\exp(c_{PA}) & 1-\mu_A - A^*\mu_A c_{PA} \end{pmatrix}.$$

The characteristic equation is given by

$$\lambda^3 + (c_{EL}L^* + \mu_A c_{PA}A^* - (1-\mu_A))\lambda^2 - c_{EL}L^*(1-\mu_A)\lambda - \left(\frac{L^*}{A^*} - c_{EA}L^*\right)(1-\mu_L)\exp(-c_{PA}A^*) = 0.$$

As of writing this edition of the book, a condition for the stability of the positive equilibrium is known only in special cases.

Case (i) If $c_{EL} = 0$, the positive equilibrium is globally attracting if

$$1 < N < e\min\{(1, (c_{EA}/c_{PA})((1-\mu_A)/\mu_A)\} \quad [83].$$

Case (ii) In several long term experiments reported in [27], the adult death rate was manipulated to equal 96% and hence $\mu_A = 0.96$. Motivated by this data, Cushing [25] assumed that $\mu_A = 1$. In this case we have $N = b(1-\mu_L)$ and equation (4.7.30) becomes

$$\begin{aligned} L(n+1) &= \frac{N}{1-\mu_L}A(n)\exp(-c_{EL}L(n) - c_{EA}A(n)), \\ P(n+1) &= (1-\mu_L)L(n), \\ A(n+1) &= P(n)\exp(-c_{PA}A(n)). \end{aligned} \tag{4.7.36}$$

Theorem 4.40 [25]. *For $N > 1$, the trivial equilibrium of equation (4.7.36) is unstable and there exists a unique positive equilibrium. This positive equilibrium, which bifurcates from the trivial equilibrium at $N = 1$, is unstable for $N = 1 + \delta$, where δ is sufficiently small.*

A subcase of Case (ii) is the case of synchronous orbits. A triple $(L(n), P(n), A(n))$ is said to be synchronous at time n if one component equals zero and at least one component is nonzero. One can see immediately from equation (4.7.36), that an orbit that is synchronous at time n_0 is synchronous for all $n \geq n_0$. Notice that a point $(L_0, P_0, 0)^T$ in the L, P-plane is mapped to the point $(0, (1-\mu_L)L_0, P_0)^T$ in the P, A-plane, which in turn is mapped to the point $\left(\frac{N}{1-\mu_L}P_0\exp(-c_{EA}P_0), 0, (1-\mu_L)L_0\exp(-c_{PA}P_0)^T\right)$ in the L, A-plane. Hence points are mapped from one nonnegative quadrant of the coordinate planes to the next in sequential order. A synchronous triplet $(L(n), P(n), A(n))^T$ is said to be fully synchronous at time n if it has two zero components. This is the case for points on the positive coordinate axes. An orbit is fully synchronous if and only if its initial point is fully

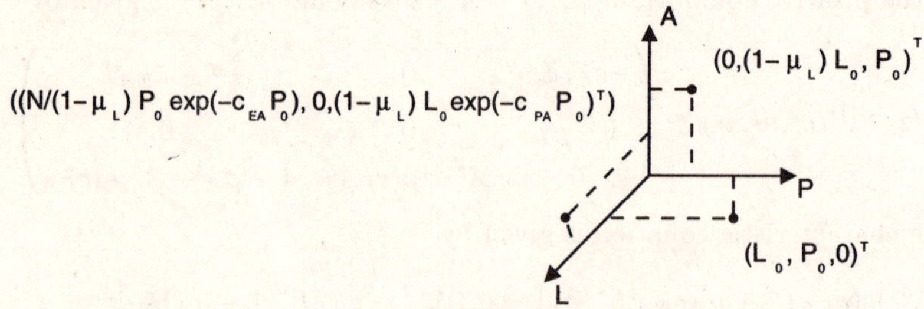

FIGURE 4.33.

synchronous. This notion is derived from the fact that the three life-cycle stages are synchronized temporarily in such a way that they never overlap.

Denote the map (4.7.36) by F,

$$\begin{pmatrix} L(n+1) \\ P(n+1) \\ A(n+1) \end{pmatrix} = F \begin{pmatrix} L(n) \\ P(n) \\ A(n) \end{pmatrix}. \tag{4.7.37}$$

Then F^3 maps the nonnegative quadrant of a coordinate plane to itself. A fixed point of F^3 corresponds to a 3-cycle of F and so on. The map F^3 is defined by the equations

$$\begin{aligned} x(n+1) = Nx(n)\exp\Big[&- c_{PA}y(n)\exp(-c_{PA}z(n)) \\ &- c_{EA}(1-\mu_L)x(n)\exp(-c_{PA}y(n)\exp(-c_{PA}z(n))) \\ &- c_{EL}\frac{N}{1-\mu_L}y(n)\exp(-c_{PA}y(n) - c_{EA}y(n)\exp(-c_{PA}z(n))) \\ &- c_{EL}\frac{N}{1-\mu_L}z(n)\exp\Big(- c_{EA}\frac{N}{1-\mu_L}z(n)\exp(-c_{EA}z(n) \\ &- c_{EL}x(n))\Big)\Big], \end{aligned} \tag{4.7.38}$$

$$\begin{aligned} y(n+1) = Ny(n)\exp\Big[&- c_{PA}z(n) - c_{EA}y(n)\exp(-c_{PA}z(n)) \\ &- c_{EA}\frac{N}{1-\mu_L}z(n)\exp(-c_{EA}z(n) - c_{EL}x(n))\Big], \end{aligned} \tag{4.7.39}$$

$$\begin{aligned} z(n+1) = Nz(n)\exp\Big[&- c_{EA}z(n) - c_{EL}x(n) \\ &- c_{PA}(1-\mu_L)x(n)\exp(-c_{PA}y(n)\exp(-c_{PA}z(n)))\Big]. \end{aligned}$$

If $(x_0, 0, z_0)^T$ is a point in the x, z-plane, then its orbit is described by the two-dimensional system

$$x(n+1) = Nx(n)\exp(-cx(n)), \tag{4.7.40}$$

$$z(n+1) = [N\exp(-\alpha x(n))]z(n)\exp(-\beta z(n)), \tag{4.7.41}$$

where $c = c_{EA}(1-\mu_L)$, $\alpha = c_{EL} + c_{PA}(1-\mu_L)$, $\beta = c_{EA}$.

The first equation (4.7.40) is the well-known Ricker's map, where $\lim_{n\to\infty} x(n) = 0$ and the convergence is exponential. Hence equation (4.7.41) may be looked at as a perturbation of (4.7.40). Hence by Corollary 8.27, $\lim_{n\to\infty} z(n) = 0$ which is consistent with what we had earlier. For $N > 1$, Ricker's map has a unique positive equilibrium $x^* = \ln N/c$. Consequently, there exists a fully synchronous 3-cycle of equation (4.7.36). As N increases, Ricker's map undergoes a period-doubling bifurcation route to chaos. If $1 < N < e^2$, then $(x^*, z^*)^T = (\frac{1}{c}\ln N, 0)^T$ is an asymptotically stable equilibrium point of equations (4.7.40) and (4.7.41) and globally attracts all positive initial conditions in the x, z-plane. This fixed point of F^3 corresponds to the fully synchronous 3-cycle of the LPA model (4.7.36)

$$\begin{pmatrix} \dfrac{\ln N}{c_{EA}(1-\mu_L)} \\ 0 \\ 0 \end{pmatrix} \to \begin{pmatrix} 0 \\ \dfrac{1}{c_{EA}}\ln N \\ 0 \end{pmatrix} \to \begin{pmatrix} 0 \\ 0 \\ \dfrac{1}{c_{EA}}\ln N \end{pmatrix}. \tag{4.7.42}$$

Thus we have the following result. Theorem 4.40 [25] for $1 < N < e^2$, the LPA model (4.7.36) has a unique, nontrivial fully synchronous 3-cycle given by (4.7.42). This 3-cycle attracts all fully synchronous orbits or equation (4.7.36). For $N > e^2$, the system has a period-doubling cascade of fully synchronous (3×2^n)-cycle attractors and, for sufficiently large N, has "fully synchronous chaotic" attractors (with respect only to fully synchronous orbits).

This is the first proof of the presence of chaos in a population model.

Details about synchronous but not fully synchronous orbits may be found in Cushing [25]. There are still many open problems that need to be tackled. We invite the reader to solve them.

Open Problem 1. Investigate the LPA model for the general case $\mu_A \neq 1$. For a starting point, try the case with $\mu_A = 0.96$.

Open Problem 2. Investigate the behavior of orbits that are not synchronous provided that $\mu_A = 1$.

5

Higher-Order Scalar Difference Equations

In Chapter 4 we investigated the qualitative behavior of systems of difference equations, both linear and nonlinear. In this chapter we turn our attention to linear and nonlinear higher-order scalar difference equations. Although one may be able to convert a scalar difference equation to a system, it is often advantageous to tackle the scalar difference equation directly. Moreover, since a system of difference equations may not be convertible to a scalar difference equation, results on the latter may not extend to the former. Every section in this chapter was written with this statement in mind. Section 5.1 gives explicit necessary and sufficient conditions for the stability of the zero solution of a kth-order scalar difference equation. This task is accomplished either via the Schur–Cohn criterion or by using special techniques that were developed by Levin and May [90], Kuruklis [86], Dannan [28], and Dannan and Elaydi [29]. Section 5.2 provides easy computable sufficient conditions for asymptotic stability using Gerschgorin's Theorem which provides a rough estimate of the location of eigenvalues of matrices.

In the first and second editions of this book I have used Rouché's Theorem from complex analysis to obtain the results in Section 5.2. However, the new approach is not only more accessible to readers with no background in complex analysis but, more importantly, it is much more intuitive. Section 5.3 treats nonlinear equations via linearization and follows closely the exposition in Section 4.4 for systems. Section 5.4 collects the main results in global stability of nonlinear scalar difference equations. It remains an open question of whether or not these results extend to nonlinear systems of difference equations. Finally, Section 5.5 presents the larval–pupal–adult

(LPA) model of flour beetles with no larval cannibalism on eggs, and a mosquito model.

5.1 Linear Scalar Equations

Consider the kth-order difference equation

$$x(n+k)+p_1x(n+k-1)+p_2x(n+k-2)+\cdots+p_kx(n)=0 \quad (5.1.1)$$

where the p_i's are real numbers.

It follows from Corollary 2.24 that the zero solution of (5.1.1) is asymptotically stable if and only if $|\lambda|<1$ for all characteristic roots λ of (5.1.1), that is, for every zero λ of the characteristic polynomial

$$p(\lambda)=\lambda^k+p_1\lambda^{k-1}+\cdots+p_k. \quad (5.1.2)$$

Furthermore, the zero solution of (5.1.1) is stable if and only if $|\lambda|\le 1$ for all characteristic roots of (5.1.1) and those characteristic roots λ with $|\lambda|=1$ are simple (not repeated). On the other hand, if there is a repeated characteristic root λ with $|\lambda|=1$, then according to Corollary 2.24 the zero solution of (5.1.1) is unstable.

One of the main tools that provides necessary and sufficient conditions for the zeros of a kth-degree polynomial, such as (5.1.2), to lie inside the unit disk is the Schur–Cohn criterion. This is useful for studying the stability of the zero solution of (5.1.1). Moreover, one may utilize the Schur–Cohn criterion to investigate the stability of a k-dimensional system of the form

$$x(n+1)=Ax(n) \quad (5.1.3)$$

where $p(\lambda)$ in (5.1.2) is the characteristic polynomial of the matrix A.

But before presenting the Schur–Cohn criterion we introduce a few preliminaries.

First let us define the *inners* of a matrix $B=(b_{ij})$. The inners of a matrix are the matrix itself and all the matrices obtained by omitting successively the first and last rows and the first and last columns. For example, the inners for the following matrices are highlighted:

A 3 × 3 matrix **A 4 × 4 matrix** **A 5 × 5 matrix**

$$\begin{pmatrix} b_{11} & b_{12} & b_{13} \\ b_{21} & \boxed{b_{22}} & b_{23} \\ b_{31} & b_{32} & b_{33} \end{pmatrix}, \quad \begin{pmatrix} b_{11} & b_{12} & b_{13} & b_{14} \\ b_{21} & b_{22} & b_{23} & b_{24} \\ b_{31} & b_{32} & b_{33} & b_{34} \\ b_{41} & b_{42} & b_{43} & b_{44} \end{pmatrix}, \quad \begin{pmatrix} b_{11} & b_{12} & b_{13} & b_{14} & b_{15} \\ b_{21} & b_{22} & b_{23} & b_{24} & b_{25} \\ b_{31} & b_{32} & \boxed{b_{33}} & b_{34} & b_{35} \\ b_{41} & b_{42} & b_{43} & b_{44} & b_{45} \\ b_{51} & b_{52} & b_{53} & b_{54} & b_{55} \end{pmatrix}.$$

A matrix B is said to be *positive innerwise* if the determinants of all of its inners are positive.

Theorem 5.1 (Schur–Cohn Criterion) [74]. *The zeros of the characteristic polynomial* (5.1.2) *lie inside the unit disk if and only if the following hold:*

(i) $p(1) > 0$,

(ii) $(-1)^k p(-1) > 0$,

(iii) *the* $(k-1) \times (k-1)$ *matrices*

$$B^{\pm}_{k-1} = \begin{pmatrix} 1 & 0 & \dots & & 0 \\ p_1 & 1 & \dots & & 0 \\ \vdots & & & & \vdots \\ p_{k-3} & & & & \\ p_{k-2} & p_{k-3} & \dots & p_1 & 1 \end{pmatrix} \pm \begin{pmatrix} 0 & 0 & \dots & 0 & p_k \\ 0 & 0 & \dots & p_k & p_{k-1} \\ \vdots & \vdots & & & \vdots \\ 0 & p_k & & & p_3 \\ p_k & p_{k-1} & \dots & p_3 & p_2 \end{pmatrix}$$

are positive innerwise.

Using the Schur–Cohn criterion (Theorem 5.1), one may obtain necessary and sufficient conditions on the coefficients p_i's such that the zero solution of (5.1.1) is asymptotically stable. Neat and compact, necessary and sufficient conditions for the zero solution of (5.1.1) to be asymptotically stable are available for lower-order difference equations. We will present these conditions for second- and third-order difference equations.

For the second-order difference equation

$$x(n+2) + p_1 x(n+1) + p_2 x(n) = 0 \tag{5.1.4}$$

the characteristic polynomial is

$$p(\lambda) = \lambda^2 + p_1 \lambda + p_2. \tag{5.1.5}$$

The characteristic roots are inside the unit disk if and only if

$$p(1) = 1 + p_1 + p_2 > 0, \tag{5.1.6}$$

$$p(-1) = 1 - p_1 + p_2 > 0, \tag{5.1.7}$$

$$B_1^{\pm} = 1 \pm p_2 > 0. \tag{5.1.8}$$

It follows from (5.1.6) and (5.1.7) that $1+p_2 > |p_1|$ and $1+p_2 > 0$. Now (5.1.8) reduces to $1 - p_2 > 0$ or $p_2 < 1$. Hence the zero solution of (5.1.4) is asymptotically stable if and only if

$$\boxed{|p_1| < 1 + p_2 < 2.} \tag{5.1.9}$$

For the third-order difference equation

$$x(n+3) + p_1 x(n+2) + p_2 x(n+1) + p_3 x(n) = 0 \tag{5.1.10}$$

the characteristic polynomial is

$$\lambda^3 + p_1\lambda^2 + p_2\lambda + p_3 = 0. \tag{5.1.11}$$

The Schur–Cohn criterion are

$$1 + p_1 + p_2 + p_3 > 0, \tag{5.1.12}$$

$$(-1)^3[-1 + p_1 - p_2 + p_3] = 1 - p_1 + p_2 - p_3 > 0 \tag{5.1.13}$$

$$|B_2^+| = \left| \begin{bmatrix} 1 & 0 \\ p_1 & 1 \end{bmatrix} + \begin{bmatrix} 0 & p_3 \\ p_3 & p_2 \end{bmatrix} \right| = \begin{vmatrix} 1 & p_3 \\ p_1 + p_3 & 1 + p_2 \end{vmatrix} > 0.$$

Thus

$$1 + p_2 - p_1 p_3 - p_3^2 > 0 \tag{5.1.14}$$

and

$$|B_2^-| = \left| \begin{bmatrix} 1 & 0 \\ p_1 & 1 \end{bmatrix} - \begin{bmatrix} 0 & p_3 \\ p_3 & p_2 \end{bmatrix} \right| = \begin{vmatrix} 1 & -p_3 \\ p_1 - p_3 & 1 - p_2 \end{vmatrix} > 0. \tag{5.1.15}$$

Hence

$$1 - p_2 + p_3 p_1 - p_3^2 > 0. \tag{5.1.16}$$

Using (5.1.12), (5.1.13), (5.1.14), and (5.1.16), we conclude that a necessary and sufficient condition for the zero solution of (5.1.10) to be asymptotically stable is

$$\boxed{|p_1 + p_3| < 1 + p_2 \quad \text{and} \quad |p_2 - p_1 p_3| < 1 - p_3^2.} \tag{5.1.17}$$

It is now abundantly clear that the higher the order of the equation, the more difficult the computation involved in applying the Schur–Cohn criterion becomes. However, Levin and May [90], using a very different technique, were able to obtain a simple criterion for the asymptotic stability of the following special equation

$$x(n+1) - x(n) + qx(n-k) = 0, \tag{5.1.18}$$

or, equivalently,

$$x(n+k+1) - x(n+k) + qx(n) = 0. \tag{5.1.19}$$

Theorem 5.2. *The zero solution of* (5.1.18) *is asymptotically stable if and only if*

$$\boxed{0 < q < 2\cos\left(\frac{k\pi}{2k+1}\right).} \tag{5.1.20}$$

The proof of this theorem is given in Appendix E. The theorem may also be obtained as a corollary of a more general theorem by Kuruklis [86] which we now state.

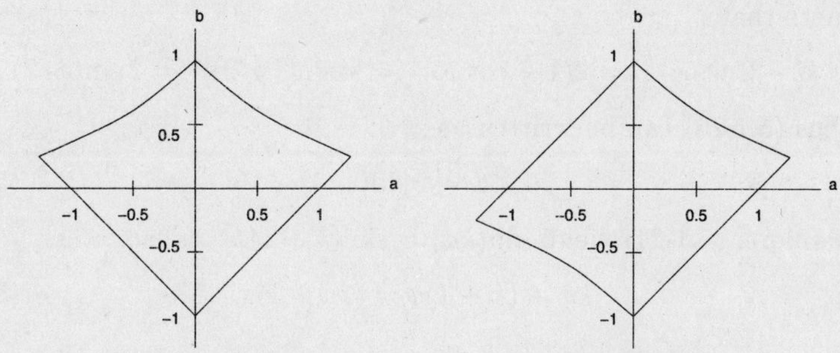

(i) k odd **(ii) k even**

FIGURE 5.1. The graphs of the domains of (a, b), for which the roots of $\lambda^{k+1} - a\lambda^k + b = 0$, with $a \neq 0$ and $k > 1$, are inside the unit disk. The curved sides are parts of $|b| = |a^2 + 1 - 2|a| \cos\phi|^{\frac{1}{2}}$, where ϕ is the solution in $(0, \pi(k+1))$ of $\sin(k\theta)/\sin(k+1)\theta = 1/|a|$.

Contemplate the equation

$$x(n+1) - ax(n) + bx(n-k) = 0, \quad n \in \mathbb{Z}^+. \tag{5.1.21}$$

Theorem 5.3. *Let a be a nonnegative real number, b an arbitrary real number, and k a positive integer. The zero solution of (5.1.21) is asymptotically stable if and only if $|a| < (k+1)/k$, and:*

(i) $|a| - 1 < b < (a^2 + 1 - 2|a| \cos\phi)^{\frac{1}{2}}$ *for k odd, or*

(ii) $|b - a| < 1$ *and* $|b| < (a^2 + 1 - 2|a| \cos\phi)^{\frac{1}{2}}$ *for k even,*

where ϕ is the solution in $(0, \pi/(k+1))$ of $\sin(k\theta)/\sin(k+1)\theta = 1/|a|$. (See Figure 5.1.)

Using the above theorem we present a simple proof of Theorem 5.2.

PROOF OF THEOREM 5.2.

From Theorem 5.3 we have that the zero solution of the difference equation (5.1.19) is asymptotically stable if and only if

$$0 < b < (2 - 2\cos\phi)^{\frac{1}{2}} \quad \text{for } k \text{ odd} \tag{5.1.22}$$

or

$$|b - 1| < 1 \text{ and } |b| < (2 - 2\cos\phi)^{\frac{1}{2}} \quad \text{for } k \text{ even}, \tag{5.1.23}$$

where ϕ is the solution in $(0, \pi/(k+1))$ of

$$\sin(k\theta)/\sin(k+1)\theta = 1. \tag{5.1.24}$$

Note that $|b - 1| < 1$ implies $b > 0$. Therefore conditions (5.1.22) and (5.1.23) are reduced to

$$0 < b < (2 - 2\cos\phi)^{\frac{1}{2}}. \tag{5.1.25}$$

Also note that

$$(2-2\cos\phi)^{\frac{1}{2}} = [2(1-\cos\phi)]^{\frac{1}{2}} = [4\sin^2(\phi/2)]^{\frac{1}{2}} = 2\sin(\phi/2)$$

and thus (5.1.25) can be written as

$$0 < b < 2\sin(\phi/2). \tag{5.1.26}$$

Furthermore, (5.1.24) yields $\sin(k\phi) = \sin[(k+1)\phi]$ and so either

$$k\phi + (k+1)\phi = (2n+1)\pi \tag{5.1.27}$$

or

$$k\phi = (k+1)\phi + 2n\pi, \tag{5.1.28}$$

where n is an integer. Since (5.1.28) cannot be valid for $0 < \phi < \pi/(k+1)$ we have that (5.1.27) holds. In fact, $0 < \phi < \pi/(k+1)$ forces $n = 0$ and so $\phi = \pi/2$ and thus condition (5.1.25) may be written as $0 < b < 2\cos[k\pi/(2k+1)]$, which is the condition of Theorem 5.3. □

Dannan [28] considered the following more general equation

$$x(n+k) + ax(n) + bx(n-l) = 0, \quad n \in \mathbb{Z}^+, \tag{5.1.29}$$

where $k \geq 1$ and $l \geq 1$ are integers.

Theorem 5.4 [28]. *Let $l \geq 1$ and $k > 1$ be relatively prime odd integers. Then the zero solution of* (5.1.29) *is asymptotically stable if and only if* $|a| < 1$ *and*

$$|a| - 1 < b \min_{\theta \in S}(1 + a^2 - 2|a|\cos k\theta)^{\frac{1}{2}} \tag{5.1.30}$$

where S is the solution set of

$$\frac{1}{|a|} = \frac{\sin l\theta}{\sin(l+k)\theta} \tag{5.1.31}$$

on the interval $(0, \pi)$.

Theorem 5.5 [28]. *Let $l \geq 1$ be an odd integer, k an even integer, with l and k relatively prime. Then the zero solution of* (5.1.29) *is asymptotically stable if and only if*

$$|b| < 1 - |a| \quad \textit{for } -1 < a < 0 \tag{5.1.32}$$

and

$$|b| < \min_{\theta \in S^*}(1 + a^2 + 2a\cos k\theta)^{\frac{1}{2}} \quad \textit{for } 0 < a < 1, \tag{5.1.33}$$

where S^* is the solution set of

$$-\frac{1}{a} = \frac{\sin l\theta}{\sin(l+k)\theta} \tag{5.1.34}$$

on the interval $(0, \pi)$.

Theorem 5.6 [28]. *Let l be an even integer and $k > 1$ an odd integer, where l and k are relatively prime. Then the zero solution of* (5.1.29) *is asymptotically stable if and only if*

$$|a| - 1 < b \min_{\theta \in S}(1 + a^2 - 2|a|\cos k\theta)^{\frac{1}{2}} \quad \text{for } -1 < a < 0 \tag{5.1.35}$$

and

$$a - 1 < -b < \min_{\theta \in S}(1 + a^2 - 2a\cos k\theta)^{\frac{1}{2}} \quad \text{for } 0 < a < 1. \tag{5.1.36}$$

where S is as in Theorem 5.4.

Remark: If l and k in Theorems 5.4, 5.5, and 5.6 are not relatively prime, then $l = s\tilde{l}$ and $k = s\tilde{k}$ for some positive integers s, $\tilde{l}$, and $\tilde{k}$, where $\tilde{l}$ and $\tilde{k}$ are relatively prime. The asymptotic stability of (5.1.29) is equivalent to the asymptotic stability of

$$x(n + \tilde{k}) + ax(n) + bx(n - \tilde{l}) = 0. \tag{5.1.37}$$

(Why?) The reader is asked to prove this in Exercises 5.1, 5.2, Problem 5.

Example 5.7. Consider the difference equation

$$x(n + 25) + ax(n) + bx(n - 15) = 0, \quad n = 0, 1, 2, \ldots.$$

The corresponding characteristic equation is

$$\lambda^{40} + a\lambda^{25} + b = 0,$$

and in the reduced form is

$$\lambda^8 + a\lambda^5 + b = 0.$$

Here we have $\tilde{l} = 5$ and $\tilde{k} = 3$. Therefore, Theorem 5.5 is applicable and the given equation is asymptotically stable if and only if $|a| < 1$ and

$$|a| - 1 < b \min_{\theta \in S}(1 + a^2 - 2|a|\cos 3\theta)^{\frac{1}{2}},$$

where S is the solution set of $\frac{1}{|a|} = \frac{\sin 5\theta}{\sin 8\theta}$ on the interval $(0, \pi)$. If we let $a = 0.6$, then $\theta = 2.007548968$ and the given equation is asymptotically stable if and only if $-0.4 < b < 0.4477703541$.

5.2 Sufficient Conditions for Stability

Clark [21] considered the equation

$$x(n + k) + px(n + k - 1) + qx(n) = 0 \tag{5.2.1}$$

where $p, q \in \mathbb{R}$. When $p = -1$, we revert back to the Levin and May equation (5.1.18). He showed that the zero solution of (5.2.1) is asymptotically stable if

$$|p| + |q| < 1. \tag{5.2.2}$$

Moreover, the zero solution is unstable if

$$|p| - |q| > 1. \tag{5.2.3}$$

Here we will extend Clark's Theorem to the general equation (5.1.1). The novelty of Clark's proof is the use of Rouché's Theorem from Complex Analysis to locate the characteristic roots of the equation. In the first two editions of this book, I followed Clark's proof. In this edition, I am going to deviate from this popular method, and give instead a simpler proof based on Gerschgorin's Theorem [111] which we now state.

Theorem 5.8. *Let A be a $k \times k$ real or complex matrix. Let S_i be the disk in the complex plane with center at a_{ii} and radius $r_i = \sum_{\substack{j=1 \\ j \neq i}}^{k} |a_{ij}|$. Then all the eigenvalues of A lie in $S = \bigcup_{i=1}^{k} S_i$.*

PROOF. Let λ be an eigenvalue of A with a corresponding eigenvector $v = (v_1, v_2, \ldots, v_k)^T$ such that $||v||_\infty = \max_i\{|v_i|\} = 1$. (Why?).

Since $Av = \lambda v$, equating the ith row in both sides yields $\sum_{j=1}^{k} a_{ij} v_j = \lambda v_i$. Hence

$$(\lambda - a_{ii}) v_i = \sum_{j \neq i}^{k} a_{ij} v_j, \quad i = 1, 2, \ldots, k.$$

Since $||v||_\infty = 1$, there exists r, $1 \leq r \leq k$, such that $||v||_\infty = |v_r| = 1$. Then

$$|\lambda - a_{rr}| = |(\lambda - a_{rr}) v_r| \leq \sum_{j \neq r}^{k} |a_{rj}||v_j| \leq r_k$$

so that λ is in the disk S_r. □

The following example illustrates the above theorem.

Example 5.9. Consider the difference equation

$$x(n+3) + \frac{1}{2}x(n+2) - \frac{1}{4}x(n+1) + \frac{1}{5}x(n) = 0. \tag{5.2.4}$$

This equation can be converted to the system

$$\begin{aligned} x_1(n+1) &= x_2(n), \\ x_2(n+1) &= x_3(n), \\ x_3(n+1) &= -\frac{1}{5}x_1(n) + \frac{1}{4}x_2(n), -\frac{1}{2}x_3(n) \end{aligned}$$

where $x_1(n) = x(n)$, $x_2(n) = x(n+1)$, $x_3(n) = x(n+2)$.

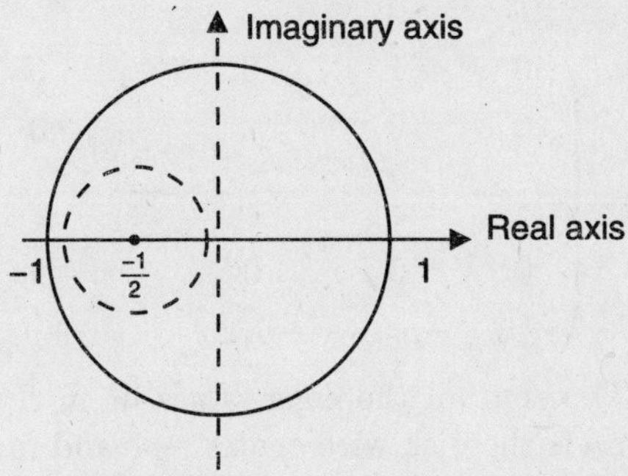

FIGURE 5.2. Gerschgorin disks.

This can be written in the compact form

$$y(n+1) = Ay(n),$$
$$y(n) = (x_1(n), x_2(n), x_3(n))^T,$$
$$A = \begin{pmatrix} 0 & 1 & 0 \\ 0 & 0 & 1 \\ -\frac{1}{5} & \frac{1}{4} & -\frac{1}{2} \end{pmatrix}.$$

The eigenvalues of A are the characteristic roots of (5.2.4). By Gerschgorin's Theorem, all the eigenvalues of A lie in the union of the disks S_1 and S_2, where S_1 is the disk centered at the origin with radius 1 and S_2 is centered at $-\frac{1}{2}$ and with radius $\frac{1}{4} + \frac{1}{5} = \frac{9}{10} = 0.45$ (Figure 5.2). Thus the spectral radius of A, $\rho(A) \leq 1$. In fact we can do better if we realize that an eigenvalue λ_0 of A is also a characteristic root of (5.2.4). Hence $\lambda_0^3 + \frac{1}{2}\lambda_0^2 - \frac{1}{4}\lambda_0 + \frac{1}{5} = 0$ or $\lambda_0^3 = -\frac{1}{2}\lambda_0^2 + \frac{1}{4}\lambda_0 - \frac{1}{5} = 0$. Now if $|\lambda_0| = 1$, $1 = |\lambda_0^3| \leq |-\frac{1}{2}\lambda_0| + |\frac{1}{4}\lambda_0| + |-\frac{1}{5}| = \frac{19}{20}$, a contradiction. Hence $\rho(A) < 1$. Thus by Corollary 2.24, the zero solution of (5.2.4) is asymptotically stable.

Theorem 5.10. *The zero solution of* (5.1.1) *is asymptotically stable if*

$$\boxed{\sum_{i=1}^{k} |p_i| < 1.} \tag{5.2.5}$$

PROOF. We first convert (5.1.1) into a system of first-order difference equations

$$x(n+1) = Ax(n) \tag{5.2.6}$$

where

$$A = \begin{pmatrix} 0 & 1 & 0 & \cdots & 0 \\ 0 & 0 & 1 & \cdots & 0 \\ \vdots & \vdots & \vdots & \vdots & \vdots \\ 0 & 0 & 0 & \cdots & 1 \\ -p_k & -p_{k-1} & -p_{k-2} & \cdots & -p_1 \end{pmatrix}. \tag{5.2.7}$$

By Gerschgorin's Theorem all the eigenvalues lie in $S_1 \cup S_2$, where S_1 is the unit disk, and S_2 is the disk with center $-p_1$ and radius $r = \sum_{i=2}^{k} |p_i|$. By assumption (5.2.5), $|\lambda| \le 1$. To eliminate the possibility that $|\lambda| = 1$, we assume that A has an eigenvalue λ_0 such that $|\lambda_0| = 1$. Now λ_0 is also a characteristic root of (5.1.1) and thus from (5.1.2), $p(\lambda_0) = \lambda_0^k + p_1\lambda_0^{k-1} + \cdots + p_k = 0$. This implies that

$$\begin{aligned} 1 = |\lambda_0^k| &\le |-p_1\lambda_0^{k-1}| + |-p_2\lambda_0^{k-2}| + \cdots + |-p_k| \\ &= \sum_{i=1}^{k} |p_i| \\ &< 1 \end{aligned}$$

which leads to a contradiction. It follows by Corollary 2.24 that the zero solution (5.1.1) is asymptotically stable. □

A partial converse of this theorem may be obtained by a more refined statement of Gerschgorin's Theorem.

Theorem 5.11. *Let S_i, $i = 1, 2, \ldots, k$, be the Gerschgorin disks of a $k \times k$ matrix A. Suppose that for some $1 \le m \le k$, $(\bigcup_{i=1}^{m} S_i) \cap (\bigcup_{i=m+1}^{k} S_i) = \emptyset$. Then there are exactly m eigenvalues (counting multiplicities) of A that lie in $\bigcup_{i=1}^{m} S_i$ and $k - m$ eigenvalues that lie in $\bigcup_{i=m+1}^{k} S_i$.*

PROOF. The proof may be found in [111]. □

We are now ready to provide the promised partial converse to Theorem 5.10.

Theorem 5.12. *The zero solution of* (5.1.1) *is unstable if*

$$\boxed{|p_1| - \sum_{i=2}^{k} |p_i| > 1.} \tag{5.2.8}$$

PROOF. We first convert (5.1.1) into system (5.2.6). Then by Gerschgorin's Theorem all the eigenvalues of A lie in the disks S_1 and S_2 where S_1 is the unit disk and S_2 is the disk centered at $-p_1$ and with radius $r = \sum_{i=2}^{k} |p_i|$. Condition (5.2.8) implies that $S_1 \cap S_2 = \emptyset$. By Theorem 5.11, S_2 must contain an eigenvalue λ_j of A. Moreover, $|\lambda_j| > 1$. Hence the zero solution of (5.1.1) is unstable. □

Exercises 5.1 and 5.2

1. Show that the zero solution of $x(n+4) + p_1 x(n+3) + p_2 x(n+2) + p_3 x(n+1) + p_4 x(n) = 0$ is asymptotically stable if and only if $|p_4| < 1, |p_3 + p_1| < 1 + p_2 + p_4$, and $|p_2(1-p_4) + p_4(1-p_4^2) + p_1(p_4 p_1 - p_3)| < p_4 p_2(1-p_4) + (1-p_4^2) + p_3(p_4 p_1 - p_3)$.

2. Extend the result in Problem 1 to the fifth-order equation

$$x(n+5) + p_1 x(n+4) + p_2 x(n+3) + p_3 x(n+2) + p_4 x(n+1) + p_5 x(n) = 0.$$

3. For what values of α is the zero solution of $x(n+3) - x(n+2) + \frac{\alpha-1}{\alpha} x(n) = 0$ asymptotically stable?

4. Consider the equation $x(n+k) - x(n+k-1) + \frac{\alpha-1}{\alpha} x(n) = 0$, $k \geq 2$.

   (i) Show that the zero solution is asymptotically stable if and only if

$$\text{(L)} \qquad 1 < \alpha < 1 + \left[2\cos\left[\frac{(k-1)\pi}{2k-1}\right] \Big/ \left(1 - 2\cos\left[\frac{(k-1)\pi}{2k-1}\right]\right)\right].$$

   (ii) Show that in (L) as k increases to ∞, α decreases monotonically to 1.

5. Prove that the zero solution of (5.1.37) is asymptotically stable if and only if the zero solution of (5.1.29) is asymptotically stable.

6. Apply Theorem 5.1 to show that the zero solution of the difference equation $\Delta x(n) = -qx(n-1)$, $q > 0$, is asymptotically stable if and only if $q < 1$.

*7. (Hard). Prove that the zero solution of the difference equation $\Delta x(n) = -qx(n-k), q > 0, k > 1$, is asymptotically stable if $qk < 1$.

*8. Consider the linear difference equation

$$x(n+1) - x(n) + \sum_{i=1}^{m} p_i x(n-k_i) = 0, \quad n \in \mathbb{Z}^+,$$

$p_1, p_2, \ldots, p_m \in (0, \infty)$ and $k_1, k_2, \ldots, k_m$ are positive integers. Show that the zero solution is asymptotically stable if $\sum_{i=1}^{m} k_i p_i < 1$.

9. Consider the difference equation

$$x(n+1) - x(n) + px(n-k) - qx(n-m) = 0, \quad n \in \mathbb{Z}^+,$$

where k and m are nonnegative integers, $p \in (0, \infty)$, and $q \in [0, \infty)$. Show that the zero solution is (globally) asymptotically stable if

$$kp < 1 \quad \text{and} \quad q < p(1-kp)/(1+kp).$$

*10. Consider the following model of haematopoiesis (blood cell production) [97], [78]

$$N(n+1) = \alpha N(n) + \frac{\beta}{1 + N^p(n-k)}, \quad n \in \mathbb{Z}^+,$$

and $\alpha \in [0, 1)$, p, $\beta \in (0, \infty)$, k is a positive integer. Here $N(n)$ denotes the density of mature cells in blood circulation.

(a) Find the positive equilibrium point N^*.

(b) Show that N^* is (locally) asymptotically stable if either:

(i) $p \leq 1$; or

(ii) $p > 1$ and $\left(\frac{\beta}{1-\alpha}\right)^p < \frac{p^p}{(p-1)^{p+1}}$.

The following problems give Clark's proof of Theorems 5.10 and 5.12. These proofs are based on Rouché's Theorem from Complex Analysis [20].

Theorem 5.13 (Rouché's Theorem). *Suppose that:*

(i) *two functions $f(z)$ and $g(z)$ are analytic inside and on a simple closed contour γ in the complex domain, and*

(ii) *$|f(z)| > |g(z)|$ at each point on γ.*

Then $f(z)$ and $f(z) + g(z)$ have the same number of zeros, counting multiplicities, inside γ.

11. Use Rouché's Theorem to prove Theorem 5.10.

12. Use Rouché's Theorem to prove Theorem 5.12.

5.3 Stability via Linearization

Let I be a subset of the real line $\mathbb{R}$. Then $I^m = I \times I \times \cdots \times I \subset \mathbb{R}^m$ is the product of m copies of I equipped with any of the norms l_1, l_2, or l_∞ as discussed in Chapter 3.

Now a function $f : I^m \to I$ is continuous at $x = (x_1, x_2, \ldots, x_m)$ if given $\varepsilon > 0$, there exists $\delta > 0$ such that if $y = (y_1, y_2, \ldots, y_m) \in I^m$ and $||x-y|| < \delta$, then $|f(x)-f(y)| < \varepsilon$. Notice that the l_1-norm gives $||x-y||_1 = \sum_{i=1}^m |x_i - y_i|$. For the l_2-norm, we have $||x - y||_2 = \sum_{i=1}^m (x_i^2 - y_i^2)$, and finally the l_∞-norm gives $||x - y||_\infty = \max_{1 \leq i \leq m} |x_i - y_i|$.

Following the work of Ladas and his collaborators [85] we will use the l_1-norm, unless otherwise noted.

Consider the following difference equation of order $k + 1$,

$$x(n + 1) = f(x(n), x(n - 1), \ldots, x(n - k)) \tag{5.3.1}$$

where $f : I^{k+1} \to I$ is a continuous function.

Given a set of $(k + 1)$ initial conditions $x_{-k}, x_{-k+1}, \ldots, x_0 \in I$, there exists a unique solution $\{x(n)\}_{n=-k}^{\infty}$ of (5.3.1) such that $x(-k) = x_k, x(-k + 1) = x_{-k+1}, \ldots, x(0) = x_0$. Of course one may convert (5.3.1) to a system of first-order difference equations of order $k+1$ as follows. Let $y_1(n) = x(n-k), y_2(n) = (x-k+1), \ldots, y_{k+1}(n) = x(n)$. Then (5.3.1) may be converted to the system

$$y(n+1) = F(y(n)), \tag{5.3.2}$$

where

$$y(n) = (y_1(n), y_2(n), \ldots, y_{k+1}(n))^T,$$
$$F(y(n)) = (y_2(n), y_3(n), \ldots, y_{k+1}(n), f(y_{k+1}(n), y_k(n), \ldots, y_1(n))^T.$$

We may write $F = (F_1, F_2, \ldots, F_{k+1})^T$, where $F_1(y_1) = y_2, F_2(y_2) = y_3, \ldots, F_{k+1}(y_{k+1}) = f(y_{k+1}, \ldots, y_1)$.

A point $x^* \in I$ is an equilibrium point of (5.3.1) if $f(x^*, x^*, \ldots, x^*) = x^*$. This corresponds to the equilibrium point $(x^*, x^*, \ldots, x^*) \in \mathbb{R}^{k+1}$ for system (5.3.2). Notions of stability of equilibrium points and periodic points of (5.3.1) may be stated via (5.3.2) and the use of proper interpretations of the notions in regards to (5.3.1). Here is a sample.

Definition 5.14. An equilibrium point x^* of (5.3.1) is stable (S) if, given $\varepsilon > 0$, there exists $\delta > 0$ such that if $\{x(n)\}_{n=-k}^{\infty}$ is a solution of (5.3.1) with

$$(|x(-k) - x^*| + |x(-k+1) - x^*| + \cdots + |x(0) - x^*|) < \delta,$$

then

$$|x(n) - x^*| < \varepsilon \quad \text{for all} \quad n \geq -k.$$

Analogous definitions can be given for the remaining notions of stability as defined in Chapter 4.

If f is continuously differentiable in some open neighborhood of $X^* = (x^*, x^*, \ldots, x^*)$, then one can linearize (5.3.1) around X^*. One way to do this is to revert to system (5.3.2) to obtain the linear system

$$z(n+1) = Az(n), \tag{5.3.3}$$

where A is the Jacobian of F at X^*, $A = DF(X^*)$. Then convert (5.3.3) to a scalar equation. However, one may also linearize (5.3.1) directly using the chain rule. Thus the linearized equation around x^* is given by

$$u(n+1) = p_0 u(n) + p_1 u(n-1) + \cdots + p_k u(n-k), \tag{5.3.4}$$

where

$$p_i = \frac{\partial f}{\partial u_i}(\bar{x}, \bar{x}, \ldots, \bar{x}),$$

with $f(u_0, u_1, \ldots, u_k)$.

The characteristic equation of (5.3.4) is given by

$$\lambda^{k+1} - p_0 \lambda^k - p_1 \lambda^{k-1} - \cdots - p_k = 0. \tag{5.3.5}$$

Using the stability results in Chapter 4 for system (5.3.2), one may easily establish the following fundamental stability result.

Theorem 5.15 (The Linearized Stability Result). *Suppose that* f *is continuously differentiable on an open neighborhood* $G \subset \mathbb{R}^{k+1}$ *of* $(x^*, x^*, \ldots, x^*)$, *where* x^* *is a fixed point of* (5.3.1). *Then the following statements hold true:*

(i) *If all the characteristic roots of* (5.3.5) *lie inside the unit disk in the complex plane, then the equilibrium point* x^* *of* (5.3.1) *is (locally) asymptotically stable.*

(ii) *If at least one characteristic root of* (5.3.5) *is outside the unit disk in the complex plane, the equilibrium point* x^* *is unstable.*

(iii) *If one characteristic root of* (5.3.5) *is on the unit disk and all the other characteristic roots are either inside or on the unit disk, then the equilibrium point* x^* *may be stable, unstable, or asymptotically stable.*

PROOF. The proofs of (i) and (ii) follow from Corollary 4.34 and Theorem 4.38.

(iii) This part may be proved by the following examples. First consider the logistic equation $x(n+1) = x(n)(1 - x(n)) = f(x(n))$. The linearized equation around the equilibrium point $x^* = 0$ is given by $u(n+1) = u(n)$ with the characteristic roots $\lambda = 1$. But we know from Section 1.6 that $x^* = 0$ is unstable.

Now we give an example that produces a different conclusion from the above example. Consider the equation

$$x(n+1) = x(n) - x^3(n) = f(x(n)).$$

The linearized equation is given by

$$u(n+1) = u(n)$$

with the characteristic root $\lambda = 1$. Now for the equilibrium point $x^* = 0$, we have $f'(0) = 1$, $f''(0) = 0$, $f'''(0) = -6 < 0$. This implies by Theorem 1.15 that $x^* = 0$ is asymptotically stable. □

Example 5.16 [15]. Consider the difference equation

$$x(n+1) = ax(n) + F(x(n-k)) \tag{5.3.6}$$

which models whale populations. Here $x(n)$ represents the adult breeding population, a, $0 \leq a \leq 1$, the survival coefficient, and $F(x(n-k))$ the recruitment to the adult stage with a delay of k years. The equilibrium point x^* of (5.3.6) is given by the equation

$$x^* = ax^* + F(x^*),$$

$$x^* = F(x^*)/(1-a). \tag{5.3.7}$$

Since $F(x^*) = (1-a)x^*$, $(1-a)$ is the annual mortality rate of the whale population. The linearized equation associated with (5.3.6) is given by

$$u(n+1) = au(n) + bu(n-k), \tag{5.3.8}$$

where $b = F'(x^*)$.

Equation (5.3.8) may be written in the form

$$u(n+1) - au(n) - bu(n-k) = 0. \tag{5.3.9}$$

By Theorem 5.10, a sufficient condition for the zero solution of (5.3.9) to be asymptotically stable is

$$\begin{aligned} |a| + |b| &< 1, \\ a + |b| &< 1. \end{aligned} \tag{5.3.10}$$

Condition (5.3.10) is a sufficient condition for the asymptotic stability of the equilibrium point x^* given by (5.3.7).

Exercises 5.3

1. Consider the delayed recruitment model

$$x(n+1) = \frac{1}{2}x(n) + F(x(n-k)).$$

Let x^* be the equilibrium point and let $b = F'(x^*)$. Assume that F is continuously differentiable in an open neighborhood of x^*. Find sufficient conditions for x^* to be asymptotically stable if:

   (i) $k = 2$,

   (ii) $k = 3$.

2. Consider the single species, age-structured population model

$$x(n+2) = x(n)\exp(r - ax(n+1) - x(n)),$$

where $x_n \geq 0$ for all $n \in \mathbb{Z}^+$, $a, r > 0$.

   (i) Show that all solutions are bounded.

   (ii) Find conditions on r and α under which the positive equilibrium is asymptotically stable.

In Subsection 4.7.5 we studied in detail the larval–pupal–adult (LPA) of the flour beetle.

$$\begin{aligned} L(n+1) &= bA(n)\exp(-c_{EA}A(n) - c_{EL}L(n)), \\ P(n+1) &= (1-\mu_L)L(n), \\ A(n+1) &= P(n)\exp(-c_{PA}A(n)) + (1-\mu_A)A(n). \end{aligned} \tag{5.3.11}$$

Kuang and Cushing [83] considered the simplified case when larval cannibalism of eggs is not present, i.e., $c_{EL} = 0$. Problems 3 though 5 refer to this simplified model.

3. Prove that (5.3.11) reduces to

$$x(n+1) - \alpha x(n) - \beta x(n-2)\exp(-c_1 x(n-2) - c_2 x(n)) = 0, \quad (5.3.12)$$

where $\alpha = 1-\mu_A$, $\beta = b(1-\mu_L)$, $c_1 = c_{EA}$, $c_2 = c_{PA}$, $x(n) = A(n+2)$, $n \geq -2$.

Then show that if $\alpha + \beta \leq 1$, equation (5.3.12) has only the trivial equilibrium $x_1^* = 0$. Furthermore, if $\alpha + \beta > 1$, then (5.3.12) has two equilibria, $x_1^* = 0$ and $x_2^* > 0$, with $x_2^* = (1/c_1 + c_2)\ln(\beta/(1-\alpha))$.

4. Show that the linearized equation of (5.3.12) around an equilibrium point x^* is given by

$$y(n+1) - [\alpha - \beta c_2 x^* \exp\{-(c_1+c_2)x^*\}]y(n) \\ - \beta(1 - c_1 x^*)\exp\{-(c_1+c_2)x^*\}y(n-2) = 0.$$

5. Prove that:

   (i) The trivial solution $x_1^* = 0$ is asymptotically stable

$$\frac{b(1-\mu_L)}{\mu_A} < 1.$$

   (ii) The positive equilibrium $x_2^* = (1/c_1 + c_2)\ln(\beta/(1-\alpha))$ is asymptotically stable if and only if

$$|A+B| < 1, \quad |A-3B| < 3, \quad \text{and} \quad B(B-A) < 1, \quad (5.3.13)$$

   where

$$A = \frac{c_2(1-\alpha)}{c_1+c_2}\left(\ln\frac{\beta}{1-\alpha}\right) - \alpha,$$

$$B = (1-\alpha)\left[\frac{c_1}{c_1+c_2}\left(\ln\frac{\beta}{1-\alpha}\right) - 1\right].$$

6. Consider the difference equation $N(n+1) - N(n) = N(n)[a + bN(n-k) - cN^2(n-k)]$, $n \in \mathbb{Z}^+$, where $a, c \in [0,\infty)$ and $b \in \mathbb{R}$, $k \in \mathbb{Z}^+$.

   (a) Prove the equation has a unique positive equilibrium N^*.

   (b) Show that N^* is (locally) asymptotically stable if

$$N^*\sqrt{b^2+4ac} < 2\cos\left[\frac{k\pi}{2k+1}\right].$$

7. Consider the rational difference equation

$$x(n+1) = \left(a + \sum_{i=0}^{k} a_i x(n-i)\right) \Bigg/ \left(b + \sum_{i=0}^{k} b_i x(n-i)\right),$$

where k is a nonnegative integer, $a_0, a_1, \ldots, a_k, b_0, b_1, \ldots, b_k \in [0,\infty)$, $a, b \in (0,\infty)$, $\sum_{i=0}^{k} a_i = 1$ and $B = \sum_{i=0}^{k} b_i > 0$. Find the positive equilibrium point x^* of the equation. Then show that x^* is asymptotically stable if $b > 1$.

Consider the rational difference equation

$$x(n+1) = [a + bx(n)]/[A + x(n-k)], \tag{5.3.14}$$

$n \in \mathbb{Z}^+$, $a, b \in [0, \infty)$, with $a + b > 0$, $A \in (0, \infty)$, and k is a positive integer. Problems 8, 9, and 10 refer to equation (5.3.14) with the above assumptions.

8. (i) Show that if either $a > 0$ or $a = 0$ and $b > A$, equation (5.3.14) has a positive equilibrium x^*. Then find x^*.

   (ii) Show that x^* is (locally) asymptotically stable if either $b = 0$ or $a = 0$ and

$$b > A > b\left(1 - 2\cos\frac{k\pi}{2k+1}\right).$$

9. Show that the positive equilibrium x^* of equation (5.3.14) is asymptotically stable if either $k = 1$ and $a > 0$ or $k \geq 2$ and $A > b$.

*10. Show that for any positive solution $x(n)$ of equation (5.3.14) there exists positive constants C and D such that $C \leq x(n) \leq D$, $n \in \mathbb{Z}^+$ provided that $b > 1$.

5.4 Global Stability of Nonlinear Equations

Results on global asymptotic stability are scarce and far from complete. In Chapter 4 we have seen how Liapunov functions can be used to establish both local and global asymptotic stability. In this section we will utilize the special nature of scalar equations to present a few results that can deal with several types of equations. Roughly speaking, we will be mainly concerned with "monotone" equations. A more general investigation on monotone discrete dynamical systems is beyond the scope of this book and the interested reader is referred to the work of Hal Smith [138]. More detailed expositions may also be found in Sedaghat [133], Kocic and Ladas [80], and Kulenovic and Ladas [85].

Consider the following difference equation of order $(k+1)$,

$$x(n+1) = f(x(n), x(n-1), \ldots, x(n-k)), \tag{5.4.1}$$

$n \in \mathbb{Z}^+$ and $k \geq 1$ is a positive integer. The main result in this section is due to Hautus and Bolis [65]. (See also Kocic and Ladas [80] and Kulenovic and Ladas [84].)

Theorem 5.17. *Consider (5.4.1) with $f \in C(I^{k+1}, \mathbb{R})$, where I is an open interval in $\mathbb{R}$ and $x^* \in I$ is an equilibrium point. Suppose that f satisfies the following assumptions:*

(i) *f is nonincreasing in each of its arguments, i.e., if $a \leq b$, then $f(\cdot, \ldots, a, \ldots, \cdot) \leq (f(\cdot, \ldots, b, \ldots, \cdot)$.*

(ii) *$(u - x^*)[f(u, u, \ldots, u) - u] < 0$ for all $u \in I/\{x^*\}$.*

Then with initial values $(x(0), x(-1), \ldots, x(-k)) \in I$, we have $x(n) \in I$ for $n \geq 0$ and $\lim_{n\to\infty} x(n) = x^*$.

PROOF. Condition (ii) ensures that x^* is the only equilibrium point in I. For if $y^* \in I$ is another equilibrium point, then $(y^* - x^*)[f(y^*, y^*, \ldots, y^*) - y^*] = 0$ which violates condition (ii). Let $x(n)$ be a solution of (5.4.1) with $x(0), x(-1), \ldots, x(-k) \in I$. Set $m = \min\{x^*, x(0), \ldots, x(-k)\}$, $M = \max\{x^*, x(0), \ldots, x(-k)\}$. By condition (ii) and since $m \leq x^*$, we have $m \leq f(m, m, \ldots, m)$. Moreover, by condition (i) we obtain $f(m, m, \ldots, m) \leq f(x(0), x(-1), \ldots, x(-k)) = x(1)$. Thus $m \leq f(m, m, \ldots, m) \leq x(1)$. Similarly, one may show that $x(1) \leq f(M, M, \ldots, M) \leq M$. By induction on n, it is easy to show that $m \leq x(n) \leq M$ for all $n \geq -k$. In particular, since $[m, M] \subset I$, it follows that $x(n) \in I$, for all $n \geq -k$. Since $x(n)$ is bounded, both $\lim_{n\to\infty} \inf x(n) = L_1$ and $\lim_{n\to\infty} \sup x(n) = L_2$ exist. Furthermore, $m \leq L_1 \leq L_2 \leq M$. Let $\varepsilon > 0$ be sufficiently small such that $[m + \varepsilon, M + \varepsilon] \subset I$. There exists a positive integer N such that $L_1 - \varepsilon < x(n-k)$ for all $n \geq N$. This implies by condition (i) that $f(L_1 - \varepsilon, L_1 - \varepsilon, \ldots, L_1 - \varepsilon) \leq f(x(n), x(n-1), \ldots, x(n-k)) = x(n+1)$, for all $n \geq N$. Consequently, $f(L_1 - \varepsilon, L_1 - \varepsilon, \ldots, L_1 - \varepsilon) \leq L_1$. Since f is continuous, and ε is arbitrary, it follows that $f(L_1, L_1, \ldots, L_1) \leq L_1$. This implies by condition (ii) that $x^* \leq L_1$. By a similar argument (Problem 12). one may show that $L_2 \leq x^*$ and, consequently, $L_2 \leq x^* \leq L_1$. Hence $x^* = L_1 = L_2$, and $\lim_{n\to\infty} x(n) = x^*$. □

Example 5.18. The Beverton–Holt model [10]

$$x(n+1) = \frac{rKx(n)}{K + (r-1)x(n)}, \quad K > 0, \quad r > 0, \tag{5.4.2}$$

has been used to model populations of bottom-feeding fish, including the North Atlantic plaice and haddock. These species have very high fertility rates and very low survivorship to adulthood. Furthermore, recruitment is essentially unaffected by fishing.

Local stability analysis (Theorem 1.13) reveals that:

(i) if $0 < r < 1$, then the zero solution $x_1^* = 0$ is asymptotically stable,

(ii) if $r > 1$, then the equilibrium $x_2^* = K$ is asymptotically stable.

Using Theorem 5.17 one can say more about x_2^*. Since $f(x) = rKx/K + (r-1)x$ is monotonically increasing for $r > 1$, condition (i) in Theorem

5.17 holds. Now for any $u \in (0, \infty)$,

$$(u-K)(f(u)-u) = (u-K)\left(\frac{rKu - u(K+(r-1)u)}{K+(r-1)u}\right)$$
$$= -u(r-1)\frac{(K-u)^2}{K+(r-1)u} < 0$$

and condition (ii) in Theorem 5.17 is satisfied. It follows that $x_2^* = K$ is globally asymptotically stable.

Cushing and Henson [26] used more elementary methods to prove the above results.

Now consider the second-order (modified) Beverton–Holt model

$$x(n+1) = \frac{rK(\alpha x(n) + \beta x(n-1))}{K+(r-1)x(n-1)} \tag{5.4.3}$$

where $r > 0$, $K > 0$, and $\alpha + \beta = 1$.

In this model, the future generation $x(n+1)$ depends not only on the present generation $x(n)$ but also on the previous generation $x(n-1)$. This model has two equilibrium points as before $x_1^* = 0$ and $x_2^* = K$.

We first investigate the local stability of these two equilibria. Using Theorem 5.15, we have the following conclusions:

(a) The zero solution $x_1^* = 0$ of (5.4.3) is locally asymptotically stable if and only if $0 < r < 1$.

(b) The equilibrium $x_2^* = K$ is locally asymptotically stable if and only if $r > 1$.

In fact, one can say more about $x_2^* = K$. Following the same analysis used for (5.4.2), one may conclude by employing Theorem 5.17 that $x_2^* = K$ is in fact globally asymptotically stable. A higher-order Beverton–Holt equation has been investigated in [81].

The second result that we will present is of a different flavor. It is much more flexible than Theorem 5.17 since it allows f to be either nondecreasing or nonincreasing in its arguments. This leads to the notion of weak monotonicity.

Definition 5.19. The function $f(u_1, u_2, \ldots, u_{k+1})$ is said to be *weakly monotonic* if f is nondecreasing or nonincreasing in each of its arguments, i.e., for a given integer j, $1 \le j \le k+1$, if $a \le b$, then either

$$f(\cdot, \ldots, a, \ldots, \cdot) \le f(\cdot, \ldots, b, \ldots, \cdot) \quad \text{or}$$
$$f(\cdot, \ldots, a, \ldots, \cdot) \ge f(\cdot, \ldots, b, \ldots, \cdot),$$

where a and b are in the jth slot, and all the other slots are filled with fixed numbers $z_1, z_2, \ldots, z_{j-1}, z_j, \ldots, z_{k+1}$.

Theorem 5.20 [60]. *Suppose that f in (5.4.1) is continuous and weakly monotonic. Assume, in addition, that whenever (m, M) is a solution of the* *system*

$$m = f(m_1, m_2, \ldots, m_{k+1}) \quad \textit{and} \quad M = f(M_1, M_2, \ldots, M_{k+1}),$$

where, for each $i = 1, 2, \ldots, k+1$,

$$m_i = \begin{cases} m & \textit{if } f \textit{ is nondecreasing in } z_i, \\ M & \textit{if } f \textit{ is nonincreasing in } z_i, \end{cases}$$

and

$$M_i = \begin{cases} M & \textit{if } f \textit{ is nondecreasing in } z_i, \\ m & \textit{if } f \textit{ is nonincreasing in } z_i, \end{cases}$$

then $m = M$.

Then (5.4.1) *has a unique equilibrium point* $x^* = m$ *which is globally attracting.*

PROOF. Let $m^0 = a$, $M^0 = b$, and for each $i = 1, 2, \ldots$ set

$$M^i = f(M_1^{i-1}, M_2^{i-1}, \ldots, M_{k+1}^{i-1})$$

and

$$m^i = f(m_1^{i-1}, m_2^{i-1}, \ldots, m_{k+1}^{i-1}),$$

where

$$m_i^0 = \begin{cases} a & \text{if } f \text{ is nondecreasing in the } i\text{th slot}, \\ b & \text{if } f \text{ is nonincreasing in the } i\text{th slot}, \end{cases}$$

and

$$M_i^0 = \begin{cases} b & \text{if } f \text{ is nondecreasing in the } i\text{th slot}, \\ a & \text{if } f \text{ is nonincreasing in the } i\text{th slot}. \end{cases}$$

For $r > 0$,

$$m_i^r = \begin{cases} m_i^{r-1} & \text{if } f \text{ is nondecreasing in the } i\text{th slot}, \\ M_i^{r-1} & \text{if } f \text{ is nonincreasing in the } i\text{th slot}, \end{cases}$$

and

$$M_i^r = \begin{cases} M_i^{r-1} & \text{if } f \text{ is nondecreasing in the } i\text{th slot}, \\ m_i^{r-1} & \text{if } f \text{ is nonincreasing in the } i\text{th slot}. \end{cases}$$

It follows from the assumptions on f that, for $i \geq 0$,

$$m^0 \leq m^1 \leq \cdots < m^i \leq \cdots \leq M^i \leq \cdots \leq M^1 \leq M^0.$$

Furthermore, $m^i \leq x(n) \leq M^i$ for $n \geq 2i+1$. Set $m = \lim_{i\to\infty} m^i$ and $M = \lim_{i\to\infty} M^i$. Then $M \geq \limsup_{n\to\infty} x(n) \geq \liminf_{n\to\infty} x(n) \geq m$. Since f is continuous, it follows that $m = f(m_1, m_2, \ldots, m_{k+1})$ and $M = f(M_1, M_2, \ldots, M_{k+1})$. Hence $x^* = m = M$ is the unique equilibrium point of (5.4.1). Moreover, $\lim_{n\to\infty} x(n) = 0$ and the proof of the theorem is now complete. □

Example 5.21 [85]. Consider the difference equation

$$y(n+1) = \frac{py(n) + y(n-1)}{p + y(n-1)}, \quad p > 0, \quad n \in \mathbb{Z}^+. \tag{5.4.4}$$

This equation has only two fixed points $y_1^* = 0$ and $y_2^* = 1$. Local stability analysis shows that y_2^* is locally asymptotically stable. Now assume that $y(n)$ is a solution of (5.4.4), such that $y(n) \geq 1$ for all $n \in \mathbb{Z}^+$. Then from (5.4.4) we have $y(n+1) - y(n) = (1 - y(n))\frac{y(n-1)}{p+y(n-1)} \leq 0$. This implies that $y(n)$ is nonincreasing and thus has a limit in $[1, \infty)$. But this leads to a contradiction since $0 < y_2^* < 1$. Hence for some positive integer N, $y(N) \in (0,1)$. Writing (5.4.4) in the form

$$y(n+1) - 1 = [y(n) - 1]\frac{p}{p + y(n-1)}$$

we conclude that $y(N+r) \in (0,1)$ for all $r \in \mathbb{Z}^+$. Now in the interval $(0,1)$, the function $f(u,v) = \frac{pv+u}{p+u}$ is increasing in both arguments and, by Theorem 5.20, y_2^* is globally asymptotically stable.

The following corollary of Theorem 5.20 is easy to apply to establish the global asymptotic stability of the zero solution.

An independent proof of this conclusion may be found in Grove et al. [61].

Corollary 5.22. *Contemplate the difference equation*

$$x(n+1) = \sum_{i=0}^{k} x(n-i) f_i(x(n), x(n-1), \ldots, x(n-k)), \quad n \in \mathbb{Z}^+, \tag{5.4.5}$$

with initial values $x(0), x(-1), \ldots, x(-k) \in [0, \infty)$ such that

(i) $k \in \mathbb{Z}^+$;

(ii) $f_0, f_1, \ldots, f_k \in C\left[[0,\infty)^{k+1}, [0,1)\right]$;

(iii) $f_0, f_1, \ldots, f_k$ *are nonincreasing in each argument;*

(iv) $\sum_{i=0}^{k} f_i(u_0, u_1, \ldots, u_k) < 1$ *for all* $(u_0, u_1, \ldots, u_k) \in (0, \infty)^{k+1}$;

(v) $f_0(u, u, \ldots, u) > 0$ *for all* $u \geq 0$.

Then the trivial solution $x^* = 0$ of (5.4.5) is globally asymptotically stable.

Proof. This will be left as Exercises 5.4, Problem 13. □

Example 5.23. Consider again (5.4.4). If $p + 1 < q$, then $y_1^* = 0$ is the only equilibrium point in $[0, \infty)$. By Theorem 5.15, $y_1^* = 0$ is locally asymptotically stable. Now we write (5.4.4) in the form

$$y(n+1) = y(n) \cdot \frac{p}{q + y(n-1)} + y(n-1) \cdot \frac{1}{q + y(n-1)}.$$

Hence

$$f_0(u, v) = \frac{p}{q+v}, \quad f_1(y, v) = \frac{1}{q+v}.$$

It is easy to show that f_0 and f_1 satisfy all the conditions (i) through (iv) in Corollary 5.22. Hence by Corollary 5.22, $y_1^* = 0$ is globally asymptotically stable.

Exercise 5.4

1. Consider a modified Beverton–Holt equation

$$x(n+1) = \frac{rK(\alpha x(n) + \beta x(n-1))}{K + (r-1)x(n-1)}, \quad \alpha + \beta = 1, \quad \alpha, \beta > 0.$$

   Show that the zero solution is globally asymptotically stable if $0 < r < 1$.

2. Show that the zero solution of the equation $x(n+1) = ax(n-k)\exp[-b(x^2(n) + \cdots + x^2(n-m))]$, $|a| \leq 1$, $b > 0$, is globally asymptotically stable.

3. Consider the LPA model of the flour beetle (5.3.11) with no larval cannibalism on eggs:

$$x(n+1) = \alpha x(n) + \beta x(n-2)\exp[-c_1 x(n-2) - c_2 x(n)]$$

   where $\alpha = 1 - \mu_A$, $\beta = b(1 - \mu_L)$, $c_1 = c_{EA}$, $c_2 = c_{PA}$, $x(n) = A(n+2)$. Show that the zero solution is globally asymptotically stable if $\alpha + \beta \leq 1$ and $\beta > 0$.

4. The following equation describes the growth of a mosquito population:

$$x(n+1) = (ax(n) + bx(n-1)e^{-x(n-1)})e^{-x(n)}, \quad n \in \mathbb{Z}^+,$$

   where $a \in (0, 1)$, $b \in [0, \infty)$. Prove that the zero solution is globally asymptotically stable if $a + b \leq 1$.

5. A variation of the mosquito model in Problem 4 is given by the equation

$$x(n+1) = (ax(n) + bx(n-1))e^{-x(n)}, \quad n \in \mathbb{Z}^+, \tag{5.4.6}$$

   where $a \in [0, 1)$, $b \in (0, \infty)$. Prove that the zero solution is globally asymptotically stable if $a + b \leq 1$ and $b < 1$.

6.[1] Show that the positive equilibrium of the equation

$$x(n+1) = \frac{p + qx(n-1)}{1 + x(n)}, \quad p, q > 0,$$

is globally asymptotically stable if $q < 1$.

7. Show that the positive equilibrium of the equation

$$x(n+1) = \frac{p + x(n-1)}{qx(n) + x(n-1)}, \quad p, q > 0,$$

is globally asymptotically stable if $q \leq 1 + 4p$.

8. Consider equation (5.4.4). Show that if $y(-1)+y(0) > 0$ and $p+1 > q$, then the positive equilibrium $y^* = p+1-q$ is globally asymptotically stable.

9. Consider the equation

$$x(n+1) = \frac{x(n) + p}{x(n) + qx(n-1)}, \quad p, q > 0, \quad n \in \mathbb{Z}^+.$$

Show that the positive equilibrium point of the equation is globally asymptotically stable if $q \leq 1 + 4p$.

10. Show that the positive equilibrium point of the equation

$$x(n+1) = \frac{p + qx(n)}{1 + x(n-1)}, \quad p, q > 0, \quad n \in \mathbb{Z}^+,$$

is globally asymptotically stable if one of the following two conditions holds:

(i) $q < 1$,

(ii) $q \geq 1$ and either $p \leq q$ or $q < p < 2(q+1)$.

11. Show that the positive equilibrium of the equation

$$x(n+1) = \frac{px(n) + x(n-1)}{qx(n) + x(n-1)}, \quad p, q > 0, \quad n \in \mathbb{Z}^+,$$

is globally asymptotically stable if $q < pq + 1 + 3p$ and $p < q$.

12. Complete the proof of the Theorem 5.17 by showing that $L_2 \leq x^*$.

13. Prove Corollary 5.22.

*14. (Term project). Consider equation (5.4.6) with the assumption $a+b > 1$ and $b < \frac{e-a}{e+1}$ (where e^x is the exponential function). Show that the positive equilibrium is globally asymptotically stable, with basin of attraction $[0, \infty) \times [0, \infty)/\{(0,0)\}$.

---

[1]Problems 6–11 are from Kulenovic and Ladas [85].

*15. Conjecture [85]. Prove that every positive solution of the equation

$$x(n+1) = \frac{\alpha + \beta x(n) + \gamma x(n-1)}{cx(n-1)},$$

where $n \in \mathbb{Z}^+$, $\alpha c > 0$, converges to the positive equilibrium of the equation.

*16. Conjecture [85]. Consider the equation

$$x(n+1) = \frac{\alpha + \gamma x(n-1)}{A + Bx(n) + Cx(n-1)},$$

where $\alpha, \gamma, A, B, C > 0$. Prove that if the equation has no solutions of prime period 2, then the positive equilibrium is globally asymptotically stable.

(Open Problems: Kulenovic and Ladas.) Assume that $p, q, r \in [0, \infty)$, $k \geq 2$ is a positive integer. Investigate the global stability of the following equations.

*17. $y(n+1) = \dfrac{p + qy(n)}{1 + y(n) + ry(n-k)}$.

*18. $y(n+1) = \dfrac{p + qy(n-k)}{1 + y(n) + ry(n-k)}$.

*19. $y(n+1) = \dfrac{py(n) + y(n-k)}{r + qy(n) + y(n-k)}$.

5.5 Applications

5.5.1 *Flour Beetles*

In this section we consider again the LPA model [83] of the flour beetle (5.3.12) with no larval cannibalism on eggs:

$$\begin{aligned} x(n+1) &= \alpha x(n) + \beta x(n-2) \exp[-c_1 x(n-2) - c_2 x(n)] \\ &= f(x(n), x(n-1), x(n-2)) \end{aligned} \tag{5.5.1}$$

where $\alpha = 1 - \mu_A$, $\beta = b(1 - \mu_L)$, $c_1 = c_{EA}$, $c_2 = c_{PA}$, $x(n) = A(n+2)$.

Using the Schur–Cohn Criterion one may show that the positive equilibrium $x_2^* = \frac{1}{c_1+c_2} \ln\left(\frac{\beta}{1-\alpha}\right)$ is (locally) asymptotically stable if and only if the following conditions hold:

$$|A + B| < 1, \quad |A - 3B| < 3, \quad B(B - A) < 1, \tag{5.5.2}$$

where

$$A = \frac{c_2(1-\alpha)}{c_1+c_2}\left(\ln \frac{\beta}{1-\alpha}\right) - \alpha, \quad B = (1-\alpha)\left[\frac{c_1}{c_1+c_2}\left(\ln \frac{\beta}{1-\alpha}\right) - 1\right].$$

In the sequel, we will go one step further and prove that the equilibrium point x_2^* is in fact globally asymptotically stable under the following assumptions:

$$A_1 : \alpha + \beta > 1 \quad \text{and} \quad \beta < \min\{e(1-\alpha), e\alpha c_1/c_2\},$$
$$A_2 : \max\{x(-2), x(-1), x(0)\} > 0.$$

The following two lemmas from Kuang and Cushing [83] will facilitate the proof of the main result and make it more transparent.

Lemma 5.24. *If $\alpha + \beta > 1$, then*

$$\limsup_{n\to\infty} x(n) \leq \frac{\beta}{c_1 e(1-\alpha)}. \tag{5.5.3}$$

PROOF. Consider the function $h(x) = c_1 e x e^{-c_1 x}$. This function has a critical point x_c, where

$$h'(x_c) = c_1 e^{1-c_1 x_c} - c_1^2 x_c e^{1-c_1 x_c} = 0.$$

Hence $x_c = \frac{1}{c_1}$, and $h\left(\frac{1}{c_1}\right) = 1$ is the maximum value of h. It follows that $c_1 e x e^{-c_1 x} \leq 1$ and, consequently,

$$x e^{-c_1 x} \leq \frac{1}{c_1 e}. \tag{5.5.4}$$

Going back to equation (5.5.1) we obtain

$$\begin{aligned} x(n+1) &\leq \alpha x(n) + \beta x(n-2)\exp(-c_1 x(n-2)) \\ &\leq \alpha x(n) + \frac{\beta}{c_1 e} \quad \text{(by (5.5.4))}. \end{aligned} \tag{5.5.5}$$

By (1.2.8), the solution of (5.5.5) is given by

$$x(n) \leq \alpha^n x(0) + \frac{\beta}{c_1 e(1-\alpha)}(1-\alpha^n)$$

and since $\alpha \in (0,1)$, it follows that

$$\limsup_{n\to\infty} x(n) \leq \frac{\beta}{c_1 e(1-\alpha)}. \qquad \square$$

The next lemma shows that (5.5.1) satisfies condition (ii) in Theorem 5.17.

Lemma 5.25. *For any $u > 0$, $u \neq x_2^*$,*

$$(u - x_2^*)[f(u,u,u) - u] < 0. \tag{5.5.6}$$

PROOF. Let $g(u) = f(u,u,u_r) - u$. Then

$$g(u) = u[\alpha + \beta\exp(-(c_1+c_2)u) - 1].$$

Clearly $g(u) = 0$ if and only if $u = 0$ or $u = x_2^*$. We now have two cases to consider.

Case (a): If $0 < u < x_2^* = \frac{1}{c_1+c_2} \ln\left(\frac{\beta}{1-\alpha}\right)$, then

$$\alpha + \beta \exp[-(c_1 + c_2)u] - 1 > \alpha + \beta e^{-(c_1+c_2)x_2^*} - 1 = 0,$$

and (5.5.6) holds true.

Case (b): If $u > x_2^*$, then

$$\alpha + \beta \exp[-(c_1 + c_2)u] - 1 < \alpha + \beta \exp[-(c_1 + c_2)x_2^*] - 1 = 0,$$

and (5.5.6) holds true. □

We are now ready to prove the following theorem.

Theorem 5.26. *If conditions A_1 and A_2 hold, then*

$$\lim_{n\to\infty} x(n) = x_2^*.$$

PROOF. By virtue of Lemma 5.25 it remains to show that condition (ii) of Theorem 5.17 holds. To accomplish this task we need to show that $\frac{\partial f}{\partial x(n)}, \frac{\partial f}{\partial x(n-1)}, \frac{\partial f}{\partial x(n-2)} \geq 0$ on a region D. Simple computations show that

$$\frac{\partial f}{\partial x(n)} = \alpha - c_2\beta x(n-2) \exp[-c_1 x(n-2) - c_2 x(n)],$$

$$\frac{\partial f}{\partial x(n-1)} = 0,$$

$$\frac{\partial f}{\partial x(n-2)} = \beta(1 - c_1 x(n-2)) \exp[-c_1 x(n-2) - c_2 x(n)].$$

Now by Lemma 5.24, $\limsup_{n\to\infty} x(n) \leq \beta[c_1 e(1-\alpha)]^{-1}$. Since $\beta < e(1-\alpha)$, this implies that $\limsup_{n\to\infty} \leq \frac{1}{c_1}$. Hence, for $n > N$, for some integer N greater than 2, $x(n-2) < \frac{1}{c_1}$. Let $I = \left(0, \frac{1}{c}\right)$ and $D = I^3$. Then $x_2^* \in I$ and $\frac{\partial f}{\partial x(n-i)} \geq 0$ for $i = 1, 2$. Furthermore, $\frac{\partial f}{\partial x(n)} \geq \alpha - c_2\beta \exp(-c_1 x(n-2)) \geq \alpha - \frac{c_2\beta}{c_1 e} \geq 0$. This shows that f is nondecreasing in each of its arguments restricted to D. Since $\max\{x(-2), x(-1), x(0)\} > 0$, there is n_0 such that for $n \geq n_0$, $0 < x(n-2) < \frac{1}{c_1}$. Let $\hat{x}(0) = x(n_0)$, $\hat{x}(-1) = x(n_0 - 1)$, $\hat{x}(-2) = x(n_0 - 2)$. Then since the equation is autonomous, $\hat{x}(n) = x(n)$ for $n \geq n_0$. By Theorem 5.17, we have $\lim_{n\to\infty} x(n) = x_2^*$. □

5.5.2 A Mosquito Model

Consider the following mosquito model [61]

$$x(n+1) = [ax(n) + bx(n-1)\exp(-x(n-1))]\exp(-x(n)), \qquad (5.5.7)$$

where

$$a \in (0,1), \quad b \in ([0,\infty), \quad n \in \mathbb{Z}^+.$$

This equation describes the growth of a mosquito population. Mosquitoes lay eggs, some of which hatch as soon as conditions are favorable, while others remain dormant for a year or two. In this model, it is assumed that eggs are dormant for one year at most.

Clearly $x_1^* = 0$ is an equilibrium point. The positive equilibrium point may be obtained by solving the equation

$$(a + be^{-x})e^{-x} = 1.$$

Let $g(x) = (a+be^{-x})e^{-x} - 1$. Then since $g'(x) = -e^{-x}[a+2be^{-x}] < 0$ and $g(0) = a + b - 1$, a positive root of g exists if and only if

$$a + b > 1. \tag{5.5.8}$$

Moreover, the positive equilibrium point is given by

$$x_2^* = \ln\left(\frac{a + \sqrt{a^2+4b}}{2}\right).$$

Note also that if $x(n)$ is a solution of (5.5.7) with $x(-1) + x(0) > 0$, then $x(n) > 0$ for all $n \geq 1$.

Our main objective in this subsection is to prove the following result.

Theorem 5.27. *Suppose that*

$$1 - a < b \leq \frac{a(e-a)}{a+1}. \tag{5.5.9}$$

Then the positive equilibrium x_2^* of (5.5.7) is globally asymptotically stable whose basin of attraction is $(0, \infty)$.

PROOF. The proof is divided into two parts. The first part establishes local asymptotic stability and the second part establishes global attractivity. Here we denote x_2^* by x^*.

Part I: Local stability analysis.

By criterion (5.1.9), it follows that the equilibrium point x_2^* is locally asymptotically stable if and only if the following three inequalities hold:

$$b - bx^* - e^{2x^*} < ae^{x^*} - ax^*e^{x^*} - bx^*, \tag{5.5.10}$$

$$ae^{x^*} - ax^*e^{x^*} - bx^* < e^{2x^*} - b + bx^*, \tag{5.5.11}$$

$$bx^* - b < e^{2x^*}. \tag{5.5.12}$$

Observe that x^* satisfies the equation

$$x^* = (ax^* + bx^*e^{-x^*})e^{-x^*}$$

or, equivalently, if $x_2^* \neq 0$,

$$e^{2x^*} = ae^{x^*} + b. \tag{5.5.13}$$

Inequality (5.5.11) is satisfied since by using (5.5.13) we obtain

$$e^{2x^*} - b + 2bx^* + ax^*e^{x^*} - ae^{x^*} = 2bx^* + ax^*e^{x^*} > 0.$$

Inequalities (5.5.10) and (5.5.12) both hold using (5.5.13) if $0 < x^* < 2$.

But this is true in general for any positive solution $x(n)$ since from (5.5.7) it follows that

$$x(n+1) < ax(n)e^{-x(n)} + bx(n-1)e^{-x(n-1)} \leq \frac{a}{e} + \frac{b}{e} = \frac{a+b}{e}.$$

(Why?) (The function $\frac{x}{e^x}$ attains its maximum at $x = 1$.) But by assumption (5.5.9), $b \leq \frac{a(e-a)}{a+1} < e - a$, and so $\frac{a+b}{e} < 1$. Hence $x^* < 2$ and x^* is thus locally asymptotically stable.

Part II: Global attractivity.

We set $I = \left(0, \frac{a+b}{e}\right)$ and $D = I \times I$ and let our function be $f : D \to (0, \infty)$ defined as

$$f(u,v) = (au + bve^{-v})e^{-u}.$$

Then clearly $x^* \in I$.

Now $\frac{\partial f}{\partial v}(u,v) = be^{-u}\frac{\partial(ve^{-v})}{\partial v} \geq 0$ since $v \to ve^{-v}$ is increasing on $[0,1]$. Hence f is nondecreasing in v for $v \in I$. Moreover,

$$\frac{\partial f}{\partial v}(u,v) = e^{-u}(a - au - bve^{-v}) \geq 0$$

if $(au + bve^{-v}) \leq a$ for all $u, v \in \left(0, \frac{a+b}{e}\right)$. But

$$\begin{aligned} au + bve^{-v} &\leq a\left(\frac{a+b}{e}\right) + \frac{b}{e} \\ &\leq \frac{a^2(a+1) + a^2(e-a) + a(e-a)}{e(a+1)} = a. \end{aligned}$$

Hence f is nondecreasing in u, for $u \in I$.

This shows that condition (i) in Theorem 5.17 is fulfilled. Condition (ii) in Theorem 5.17 can easily be established and will be left to the reader to verify. The proof of the theorem is now complete. □

6

The Z-Transform Method and Volterra Difference Equations

In the last five chapters we used the so-called *time domain analysis.* In this approach we investigate difference equations as they are, that is, without transforming them into another domain. We either find solutions of the difference equations or provide information about their qualitative behavior.

An alternative approach will be developed in this chapter. The new approach is commonly known as the *transform method.* By using a suitable transform, one may reduce the study of a linear difference or differential equation to an examination of an associated complex function. For example, the Laplace transform method is widely used in solving and analyzing linear differential equations and continuous control systems, while the Z-transform method is most suitable for linear difference equations and discrete systems. It is widely used in the analysis and design of digital control, communication, and signal processing.

The Z-transform technique is not new and may be traced back to De Moivre around the year 1730. In fact, De Moivre introduced the more general concept of "generating functions" to probability theory.

6.1 Definitions and Examples

The Z-transform of a sequence $x(n)$, which is identically zero for negative integers n (i.e., $x(n) = 0$ for $n = -1, -2, \ldots$), is defined by

$$\boxed{\tilde{x}(z) = Z(x(n)) = \sum_{j=0}^{\infty} x(j)z^{-j},} \tag{6.1.1}$$

where z is a complex number.

The set of numbers z in the complex plane for which series (6.1.1) converges is called the region of convergence of $x(z)$. The most commonly used method to find the region of convergence of the series (6.1.1) is the ratio test. Suppose that

$$\lim_{j\to\infty}\left|\frac{x(j+1)}{x(j)}\right| = R.$$

Then by the ratio test, the infinite series (6.1.1) converges if

$$\lim_{j\to\infty}\left|\frac{x(j+1)z^{-j-1}}{x(j)z^{-j}}\right| < 1$$

and diverges if

$$\lim_{j\to\infty}\left|\frac{x(j+1)z^{-j-1}}{x(j)z^{-j}}\right| > 1.$$

Hence the series (6.1.1) converges in the region $|z| > R$ and diverges for $|z| < R$. This is depicted in Figure 6.1, where Re z denotes the real axis and Im z represents the imaginary axis.

The number R is called the radius of convergence of series (6.1.1). If $R = 0$, the Z-transform $\tilde{x}(z)$ converges everywhere with the possible exception of the origin. On the other hand, if $R = \infty$, the Z-transform diverges everywhere.

We now compute the Z-transform of some elementary functions.

Example 6.1. Find the Z-transform of the sequence $\{a^n\}$, for a fixed real number a, and its region of convergence.

Solution $Z(a^n) = \sum_{j=0}^{\infty} a^j z^{-j}$. The radius of convergence R of $Z(a^n)$ is given by

$$R = \lim_{j\to\infty}\left|\frac{a^{j+1}}{a^j}\right| = |a|.$$

Hence (Figure 6.2)

$$Z(a^n) = \sum_{j=0}^{\infty}\left(\frac{a}{z}\right)^j = \frac{1}{1-(a/z)} = \frac{z}{z-a} \quad \text{for } |z| > |a|. \tag{6.1.2}$$

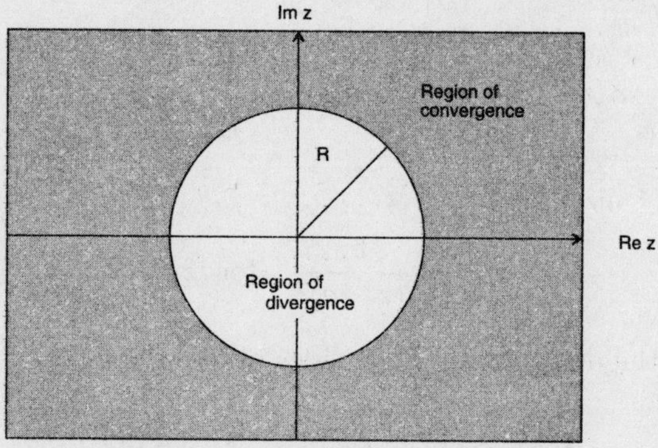

FIGURE 6.1. Regions of convergence and divergence for $\tilde{x}(z)$.

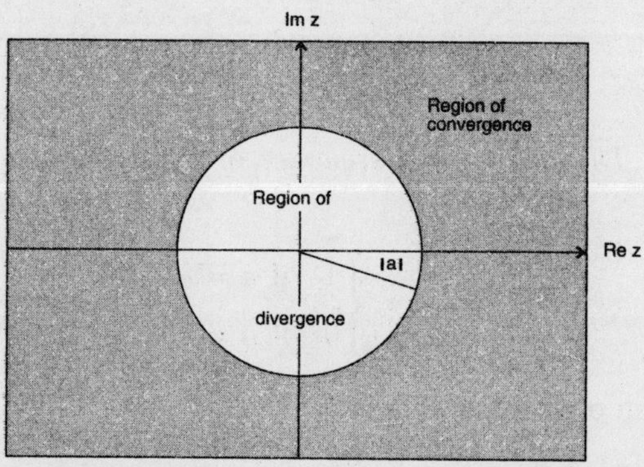

FIGURE 6.2. Regions of convergence and divergence for $Z(a^n)$.

A special case of the above result is that of $a = 1$. In this case we have

$$Z(1) = \frac{z}{z-1} \quad \text{for } |z| > 1.$$

Example 6.2. Find the Z-transform of the sequences $\{na^n\}$ and $\{n^2a^n\}$.

Solution Recall that an infinite series $Z(a^n)$ may be differentiated, term by term, any number of times in its region of convergence [20]. Now,

$$\sum_{j=0}^{\infty} a^j z^{-j} = \frac{z}{(z-a)} \quad \text{for } |z| > |a|.$$

Taking the derivative of both sides yields

$$\sum_{j=0}^{\infty} -ja^j z^{-j-1} = \frac{-a}{(z-a)^2} \quad \text{for } |z| > |a|.$$

Hence

$$Z(na^n) = \sum_{j=0}^{\infty} ja^j z^{-j} = -z \sum_{j=0}^{\infty} -ja^j z^{-j-1}.$$

Therefore,

$$Z(na^n) = \frac{az}{(z-a)^2} \quad \text{for } |z| > 1. \tag{6.1.3}$$

Again taking the derivative of both sides of the identity

$$\sum_{j=0}^{\infty} ja^j z^{-j} = \frac{az}{(z-a)^2} \quad \text{for } |z| > |a|$$

yields

$$Z(n^2 a^n) = \frac{az(z+a)}{(z-a)^3} \quad \text{for } |z| > |a|. \tag{6.1.4}$$

Example 6.3. *The unit impulse sequence*, or the *Kronecker delta sequence*, is defined by

$$\delta_k(n) = \begin{cases} 1 & \text{if } n = k, \\ 0 & \text{if } n \neq k. \end{cases}$$

The Z-transform of this function is

$$Z(\delta_k(n)) = \sum_{j=0}^{\infty} \delta_k(j) z^{-j} = z^{-k}.$$

If $k = 0$, we have the important special case

$$Z(\delta_0(n)) = 1. \tag{6.1.5}$$

Notice that the radius of convergence of $Z(\delta_k(n))$ is $R = 0$.

Example 6.4. Find the Z-transform of the sequence $\{\sin(\omega n)\}$.

Solution Recall that the Euler identity gives $e^{i\theta} = \cos\theta + i\sin\theta$ for any real number θ. Hence $e^{-i\theta} = \cos\theta - i\sin\theta$. Both identities yield

$$\cos\theta = \frac{e^{i\theta} + e^{-i\theta}}{2} \quad \text{and} \quad \sin\theta = \frac{e^{i\theta} + e^{-i\theta}}{2i}.$$

Thus

$$Z(\sin\omega n) = \frac{1}{2i}[Z(e^{i\omega n}) - Z(e^{-i\omega n})].$$

Using formula (6.1.2) we obtain

$$\begin{aligned} Z(\sin\omega n) &= \frac{1}{2i}\left[\frac{z}{z-e^{i\omega}} - \frac{z}{z-e^{-i\omega}}\right] \quad \text{for } |z|>1 \\ &= \frac{z\sin\omega}{(z-e^{i\omega})(z-e^{-i\omega})} \\ &= \frac{z\sin\omega}{z^2-(e^{i\omega}+e^{-i\omega})z+1}, \end{aligned}$$

or

$$Z(\sin\omega n) = \frac{z\sin\omega}{z^2-2z\cos\omega+1} \quad \text{for } |z|>1. \tag{6.1.6}$$

6.1.1 *Properties of the Z-Transform*

We now establish some useful properties of the Z-transform that will be needed in the sequel.

(i) *Linearity.*
Let $\tilde{x}(z)$ be the Z-transform of $x(n)$ with radius of convergence R_1, and let $\tilde{y}(z)$ be the Z-transform of $y(n)$ with radius of convergence R_2. Then for any complex numbers α, β we have

$$Z[\alpha x(n)+\beta y(n)] = \alpha\tilde{x}(z)+\beta\tilde{y}(z) \quad \text{for} \quad |z|>\max(R_1,R_2). \tag{6.1.7}$$

The proof of property (6.1.7) is left to the reader as Exercises 6.1, Problem 18.

(ii) *Shifting.* Let R be the radius of convergence of $\tilde{x}(z)$.

(a) Right-shifting: If $x(-i)=0$ for $i=1,2,\dots,k$, then

$$\boxed{Z[x(n-k)] = z^{-k}\tilde{x}(z) \quad \text{for } |z|>R.} \tag{6.1.8}$$

(b) Left-shifting:

$$\boxed{Z[x(n+k)] = z^k\tilde{x}(z) - \sum_{r=0}^{k-1} x(r)z^{k-r} \quad \text{for } |z|>R.} \tag{6.1.9}$$

The proofs are left as Exercises 6.1, Problem 16. The most commonly used cases of formula (6.1.9) are

$$\begin{aligned} Z[x(n+1)] &= z\tilde{x}(z) - zx(0) \quad \text{for } |z|>R, \\ Z[x(n+2)] &= z^2\tilde{x}(z) - z^2x(0) - zx(1) \quad \text{for } |z|>R. \end{aligned}$$

(iii) *Initial and final value.*

(a) *Initial value theorem:*

$$\boxed{\lim_{|z|\to\infty} \tilde{x}(z) = x(0).} \tag{6.1.10}$$

(b) *Final value theorem:*

$$\boxed{x(\infty) = \lim_{n\to\infty} x(n) = \lim_{z\to 1}(z-1)\tilde{x}(z).} \tag{6.1.11}$$

The proof of formula (6.1.10) follows immediately from the definition of $\tilde{x}(z)$. To prove formula (6.1.11) we first observe that

$$Z[x(n+1)-x(n)] = \sum_{j=0}^{\infty}[x(j+1)-x(j)]z^{-j}.$$

Using formula (6.1.9) on the left-hand side of the above identity leads to

$$(z-1)\tilde{x}(z) = zx(0) + \sum_{j=0}^{\infty}[x(j+1)-x(j)]z^{-j}.$$

Thus

$$\lim_{z\to 1}(z-1)\tilde{x}(z) = x(0) + \sum_{j=0}^{\infty}[x(j+1)-x(j)] = \lim_{n\to\infty} x(n).$$

(iv) *Convolution.*
The convolution* of two sequences $x(n), y(n)$ is defined by

$$\boxed{x(n) * y(n) = \sum_{j=0}^{n} x(n-j)y(j) = \sum_{j=0}^{n} x(n)y(n-j).}$$

Now,

$$Z[x(n) * y(n)] = \sum_{m=0}^{\infty}\left[\sum_{j=0}^{m} x(m-j)y(j)\right] z^{-m}.$$

Interchanging the summation signs yields

$$Z[x(n) * y(n)] = \sum_{j=0}^{\infty} y(j) \sum_{m=j}^{\infty} x(m-j)z^{-m}.$$

And if we put $m - i = s$, we obtain

$$Z[x(n) * y(n)] = \left(\sum_{j=0}^{\infty} y(j)z^{-j}\right)\left(\sum_{s=0}^{\infty} x(s)z^{-s}\right),$$

$$\boxed{Z[x(n) * y(n)] = \tilde{x}(z)\tilde{y}(z).} \tag{6.1.12}$$

It is interesting to know that one may obtain formula (6.1.12) if the convolution is defined as

$$x(n) * y(n) = \sum_{j=0}^{\infty} x(n-j)y(j).$$

(v) *Multiplication by a^n property.*
Suppose that $\tilde{x}(z)$ is the Z-transform of $x(n)$ with radius of convergence R. Then

$$\boxed{Z[a^n x(n)] = \tilde{x}\left(\frac{z}{a}\right),} \qquad \text{for } |z| > |a|R. \tag{6.1.13}$$

The proof of (6.1.13) follows easily from the definition and will be left to the reader as Exercises 6.1, Problem 19.

Example 6.5. Determine the Z-transform of

$$g(n) = a^n \sin \omega n, \quad n = 0, 1, 2, \ldots.$$

Using Example 6.4 and formula (6.1.13) we have

$$\begin{aligned} \tilde{g}(z) = Z(a^n \sin \omega n) &= \frac{(z/a)\sin\omega}{(z/a)^2 - 2(z/a)\cos\omega + 1} \\ &= \frac{az\sin\omega}{z^2 - 2az\cos\omega + a^2}, \qquad \text{for } |z| > |a|. \end{aligned} \tag{6.1.14}$$

(vi) *Multiplication by n^k.*
In Example 6.2 it was shown that $Z(na^n) = \frac{az}{(z-a)^2}$, which may be written in the form

$$Z(na^n) = -z\frac{d}{dz}Z(a^n).$$

Similarly, formula (6.1.4) may be written in the form

$$Z(n^2a^n) = -z\frac{d}{dz}\left[-z\frac{d}{dz}Z(a^n)\right].$$

This may be written in the compact form

$$Z(n^2a^n) = \left(-z\frac{d}{dz}\right)^2 Z(a^n).$$

Generally speaking, we write

$$\left(-z\frac{d}{dz}\right)^k \tilde{x}(z) = \left(-z\frac{d}{dz}\left(-z\frac{d}{dz}\left(\cdots\left(-z\frac{d}{dz}\tilde{x}(z)\right)\cdots\right)\right)\right).$$

It may be shown (Exercises 6.1, Problem 7) that

$$Z[n^k x(n)] = \left(-z\frac{d}{dz}\right)^k Z(x(n)). \tag{6.1.15}$$

Exercises 6.1

1. Find the Z-transform and its region of convergence of the given sequence $\{x(n)\}$.

   (a) $\cos \omega n$. (b) $n \sin 2n$. (c) n.

2. Find the Z-transform and its region of convergence of the sequence

$$f(n) = \begin{cases} 1 & \text{for } n = 1, 3, 5, \\ 0 & \text{for all other values of } n. \end{cases}$$

3. Find the Z-transform and its region of convergence of the sequence

$$f(n) = \begin{cases} 0 & \text{for } n = 0, -1, -2, \ldots, \\ -1 & \text{for } n = 1, \\ a^n & \text{for } n = 2, 3, 4, \ldots. \end{cases}$$

4. Let $x(n)$ be a periodic sequence of period N, i.e., $x(n+N) = x(n)$ for all $n \in \mathbb{Z}^+$. Prove that $\tilde{x}(z) = [z^n/(z^n - 1)]\tilde{x}_1(z)$ for $|z| > 1$, where $\tilde{x}_1(z) = \sum_{j=0}^{N-1} x(j)z^{-j}(\tilde{x}_1)(z)$ is called the Z-transform of the first period.

5. Determine the Z-transform of the periodic sequence shown in Figure 6.3.

6. Use Problem 4 to find the Z-transform and its radius of convergence for the periodic sequence of period 4

$$f(n) = \begin{cases} 1 & \text{for } n = 0, 1, \\ -1 & \text{for } n = 2, 3. \end{cases}$$

7. Let R be the radius of convergence of $\tilde{x}(z)$. Show that

$$Z[n^k x(n)] = \left(-z\frac{d}{dz}\right)^k \tilde{x}(z) \quad \text{for } |z| > R.$$

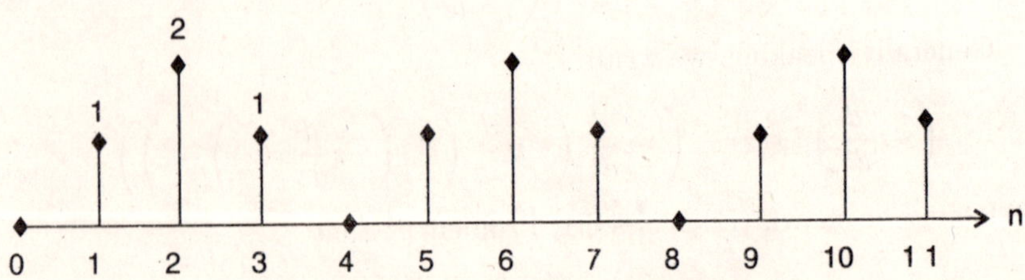

FIGURE 6.3. A periodic sequence.

8. Prove that the Z-transform of the sequence

$$x(n) = \begin{cases} (n-1)a^{n-2}, & n = 0, 1, 2, \ldots, \\ 0 & n = -1, -2, \ldots, \end{cases}$$

is $\tilde{x}(z) = \frac{1}{z-a^2}$ for $|z| > |a|$.

9. Find the Z-transform and its region of convergence of the sequence defined by

$$x(n) = \begin{cases} \dfrac{(n-1)(n-2)}{2} a^{n-3}, & n = 0, 1, 2, \ldots, \\ 0 & n = -1, -2, \ldots. \end{cases}$$

The first backward difference for a sequence $x(n)$ is defined by $\nabla x(n) = x(n) - x(n-1)$.

10. Find $Z[\nabla x(n)], Z[\nabla^2 x(n)]$.

11. Generalize the results of Problem 10 and show that $Z[\nabla^k x(n)] = \left(\frac{z-1}{z}\right)^k \tilde{x}(z)$.

12. Find $Z[\Delta x(n)], Z[\Delta^2 x(n)]$.

13. Show that $Z[\Delta^k x(n)] = (z-1)^k \tilde{x}(z) - z \sum_{j=0}^{k-1} (z-1)^{k-j-1} \Delta^j x(0)$.

14. Let $y(n) = \sum_{i=1}^{n} x(i), n \in \mathbb{Z}^+$. Show that $\tilde{y}(z) = \frac{z}{z-1}\tilde{x}(z)$ for $|z| > \max\{1, R\}$, where R is the radius of convergence of $\tilde{x}(z)$.

15. Let $y(n) = \sum_{i=0}^{n} i x(i)$. Prove that $\tilde{y}(z) = \frac{-z^2}{z-1} \frac{d}{dz} \tilde{x}(z)$.

16. Prove formulas (6.1.8) and (6.1.9).

17. Find the Z-transform of:

(a) $x(n) = \sum_{r=0}^{n} a^{n-r} \sin(\omega r)$.

(b) $\sum_{r=0}^{n} \cos \omega(n-r)$.

18. Prove expression (6.1.7).

19. Show that $Z[a^n x(n)] = \tilde{x}\left(\frac{z}{a}\right)$ for $|z| > |a|R$, where R is the radius of convergence of $\tilde{x}(z)$.

20. Find the Z-transform and its radius of convergence of the sequence $g(n) = a^n \cos(\omega n)$.

21. Use the initial value theorem to determine $x(0)$ for the sequence $\{x(n)\}$ whose Z-transform is given by:

(a) $\frac{2}{z-a}$, for $|z| > a$.

(b) $\frac{3z}{z-6}$, for $|z| > 3$.

22. Extend the initial value theorem to finding $x(1), x(2)$ by proving:

    (i) $x(1) = \lim_{|z|\to\infty} [z(\tilde{x}(z) - x(0))]$,

    (ii) $x(2) = \lim_{|z|\to\infty} [z(\tilde{x}(z) - zx(0) - x(1))]$.

6.2 The Inverse Z-Transform and Solutions of Difference Equations

As we have mentioned in the introduction to this chapter, the Z-transform transforms a difference equation of an unknown sequence $x(n)$ into an algebraic equation in its Z-transform $\tilde{x}(z)$. The sequence $x(n)$ is then obtained from $\tilde{x}(z)$ by a process called the inverse Z-transform. This process is symbolically denoted by

$$Z^{-1}[\tilde{x}(z)] = x(n). \tag{6.2.1}$$

The uniqueness of the inverse Z-transform may be established as follows: Suppose that there are two sequences $x(n), y(n)$ with the same Z-transform, that is,

$$\sum_{i=0}^{\infty} x(i)z^{-i} = \sum_{i=0}^{\infty} y(i)z^{-i}, \qquad \text{for } |z| > R.$$

Then

$$\sum_{i=0}^{\infty} [x(i) - y(i)]z^{-i} = 0, \qquad \text{for } |z| > R.$$

It follows from Laurent's theorem [20] that $x(n) \equiv y(n)$. The most commonly used methods for obtaining the inverse Z-transform are:

1. power series method;
2. partial fractions method;
3. inversion integral method.

It is imperative to remind the reader that when finding the inverse Z-transform, it is always assumed that for any sequence $x(n)$, $x(k) = 0$ for $k = -1, -2, \ldots$.

6.2.1 *The Power Series Method*

In this method we obtain the inverse Z-transform by simply expanding $\tilde{x}(z)$ into an infinite power series in z^{-1} in its region of convergence: $\tilde{x}(z) = \sum_{i=0}^{\infty} a_i z^{-i}$ for $|z| > R$. Then by comparing this with $Z[x(n)] = \sum_{i=0}^{\infty} x(i)z^{-i}$ for $|z| > R$, one concludes that $x(n) = a_n$, $n = 0, 1, 2, \ldots$.

If $\tilde{x}(z)$ is given in the form of a rational function $\tilde{x}(z) = g(z)/h(z)$, where $g(z)$ and $h(z)$ are polynomials in z, then we simply divide $g(z)$ by $h(z)$ to obtain a power series expansion $\tilde{x}(z)$ in z^{-1}. The only possible drawback of this method is that it does not provide us with a closed-form expression of $x(n)$.

Example 6.6. Obtain the inverse Z-transform of

$$\tilde{x}(z) = \frac{z(z+1)}{(z-1)^2}.$$

Solution We first write $x(z)$ as a ratio of two polynomials in z^{-1}:

$$\tilde{x}(z) = \frac{1+z^{-1}}{1-2z^{-1}+z^{-2}}.$$

Dividing the numerator by the denominator, we have

$$\tilde{x}(z) = 1 - 3z^{-1} + 5z^{-2} + 7z^{-3} + 9z^{-4} + 11z^{-5} + \cdots .$$

Thus

$$x(0) = 1, \quad x(2) = 3, \quad x(3) = 5, \quad x(4) = 7, \ldots, \quad x(n) = 2n+1.$$

6.2.2 The Partial Fractions Method

This method is used when the Z-transform $\tilde{x}(z)$ is a rational function in z, analytic at ∞, such as

$$\tilde{x}(z) = \frac{b_0 z^m + b_1 z^{m-1} + \cdots + b_{m-1}z + b_m}{z^n + a_1 z^{n-1} + \cdots + a_{n-1}z + b_n}, \quad m \leq n. \tag{6.2.2}$$

If $\tilde{x}(z)$ in expression (6.2.2) is expressed by a partial fraction expression,

$$\tilde{x}(z) = \tilde{x}_1(z) + \tilde{x}_2(z) + \tilde{x}_3(z) + \cdots ,$$

then by the linearity of the inverse Z-transform one obtains

$$x(n) = Z^{-1}[\tilde{x}_1(z)] + Z^{-1}[\tilde{x}_2(z)] + Z^{-1}[\tilde{x}_3(z)] + \cdots .$$

Then a Z-transform table (Table 6.1; see the end of this chapter) is used to find $Z^{-1}[\tilde{x}_i(z)]$, $i = 1, 2, 3, \ldots$.

Before giving some examples to illustrate this method we remind the reader that the zeros of the numerator of expression (6.2.2) are called *zeros* of $\tilde{x}(z)$, and zeros of the denominator of expression (6.2.2) are called *poles* of $\tilde{x}(z)$.

Remark: Since $\tilde{x}(z)$ is often an improper fraction, it is more convenient to expand $\tilde{x}(z)/z$ rather than $\tilde{x}(z)$ into sums of partial fractions.

Example 6.7. Simple Poles

Solve the difference equation

$$x(n+2)+3x(n+1)+2x(n)=0, \quad x(0)=1, \quad x(1)=-4.$$

Solution Taking the Z-transform of both sides of the equation, we get

$$\tilde{x}(z)=\frac{z(z-1)}{(z+1)(z+2)}.$$

We expand $\tilde{x}(z)/z$ into partial fractions as follows:

$$\tilde{x}(z)/z=\frac{(z-1)}{(z+1)(z+2)}=\frac{a_1}{z+1}+\frac{a_2}{z+2}.$$

Clearing fractions, we obtain

$$z-1=a_1(z+2)+a_2(z+1).$$

This reduces to

$$z-1=(a_1+a_2)z+(2a_1+a_2).$$

Comparing coefficients of like powers of z, we get

$$a_1+a_2=1,$$
$$2a_1+a_2=-1.$$

Hence $a_1=-2$, $a_2=3$. Consequently,

$$\tilde{x}(z)=\frac{-2z}{z+1}+\frac{3z}{z+2}.$$

Thus

$$x(n)=-2(-1)^n+3(-2)^n.$$

Remark: If $\tilde{x}(z)$ has a large number of poles, a computer may be needed to determine the constants $a_1, a_2, \ldots$.

Example 6.8. Repeated Poles

Solve the difference equation

$$x(n+4)+9x(n+3)+30x(n+2)+44x(n+1)+24x(n)=0,$$
$$x(0)=0, \quad x(1)=0, \quad x(2)=1, \quad x(3)=10.$$

Solution Taking the Z-transform, we get

$$\tilde{x}(z)=\frac{z(z-1)}{(z+2)^3(z+3)}.$$

It is convenient here to expand $\tilde{x}(z)/z$ into partial fractions as follows:

$$\tilde{x}(z)/z=\frac{z-1}{(z+2)^3(z+3)}=\frac{b}{z+3}+\frac{a_1}{(z+2)^3}+\frac{a_2}{(z+2)^2}+\frac{a_3}{z+2}. \quad (6.2.3)$$

This time we use a smarter method to find a_1, a_2, a_3, and a_4. To find b we multiply (6.2.3) by $(z+3)$ and then evaluate at $z = -3$. This gives

$$b = \left.\frac{(z-1)}{(z+2)^3}\right|_{z=-3} = 4.$$

To find a_1 we multiply (6.2.3) by $(z+2)^3$ to get

$$\frac{z-1}{z+3} = a_3(z+2)^2 + a_2(z+2) + a_1 + 4\frac{(z+2)^3}{(z+3)} \tag{6.2.4}$$

and evaluate at $z = -2$. This gives

$$a_1 = \left.\frac{z-1}{z-3}\right|_{z=-2} = -3.$$

To find a_2 we differentiate (6.2.4) with respect to z to get

$$\frac{4}{(z+3)^2} = 2a_3(z+2) + a_2 + \frac{r(2z+7)(z+2)^2}{(z+3)^2} \tag{6.2.5}$$

and again evaluate at $z = -2$. This gives

$$a_2 = \left.\frac{d}{dz}\left(\frac{z-1}{z+3}\right)\right|_{z=-2} = 4.$$

Finally, to find a_3 we differentiate (6.2.5) to obtain

$$\frac{-8}{(z+3)^3} = 2a_3 + 4\frac{d^2}{dz^2}\frac{(z+2)^3}{(z+3)},$$

and if we let $z = -2$, then we have

$$a_3 = \left.\frac{1}{2}\frac{d^2}{dz^2}\left(\frac{z-1}{z+3}\right)\right|_{z=-2} = -4.$$

Hence

$$\tilde{x}(z) = \frac{-4z}{z+2} + \frac{4z}{(z+2)^2} - \frac{3z}{(z+2)^3} + \frac{4z}{z+3}.$$

The corresponding sequence is (Table 6.1, at the end of this chapter)

$$\begin{aligned} x(n) &= -4(-2)^n - 2n(-2)^n + \frac{3}{4}n(n-1)(-2)^n + 4(-3)^n \\ &= \left(\frac{3}{4}n^2 - \frac{11}{4}n - 4\right)(-2)^n + 4(-3)^n. \end{aligned}$$

Remark: The procedure used to obtain a_1, a_2, and a_3 in the preceding example can be generalized. If $\tilde{x}(z)/z$ has a pole of multiplicity m at $z = z_0$, then the corresponding terms in the partial fraction expansion can be written

$$\cdots + \frac{a_1}{(z-z_0)^m} + \cdots + \frac{a_m}{z-z_0} + \cdots,$$

and $a_1, a_2, \ldots, a_m$ can be found using the formula

$$a_r = \frac{1}{(r-1)!} \frac{d^{r-1}}{dz^{r-1}} \left[(z - z_0)^m \frac{\tilde{x}(z)}{z} \right] \Bigg|_{z=z_0}.$$

Example 6.9. Complex Poles

Solve the difference equation

$$x(n+3) - x(n+2) + 2x(n) = 0, \quad x(0) = 1, \quad x(1) = 1.$$

Solution Taking the Z-transform of the equation, we get

$$\tilde{x}(z) = \frac{z^3}{(z^2 - 2z + 2)(z+1)}.$$

Next we expand $\tilde{x}(z)/z$ as a sum of the partial fraction in the form

$$\tilde{x}(z)/z = \frac{z^2}{(z^2 - 2z + 2)(z+1)} = \frac{a_1}{[z - (1+i)]} + \frac{a_2}{[z - (1-i)]} + \frac{a_3}{(z+1)}.$$

Using the method of the preceding example we obtain

$$a_3 = \frac{z^2}{z^2 - 2z + 2} \Bigg|_{z=-1} = \frac{1}{5},$$

$$a_1 = \frac{z^2}{[z - (1-i)](z+1)} \Bigg|_{z=1+i} = \frac{1}{2+i} = \frac{2}{5} - \frac{1}{5}i,$$

$$a_2 = \bar{a}_1 = \frac{2}{5} + \frac{1}{5}i.$$

Hence

$$\tilde{x}(z) = \frac{\frac{1}{5}z}{z+1} + \frac{a_1 z}{z - \lambda} + \frac{\bar{a}_1 z}{z - \bar{\lambda}},$$

where $\lambda = 1 + i$. Thus

$$x(n) = \frac{1}{5}(-1)^n + a_1 \lambda^n + \bar{a}_1 \bar{\lambda}^n.$$

But

$$a_1 \lambda^n + \bar{a}_1 \bar{\lambda}^n = 2\,\mathrm{Re}(a_1 \lambda^n) = 2|\bar{a}_1| (\sqrt{2})^n \cos\left(\frac{n\pi}{4} + \arg a_1\right),$$

where $|a_1| = \frac{1}{5}\sqrt{5}$ and $\arg a_1 = \tan^{-1}(\frac{1}{2}) = 0.46$ radians. Thus

$$x(n) = \frac{1}{5}(-1)^n + \frac{2}{5}\sqrt{5}(\sqrt{2})^n \cos\left(\frac{n\pi}{4} + 0.46\right).$$

6.2.3 The Inversion Integral Method[1]

From the definition of the Z-transform, we have

$$\tilde{x}(z) = \sum_{i=0}^{\infty} x(i) z^{-i}.$$

Multiplying both sides of the above equation by z^{n-1}, we get

$$\begin{aligned}\tilde{x}(z)z^{n-1} &= \sum_{i=0}^{\infty} x(i) z^{n-i-1} \\ &= x(0)z^{n-1} + x(1)z^{n-2} + \cdots + x(n)z^{-1} + x(n+1)z^{-2} + \cdots .\end{aligned} \tag{6.2.6}$$

Equation (6.2.6) gives the Laurent series expansion of $\tilde{x}(z)z^{n-1}$ around $z = 0$.

Consider a circle C, centered at the origin of the z-plane, that encloses all poles of $\tilde{x}(z)z^{n-1}$. Since $x(n)$ is the coefficient of z^{-1}, it follows by the Cauchy integral formula [20] that

$$\boxed{x(n) = \frac{1}{2\pi i} \oint_c \tilde{x}(z) z^{n-1}\, dz,} \tag{6.2.7}$$

and by the residue theorem [20] we obtain

$$x(n) = \text{sum of residues of} \tilde{x}(z) z^{n-1}. \tag{6.2.8}$$

Suppose that

$$\tilde{x}(z) z^{n-1} = \frac{h(z)}{g(z)}.$$

In evaluating the residues of $\tilde{x}(z)z^{n-1}$, there are two cases to consider:

(i) $g(z)$ has simple zeros (i.e., $\tilde{x}(z)z^{n-1}$ has simple poles) (see Figure 6.4). In this case the residue K_i at a pole z_i is given by

$$\boxed{K_i = \lim_{z \to z_i} \left[(z - z_i) \frac{h(z)}{g(z)} \right].} \tag{6.2.9}$$

(ii) $g(z)$ has multiple zeros (i.e., $\tilde{x}(z)z^{n-1}$ has multiple poles). If $g(z)$ has a multiple zero z_i of order r, then the residue K_i at z_i is given by

$$\boxed{K_i = \frac{1}{(r-1)!} \lim_{z \to z_i} \frac{d^{r-1}}{dz^{r-1}} \left[(z - z_i)^r \frac{h(z)}{g(z)} \right].}$$

[1]Requires some knowledge of residues in complex analysis [20].

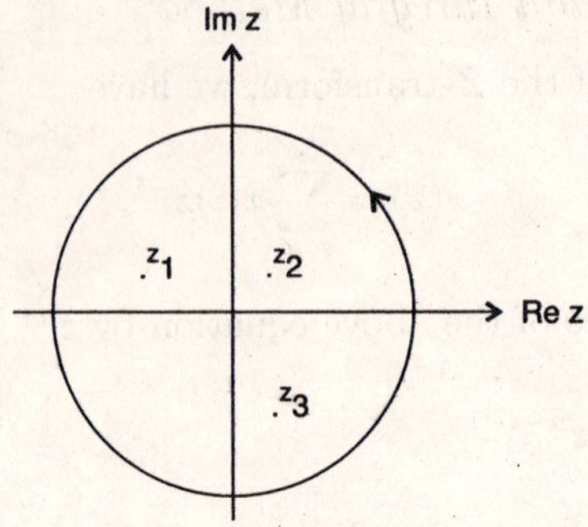

FIGURE 6.4. Poles of $\tilde{x}(z)$.

Example 6.10. Obtain the inverse Z-transform of

$$\tilde{x}(z) = \frac{z(z-1)}{(z-2)^2(z+3)}.$$

Solution Notice that

$$\tilde{x}(z)z^{n-1} = \frac{(z-1)z^n}{(z-2)^2(z+3)}.$$

Thus $\tilde{x}(z)z^{n-1}$ has a simple pole at $z_1 = -3$ and a double pole at $z_2 = 2$. Thus from formula (6.2.8), we get $x(n) = K_1 + K_2$, where K_1, K_2 are the residues of $x(z)z^{n-1}$ at z_1, z_2, respectively. Now,

$$K_1 = \lim_{z\to -3}\left[\frac{(z+3)(z-1)z^n}{(z-2)^2(z+3)}\right] = \frac{-4}{25}(-3)^n,$$

$$K_2 = \frac{1}{(2-1)!}\lim_{z\to 2}\frac{d}{dz}\left[\frac{(z-2)^2(z-1)z^n}{(z-2)^2(z+3)}\right]$$

$$= \lim_{z\to 2}\frac{z^{n-1}[(z+3)(z+nz-n) - z(z-1)]}{(z+3)^2}$$

$$= \frac{(8+5n)}{25}(2)^{n-1}.$$

Thus

$$x(n) = \frac{-4}{25}(-3)^n + \frac{(8+5n)}{25}(2)^{n-1}, \quad n = 0, 1, 2, \ldots.$$

Example 6.11. Electric Circuits or a Ladder Network

Consider the electric network shown in Figure 6.5. Here $i(n)$ is the current in the nth loop; R is the resistance, which is assumed to be constant in every loop; and V is the voltage. By Ohm's law, the voltage (or electric potential) between the ends of a resistor R may be expressed as $V = iR$.

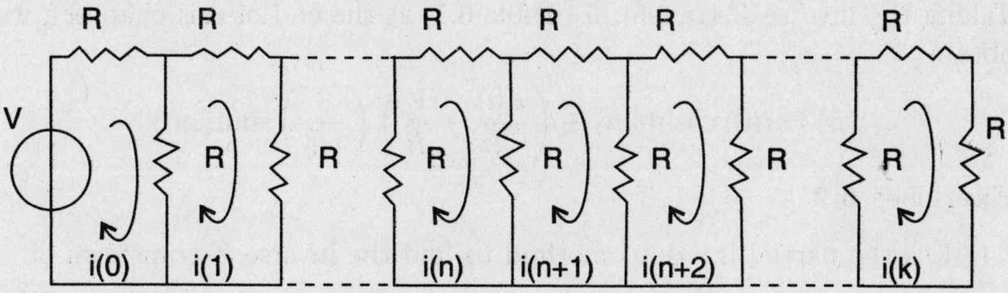

FIGURE 6.5. A ladder network.

Now, Kirchhoff's[2] second law states that "in a closed circuit the impressed voltage is equal to the sum of the voltage drops in the rest of the circuit." By applying Kirchhoff's law to the loop corresponding to $i(n+1)$ we obtain

$$R[i(n+1) - i(n+2)] + R[i(n+1) - i(n)] + Ri(n+2) = 0,$$

or

$$i(n+2) - 3i(n+1) + i(n) = 0. \tag{6.2.10}$$

For the first loop on the left we have

$$V = Ri(0) + R(i(0) - i(1)),$$

or

$$i(1) = 2i(0) - \frac{V}{R}. \tag{6.2.11}$$

Taking the Z-transform of (6.2.10) with the data (6.2.11) yields the equation

$$\tilde{i}(z) = \frac{z[zi(0) - 3i(0) + i(1)]}{z^2 - 3z + 1} = \left[\frac{z^2 - \left(1 + \frac{V}{Ri(0)}\right) z}{z^2 - 3z + 1}\right] i(0). \tag{6.2.12}$$

Let $\omega > 0$ be such that $\cosh\omega = \frac{3}{2}$. Then $\sinh\omega = \frac{\sqrt{5}}{2}$. Then expression (6.2.12) becomes

$$\tilde{i}(z) = i(0)\left[\frac{z^2 - z\cosh\omega}{z^2 - 2z\cosh\omega + 1}\right] + \left(\frac{i(0)}{2} + \frac{V}{R}\right)\left(\frac{2}{\sqrt{5}}\right)\left[\frac{z\sinh\omega}{z^2 - 2z\cosh\omega + 1}\right].$$

[2]Gustav Kirchhoff, a German physicist (1824–1887), is famous for his contributions to electricity and spectroscopy.

Taking the inverse Z-transform (Table 6.1, at the end of this chapter), we obtain

$$i(n) = i(0)\cosh(\omega n) + \left(\frac{i(0)}{2} + \frac{V}{R}\right)\left(\frac{2}{\sqrt{5}}\right)\sinh(\omega n).$$

Exercises 6.2

1. Use the partial fractions method to find the inverse Z-transform of:
   (a) $\frac{z}{(z-\frac{1}{2})(z+1)}$.
   (b) $\frac{z(z+1)}{(z+2)^2(z-1)}$.
2. Use the power series method to find the inverse Z-transform of:
   (a) $\frac{z-2}{(z-1)(z+3)}$.
   (b) $\frac{e^{-a}z}{(z-e^{-a})^2}$.
3. Use the inversion integral method to find the inverse Z-transform of:
   (a) $\frac{z(z-1)}{(z+2)^3}$.
   (b) $\frac{z(z+2)}{(z-\frac{1}{2})(z+i)(z-i)}$.
4. Use the partial fractions method and the inversion integral method to find the inverse Z-transform of:
   (a) $\frac{z(z+1)}{(z-2)^2}$.
   (b) $\frac{z^2+z+1}{(z-1)(z^2-z+1)}$.

In Problems 5 through 7, use the Z-transform method to solve the given difference equation.

5. (The Fibonacci Sequence). $x(n+2) = x(n+1) + x(n)$, $x(0) = 0$, $x(1) = 1$.
6. $x(n+2) - 3x(n+1) + 2x(n) = \delta_0(n)$, $x(0) = x(1) = 0$.
7. $(n+1)x(n+1) - nx(n) = n+1$, $x(0) = 0$.
8. Consider the continued fraction

$$K = K\left(\frac{a_n}{b_n}\right) = \frac{a_0}{b_0} + \frac{a_1}{b_1 + \frac{a_2}{b_2 + \frac{a_3}{b_3 + \cdots}}}$$

$$= \frac{a_0}{b_0+}\,\frac{a_1}{b_1+}\,\frac{a_2}{b_2+}\cdots.$$

Let $a_i = b_i = 1$ for all $i \in \mathbb{Z}^+$, and $x(n) = \frac{a_1}{b_1+}\,\frac{a_2}{b_2+}\cdots\frac{a_n}{b_n+}$.

(a) Show that $x(n+1) = 1 + \frac{1}{x(n)}$. Find $x(n)$.

(b) Find $K = 1 + \lim_{n\to\infty} x(n)$.

9. Prove that the convolution product is commutative and associative (i.e., $x * y = y * x; x * (y * f) = (x * y) * f$).

10. Solve, using convolution, the equation $x(n+1) = 2+4\sum_{r=0}^{n}(n-r)x(r)$.

11. Solve the equation $x(n) = 1 - \sum_{r=0}^{n-1} e^{n-r-1}x(r)$.

6.3 Volterra Difference Equations of Convolution Type: The Scalar Case[3]

Volterra difference equations of convolution type are of the form

$$x(n+1) = Ax(n) + \sum_{j=0}^{n} B(n-j)x(j), \tag{6.3.1}$$

where $A \in \mathbb{R}$ and $B : Z^+ \to \mathbb{R}$ is a discrete function. This equation may be considered as the discrete analogue of the famous Volterra integrodifferential equation

$$x'(t) = Ax(t) + \int_0^t B(t-s)x(s)\,ds. \tag{6.3.2}$$

Equation (6.3.2) has been widely used as a mathematical model in population dynamics. Both (6.3.1) and (6.3.2) represent systems in which the future state $x(n+1)$ does not depend only on the present state $x(n)$ but also on all past states $x(n-1), x(n-2), \ldots, x(0)$. These systems are sometimes called *hereditary.* Given the initial condition $x(0) = x_0$, one can easily generate the solution $x(n, x_0)$ of (6.3.1). If $y(n)$ is any other solution of (6.3.1) with $y(0) = x_0$, then it is easy to show that $y(n) = x(n)$ for all $n \in \mathbb{Z}^+$ (Exercises 6.3, Problem 8).

One of the most effective methods of dealing with (6.3.1) is the Z-transform method. Let us rewrite (6.3.1) in the convolution form

$$x(n+1) = Ax(n) + B * x. \tag{6.3.3}$$

Taking formally the Z-transform of both sides of (6.3.3), we get

$$z\tilde{x}(z) - zx(0) = A\tilde{x}(z) + \tilde{B}(z)\tilde{x}(z),$$

which gives

$$[z - A - \tilde{B}(z)]\tilde{x}(z) = zx(0),$$

or

$$\tilde{x}(z) = zx(0)/[z - A - \tilde{B}(z)]. \tag{6.3.4}$$

[3]This section requires some rudiments of complex analysis [20].

Let

$$g(z) = z - A - \tilde{B}(z). \tag{6.3.5}$$

The complex function $g(z)$ will play an important role in the stability analysis of (6.3.1). Before embarking on our investigation of $g(z)$ we need to present a few definitions and preliminary results.

Definition 6.12. Let E be the space of all infinite sequences of complex numbers (or real numbers) $x = (x(0), x(1), x(2), \ldots)$. There are three commonly used norms that may be defined on subsets of E. These are

(i) the l_1 norm: $\|x\|_1 = \sum_{i=0}^{\infty} |x(i)|$;

(ii) the l_2, or, Euclidean norm: $\|x\|_2 = \left[\sum_{i=0}^{\infty} |x(i)|^2\right]^{1/2}$;

(iii) the l_∞ norm: $\|x\|_\infty = \sup_{i \geq 0} |x(i)|$.

The corresponding normed spaces are called l_1, l_2, and l_∞, respectively. One may show easily that (Exercises 6.3, Problem 6)

$$l_1 \subset l_2 \subset l_\infty.$$

Definition 6.13. A complex function $g(z)$ is said to be *analytic* in a region in the complex plane if it is differentiable there. The next result establishes an important property of l_1 sequences.

Theorem 6.14. *If $x(n) \in l_1$, then:*

(i) *$\tilde{x}(z)$ is an analytic function for $|z| \geq 1$;*

(ii) *$|\tilde{x}(z)| \geq \|x\|$ for $|z| \geq 1$.*

PROOF.

(i) Since $x(n) \in l_1$, the radius of convergence of $\tilde{x}(z) = \sum_{n=0}^{\infty} x(n) z^{-n}$ is $R = 1$. Hence $\tilde{x}(z)$ can be differentiated term by term in its region of convergence $|z| > 1$. Thus $\tilde{x}(z)$ is analytic on $|z| > 1$. Furthermore, since $x(n) \in l_1, \tilde{x}(z)$ is analytic for $|z| = 1$.

(ii) This is left as Exercises 6.3, Problem 9. □

We now turn our attention to the function $g(z) = z - A - \tilde{B}(z)$ in formula (6.3.5). This function plays the role of the characteristic polynomial of linear difference equations. (See Chapter 2.) In contrast to polynomials, the function $g(z)$ may have infinitely many zeros in the complex plane. The following lemma sheds some light on the location of the zeros of $g(z)$.

Lemma 6.15 [39]. *The zeros of*

$$g(z) = z - A - \tilde{B}(z)$$

all lie in the region $|z| < c$, for some real positive constant c. Moreover, $g(z)$ has finitely many zeros z with $|z| \geq 1$.

PROOF. Suppose that all the zeros of $g(z)$ do not lie in any region $|z| < c$ for any positive real number c. Then there exists a sequence $\{z_i\}$ of zeros of $g(z)$ with $|z_i| \to \infty$ as $i \to \infty$. Now,

$$|z_i - A| = |\tilde{B}(z_i)| \leq \sum_{n=0}^{\infty} |B(n)||z_i|^{-n}. \tag{6.3.6}$$

Notice that the right-hand side of inequality (6.3.6) goes to $B(0)$ as $i \to \infty$, while the left-hand side goes to ∞ as $i \to \infty$, which is a contradiction. This proves the first part of the lemma. □

To prove the second part of the lemma, we first observe from the first part of the lemma that all zeros z of $g(z)$ with $|z| \geq 1$ lie in the annulus $1 \leq |z| \leq c$ for some real number c. From Theorem 6.14 we may conclude that $g(z)$ is analytic in this annulus $(1 \leq |z| \leq c)$. Therefore, $g(z)$ has only finitely many zeros in the region $|z| \geq 1$ [39].

Next we embark on a program that will reveal the qualitative behavior of solutions of (6.3.1). In this program we utilize (6.3.4), which may be written as

$$\tilde{x}(z) = x(0)zg^{-1}(z). \tag{6.3.7}$$

Let γ be a circle that includes all the zeros of $g(z)$. The circle γ is guaranteed to exist by virtue of Lemma 6.15. By formula (6.2.7) we obtain

$$x(n) = \frac{1}{2\pi i} \oint_{\gamma} x(0)z^n g^{-1}(z)\, dz, \tag{6.3.8}$$

and by formula (6.2.8) we get

$$x(n) = \text{sum of residues of } [x(0)z^n g^{-1}(z)]. \tag{6.3.9}$$

This suggests that

$$x(n) = \sum p_r(n) z_r^n, \tag{6.3.10}$$

where the sum is taken over all the zeros of $g(z)$ and where $p_r(n)$ is a polynomial in n of degree less than $k-1$ if z_r is a multiple root of order k. To show the validity of formula (6.3.10), let z_r be a zero of $g(z)$ of order k. We write the following Laurent's series expansion [20]:

$$g^{-1}(z) = \sum_{n=-k}^{\infty} g_n(z - z_r)^n, \quad \text{for some constants } g_n,$$

$$z^n = [z_r - (z_r - z)]^n = \sum_{i=0}^{n} \binom{n}{i} z_r^{n-i}(z - z_r)^i.$$

The residue of $x(0)z^n g^{-1}$ at z_r is $x(0)$ times the coefficient of $(z-z_r)^{-1}$ in $g^{-1}(z)z^n$. The coefficient of $(z-z_r)^{-1}$ in $g^{-1}(z)z^n$ is given by

$$g_{-k}\binom{n}{k-1}z_r^{n-k+1}+g_{-k+1}\binom{n}{k-2}z_r^{n-k+2}+\cdots+g_{-1}\binom{n}{0}z_r^n. \quad (6.3.11)$$

It follows from formula (6.3.9) that $x(n)$ may be given by formula (6.3.10).

Formula (6.3.10) has the following important consequences.

Theorem 6.16 [39]. *The zero solution of* (6.3.1) *is uniformly stable if and only if:*

(a) $z - A - \tilde{B}(z) \neq 0$ *for all* $|z| > 1$, *and*

(b) *if* z_r *is a zero of* $g(z)$ *with* $|z_r| = 1$, *then the residue of* $z^n g^{-1}(z)$ *at* z_r *is bounded as* $n \to \infty$.

PROOF. Suppose that conditions (a) and (b) hold. If z_r is a zero of $g(z)$ with $|z_r| < 1$, then from formula (6.3.10) its contribution to the solution $x(n)$ is bounded. On the other hand, if z_r is a zero of $g(z)$ with $|z_r| = 1$ at which the residue of $x(0)z^n g^{-1}(z)$ is bounded as $n \to \infty$, then from formula (6.3.9) its contribution to the solution $x(n)$ is also bounded. This shows that $|x(n)| \leq L|x(0)|$ for some $L > 0$, and thus we have uniform stability. The converse is left to the reader as Exercises 6.3, Problem 10. □

We observe here that a necessary and sufficient condition for condition (b) is that each zero z of $g(z)$ with $|z| = 1$ must be simple (Exercises 6.3, Problem 11).

The next result addresses the question of asymptotic stability.

Theorem 6.17 [39]. *The zero solution of* (6.3.1) *is uniformly asymptotically stable if and only if*

$$\boxed{z - A - \tilde{B}(z) \neq 0, \quad \text{for all } |z| \geq 1.} \quad (6.3.12)$$

PROOF. The proof follows easily from formula (6.3.10) and is left to the reader as Exercises 6.3, Problem 12. □

Exercises 6.3

1. Solve the Volterra difference equation $x(n+1) = 2x(n)+\sum_{r=0}^{n} 2^{n-r}x(r)$, and then determine the stability of its zero solution.

2. Solve the Volterra difference equation $x(n+1) = -\frac{1}{2}x(n) + \sum_{r=0}^{n} 3^{r-n}\, x(r)$, and then determine the stability of its zero solution.

3. Use Theorems 6.16 and 6.17 to determine the stability of the zero solutions of the difference equations in Problems 1 and 2.

4. Without finding the solution of the equation

$$x(n+1) = -\frac{1}{4}x(n) + \sum_{r=0}^{n} \left(\frac{1}{2}\right)^{r-n} x(r),$$

determine the stability of its zero solution.

5. Determine the stability of the zero solution of $x(n+1) = 2x(n) - 12\sum_{r=0}^{n}(n-r)x(r)$, using Theorem 6.16 or 6.17.

6. Prove that $l_1 \subset l_2 \subset l_\infty$.

7. Let $x = \{x_n\}$ and $y = \{y_n\}$ be two l_1 sequences. Prove that $x * y \in l_1$ by following these steps:

   (i) If $\sum_{i=0}^{\infty} x(i) = a, \sum_{i=0}^{\infty} y(i) = b,$ and $c(n) = \sum_{i=1}^{n} x(n-i)y(i)$, show that $\sum_{i=0}^{\infty} c(i) = ab$.

   (ii) Prove that $\sum_{n=0}^{\infty} |c(n)| \le (\sum_{i=0}^{\infty} |x(i)|) \left(\sum_{j=0}^{\infty} |y(j)|\right)$.

8. Prove the uniqueness of solutions of (6.3.1), that is, if $x(n)$ and $y(n)$ are solutions of (6.3.1) with $x(0) = y(0)$, then $x(n) = y(n)$ for all $n \in \mathbb{Z}^+$.

9. If $x(n) \in l_1$ show that $|\tilde{x}(z)| \le \|x\|_1$ for $|z| \ge 1$.

*10. Suppose that the zero solution of (6.3.1) is uniformly stable. Prove that:

   (a) $g(z) = z - A - \hat{B}(z) \ne 0$ for all $|z| > 1$.

   (b) If z_r is a zero of $g(z)$ with $|z_r| = 1$, then the residue of $z^n g^{-1}(z)$ at z_r is bounded.

11. Prove that a necessary and sufficient condition for condition (b) in Theorem 6.16 is that z_r be a simple root of $g(z)$.

*12. Prove Theorem 6.17.

6.4 Explicit Criteria for Stability of Volterra Equations

The stability results in Section 6.3 are not very practical, since locating the zeros of $g(z)$ is more or less impossible in most problems. In this section we provide explicit conditions for the stability of (6.3.1). The main tools in this study are Theorems 6.17 and Rouché's Theorem (Theorem 5.13).

Theorem 6.18 [39]. *Suppose that $B(n)$ does not change sign for $n \in \mathbb{Z}^+$. Then the zero solution of* (6.3.1) *is asymptotically stable if*

$$|A| + \left|\sum_{n=0}^{\infty} B(n)\right| < 1. \tag{6.4.1}$$

PROOF. Let $\beta = \sum_{n=0}^{\infty} B(n)$ and $D(n) = \beta^{-1}B(n)$. Then $\sum_{n=0}^{\infty} D(n) = 1$. Furthermore, $\tilde{D}(1) = 1$ and $|\tilde{D}(z)| \leq 1$ for all $|z| \geq 1$. Let us write $g(z)$ in the form

$$g(z) = z - A - \beta\tilde{D}(z). \tag{6.4.2}$$

To prove uniform asymptotic stability of the zero solution of (6.3.1), it suffices to show that $g(z)$ has no zero z with $|z| \geq 1$. So assume that there exists a zero z_r of $g(z)$ with $|z_r| \geq 1$. Then by (6.4.2) we obtain $|z_r - A| = |\beta\tilde{D}(z)| \leq |\beta|$. Using condition (6.4.1) one concludes that $|z_r| \leq |A| + |\beta| < 1$, which is a contradiction. This concludes the proof of the theorem. □

Unfortunately, we are not able to show that condition (6.4.1) is a necessary condition for asymptotic stability. However, in the next result we give a partial converse to the above theorems.

Theorem 6.19 [39]. *Suppose that $B(n)$ does not change sign for $n \in \mathbb{Z}^+$. Then the zero solution of* (6.3.1) *is not asymptotically stable if any one of the following conditions holds:*

(i) $A + \sum_{n=0}^{\infty} B(n) \geq 1$.

(ii) $A + \sum_{n=0}^{\infty} B(n) \leq -1$ *and* $B(n) > 0$ *for some* $n \in \mathbb{Z}^+$.

(iii) $A + \sum_{n=0}^{\infty} B(n) < -1$ *and* $B(n) < 0$ *for some* $n \in \mathbb{Z}^+$, *and* $\sum_{n=0}^{\infty} B(n)$ *is sufficiently small.*

PROOF. Let β and $D(n)$ be as defined in the proof of Theorem 6.18.

(i) Assume condition (i). If $A + \beta = 1$, then clearly $z = 1$ is a root of $g(z)$ defined in (6.4.2). Hence by Theorem 5.17 the zero solution of (6.3.1) is not asymptotically stable. If $A + \beta > 1$, say $A + \beta = 1 + \delta$, then there are two areas to consider.

   (a) If $\beta < 0$, then we let γ be the circle in the complex plane with center at A and radius equal to $|\beta| + \frac{1}{2}\delta$. Then on γ (Figure 6.6) we have $|z| > 1$ and thus

$$|\beta\tilde{D}(z)| \leq |\beta| < |z - A|. \tag{6.4.3}$$

   Let $h(z) = -\beta\tilde{D}(z)$, $f(z) = z - A$. Then from inequality (6.4.3) $|h(z)| < |f(z)|$ on γ. Hence by Rouché's Theorem (Theorem 5.13), $g(z) = f(z) + h(z)$ and $f(z)$ have the same number of zeros inside γ. Since A is the only zero of $f(z)$ inside γ, then $g(z)$ has exactly

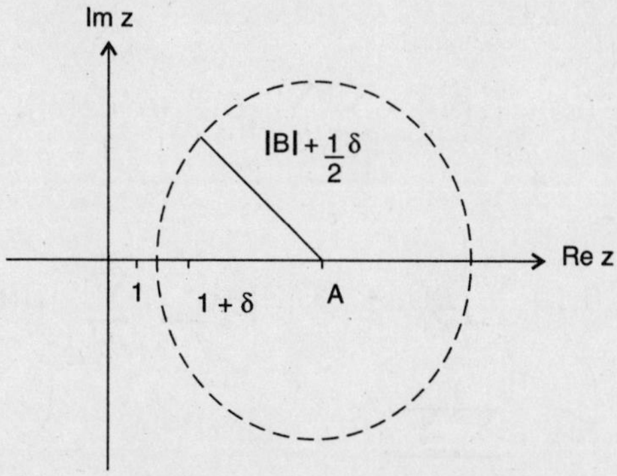

FIGURE 6.6. A circle with center A and radius $|\beta| + \frac{\delta}{2}$.

one zero z_0 inside γ with $|z_0| > 1$. Again by using Theorem 6.16, the zero solution of (6.3.1) is not asymptotically stable.

(b) Suppose that $\beta > 0$. Since $A + \beta > 1$, it follows that $g(z) = 1 - A - \beta < 0$. Moreover, $|\tilde{D}(A + \beta)| \leq 1$. Thus $g(A + \beta) = \beta[1 - \tilde{D}(A + \beta)] \geq 0$. Therefore, $g(z)$ has a zero between 1 and $A + \beta$ and, consequently, by Theorem 6.17, the zero solution of (6.3.1) is not asymptotically stable. This completes the proof of condition (i).

Parts (ii) and (iii) are left to the reader as Exercises 6.4, Problems 7 and 8. □

The above techniques are not expendable to uniform stability. This is mainly due to the lack of easily verifiable criteria for condition (b) of Theorem 6.16. Therefore, new techniques are needed to tackle the problem of uniform stability. These techniques involve the use of Liapunov functionals (functions), which we have encountered in Chapter 4.

Let E by the space of all infinite sequences of complex numbers as defined in Definition 6.12. Then a function $V : E \to R$ is said to *a Liapunov functional* if, for $x = \{x(n)\} \in E$,

(i) $V(x)$ is positive definite (Chapter 4),

(ii) $\Delta V(x) \leq 0$,

where $\Delta V(x) = V(\hat{x}) - V(x)$ and $\hat{x}(n) = x(n+1)$ for all $n \in \mathbb{Z}^+$.

The next result illustrates the use of Liapunov functionals in stability theory.

Theorem 6.20 [39]. *The zero solution of* (6.3.1) *is uniformly stable if*

$$|A| + \sum_{j=0}^{n} |B(j)| \leq 1 \quad \text{for all } n \in \mathbb{Z}^+. \tag{6.4.4}$$

PROOF. For $x \in E$, let

$$V(x) = |x(n)| + \sum_{r=0}^{n-1} \sum_{s=n}^{\infty} |B(s-r)||x(r)|. \tag{6.4.5}$$

Then

$$\Delta V(x) = \left| Ax(n) + \sum_{j=0}^{n} B(n-j)x(j) \right| + \sum_{r=0}^{n} \sum_{s=n+1}^{\infty} |B(s-r)||x(r)|$$

$$- |x(n)| - \sum_{r=0}^{n-1} \sum_{s=n}^{\infty} |B(s-r)||x(r)| \tag{6.4.6}$$

$$\leq \left(|A| + \sum_{j=0}^{\infty} |B(j)| - 1 \right) |x(n)|. \tag{6.4.7}$$

By assumption (6.4.4) we thus have

$$\Delta V(x) \leq 0. \tag{6.4.8}$$

From (6.4.5) we obtain $|x(n)| \leq V(x)$. Using inequality (6.4.8) and expression (6.4.5) again we obtain

$$|x(n)| \leq V(x) \leq |x(0)|.$$

Consequently, the zero solution is uniformly stable (Chapter 4). □

Exercises 6.4

Use Theorem 6.19 to determine the stability and instability of the zero solution of the equations in Problems 1, 2, and 3.

1. $x(n+1) = -\frac{1}{4}x(n) + \sum_{r=0}^{n} \left(\frac{1}{3}\right)^{n+1-r} x(r)$.
2. $x(n+1) = \frac{1}{2}x(n) + \sum_{r=0}^{n} (n-r)x(r)$.
3. $x(n+1) = \frac{1}{3}x(n) + \sum_{r=0}^{n} e^{r-n}x(r)$.
4. Find the values of a for which the zero solution of the equation $x(n) = \sum_{r=0}^{n-1} (n-r-1)a^{n-r-1}x(r)$ is:
   (i) uniformly stable,
   (ii) asymptotically stable,
   (iii) not asymptotically stable.
5. Determine the values of a for which the zero solution of the equation $\Delta x(n) = -\frac{2}{3}x(n) + \sum_{r=0}^{n} (n-r)^2 a^{n-r} x(r)$ is asymptotically stable.
6. Prove Theorem 6.18 using the method of Liapunov functionals used in the proof of Theorem 6.20.

7. Prove part (ii) of Theorem 6.19.

8. Prove part (iii) of Theorem 6.19.

9. Provide details of how inequality (6.4.7) is obtained from inequality (6.4.6).

10. (Open problem). Discuss the stability of the zero solution of (3.5.1) under the condition $A + \sum_{n=0}^{\infty} B(n) = -1$ and $\sum_{n=0}^{\infty} B(n) < 0$.

11. (Open problem). Can we omit the assumption that $\sum_{n=0}^{\infty} B(n)$ is sufficiently small in Theorem 6.19, part (iii)?

12. (Open problem). Develop a necessary and sufficient condition for the asymptotic stability of the zero solution of (6.3.1).

6.5 Volterra Systems

In this section we are mainly interested in the following Volterra system of convolution type:

$$x(n+1) = Ax(n) + \sum_{j=0}^{n} B(n-j)x(j), \tag{6.5.1}$$

where $A = (a_{ij})$ is a $k \times k$ real matrix and $B(n)$ is a $k \times k$ real matrix defined on $\mathbb{Z}^+$. It is always assumed that $B(n) \in l_1$, i.e., $\sum_{j=0}^{\infty} |B(j)| < \infty$. The Z-transform for sequences in R^k and matrices $R^{k\times k}$ is defined in the natural way, that is,

$$\begin{aligned} Z[x(n)] &= \left(Z(x_1(n)), Z(x_2(n)), \ldots, Z(x_k(n))\right)^T, \\ Z[B(n)] &= (Z(b_{ij}(n))). \end{aligned}$$

Thus all the rules and formulas for the Z-transform of scalar sequences hold for vector sequences and matrices.

Taking the Z-transform of both sides of (6.5.1), one obtains

$$z\tilde{x}(z) - zx(0) = A\tilde{x}(z) + \tilde{B}(z)\tilde{x}(z), \quad |z| > R,$$

which yields

$$\tilde{x}(z) = [zI - A - \tilde{B}(z)]^{-1} zx(0), \quad |z| > R. \tag{6.5.2}$$

Theorem 6.17 for scalar equations has the following counterpart for systems.

Theorem 6.21. *A necessary and sufficient condition for uniform asymptotic stability is*

$$\boxed{\det(zI - A - \tilde{B}(z)) \neq 0, \quad \text{for all } |z| \geq 1.} \tag{6.5.3}$$

PROOF. See [39]. □

An application of the preceding theorem will be introduced next. This will provide explicit criteria for asymptotic stability. But before introducing our result we need the following lemma concerning eigenvalues of matrices.

Lemma 6.22 [14]. *Let $G = (g_{ij})$ be a $k \times k$ matrix. If z_0 is an eigenvalue of G, then:*

(i) $|z_0 - g_{ii}||z_0 - g_{jj}| \le \sum_r' |g_{ir}| \sum_r' |g_{jr}|$, *for some $i, j, i \ne j$, and*

(ii) $|z_0 - g_{tt}||z_0 - g_{ss}| \le \sum_r' |g_{rt}| \sum_r' |g_{rs}|$, *for some $t, s, t \ne s$,*

where $\sum_r' g_{ir}$ means $\left(\sum_{r=1}^k g_{ir}\right) - g_{ii}$.

Using the above lemma we can prove the next result. Let

$$\beta_{ij} = \sum_{n=0}^{\infty} |b_{ij}(n)|, \quad 1 \le i, j \le k.$$

Theorem 6.23 [39]. *The zero solution of (6.5.1) is uniformly asymptotically stable if either one of the following conditions holds:*

(i) $\sum_{j=1}^k (|a_{ij}| + \beta_{ij}) < 1$, *for each $i, 1 \le i \le k$, or*

(ii) $\sum_{i=1}^k (|a_{ij}| + \beta_{ij}) < 1$, *for each $j, 1 \le j \le k$.*

PROOF. (i) To prove uniform asymptotic stability under condition (i) we need to show that condition (6.5.3) holds. So assume the contrary, that is,

$$\det(z_0 I - A - \tilde{B}(z_0)) = 0 \quad \text{for some } z_0 \text{ with } |z_0| \ge 1.$$

Then z_0 is an eigenvalue of the matrix $A + \tilde{B}(z_0)$. Hence by condition (i) in Lemma 6.22, we have

$$|z_0 - a_{ii} - \tilde{b}_{ii}(z_0)||z_0 - a_{jj} - \tilde{b}_{jj}(z_0)| \le \sum_r' |a_{ir} + \tilde{b}_{ir}(z_0)| \sum_r' |a_{jr} + \tilde{b}_{jr}(z_0)|. \tag{6.5.4}$$

But

$$\begin{aligned} |z_0 - a_{ii} - \tilde{b}_{ii}(z_0)| &\ge |z_0| - |a_{ii}| - |\tilde{b}_{ii}(z_0)| \ge 1 - |a_{ii}| - |\tilde{b}_{ii}(z_0)| \\ &> \sum_r' (|a_{ir}| + |\beta_{ir})| \quad \text{(by condition (i))}. \end{aligned}$$

Similarly,

$$|z_0 - a_{jj} - \tilde{b}_{jj}(z_0)| > \sum_r' (|a_{jr}| + \beta_{jr}).$$

Combining both inequalities, we get

$$|z_0 - a_{ii} - \tilde{b}_{ii}(z_0)||z_0 - a_{jj} - \tilde{b}_{jj}(z_0)| > \sum_r{}'(|a_{ir}| + \beta_{ir}) \sum_r{}'(|a_{jr}| + \beta_{jr}).$$

It is clear that this contradicts inequality (6.5.4) if one notes that for any $1 \leq s, m \leq k$,

$$|a_{st}| + \beta_{st} \geq |a_{st}| + |\tilde{b}_{st}(z_0)| \geq |a_{st} + \tilde{b}_{st}(z_0)|. \qquad \square$$

As in the scalar case, the above method may be extended to provide criteria for uniform stability. Again, the method of Liapunov functionals will come to the rescue.

Theorem 6.24 [39]. *The zero solution of* (6.5.1) *is uniformly stable if*

$$\boxed{\sum_{i=1}^{k} |a_{ij}| + \beta_{ij} \leq 1} \tag{6.5.5}$$

for all $j = 1, 2, \ldots, k$.

PROOF. Define the Liapunov functional

$$V(x) = \sum_{i=1}^{k} \left[|x_i(n)| + \sum_{j=1}^{k} \sum_{r=0}^{n-1} \sum_{s=n}^{\infty} |b_{ij}(s-r)||x_j(r)| \right].$$

Then

$$\Delta V_{(6.5.1)}(x) \leq \sum_{i=1}^{k} \left[\sum_{j=1}^{k} |a_{ij}||x_j(n)| - |x_i(n)| + \sum_{j=1}^{k} \sum_{s=n}^{\infty} |b_{ij}(s-n)||x_j(n)| \right]. \tag{6.5.6}$$

A crucial but simple step is now in order. Observe that

$$\sum_{i=1}^{k} \sum_{j=1}^{k} |a_{ij}||x_j(n)| = \sum_{i=1}^{k} \sum_{j=1}^{k} |a_{ji}||x_i(n)|,$$

and

$$\sum_{i=1}^{k} \sum_{j=1}^{k} \sum_{s=n}^{\infty} |b_{ij}(s-n)||x_j(n)| = \sum_{i=1}^{k} \sum_{j=1}^{k} \sum_{s=n}^{\infty} |b_{ij}(s-n)||x_i(n)|$$

(Exercises 6.5, Problem 1).

Hence inequality (6.5.6) now becomes

$$\begin{aligned}\Delta V_{(6.5.1)}(x) &\leq \sum_{i=1}^{k}\left[\sum_{j=1}^{k}|a_{ji}| + b_{ji} - 1\right]|x_i(n)| \\ &\leq 0 \quad \text{(by condition (6.5.5)).}\end{aligned}$$

This implies that

$$|x(n)| \leq V(x) \leq \sum_{i=1}^{k}|x_i(0)| = \|x(0)\|,$$

which proves uniform stability. □

Example 6.25. An Epidemic Model [89]

Let $x(n)$ denote the fraction of susceptible individuals in a certain population during the nth day of an epidemic, and let $a(k) > 0$ be the measure of how infectious the infected individuals are during the kth day. Then the spread of an epidemic may be modeled by the equation

$$\ln \frac{1}{x(n+1)} = \sum_{j=0}^{n}(1+\varepsilon - x(n-j))a(j), \tag{6.5.7}$$

where ε is a small positive number, $n \in \mathbb{Z}^+$.

To transform (6.5.7) into a Volterra-type equation, we put $x(n) = \bar{e}^{y(n)}$. Then we obtain

$$y(n+1) = \sum_{j=0}^{n} a(n-j)(1+\varepsilon - \bar{e}^{y(j)}). \tag{6.5.8}$$

Since $x(n) \in [0,1]$, we have $y(n) \geq 0$ for all solutions of (6.5.8). Observe that during the early stages of the epidemic $x(n)$ is close to 1 and, consequently $y(n)$ is close to zero. Hence it is reasonable to linearize (6.5.8) around zero. So if we replace $\bar{e}^{y(j)}$ by $1 - y(j)$, (6.5.8) becomes

$$y(n+1) = \sum_{j=0}^{n} a(n-j)(\varepsilon + y(j)), \quad y(0) = 0. \tag{6.5.9}$$

Taking the Z-transform of both sides of the equation yields

$$\begin{aligned} z\tilde{y}(z) &= \tilde{a}(z)\frac{\varepsilon z}{z-1} + \tilde{a}(z)\tilde{y}(z), \\ \tilde{y}(z) &= \frac{\varepsilon z\, \tilde{a}(z)}{(z+1)(z-\tilde{a}(z))} \end{aligned} \tag{6.5.10}$$

If $a(n)$ has a simple form, one may be able to compute $y(n)$. For example, if $a(n) = ca^n$, then $\tilde{a} = \frac{c\,z}{z-a}$. Hence

$$\tilde{y}(z) = \frac{\varepsilon c\, z}{(z-1)(z-(a+c))} = \frac{\varepsilon c}{1-a-c}\left[\frac{1}{z-1} - \frac{a+c}{z-a-c}\right].$$

Thus

$$y(n) = \frac{\varepsilon c}{1-(a+c)}[1-(a+c)^n]. \tag{6.5.11}$$

Now if $0 < a+c < 1$, then $\lim_{n\to\infty} y(n) = \dfrac{\varepsilon c}{1-(a+c)}$, and hence the spread of the disease will not reach epidemic proportions.

Let us now turn our attention to the nonlinear equation (6.5.8). Let $\hat{y}(n)$ be the solution of (6.5.8) with $\hat{y}(0) = 0$. Then the global stability of $\hat{y}(0)$ is established by the following result (Kocic and Ladas [80]).

Theorem 6.26. *Assume that*

$$\sum_{n=0}^{\infty} a(n) < 1.$$

Then $\hat{y}(n)$ *is a globally asymptotically stable solution of* (6.5.8).

PROOF. We first make the change of variable $u(n) = y(n) - \hat{y}(n)$ in (6.5.8). Then we have

$$u(n+1) = \sum_{j=0}^{n} a(n-j)\bar{e}^{\hat{y}(j)}(1-\bar{e}^{u(j)}). \tag{6.5.12}$$

By induction on n, one may show that $u(n) = y(n) - \hat{y}(n) \geq 0$.

Consider the Liapunov function

$$V(n,u(n)) = (1-a)^{-1}(u(n) + \sum_{r=0}^{n-1}\sum_{s=n}^{\infty} a(s-r)u(r). \tag{6.5.13}$$

Then

$$\begin{aligned}
\Delta V(n,u(n)) &= V(n+1,u(n+1)) - V(n,u(n)) \\
&= (1-a)^{-1}\left[\sum_{j=0}^{n} a(n-j)\bar{e}^{\hat{y}(j)}(1-\bar{e}^{u(j)})\right. \\
&\quad \left. + \sum_{r=0}^{n}\sum_{s+n+1}^{\infty} a(s-r)u(r)\right] \\
&\quad - (1-a)^{-1}\left[u(n) + \sum_{r=0}^{n-1}\sum_{s=n}^{\infty} a(s-r)u(r)\right]
\end{aligned}$$

$$\leq (1-a)^{-1}\left[\sum_{j=0}^{n} a(n-j)u(j) + \sum_{r=0}^{n-1}\sum_{s=n+1}^{\infty} a(s-r)u(r) + \sum_{s=n+1}^{\infty} a(s-n)u(n) - \hat{y}(n) - \sum_{r=0}^{n-1}\sum_{s=n+1}^{\infty} a(s-r)u(r) - \sum_{r=0}^{n-1} a(n-r)u(r)\right]$$

$$\leq (1-a)^{-1}\left(\sum_{s=n}^{\infty} a(s-n) - 1\right)u(n)$$

$$\leq (1-a)^{-1}(a-1)u(n) \leq -u(n).$$

Hence by Theorem 4.20,[4] the zero solution of (6.5.8) is globally asymptotically stable. □

Exercises 6.5

1. Prove that

$$\sum_{i=1}^{k}\sum_{j=1}^{k}\sum_{s=n}^{\infty} |b_{ij}(s-n)||x_j(n)| = \sum_{i=1}^{k}\sum_{j=1}^{k}\sum_{s=n}^{\infty} |b_{ji}(s-n)||x_i(n)|.$$

In Problems 2 through 6 determine whether the zero solution of the given equation is uniformly stable or uniformly asymptotically stable.

2. $x(n+1) = \sum_{j=0}^{n} B(n-j)x(j)$, where $B(n) = \begin{pmatrix} e^{-n} & 1 \\ 0 & e^{-n} \end{pmatrix}$.

3. $x(n+1) = Ax(n) + \sum_{j=o}^{n} B(n-j)x(j)$, where $A = \begin{pmatrix} 0 & \frac{1}{5} \\ \frac{1}{3} & \frac{1}{4} \end{pmatrix}$,

$$B(n) = \begin{pmatrix} 4^{-n-1} & 0 \\ 0 & 3^{-n-1} \end{pmatrix}.$$

4. $x(n+1) = \sum_{j=0}^{n} B(n-j)x(j)$, where $B(n) = \begin{pmatrix} -1 & 4^{-n-1} \\ 0 & 3^{-n-1} \end{pmatrix}$.

5. $x(n+1) = \sum_{j=0}^{n} B(n-j)x(j)$, where $B(n) = \begin{pmatrix} 2^{-n-1} & e^{-n-1} \\ 0 & 5^{-n-1} \end{pmatrix}$.

---

[4]Observe that Theorem 4.20 is valid also for nonautonomous equations as well as Volterra difference equations. The proof is a slight modification of the proof in the text.

*6. Theorem ([14]). Let $G = (g_{ij})$ be a real $k \times k$ matrix. Then $\det G > 0$ if $g_{ii} > 0$, $g_{ii}g_{jj} > \sum_r' |g_{ir}| \sum_r' |g_{jr}|$, for all $1 \le i,\ j \le k$, $i \ne j$. Assume that $\nu_{ij} = \sum_{n=0}^{\infty} b_{ij}(n) < \infty$ for $1 \le i,\ j \le k$. Suppose that the following two conditions hold:

   (i) $a_{ii} + \nu_{ii} > 1, \quad 1 \le i \le k.$

   (ii) $(a_{ii}+\nu_{ii}-1)(a_{jj}+\nu_{jj}-1) > \sum_r' |a_{ir}+\nu_{ir}|$ for $1 \le i, \quad j \le k, i \ne j.$
   Prove that:

   (a) If k is odd, then the zero solution of (6.5.1) is not asymptotically stable.

   (b) If k is even, then the zero solution of (6.5.1) may or may not be asymptotically stable.

*7. (Open problem). Discuss the stability of the zero solution of (6.5.1) under the condition $a_{ii} + \nu_{ii} \le 1$, $1 \le i \le k$. Consider the Volterra system with infinite delay

$$x(n+1) = Ax(n) + \sum_{j=0}^{\infty} B(n-j)x(j). \tag{6.5.14}$$

8. Mimic the proof of Theorem 6.23 to find criteria for the asymptotic stability of the zero solution of (6.5.7).

9. Mimic the proof of Theorem 6.24 to find criteria for the uniform stability of the zero solution of (6.5.7).

10. Prove Theorem 6.23 using the method of Liapunov functionals (as in Theorem 6.24).

6.6 A Variation of Constants Formula

Associated with the homogeneous system (6.5.1) we contemplate the following nonhomogeneous system:

$$y(n+1) = Ay(n) + \sum_{j=0}^{n} B(n-j)y(j) + g(n), \tag{6.6.1}$$

where $g(n) \in \mathbb{R}^k$.

The existence and uniqueness of solutions of system (6.5.1) may be established by a straightforward argument (Exercises 6.6, Problem 13). Let $e_i = (0, \ldots, 1, \ldots, 0)^T$ be the standard ith unit vector in $\mathbb{R}^k$, $1 \le i \le k$. Then there are k vector solutions $x_1(n), x_2(n), x_3(n), \ldots, x_k(n)$ of system (6.5.1) with $x_i(n) = e_i, 1 \le i \le k$. The solutions are linearly independent on $\mathbb{Z}^+$. For if there is a nontrivial linear relation $c_1x_1(n) + c_2x_2(n) + \cdots + c_kx_k(n) = 0$ on $\mathbb{Z}^+$, then at $n = 0$ we have $c_1e_1 + c_2e_2 + \cdots + c_ke_k = 0$.

This proves that $c_1 = c_2 = \cdots = c_k = 0$, which is a contradiction. The $k \times k$ matrix $X(n)$ whose ith column is $x_i(n)$, is called the *fundamental matrix* of system (6.5.1). Notice that $X(n)$ is a nonsingular matrix with $X(0) = I$. Moreover, $x(n) = X(n)x_0$ is a solution of system equations (6.5.1) with $x(0) = x_0$ (Exercises 6.6, Problem 1). Furthermore, the fundamental matrix $X(n)$ satisfies the matrix equation (Exercises 6.6, Problem 2)

$$X(n+1) = AX(n) + \sum_{j=0}^{n} B(n-j)X(j). \tag{6.6.2}$$

It should be pointed out that the fundamental matrix $X(n)$ enjoys all the properties possessed by its counterpart in ordinary difference equations (Chapter 3).

Next we give the variation of constants formula.

Theorem 6.27. *Suppose that the Z-transforms of $B(n)$ and $g(n)$ exist. Then the solution $y(n)$ of system (6.6.1) with $y(n_0) = y_0$ is given by*

$$\boxed{y(n, 0, y_0) = X(n)y_0 + \sum_{r=0}^{n-1} X(n-r-1)g(r).} \tag{6.6.3}$$

PROOF. We first observe that (Why?)

$$X(n+1) = AX(n) + \sum_{r=0}^{n} B(n-r)X(r). \tag{6.6.4}$$

Taking the Z-transform of both sides of (6.6.4), we obtain, for some $R > 0$,

$$z\tilde{X}(z) - zX(0) = A\tilde{X}(z) + \tilde{B}(z)\tilde{X}(z), \quad |z| > R.$$

This yields

$$[zI - A - \tilde{B}(z)]\tilde{X}(z) = zI, \quad |z| > R. \tag{6.6.5}$$

Since the right-hand side of (6.6.5) is nonsingular, it follows that the matrix $zI - A - \tilde{B}(z)$ is also nonsingular. (Why?) This implies that

$$\tilde{X}(z) = z[zI - A - \tilde{B}(z)]^{-1}, \quad |z| > R. \tag{6.6.6}$$

In the next step of the proof we take the Z-transform of both sides of system (6.6.1) to obtain

$$\tilde{y}(z) = [zI - A - \tilde{B}(z)]^{-1}[zy_0 + \tilde{g}(z)], \quad |z| > R_1,$$

for some $R_1 \geq R$, and by using formula (6.6.6) this gives

$$\tilde{y}(z) = \tilde{X}(z)y_0 + \frac{1}{2}\tilde{X}(z)\tilde{g}(z), \quad |z| > R_1.$$

Thus

$$y(n) = Z^{-1}[\tilde{X}(z)y_0] + Z^{-1}\left[\frac{1}{2}\tilde{X}(z)\tilde{g}(z)\right]$$

$$= X(n)y_0 + \sum_{r=0}^{n-1} X(n-r-1)g(r)$$

(using formulas (6.1.4) and (6.1.8)). □

Exercises 6.6

1. Let $X(n)$ be the fundamental matrix of system equation (6.5.1). Prove that $x(n) = X(n)x_0$ is a solution of (6.5.1) for any vector $x_0 \in \mathbb{R}^k$.
2. Prove that the fundamental matrix $X(n)$ satisfies (6.6.2).
3. Prove that the zero solution of (6.5.1) is uniformly stable if and only if $|x(n, n_0, x_0)| \le M|x_0|$ for some $M > 0$.
4. Prove that the zero solution of (6.5.1) is uniformly asymptotically stable if and only if there exist $M > 0$, $\nu \in (0,1)$ such that $|x(n, n_0, x_0)| \le M\nu^{n-n_0}$.
5. Solve the equation

$$x(n+1) = -2\sqrt{3}x(n) + \sum_{r=0}^{n} 2^{n-r}\left(3^{\frac{1}{2(n-r+1)}}\right)x(r) + 2^n(3^{n/2}),\ x(0) = 0:$$

   (a) by the Z-transform method,

   (b) by using Theorem 6.27.
6. Solve the equation $x(n+1) = \frac{1}{2}x(n) + \sum_{r=0}^{n}(n-r)x(r) + n$:

   (a) by the Z-transform method,

   (b) by using Theorem 6.27.
7. Consider the planar system $x(n+1) = Ax(n) + \sum_{j=0}^{n} B(n-j)x(j) + g(n)$, $x(0) = 0$, where

$$A = \begin{pmatrix} -\sqrt{2} & 0 \\ 0 & -\sqrt{6} \end{pmatrix}, \quad B(n) = \begin{pmatrix} 2^{-n/2} & 0 \\ 0 & 6^{-n/2} \end{pmatrix}.$$

   (a) Find the fundamental matrix $X(n)$ of the homogeneous equation.

   (b) Use Theorem 6.27 to solve the equation when $g(n) = \begin{pmatrix} n \\ 0 \end{pmatrix}$.
8. Consider the system $\Delta x(n) = \sum_{j=0}^{n} B(n-j)x(j) + g(n)$, where $B(n) = \begin{pmatrix} 1 & 0 \\ 0 & 2^n \end{pmatrix}$.

(a) Solve the homogeneous part when $g(n) = 0$.

(b) Use Theorem 6.27 to find the solution of the nonhomogeneous equation when $g(n) = \begin{pmatrix} a \\ 0 \end{pmatrix}$, where a is a constant.

9. Consider the system

$$y(n+1) = Ay(n) + g(n), \quad y(0) = y_0. \tag{6.6.7}$$

Use the Z-transform to show that:

(a) $A^n = Z^{-1}[z(zI - A)^{-1}]$.

(b) $\sum_{r=0}^{n-1} A^{n-r-1}g(r) = Z^{-1}[(zI - A)^{-1}\tilde{g}(z)]$.

(c) Conclude that the solution of the given equation is given by $y(n) = Z^{-1}[z(zI - A)^{-1}]y_0 + Z^{-1}[(zI - A)^{-1}\tilde{g}(z)]$.

10. Use (6.6.5) to show that for some $R > 0, \det(zI - A - \tilde{B}(z)) \neq 0$ for $|z| > R$.

Apply the method of Problem 9 to solve equation (6.6.7) in Problem 9 if A and $g(n)$ are given as follows:

11. $A = \begin{pmatrix} 3 & -2 \\ 1 & 0 \end{pmatrix}, \quad g(n) = \begin{pmatrix} n \\ 3 \\ 0 \end{pmatrix}, \quad y(0) = 0.$

12. $A = \begin{pmatrix} 0.5 & 1 \\ 0 & 0.5 \end{pmatrix}, \quad g(n) = 0.$

13. Prove the existence and uniqueness of the solutions of (6.6.1).

6.7 The Z-Transform Versus the Laplace Transform[5]

The Laplace transform plays the same role in differential equations as does the Z-transform in difference equations. For a continuous function $f(t)$, the Laplace transform is defined by

$$\hat{f}(s) = L(f(t)) = \int_0^\infty e^{-st} f(t)\, dt.$$

If we discretize this integral we get $\sum_{n=0}^{\infty} e^{-sn} f(n)$. If further we let $z = e^s$, we get the Z-transform of $f(n)$, namely $\sum_{n=0}^{\infty} f(n)z^{-n}$. Hence

[5]This section may be skipped by readers who are not familiar with the Laplace transform.

given $s = \alpha + i\beta$ in the s-plane (commonly called the frequency domain in engineering), then

$$z = e^{\alpha + i\beta} = e^{\alpha} \cdot e^{i\beta} = e^{\alpha} \cdot e^{i(\beta + 2n\pi)}, \quad n \in \mathbb{Z}.$$

Hence a point in the z-plane corresponds to infinitely many points in the s-plane. Observe that the left half of the s-plane corresponds to the interior of the unit disk $|z| < 1$ in the z-plane. Thus asymptotic stability of a differential equation is obtained if all the roots of its characteristic equation have negative real parts. In difference equations this corresponds to the condition that all the roots of the characteristic equation lie inside the unit disk.

There is another method that enables us to carry the stability analysis from the s-plane to the z-plane, i.e., from differential equations to difference equations. Suppose that the characteristic equation of a difference equation is given by

$$P(z) = a_0 z^n + a_1 z^{n-1} + \cdots + a_n = 0.$$

The bilinear transformation defined by

$$z = \frac{s+1}{s-1}$$

maps the interior of the unit disk to the left half-plane in the complex plane (Figure 6.7). To show this we let $s = \alpha + i\beta$. Then

$$|z| = \left| \frac{\alpha + i\beta + 1}{\alpha + i\beta - 1} \right| < 1,$$

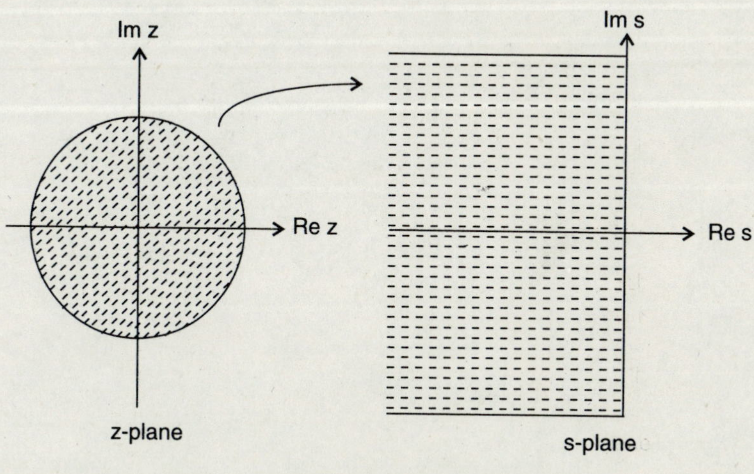

FIGURE 6.7. A bilinear transformation.

or

$$\frac{(\alpha+1)^2+\beta^2}{(\alpha-1)^2+\beta^2}<1,$$

which gives $\alpha < 0$.

Now substituting $z = \frac{s+1}{s-1}$ into $P(z)$ we obtain

$$a_0\left(\frac{s+1}{s-1}\right)^n + a_1\left(\frac{s+1}{s-1}\right)^{n-1} + \cdots + a_n = 0,$$

or

$$Q(s) = b_0 s^n + b_1 s^{n-1} + \cdots + b_n = 0.$$

We now can apply the Routh stability criterion [102] to $Q(s)$ to check whether all the zeros of $Q(s)$ are in the left half-plane. If this is the case, then we know for sure that the zeros of $P(z)$ all lie inside the unit disk. We are not going to pursue this approach here, since the computation involved is horrendous.

TABLE 6.1. Z-transform pairs.

No.	$x(n)$ for $n = 0, 1, 2, 3, \ldots$ $x(n) = 0$ for $n = -1, -2, -3, \ldots$	$\tilde{x}(z) = \sum_{n=0}^{\infty} x(n) z^{-n}$
1.	1	$z/z-1$
2.	a^n	$z/z-a$
3.	a^{n-1}	$\frac{1}{z-a}$
4.	n	$z/(z-1)^2$
5.	n^2	$z(z+1)/(z-1)^3$
6.	n^3	$z(z^2+4z+1)/(z-1)^4$
7.	n^k	$(-1)^k D^k \left(\frac{z}{z-1}\right); D = z\,\frac{d}{dz}$
8.	na^n	$az/(z-a)^2$
9.	$n^2 a^n$	$az(z+a)/(z-a)^3$
10.	$n^3 a^n$	$az(z^2+4az+a^2)/(z-a)^4$
11.	$n^k a^n$	$(-1)^k D^k \left(\frac{z}{z-a}\right); D = z\,\frac{d}{dz}$
12.	$\sin n\omega$	$z \sin \omega/$ $(z^2 - 2z \cos \omega + 1)$
13.	$\cos n\omega$	$z(z - \cos \omega)/$ $(z^2 - 2z \cos \omega + 1)$
14.	$a^n \sin n\omega$	$az \sin n\omega/$ $(z^2 - 2az \cos \omega + a^2)$
15.	$a^n \cos n\omega$	$z(z - a \cos \omega)/$ $(z^2 - 2az \cos \omega + a^2)$
16.	$\delta_0(n)$	1
17.	$\delta_m(n)$	z^{-m}
18.	$a^n/n!$	$e^{a/z}$
19.	$\cosh n\omega$	$z(z - \cosh \omega)/$ $(z^2 - 2z \cosh \omega + 1)$
20.	$\sinh n\omega$	$z \sinh \omega/$ $(z^2 - 2z \cosh \omega + 1)$
21.	$\frac{1}{n},\ n > 0$	$\ln\,(z/z-1)$
22.	$e^{-\omega n} x(n)$	$\tilde{x}\,(e^{\omega} z)$
23.	$n^{(2)} = n(n-1)$	$2z/(z-1)^3$
24.	$n^{(3)} = n(n-1)(n-2)$	$3!z/(z-1)^4$
25.	$n^{(k)} = n(n-1)\cdots(n-k+1)$	$k!z/(z-1)^{k+1}$
26.	$x(n-k)$	$z^{-k}\tilde{x}(z)$
27.	$x(n+k)$	$z^k \tilde{x}(z) - \sum_{r=0}^{k-1} x(r) z^{k-r}$

7

Oscillation Theory

In previous chapters we were mainly interested in the asymptotic behavior of solutions of difference equations both scalar and nonscalar. In this chapter we will go beyond the question of stability and asymptoticity. Of particular interest is to know whether a solution $x(n)$ oscillates around an equilibrium point x^*, regardless of its asymptotic behavior. Since we may assume without loss of generality that $x^* = 0$, the question that we will address here is whether solutions oscillate around zero or whether solutions are eventually positive or eventually negative.

Sections 7.1 and 7.3 follow closely the paper of Erbe and Zhang [53] and the book of Gyori and Ladas [63]. In Section 7.2 we follow the approach in the paper of Hooker and Patula [67]. For more advanced treatment of oscillation theory the reader is referred to [3], [63], [64], [79].

7.1 Three-Term Difference Equations

In this section we consider the three-term difference equation (of order $k+1$)

$$x(n+1) - x(n) + p(n)x(n-k) = 0, \quad n \in \mathbb{Z}^+, \tag{7.1.1}$$

where k is a positive integer and $p(n)$ is a sequence defined for $n \in \mathbb{Z}^+$.

A nontrivial solution $x(n)$ is said to be *oscillatory* (around zero) if for every positive integer N there exists $n \geq N$ such that $x(n)x(n+1) \leq 0$. Otherwise, the solution is said to be *nonoscillatory*. In other words, a solution $x(n)$ is oscillatory if it is neither eventually positive nor eventually negative. The solution $x(n)$ is said to be oscillatory around an equilibrium point x^* if $x(n) - x^*$ is oscillatory around zero. The special case, where $k = 1$ and $p(n) = p$ is a constant real number, has been treated previously in Section 2.5. In this case (7.1.1) may be written in the more convenient form

$$x(n+2) - x(n+1) + px(n) = 0. \tag{7.1.2}$$

The characteristic roots of (7.1.2) are given by

$$\lambda_{1,2} = \frac{1}{2} \pm \frac{1}{2}\sqrt{1-4p}.$$

Recall from Section 2.5 that all solutions of (7.1.2) oscillate if and only if λ_1 and λ_2 are not positive real numbers. Hence every solution of (7.1.2) oscillates if and only if $p > \frac{1}{4}$.

Let us now turn our attention back to (7.1.1). This equation is the discrete analogue of the delay differential equation

$$x'(t) + p(t)x(t-k) = 0. \tag{7.1.3}$$

The oscillatory behavior of (7.1.3) is remarkably similar to that of its discrete analogue (7.1.1), with one exception, when $k = 0$. In this case, the equation

$$x'(t) + p(t)x(t) = 0$$

has the solution

$$x(t) = x(t_0)\, \exp\left(-\int_{t_0}^{t} p(s)\, ds\right),$$

which is never oscillatory. However, the discrete analogue

$$x(n+1) = (1-p(n))x(n)$$

has the solution $x(n) = \left[\prod_{j=n_0}^{n-1}(1-p(j))\right]x(n_0)$, which oscillates if $1 - p(j) < 0$ for all $j \geq n_0$.

To prepare for the study of the oscillatory behavior of (7.1.1) we first investigate the solutions of the following associated difference inequalities:

$$x(n+1) - x(n) + p(n)x(n-k) \leq 0, \tag{7.1.4}$$

$$x(n+1) - x(n) + p(n)x(n-k) \geq 0. \tag{7.1.5}$$

In the sequel we make use of the notions of the *limit superior* and the *limit inferior* of a sequence $\{a(n)\}$, denoted by $\limsup_{n\to\infty} a(n)$ and $\liminf_{n\to\infty} a(n)$, respectively.

Definition 7.1. Let $\{a(n)\}$ be a sequence of real numbers. Let $\beta(n)$ be the least upper bound of the set $\{a(n), a(n+1), a(n+2), \ldots\}$. Then either

$\beta(n) = \pm\infty$ for every n, or the sequence $\{\beta(n)\}$ is a monotonically decreasing sequence of real numbers, and thus $\lim_{n\to\infty}\beta(n)$. Similarly, let $\alpha(n)$ be the greatest lower bound of the set $\{a(n), a(n+1), a(n+2), \ldots\}$. Then:

(i) $\limsup\limits_{n\to\infty} a(n) = \lim_{n\to\infty}\beta(n)$.

(ii) $\liminf\limits_{n\to\infty} a(n) = \lim_{n\to\infty}\alpha(n)$.

Note that $\lim_{n\to\infty} a(n)$ exists if and only if $\limsup\limits_{n\to\infty} a(n) = \liminf\limits_{n\to\infty} a(n) = \lim\limits_{n\to\infty} a(n)$.

Example 7.2. Find the limit superior and the limit inferior for the following sequences:

$$
\begin{aligned}
&S_1 : 0, 1, 0, 1, \ldots.\\
&S_2 : 1, -2, 3, -4, \ldots, (-1)^{n+1}n, \ldots.\\
&S_3 : \frac{3}{2}, -\frac{1}{2}, \frac{4}{3}, -\frac{1}{3}, \frac{5}{4}, -\frac{1}{4}, \frac{6}{5}, -\frac{1}{5}, \ldots.
\end{aligned}
$$

Solution

$$
\begin{aligned}
\limsup_{n\to\infty} S_1 &= 1, & \liminf_{n\to\infty} S_2 &= 0,\\
\limsup_{n\to\infty} S_2 &= \infty, & \liminf_{n\to\infty} S_2 &= -\infty,\\
\limsup_{n\to\infty} S_3 &= 1, & \liminf_{n\to\infty} S_3 &= 0.
\end{aligned}
$$

Theorem 7.3 [53]. *Suppose that*

$$
\boxed{\liminf_{n\to\infty} p(n) = p > \frac{k^k}{(k+1)^{k+1}}.} \tag{7.1.6}
$$

Then the following statements hold:

(i) *Inequality* (7.1.4) *has no eventually positive solution.*

(ii) *Inequality* (7.1.5) *has no eventually negative solution.*

PROOF. (i) To prove statement (i), assume the contrary, that is, there exists a solution $x(n)$ of inequality (7.1.4) that is eventually positive. Hence there exists a positive integer N_1 such that $x(n) > 0$ for all $n \geq N_1$. Dividing inequality (7.1.4) by $x(n)$, we get, for $n \geq N_1$,

$$
\frac{x(n+1)}{x(n)} \leq 1 - p(n)\frac{x(n-k)}{x(n)}. \tag{7.1.7}
$$

If we let $z(n) = \frac{x(n)}{x(n+1)}$, then

$$\begin{aligned}\frac{x(n-k)}{x(n)} &= \frac{x(n-k)}{x(n-k+1)}\frac{x(n-k+1)}{x(n-k+2)},\dots,\frac{x(n-1)}{x(n)}\\ &= z(n-k)z(n-k+1),\dots,z(n-1).\end{aligned}$$

Substituting into inequality (7.1.7) yields

$$\frac{1}{z(n)} \le 1 - p(n)z(n-k)z(n-k+1)\cdots z(n-1), \quad n \ge N_1 + k. \tag{7.1.8}$$

Now, condition (7.1.6) implies that there exists a positive integer N_2 such that $p(n) > 0$ for all $n \ge N_2$. Put $N = \max\{N_2, N_1 + k\}$. Then for $n \ge N, x(n+1) - x(n) \le -p(n)x(n-k) \le 0$. Consequently, $x(n)$ is nonincreasing, and thus $z(n) \ge 1$. Let $\liminf\limits_{n\to\infty} z(n) = q$. Then from inequality (7.1.8) we have

$$\begin{aligned}\limsup_{n\to\infty}\frac{1}{z(n)} &= \frac{1}{\liminf\limits_{n\to\infty} z(n)} = 1/q\\ &\le 1 - \liminf_{n\to\infty}[p(n)z(n-k)z(n-k-1),\dots,z(n-1)],\end{aligned}$$

or

$$\frac{1}{q} \le 1 - pq^k,$$

which yields

$$p \le \frac{q-1}{q^{k+1}}. \tag{7.1.9}$$

Let $h(q) = (q-1)/q^{k+1}$. Then $h(q)$ attains its maximum at $q = (k+1)/k$. Hence $\max_{q\ge1} h(q) = (k^k)/(k+1)^{k+1}$. Hence from inequality (7.1.9) we have $p \le (k^k)/(k+1)^{k+1}$, a contradiction. This completes the proof of part (i) of the theorem. The proof of part (ii) is left to the reader as Exercises 7.1, Problem 6. □

Corollary 7.4. *If condition* (7.1.6) *holds, then every solution of* (7.1.1) *oscillates.*

PROOF. Suppose the contrary and let $x(n)$ be an eventually positive solution of (7.1.1). Then inequality (7.1.4) has an eventually positive solution, which contradicts Theorem 7.3. On the other hand, if (7.1.1) has an eventually negative solution, then so does inequality (7.1.5), which again violates Theorem 7.3. □

The above corollary is sharp, as may be evidenced by the following example, where we let

$$p(n) = \frac{k^k}{(k+1)^{k+1}}.$$

Example 7.5. Consider the difference equation

$$x(n+1) - x(n) + (k^k/(k+1)^{k+1})x(n-k) = 0.$$

Then $x(n) = \left(\frac{k}{k+1}\right)^{n-1}$, $n > 1$, is a nonoscillatory solution of the equation.

Next we give a partial converse of Corollary 7.4.

Theorem 7.6 [53]. *Suppose that $p(n) \geq 0$ and*

$$\sup p(n) < \frac{k^k}{k+1^{k+1}}. \tag{7.1.10}$$

Then (7.1.1) has a nonoscillatory solution.

PROOF. As in the proof of Theorem 7.3, we let $z(n) = x(n)/x(n+1)$ in (7.1.1) to obtain

$$1/z(n) = 1 - p(n)z(n-k)z(n-k+1)\cdots z(n-1). \tag{7.1.11}$$

To complete the proof, it suffices to show that (7.1.11) has a positive solution. To construct such a solution we define

$$z(1-k) = z(2-k) = \cdots = z(0) = a = \frac{k+1}{k} > 1 \tag{7.1.12}$$

and

$$z(1) = [1 - p(1)z(1-k)z(2-k)\cdots z(0)]^{-1}. \tag{7.1.13}$$

Then $z(1) > 1$ also. We claim that $z(1) < a$. To show this, we have

$$\begin{aligned} \frac{z(1)}{a} &= \frac{1}{a[1-p(1)z(1-k)\cdots z(0)]} \\ &\leq \frac{k}{(k+1)\left[1 - \frac{k^k}{(k+1)^{k+1}} \cdot \left(\frac{k+1}{k}\right)^k\right]} = 1. \end{aligned}$$

Hence by induction, we may show that $1 < z(n) < a$, with $n = 1, 2, 3, \ldots$. Moreover, $z(n)$ is a solution of (7.1.11). Now let $x(1) = 1$, $x(2) = x(1)/z(1)$, $x(3) = x(2)/z(2)$, and so on. Then $x(n)$ is a nonoscillatory solution of (7.1.1). □

For the special case where $p(n) = p$ is a constant real number we have the following stronger result.

Theorem 7.7. *Consider the equation*

$$x(n+1) - x(n) + px(n-k) = 0, \tag{7.1.14}$$

where k is a positive integer and p is a nonnegative real number. Then every solution of (7.1.14) oscillates if and only if $p > k^k/(k+1)^{k+1}$.

PROOF. Combining the results of Corollary 7.4, Example 7.5, and Theorem 7.6 yields the proof. □

Remark: Gyori and Ladas [63] showed that every solution of the kth-order equation

$$x(n+k) + p_1 x(n+k-1) + \cdots + p_k x(n) = 0 \tag{7.1.15}$$

oscillates if and only if its characteristic equation has no positive roots. Based on this theorem (Exercises 7.1, Problem 8), they were able to show that every solution of the three-term equation (7.1.14), where $k \in \mathbb{Z} - \{-1, 0\}$, oscillates if and only if $p > k^k/(k+1)^{k+1}$.

Exercises 7.1

1. Find the limit superior and limit inferior of the following sequences:
   (a) $S_1 : \frac{2}{3}, \frac{1}{3}, \frac{3}{4}, \frac{1}{4}, \frac{4}{5}, \frac{1}{5}, \frac{5}{6}, \frac{1}{6}, \ldots$.
   (b) $S_2 : (-1)^{n+1}$.
   (c) $S_3 : \alpha n/(1 + \beta n)$.
   (d) $S_4 : 1 + (-1)^{n+1}$.
2. Prove the following statements:
   (a) $\limsup_{n\to\infty}(1/a(n)) = 1/\liminf_{n\to\infty} a(n)$.
   (b) If $a(n) > 0$, then $\limsup_{n\to\infty}(-a(n)) = -\liminf_{n\to\infty} a(n)$.
   (c) $\liminf_{n\to\infty} a(n) \leq \limsup_{n\to\infty} a(n)$.
3. Show that the difference equation
$$\Delta^2 x(n-1) + \frac{1}{n} x(n) = 0, \quad n \geq 1,$$
   is oscillatory on $[0, \infty)$.
4. (a) Show that every solution of the equation
$$x(n+1) - x(n) + px(n) = 0$$
   oscillates if and only if $p > 1$, where $p \in \mathbb{R}$.
   (b) Show that every solution of the equation
$$x(n+1) - x(n) + px(n-1) = 0$$
   oscillates if and only if $p > \frac{1}{4}$.
5. Consider the difference equation
$$\Delta^2 x(n) + p(n) x(n+1) = 0,$$
   where $p(n) > a > 0$ for $n \in \mathbb{Z}^+$. Show that every nontrivial solution of the equation is oscillatory.
6. Prove part (ii) of Theorem 7.3.

7. The characteristic equation of (7.1.14) is given by

$$\lambda^{k+1} - \lambda^k + p = 0, \quad \text{where } p \geq 0.$$

Show that the characteristic equation has no positive roots if and only if $p > k^k/(k+1)^{k+1}$. Then give a proof of Theorem 7.7.

8. Show that every solution of (7.1.15) oscillates if and only if its characteristic equation has no positive real roots.

*9. [53] Assume that

$$\liminf_{n\to\infty} p(n) = q > 0,$$

and

$$\limsup_{n\to\infty} p(n) > 1 - q.$$

Prove that all conclusions of Theorem 7.3 hold.

In Problems 10 through 12 consider the equation with several delays

$$x(n+1) - x(n) + \sum_{j=1}^{m} p_j(n)x(n-k_j) = 0, \tag{7.1.16}$$

where k_j are positive integers.

10. Suppose that $p_i(n) \geq 0$ and

$$\sum_{i=1}^{m} \liminf_{n\to\infty} p_i(n) \left[\frac{(k_i+1)^{k_i+1}}{(k_i)^{k_i}}\right] > 1.$$

Show that every solution of (7.1.16) oscillates.

11. Suppose that $p_i(n) \geq 0$ and

$$\liminf_{n\to\infty} \left(\sum_{j=1}^{m} p_i(n)\right) > \frac{(\bar{k})^{\bar{k}}}{(\bar{k}+1)^{\bar{k}+1}},$$

where $\bar{k} = \min\{k_1, k_2, \ldots, k_m\} \geq 1$. Show that every solution of (7.1.16) oscillates.

*12. Suppose that

$$\liminf_{n\to\infty} \sum_{i=1}^{m} p_i(n) = c > 0$$

and

$$\limsup_{n\to\infty} \sum_{i=1}^{m} p_i(n) = 1 - c.$$

Prove that every solution of (7.1.16) oscillates.

7.2 Self-Adjoint Second-Order Equations

In this section we consider second-order difference equations of the form

$$\Delta[p(n-1)\Delta x(n-1)] + q(n)x(n) = 0, \tag{7.2.1}$$

where $p(n) > 0, n \in \mathbb{Z}^+$. Equation (7.2.1) is called *self-adjoint*, a name borrowed from its continuous analogue

$$[p(t)x'(t)]' + q(t)x(t) = 0.$$

Equation (7.2.1) may be written in the more familiar form

$$p(n)x(n+1) + p(n-1)x(n-1) = b(n)x(n), \tag{7.2.2}$$

where

$$b(n) = p(n-1) + p(n) - q(n). \tag{7.2.3}$$

As a matter of fact, any equation of the form

$$p_0(n)x(n+1) + p_1(n)x(n) + p_2(n)x(n-1) = 0, \tag{7.2.4}$$

with $p_0(n) > 0$, and $p_2(n) > 0$, can be written in the self-adjoint form (7.2.1) or (7.2.2). To find $p(n)$ and $q(n)$ from $p_0(n), p_1(n)$, and $p_2(n)$, multiply both sides of (7.2.4) by a positive sequence $h(n)$. This yields

$$p_0(n)h(n)x(n+1) + p_1(n)h(n)x(n) + p_2(n)h(n)x(n-1) = 0. \tag{7.2.5}$$

Comparing (7.2.5) with (7.2.2), we obtain

$$p(n) = p_0(n)h(n),$$
$$p(n-1) = p_2(n)h(n).$$

Thus

$$p_2(n+1)h(n+1) = p_0(n)h(n),$$

or

$$h(n+1) = \frac{p_0(n)}{p_2(n+1)}h(n). \tag{7.2.6}$$

Hence

$$h(n) = \prod_{j=n_0}^{n-1} \frac{p_0(j)}{p_2(j+1)}$$

is a solution of (7.2.6). This gives us

$$p(n) = p_0(n) \prod_{j=n_0}^{n-1} \frac{p_0(j)}{p_2(j+1)}.$$

Also from (7.2.3) we obtain

$$q(n) = p_1(n)h(n) + p(n) + p(n-1).$$

In [64] Hartman introduced the notion of *generalized* zeros in order to obtain a discrete analogue of Sturm's separation theorem in differential equations. Next we give Hartman's definition.

Definition 7.8. A solution $x(n), n \geq n_0 \geq 0$, of (7.2.1) has a generalized zero at $r > n_0$ if either $x(r) = 0$ or $x(r-1)x(r) < 0$. In other words, a generalized zero of a solution is either an actual zero or where the solution changes its sign.

Theorem 7.9 (Sturm Separation Theorem). *Let $x_1(n)$ and $x_2(n)$ be two linearly independent solutions of (7.2.1). Then the following statements hold:*

(i) *$x_1(n)$ and $x_2(n)$ cannot have a common zero, that is, if $x_1(r) = 0$, then $x_2(r) \neq 0$.*

(ii) *If $x_1(n)$ has a zero at n_1 and a generalized zero at $n_2 > n_1$, then $x_2(n)$ must have a generalized zero in $(n_1, n_2]$.*

(iii) *If $x_1(n)$ has generalized zeros at n_1 and $n_2 > n_1$, then $x_2(n)$ must have a generalized zero in $[n_1, n_2]$.*

PROOF.

(i) Assume that $x_1(r) = x_2(r) = 0$. Then the Casoratian

$$W(r) = \begin{vmatrix} x_1(r) & x_2(r) \\ x_1(r+1) & x_2(r+1) \end{vmatrix} = 0.$$

It follows from Corollary 2.14 that $x_1(n)$ and $x_2(n)$ are linearly dependent, a contradiction.

(ii) Assume that $x_1(n_1) = 0, x_1(n_2-1)x(n_2) < 0$ (or $x_1(n_2) = 0$). We may assume that n_2 is the first generalized zero greater than n_1. Suppose that $x_1(n) > 0$ for $n_1 < n < n_2$ and $x_1(n_2) \leq 0$.

Now, if $x_2(n)$ has no generalized zeros in $(n_1, n_2]$, then $x_2(n)$ is either positive in $[n_1, n_2]$ or negative in $[n_1, n_2]$. Without loss of generality let $x_2(n) > 0$ on $[n_1, n_2]$. Now pick a positive real number M and $r \in (n_1, n_2)$ such that $x_2(r) = Mx_1(r)$ and $x_2(n) \geq Mx_1(n)$ in $[n_1, n_2]$. By the principle of superposition, the sequence $x(n) = x_2(n) - Mx_1(n)$ is also a solution of (7.2.1) with $x(r) = 0$ and $x(r-1)x(r+1) \geq 0$, with $r > n_1$. Letting $n = r$ in (7.2.1) we obtain

$$\Delta[p(r-1)\Delta x(r-1)] + q(r)x(r) = 0.$$

Since $x(r) = 0$, we have

$$p(r-1)\Delta^2 x(r-1) + \Delta x(r)\Delta p(r-1) = 0,$$

or

$$p(r)x(r+1) = -p(r-1)x(r-1). \tag{7.2.7}$$

Since $x(r+1) \neq 0, x(r-1) \neq 0$, and $p(n) > 0$, equation (7.2.7) implies that $x(r-1)x(r+1) < 0$, which is a contradiction. This completes the proof of part (ii). The proof of part (iii) is left to the reader as Exercises 7.2, Problem 6. □

Remark: Based on the notion of generalized zeros, we can give an alternative definition of oscillation. A solution of a difference equation is oscillatory on $[n_2, \infty)$ if it has infinitely many generalized zeros on $[n_0, \infty)$. An immediate consequence of the Sturm separation theorem (Theorem 7.9) is that if (7.2.1) has an oscillatory solution, then all its solutions are oscillatory. We caution the reader that the above conclusion does not hold in general for non-self-adjoint second-order difference equations. For example, the difference equation $x(n+1) - x(n-1) = 0$ has a nonoscillatory solution $x_1(n) = 1$ and an oscillatory solution $x_2(n) = (-1)^n$. Observe that this equation is not self-adjoint. We are now ready to give some simple criteria for oscillation.

Lemma 7.10. *If there exists a subsequence $b(n_k) \leq 0$, with $n_k \to \infty$ as $k \to \infty$, then every solution of* (7.2.2) *oscillates.*

PROOF. Assume the contrary, that there exists a nonoscillatory solution $x(n)$ of (7.2.2). Without loss of generality, suppose that $x(n) > 0$ for $n \geq N$. Then

$$p(n_k)x(n_k+1) + p(n_k-1)x(n_k-1) - b(n_k)x(n_k) > 0, \quad \text{for } n_k > N,$$

which is a contradiction. □

One of the most useful techniques in oscillation theory is the use of the so-called *Riccati transformations.* We will introduce only one transformation that is needed in the development of our results. Two more transformations will appear in the exercises. In (7.2.2) let

$$z(n) = \frac{b(n+1)x(n+1)}{p(n)x(n)}. \tag{7.2.8}$$

Then $z(n)$ satisfies the equation

$$c(n)z(n) + \frac{1}{z(n-1)} = 1, \tag{7.2.9}$$

where

$$c(n) = \frac{p^2(n)}{b(n)b(n+1)}. \tag{7.2.10}$$

Next we give a crucial result that relates (7.2.2) with (7.2.9).

Lemma 7.11. *Suppose that $b(n) > 0$ for $n \in \mathbb{Z}^+$. Then every solution $x(n)$ of* (7.2.2) *is nonoscillatory if and only if every solution $z(n)$ of* (7.2.9) *is positive for $n \geq N$, for some $N > 0$.*

PROOF. Suppose that $x(n)$ is a nonoscillatory solution of (7.2.2). Then $x(n)x(n+1) > 0$ for $n \geq N$. Equation (7.2.8) then implies that $z(n) > 0$. Conversely, assume that $z(n)$ is a positive solution of (7.2.9). Using this solution we construct inductively a nonoscillatory solution $x(n)$ of (7.2.2) as follows: Let $x(N) = 1, x(n+1) = (p(n)/b(n+1))z(n)x(n)$, with $n > N$. Then one may verify that $x(n)$, with $n \geq N$, is indeed a solution of (7.2.2) that is nonoscillatory. By the Sturm separation theorem, every solution of (7.2.2) is thus nonoscillatory. We need a comparison result concerning (7.2.9) that will be needed to establish the main result of this section. □

Lemma 7.12. *If $c(n) \geq a(n) > 0$ for all $n > 0$ and $z(n) > 0$ is a solution of the equation*

$$c(n)z(n) + \frac{1}{z(n-1)} = 1, \tag{7.2.11}$$

then the equation

$$a(n)y(n) + \frac{1}{y(n-1)} = 1 \tag{7.2.12}$$

has a solution $y(n) \geq z(n) > 1$ for all $n \in \mathbb{Z}^+$.

PROOF. Since $c(n) > 0$ and $z(n) > 0$, it follows from (7.2.1) that $1/(z(n-1)) < 1$. This implies that $z(n-1) > 1$ for all $n \geq 1$. We now define inductively a solution $y(n)$ of (7.2.12). Choose $y(0) \geq z(0)$ and let $y(n)$ satisfy (7.2.11). Now, from (7.2.12) and (7.2.12), we have

$$a(n)y(n) + \frac{1}{y(n-1)} = c(n)z(n) + \frac{1}{z(n-1)}.$$

So

$$a(1)y(1) + \frac{1}{y(0)} = c(1)z(1) + \frac{1}{z(0)}.$$

Since $y(0) \geq z(0)$, we have $1/y(0) \leq 1/z(0)$, and hence $a(1)y(1) \geq c(1)z(1)$, or

$$y(1) \geq \frac{c(1)}{a(1)}z(1) \geq z(1) > 1.$$

Inductively, one may show that

$$y(n) \geq z(n) > 1.$$ □

Theorem 7.13. *If $b(n)b(n+1) \leq (4-\varepsilon)p^2(n)$ for some $\varepsilon > 0$ and for all $n \geq N$, then every solution of (7.2.2) is oscillatory.*

PROOF. If $b(n)b(n-1) \leq (4-\varepsilon)p^2(n)$ for some $\varepsilon \geq 4$, then $b(n)b(n-1) \leq 0$. The conclusion of the theorem then follows from Lemma 7.10. Hence we may assume that $0 < \varepsilon < 4$. Now assume that (7.2.2) has a nonoscillatory solution. Then by Lemma 7.11, (7.2.9) has a positive solution $z(n)$ for $n \geq N$. Using the assumption of the theorem in formula (7.2.10) yields

$$c(n) = \frac{p^2(n)}{b(n)b(n+1)} \geq \frac{p^2(n)}{(4-\varepsilon)p^2(n)} = \frac{1}{4-\varepsilon}.$$

Then it follows from Lemma 7.12 that the equation

$$\frac{1}{4-\varepsilon}y(n) + \frac{1}{y(n-1)} = 1 \tag{7.2.13}$$

has a solution $y(n), n \geq N$, such that $y(n) \geq z(n) > 1$ for all $n \geq N$. Define a positive sequence $x(n)$ inductively as follows: $x(N) = 1, x(n+1) = (1/\sqrt{4-\varepsilon})y(n)x(n)$ for $n \geq N$. Then

$$y(n) = \sqrt{4-\varepsilon}\left(\frac{x(n+1)}{x(n)}\right). \tag{7.2.14}$$

Substituting for $y(n)$ in (7.2.14) into (7.2.13) yields $x(n+1) - \sqrt{4-\varepsilon}x(n) + x(n-1) = 0$, $n \geq N$, whose characteristic roots are

$$\lambda_{1,2} = \frac{\sqrt{4-\varepsilon}}{2} \pm i\frac{\sqrt{\varepsilon}}{2}.$$

Thus its solutions are oscillatory, which gives a contradiction. The proof of the theorem is now complete. □

It is now time to give some examples.

Example 7.14. Consider the difference equation

$$y(n+1) + y(n-1) = \left(2 + \frac{1}{2}(-1)^n\right)y(n).$$

Here $p(n) = 1$ and $b(n) = (2 + \frac{1}{2}(-1)^n)$:

$$\begin{aligned} b(n)\,b(n+1) &= \left(2 + \frac{1}{2}(-1)^n\right)\left(2 + \frac{1}{2}(-1)^{n+1}\right) \\ &= 3\frac{3}{4}. \end{aligned}$$

Thus $b(n)\,b(n+1) \leq \left(4 - \frac{1}{5}\right)p^2(n)$. By Theorem 7.13, we conclude that every solution is oscillatory.

The following example will show the sharpness of Theorem 7.13 in the sense that if ε is allowed to be a sequence tending to zero, then the theorem fails.

Example 7.15 [67]. Consider the equation

$$x(n+1) + x(n) = b(n)\,x(n-1), \quad n = 1, 2, 3, \ldots,$$

where

$$b(n) = \frac{\sqrt{n+1} + \sqrt{n-1}}{\sqrt{n}}.$$

Now,

$$b(n)b(n+1) = \frac{\sqrt{(n+1)(n+2)} + \sqrt{(n-1)(n+2)} + \sqrt{n(n+1)} + \sqrt{n(n-1)}}{\sqrt{n(n+1)}}.$$

But $\lim_{n\to\infty} b(n)b(n+1) = 4$. Hence if one takes $\varepsilon_n = 4 - b(n)b(n+1)$, then $\varepsilon_n \to 0$ as $n \to \infty$. However, Theorem 7.13 fails, since $x(n) = \sqrt{n}, n \geq 1$, is a nonoscillatory solution of the equation.

A partial converse of Theorem 7.13 now follows.

Theorem 7.16. *If $b(n)b(n+1) \geq 4p^2(n)$ for $n \geq N$, then every solution of (7.2.2) is nonoscillatory.*

Proof. From formula (7.2.10) and the assumption we obtain $c(n) \leq \frac{1}{4}$. We now construct inductively a solution $z(n)$ of (7.2.9) as follows: Put $z(N) = 2$, and

$$z(n) = \frac{1}{c(n)}\left(1 - \frac{1}{z(n-1)}\right), \quad n > N.$$

Observe that

$$z(N+1) = \frac{1}{c(N+1)}\left(1 - \frac{1}{z(N)}\right) \geq 4\left(1 - \frac{1}{2}\right) = 2.$$

Similarly, one may show that $z(n) \geq 2$ for $n \geq N$. Hence by Lemma 7.11, we conclude that every solution of (7.2.2) is nonoscillatory. □

Example 7.17. Consider the difference equation

$$\Delta(n\Delta x(n-1) - \frac{1}{n}x(n) = 0.$$

Here $p(n) = n+1$ and $q(n) = \frac{1}{n}$. Using formula (7.2.3) we obtain

$$b(n) = 2n + 1 + \frac{1}{n}.$$

Now,

$$\begin{aligned} b(n)b(n+1) &= \left(2n+1+\frac{1}{n}\right)\left(2n+3+\frac{1}{n+1}\right) \\ &= 4n^2 + 8n + 7 + \frac{2n+4}{n(n+1)} \\ &\geq 4p^2(n) \quad \text{for all } n \geq 1. \end{aligned}$$

Hence by Theorem 7.16, every solution is nonoscillatory.

Exercises 7.2

In Problems 1 through 5 determine the oscillatory behavior of all solutions.

1. $\Delta[(n-1)x(n-1)] + \frac{1}{n}x(n) = 0$.

2. $x(n+1)+x(n-1)=\left(2-\frac{1}{n}\right)x(n)$.
3. $x(n+1)+x(n-1)=\left(2+\frac{1}{n}\right)x(n)$.
4. $\Delta^2[x(n-1)]+\frac{1}{n\ln(n)}x(n)=0, \quad n>1$.
5. $\Delta[(n-1)x(n-1)]+x(n)=0$.
6. Prove part (iii) of Theorem 7.9.
7. [112] Show that if $b(n) \le \min\{p(n), p(n-1)\}$ for $n \ge N$, for some positive integer N, then every solution of (7.2.2) is oscillatory.
8. Show that if $b(n) \le p(n)$ and $p(n)$ is eventually nonincreasing, then every solution of (7.2.2) is oscillatory.
9. Show that if $b(n) \le p(n-1)$ and $p(n)$ is eventually nondecreasing, then every solution of (7.2.2) is oscillatory.
10. (A second Riccati transformation). Let $z(n)=x(n+1)/x(n)$ in (7.2.2).

    (i) Show that $z(n)$ satisfies the equation

    $$p(n)z(n)+\frac{p(n-1)}{z(n-1)}=b(n). \tag{7.2.15}$$

    (ii) Assuming $p(n)>0$, show that every solution of (7.2.2) is nonoscillatory if and only if (7.2.15) has a positive solution $z(n), n \ge N$, for some $N>0$.

*11. Use the second Riccati transformation in Problem 10 to show that if $b(n) \le p(n-1)$ and $\limsup_{n\to\infty}(p(n))/p(n-1) > \frac{1}{2}$, then every solution of (7.2.2) oscillates.

12. [67] Show that if $b(n) \ge \max\{p(n-1), 4p(n)\}$, for all $n \ge N$, for some $N>0$, then every solution of (7.2.2) is nonoscillatory.
13. Show that if $p(n_k) \ge b(n_k)b(n_k+1)$ for a sequence $n_k \to \infty$, then every solution of (7.2.2) is oscillatory.
14. As in formula (7.2.10), let

    $$c(n)=\frac{p^2(n)}{b(n)b(n+1)}, \quad n \ge 0.$$

    Show that either one of the following implies that every solution of (7.2.2) oscillates:

    (i) $\limsup_{n\to\infty} c(n) > 1$.

    (ii) $\limsup_{n\to\infty} \frac{1}{n}\sum_{j=1}^{n} c(j) > 1$.

15. Show that if $p(n)$ is bounded above on $[a, \infty)$ and (7.2.1) is nonoscillatory on $[a, \infty)$, then either $\sum_{n=a+1}^{\infty} q(n)$ exists and is finite or it is equal to $-\infty$.

16. Use Problem 15 to prove that (7.2.1) is oscillatory if either one of the following conditions hold:

    (i) $p(n)$ is bounded on $[a, \infty)$ and $\sum_{n=a+1}^{\infty} q(n) = \infty$, or

    (ii) $p(n)$ is bounded on $[a, \infty)$ and

$$-\infty \le \liminf_{n\to\infty} \sum_{s=a+1}^{n} q(s) \le \limsup_{n\to\infty} \sum_{s=a+1}^{n} q(s) \le \infty.$$

7.3 Nonlinear Difference Equations

In this section we will investigate the oscillatory behavior of the nonlinear difference equation

$$x(n+1) - x(n) + p(n) f(x(n-k)) = 0, \tag{7.3.1}$$

where $k \in \mathbb{Z}^+$ and $N \in \mathbb{Z}^+$. The first theorem is due to Erbe and Zhang [53].

Theorem 7.18. *Suppose that f is continuous on $\mathbb{R}$ and satisfies the following assumptions:*

(i) $xf(x) > 0, \quad x \neq 0,$

(ii) $\liminf_{x\to 0} \frac{f(x)}{x} = L, \quad 0 < L < \infty,$

(iii) $pL > \frac{k^k}{(k+1)^{k+1}}$ if $k \ge 1$ and $pL > 1$ if $k = 0$, *where* $p = \liminf_{n\to\infty} p(n) > 0$.

Then every solution of (7.3.1) *oscillates.*

Proof. Assume the contrary and let $x(n)$ be a nonoscillatory solution of (7.3.1). Suppose that $x(n) > 0$ for $n \ge N$. This implies by assumption (i) that $f(x(n)) > 0$. Hence $x(n+1) - x(n) = -p(n)f(x(n-k)) < 0$, and thus $x(n)$ is decreasing. Hence $\lim_{n\to\infty} x(n) = c \ge 0$.

Taking the limit of both sides of (7.3.1) yields $f(c) = 0$, which by assumption (i) gives $c = 0$. Hence $\lim_{n\to\infty} x(n) = 0$. Dividing (7.3.1) by $x(n)$ and letting $z(n) = x(n)/x(n+1) \ge 1$ yields

$$\frac{1}{z(n)} = 1 - p(n) z(n-1) \cdots z(n-k) \frac{f(x(n-k))}{x(n-k)}. \tag{7.3.2}$$

Let $\liminf_{n\to\infty} z(n) = r$. By taking the limit superior in (7.3.2) we obtain

$$\frac{1}{r} \le 1 - pLr^k,$$

or

$$pL \le \frac{r-1}{r^{k+1}}. \tag{7.3.3}$$

It is easy to see that the function $h(r) = (r-1)/r^{k+1}$ attains its maximum at $r = (k+1)/k$, and its maximum value is thus $k^k/(k+1)^{k+1}$. Hence inequality (7.3.3) becomes

$$\boxed{pL \le \frac{k^k}{(k+1)^{k+1}},}$$

which contradicts assumption (iii). □

Remark: If we let $\liminf_{n\to\infty} f(x)/x = 1$, then the linearized equation associated with (7.3.1), where $p(n)$ is equal to a constant real number p, is given by

$$y(n+1) - y(n) + py(n-k) = 0, \tag{7.3.4}$$

which has been studied in Section 7.1. We may now rephrase Theorem 7.18 as follows: Suppose that assumptions (i) and (ii) hold with $L = 1$ and that $p(n)$ is constant. If every solution of (7.3.4) oscillates, then so does every solution of (7.3.1). Gyori and Ladas [63] considered the more general equation with several delays

$$x(n+1) - x(n) + \sum_{i=1}^{m} p_i f_i(x(n-k_i)) = 0, \tag{7.3.5}$$

where $p_i > 0$, k_i is a positive integer, and f_i is a continuous function on $\mathbb{R}$, with $1 \le i \le m$. They obtained the following result.

Theorem 7.19. *Suppose that the following hold:*

(i) $p_i > 0$, $k_i \in \mathbb{Z}^+$, *and* $\sum_{i=1}^{m} (p_i + k_i) \ne 1$, $1 \le i \le m$,

(ii) f *is continuous on* $\mathbb{R}$, *and* $xf_i(x) > 0$, *for* $x \ne 0$, $1 \le i \le m$,

(iii) $\liminf_{x\to 0} \frac{f_i(x)}{x} \ge 1$, $1 \le i \le m$,

(iv) $\sum_{i=1}^{m} p_i \frac{(k_i+1)^{k_i+1}}{k_i^{k_i}} > 1$.

Then every solution of (7.3.5) *oscillates.*

To facilitate the proof of this theorem we present the following lemma.

Lemma 7.20 [63]. *Suppose that condition* (i) *in Theorem* 7.19 *holds and let* $\{q_i(n) : 1 \le i \le m\}$ *be a set of sequences of real numbers such that*

$$\liminf_{n\to\infty} q_i(n) \ge p_i, \quad 1 \le i \le m. \tag{7.3.6}$$

If the linear difference inequality

$$x(n+1) - x(n) + \sum_{i=1}^{m} q_i(n)x(n-k_i) \le 0, \quad n \in \mathbb{Z}^+, \tag{7.3.7}$$

has an eventually positive solution $x(n)$, then the corresponding limiting equation

$$y(n+1) - y(n) + \sum_{i=1}^{m} p_i y(n-k_i) = 0 \tag{7.3.8}$$

also has an eventually positive solution.

PROOF. There are two distinct cases to consider

Case (a): Assume that $k_i = 0,\ 1 \le i \le m$. Then (7.3.7) and (7.3.8) simplify to

$$x(n+1) \le \left(1 - \sum_{i=1}^{m} q_i(n)\right) x(n), \tag{7.3.9}$$

$$y(n+1) = \left(1 - \sum_{i=1}^{m} p_i\right) y(n). \tag{7.3.10}$$

Let $x(n)$ be an eventually positive solution of (7.3.9). Then from (7.3.9) it follows that, for sufficiently large n,

$$\sum_{i=1}^{m} q_i(n) < 1. \tag{7.3.11}$$

Now, from assumption (7.3.6) for any $\varepsilon > 0$ there exists $N > 0$ such that

$$0 < p_i \le q_i(n) + \varepsilon/m \quad \text{for } n \ge N. \tag{7.3.12}$$

This implies that

$$0 < \sum_{i=1}^{m} p_i \le \sum_{i=1}^{m} q_i(n) + \varepsilon < 1 + \varepsilon \quad \text{for } n \ge N.$$

Since ε was arbitrarily chosen,

$$0 < \sum_{i=1}^{m} p_i \le 1.$$

But from assumption (i) in Theorem 7.19, we have $\sum_{i=1}^{m} p_i \ne 1$. Hence

$$0 < \sum_{i=1}^{m} p_i < 1,$$

and consequently, (7.3.10) has a positive solution.

Case (b): Assume that $k = \max\{k_1, k_2, \ldots, k_m\} \geq 1$. Let $u(n) = x(n)/x(n-1)$. Then

$$\frac{x(n-k_i)}{x(n-1)} = \frac{x(n-k_i)}{x(n-k_i+1)} \cdot \frac{x(n-k_i+1)}{x(n-k_i+2)} \cdots \frac{x(n-2)}{x(n-1)} = \prod_{j=1}^{k_i-1} \frac{1}{u(n-j)}.$$

Making this change of variables in (7.3.7) yields

$$u(n+1) \leq 1 - \sum_{i=1}^{m} q_i(n) \prod_{j=1}^{k_i-1} \frac{1}{u(n-j)}. \tag{7.3.13}$$

Define $u = \limsup_{n\to\infty} u(n)$. Then it follows from (7.3.13) that $0 < u(n) < 1$ and $0 < u < 1$. We claim that

$$u - 1 + \sum_{i=1}^{m} p_i u^{-k_i} \leq 0. \tag{7.3.14}$$

Now, from (7.3.6), for every $\varepsilon \in (0,1)$ there exists $N_\varepsilon > 0$ such that $P_i(n) \geq (1-\varepsilon)p_i$, $1 \leq i \leq m$, and $n \geq N_\varepsilon$. Substituting in (7.3.13) yields

$$u(n+1) \leq 1 - (1-\varepsilon) \sum_{i=1}^{m} p_i \left(\prod_{j=1}^{k_i-1} \frac{1}{u(n-j)} \right) \quad \text{for } n \geq N_\varepsilon.$$

Choose $\tilde{N}_\varepsilon$ such that $\tilde{N}_\varepsilon \geq N_\varepsilon + k$ and

$$u(n) \leq (1-\varepsilon)u \quad \text{for } n \geq \tilde{N}_\varepsilon - k.$$

Then, for $n \geq \tilde{N}_\varepsilon - k$,

$$u(n+1) \leq 1 - (1-\varepsilon) \sum_{i=1}^{m} p_i u^{-k_i} (1+\varepsilon)^{-k_i}.$$

Consequently,

$$u \leq 1 - (1-\varepsilon) \sum_{i=1}^{m} p_i u^{-k_i} (1+\varepsilon)^{-k_i}.$$

Since ε was arbitrarily chosen, it follows that

$$u \leq 1 - \sum_{i=1}^{m} p_i u^{-k_i},$$

which proves claim (7.3.14).

To complete the proof of the lemma we consider the characteristic polynomial of (7.3.8), $p(\lambda) = \lambda - 1 + \sum_{i=1}^{m} p_i \lambda^{-k_i}$.

Observe that $p(0^+) = \infty$ and $p(u) \leq 0$. This implies by the intermediate value theorem that $p(\lambda)$ has a positive root and, consequently, (7.3.8) has a positive solution. $\square$

PROOF OF THEOREM 7.19. Assume that (7.3.5) has a nonoscillatory solution $x(n)$. Without loss of generality, assume that $x(n)$ is eventually positive. Then it follows from assumption (iii) that $x(n+1) \leq x(n) - \sum_{i=1}^{m} p_i x(n-k_i)$. As in the proof of Theorem 7.18, one may show that $\lim_{n\to\infty} x(n) = 0$. We now need to put (7.3.5) in the form of (7.3.7). This can be accomplished by setting

$$q_i(n) = \frac{p_i \ f(x(n-k_i))}{x(n-k_i)}.$$

Thus from assumption (iii) we obtain

$$\liminf_{n\to\infty} q_i(n) \geq p_i.$$

By Lemma 7.20, it follows that the limiting equation (7.3.8) also has an eventually positive solution, which is a contradiction. This completes the proof of the theorem. □

In fact, under additional conditions it was proved ([63, , Corollary 7.4.1]) that every solution of the nonlinear equation oscillates if and only if every solution of the corresponding linearized equation oscillates.

We now apply the obtained results to study the oscillatory behavior of the Pielou logistic delay equation. The stability of this equation has been determined previously in Example 4.37.

Example 7.21 [63]. Consider the Pielou logistic delay equation

$$y(n+1) = \frac{\alpha y(n)}{1+\beta y(n-k)}, \quad \alpha > 1, \beta > 0, \quad k \text{ a positive integer.} \tag{7.3.15}$$

Show that every positive solution of (7.3.15) oscillates about its positive equilibrium point $y^* = (\alpha - 1)/\beta$ if

$$\frac{\alpha - 1}{\alpha} > \frac{k^k}{(k+1)^{k+1}}. \tag{7.3.16}$$

Solution We follow Method 2 in Example 4.37 by letting $y(n) = ((\alpha - 1)/\beta)e^{x(n)}$ in (7.3.15). We obtain the equation

$$x(n+1) - x(n) + \frac{\alpha - 1}{\alpha} f(x(n-k)) = 0, \tag{7.3.17}$$

where

$$f(x) = \frac{\alpha}{\alpha - 1} \ln\left(\frac{(\alpha-1)e^x + 1}{\alpha}\right).$$

It may be shown that the function f satisfies conditions (i) and (ii) in Theorem 7.18 with $L = 1$. Hence by Theorem 7.18 every solution of (7.3.17) oscillates about 0. This implies that every solution of (7.3.15) oscillates about the equilibrium point $y^* = (\alpha - 1)/\beta$.

Exercises 7.3

1. Consider the difference equation

$$\Delta x(n) + e^{x(n-1)} - 1 = 0.$$

Determine the oscillatory behavior of all solutions.

2. Consider the difference equation

$$x(n+1) = x(n)\exp\left[r\left(1 - \frac{x(n)}{\alpha}\right)\right], \quad r > 0, \quad \alpha > 0, \quad x(0) > 0.$$

(a) Show that $x^* = \alpha$ is the only positive equilibrium point.

(b) Show that every solution oscillates about α if $r > 1$.

(c) Show that if $r = 1$, every solution converges monotonically to α.

3. Consider the difference equation

$$x(n+1) = x(n)\exp\left[r\left(1 - \frac{x(n-1)}{\alpha}\right)\right], \quad r > 0, \quad \alpha > 0, \quad x(0) > 0.$$

Show that every solution oscillates about $x^* = \alpha$ if $r > \frac{1}{4}$.

4. Consider the difference equation

$$x(n+1) = x(n)\exp\left[r\left(1 - \frac{x(n-1)}{\alpha} - \frac{x(n-2)}{\beta}\right)\right],$$

$$r > 0, \quad \alpha > 0, \quad \beta > 0, \quad x(0) > 0.$$

Show that every solution oscillates about $x^* = (\alpha\beta)/(\alpha + \beta)$ if $r > 4(\alpha + \beta)/(27\alpha + 16\beta)$.

5. Consider the difference equation

$$\Delta x(n) + p(1 + x(n))x(n) = 0, \quad p > 0, \quad 1 + x(n) > 0.$$

Show that every solution oscillates if $p > 1$.

6. Consider the difference equation

$$\Delta x(n) + p(1 + x(n))x(n-1) = 0, \quad p > 0, \quad 1 + x(n) > 0.$$

Show that every solution oscillates if $p > \frac{1}{4}$.

7. [63] Consider the difference equation

$$\Delta x(n) + p(n)[1 + x(n)]\, x(n-k) = 0, \quad p(n) > 0,$$

for $n \geq 1$, $x(n) + 1 > 0$ for $n \geq -k$. Prove that every solution oscillates if $\liminf_{n\to\infty} p(n) = c > k^k/((k+1)^{k+1})$.

8. Consider the difference equation

$$x(n+1) = \frac{\alpha x(n)}{1 + \beta x(n-k) + \gamma x(n-1)}$$

with $\alpha > 1$, $\beta > 0$, $\gamma > 0$, $k \in \mathbb{Z}^+$. Find conditions under which all solutions oscillate.

8

Asymptotic Behavior of Difference Equations

In Chapters 4 and 5 we were mainly interested in stability questions. In other words, we wanted to know whether solutions of a difference equation converge to zero or to an equilibrium point. In asymptotic theory, we are concerned rather with obtaining asymptotic formulas for the manner in which solutions tend to zero or a constant. We begin this chapter by introducing the reader to the tools of the trade.

8.1 Tools of Approximation

The symbols $\sim$, o, and O are the main tools of approximating functions and are widely used in all branches of science. For the benefit of our readers, we shall give our definitions for functions defined on the real or complex numbers. Hence sequences will be treated as a special case of the general theory.

We start with the symbol O (big oh).

Definition 8.1. Let $f(t)$ and $g(t)$ be two functions defined on R or C. Then we say that $f(t) = O(g(t))$, $t \to \infty$, if there is a positive constant M such that

$$\boxed{|f(t)| \leq M|g(t)| \quad \text{for all} \quad t \geq t_0.}$$

Equivalently, $f(t) = O(g(t))$ if $\left|\frac{f(t)}{g(t)}\right|$ is bounded for $t \geq t_0$. In other words, $f = O(g)$ if f is of order not exceeding g.

Example 8.2.

(a) Show that

$$\left(\frac{n}{t^2+n^2}\right)^n = O\left(\frac{1}{t^n}\right), \qquad n \to \infty, \text{ for } n \in \mathbb{Z}^+.$$

Solution Without loss of generality we assume $t > 1$. We have $t^2+n^2 = (t-n)^2 + 2nt \geq 2nt$. Hence

$$\left(\frac{n}{t^2+n^2}\right)^n \leq \frac{1}{(2t)^n} = \frac{1}{2^n}\left(\frac{1}{t^n}\right) \leq \frac{1}{t^n}, \qquad \text{for } n \in \mathbb{Z}^+, \approx > \nVdash.$$

It follows that

$$\left(\frac{n}{t^2+n^2}\right)^n = O\left(\frac{1}{t^n}\right)$$

with the constant $M = 1$ being independent of n.

(b) Show that

$$\sin\left(n\pi + \frac{1}{n}\right) = O\left(\frac{1}{n}\right), \qquad n \to \infty.$$

Solution Recall that $\sin\left(n\pi + \frac{1}{n}\right) = (-1)^n \sin\frac{1}{n}$. Thus

$$\left|\frac{\sin\left(n\pi + \frac{1}{n}\right)}{1/n}\right| = \left|\frac{\sin\frac{1}{n}}{1/n}\right|.$$

If we let $u = \frac{1}{n}$, then $\lim_{n\to\infty}\left|\frac{\sin\frac{1}{n}}{1/n}\right| = \lim_{u\to 0}\left|\frac{\sin u}{u}\right| = 1$.

Hence we conclude that $\left|\left(\sin\frac{1}{n}\right)/(1/n)\right|$ is bounded, which gives the required result.

(c) Show that $t^2 \log t + t^3 = O(t^3), t \to \infty$.

Solution $\left|\frac{t^2\log t + t^3}{t^3}\right| = 1 + \left|\frac{\log t}{t}\right|$.

Using the first derivative test one may show that the function $y = \log t/t$ attains its maximum value $\frac{1}{e}$ as $t = e$. Hence $|\log t/t| \leq \frac{1}{e} < 1$, and thus $\left|(t^2\log t + t^3)/t^3\right| \leq 2$. This proves the required result.

Remark: We would like to point out here that the relation defined by O is not symmetric, i.e., if $f = O(g)$, then it is not necessarily true that $g = O(f)$. To illustrate this point we cite some simple examples such as $x = O(x^2), x \to \infty$, but $x^2 \neq O(x), x \to \infty$, or $e^{-x} = O(1), x \to \infty$, but $1 \neq O(e^{-x}), x \to \infty$, since $1/e^{-x} \to \infty$, $x \to \infty$.

However, it is true that the relation O is transitive, that is to say if $f = O(g)$ and $g = O(h)$, then $f = O(h)$ (Exercises 8.1, Problem 1). In this case we say that $f = O(h)$ is a better approximation of f than $f = O(g)$.

Next we give the definition of the symbol o (little oh).

Definition 8.3. If $\lim_{t\to\infty} \frac{f(t)}{g(t)} = 0$, then we say that

$$f(t) = o(g(t)), \qquad t \to \infty.$$

Example 8.4.

(a) Show that $t^2 \log t + t^3 = o(t^4), t \to \infty$.

Solution $\lim_{t\to\infty} \dfrac{t^2 \log t + t^3}{t^4} = \lim_{t\to\infty} \dfrac{\log t}{t^2} + \lim_{t\to\infty} \dfrac{1}{t}$.
Using L'Hôpital's rule we have

$$\lim_{t\to\infty} \frac{\log t}{t^2} = \lim_{t\to\infty} \frac{1}{2t^2} = 0.$$

Hence

$$\lim_{t\to\infty} \frac{t^2 \log t + t^3}{t^4} = 0,$$

and the required conclusion follows.

(b) Show that $o(g(t)) = g(t)o(1), t \to \infty$.

Solution Let $f(t) = o(g(t)), t \to \infty$. Then

$$\lim_{t\to\infty} \frac{f(t)}{g(t)} = 0,$$

which implies that $\frac{f(t)}{g(t)} = o(1), t \to \infty$. Consequently, $f(t) = g(t)o(1)$, $t \to \infty$.

The reader may sense correctly that the symbol o plays a much less important role than the symbol O.

Finally, we introduce the asymptotic equivalence relation $\sim$.

Definition 8.5. If $\lim_{t\to\infty} \frac{f(t)}{g(t)} = 1$, then we say that f is asymptotic to g, $t \to \infty$, and we write $f \sim g, t \to \infty$.

Notice that if $f \sim g$ as $t \to \infty$, then

$$\boxed{\lim_{t\to\infty} \frac{f(t) - g(t)}{g(t)} = 0.}$$

This implies from Definition 8.3 that $f(t) - g(t) = o(g(t)) = g(t)o(1)$ (Example 8.4). Hence we have

$$f(t) = g(t)[1 + o(1)].$$

Thus, it appears that the symbol $\sim$ is superfluous, since, as has been demonstrated above, $f \sim g$ can be conveniently written as $f = g(1 + o(1))$.

Example 8.6.

(a) Show that $\sinh t \sim \frac{1}{2}e^t, t \to \infty$.

Solution $\lim_{t\to\infty} \frac{\sinh t}{\frac{1}{2}e^t} = \lim_{t\to\infty} \frac{\frac{1}{2}(e^t - e^{-t})}{\frac{1}{2}e^t} = 1.$

(b) Show that $t^2 \log t + t^3 \sim t^3, t \to \infty$.
Solution

$$\begin{aligned}\lim_{t\to\infty} \frac{t^2 \log t + t^3}{t^3} &= 1 + \lim_{t\to\infty} \frac{\log t}{t} \\ &= 1 + 0 \text{ (using L'Hôpital's rule)} \\ &= 1.\end{aligned}$$

Notice that from Examples 8.2(c) and 8.6(b) we have $t^3 \sim t^2 \log t + t^3$ and $t^2 \log t + t^3 = O(t^3)$. It is also true that $t^2 \log t + 2t^3 = O(t^3)$, but $t^2 \log t + 2t^3$ is not asymptotic to t^3, since

$$\lim_{t\to\infty} \frac{t^2 \log t + 2t^3}{t^3} = 2.$$

Before ending this section we will entertain the curious reader by introducing the prime number theorem, well known in the discipline of number theory. It says that the number of primes $\pi(t)$ that are less than the real number t is asymptotic to $t/(\log t)$, $t \to \infty$, that is,

$$\pi(t) \sim \frac{t}{\log t}, \quad t \to \infty.$$

For a proof of this result the reader may consult [144].

Another interesting asymptotic result is Stirling's formula

$$n! \sim n^n \sqrt{2\pi n}\, e^{-n}, \quad n \to \infty.$$

A proof of this formula may be found in [127].

Exercises 8.1

1. Show that $\frac{t^2}{1+t^3} + \log(1+t^2) = O(\log t), \quad t \to \infty.$

2. Show that $\sinh t = O(e^t), \quad t \to \infty.$

3. Show that $O(g(t)) = g(t)O(1), \quad t \to \infty.$

4. Show that:

   (i) $\frac{1}{t-1} = \frac{1}{t}\left[1 + \frac{1}{t} + O\left(\frac{1}{t^2}\right)\right], \quad t \to \infty,$

   (ii) $\frac{1}{t-1} = \frac{1}{t}\left[1 + \frac{1}{t} + o\left(\frac{1}{t}\right)\right], \quad t \to \infty.$

5. Show that $\sinh\left(\frac{1}{t}\right) = o(1), \qquad t \to \infty.$

6. Show that:

   (i) $[O(t)]^2 = O(t^2) = o(t^3)$,

   (ii) $t + o(t) = O(t)$.

7. Show that:

   (i) $\sin(O(t^{-1})) = O(t^{-1})$,

   (ii) $\cos(t + \alpha + o(1)) = \cos(t + \alpha) + o(1)$, for any real number α.

8. Prove that $\sim$ is an equivalence relation.

9. Prove that both relations o and O are transitive.

10. Suppose that $f(t) = O(t), t \to \infty$, and $g(t) = O(t^2), t \to \infty$. Show that for any nonzero constants $a, b, af(t) + bg(t) = O(g(t)), t \to \infty$.

11. If $f = O(g), t \to \infty$, show that:

    (i) $O(o(f)) = o(O(f)) = o(g)$,

    (ii) $O(f)o(g) = o(f)o(g) = o(fg)$.

12. Let f be a positive nonincreasing function of t, and let $f(t) \sim g(t)$, $t \to \infty$. Prove that $\sup_{s>t} f(s) \sim g(t)$, $t \to \infty$.

13. Suppose that the functions f and g are continuous and have convergent integrals on $[1, \infty)$. If $f(t) \sim g(t), t \to \infty$, prove that

$$\int_t^\infty f(s)\, ds \sim \int_t^\infty g(s)\, ds, \quad t \to \infty.$$

14. Consider the exponential integral $E_n(x)$ defined by

$$E_n(x) = \int_1^\infty \frac{e^{-xt}}{t^n} dt, \quad (x > 0), \text{ where } n \text{ is a positive integer.}$$

    (a) Show that $E_n(x)$ satisfies the difference equation

$$E_{n+1}(x) = \frac{1}{n}\left[e^{-x} - xE_n(x)\right].$$

    (b) Use integration by parts to show that

$$E_n(x) = \frac{e^{-x}}{x}\left(1 + 0\left(\frac{1}{x}\right)\right), \quad x \to \infty.$$

    (c) Show that

$$E_n(x) = \frac{e^{-x}}{n-1}\left[1 + O\left(\frac{1}{n-2}\right)\right], \quad n \to \infty.$$

15. Show that

$$\int_0^\infty \frac{e^{-1}}{x+t} = \frac{1}{x}\left[1 - \frac{1}{x} + O\left(\frac{1}{x^2}\right)\right], \quad x \to \infty.$$

16. Show that

$$\sum_{k=1}^{n} k^k = n^n \left[1 + O\left(\frac{1}{n}\right)\right], \qquad n \to \infty.$$

8.2 Poincaré's Theorem

In this section we introduce to the reader the theorems of Poincaré and Perron. Both theorems deal with the asymptotic behavior of linear difference equations with nonconstant coefficients. It is widely accepted among researchers in difference equations that the theorem of Poincaré [123] marks the beginning of research in the qualitative theory of linear difference equations. Thirty-six years later, Perron [117] made some significant improvements to Poincaré's theorem.

To motivate our study we will take the reader on a short excursion to the much simpler linear equations with constant coefficients of the form

$$x(n+k) + p_1 x(n+k-1) + \cdots + p_k x(n) = 0, \tag{8.2.1}$$

where the p_i's are real or complex numbers. The characteristic equation of (8.2.1) is given by

$$\lambda^k + p_1 \lambda^{k-1} + \cdots + p_k = 0. \tag{8.2.2}$$

Let $\lambda_1, \lambda_2, \ldots, \lambda_k$ be the characteristic roots of (8.2.2). Then there are two main cases to consider.

Case 1. Suppose that distinct characteristic roots have distinct moduli, i.e., if $\lambda_i \neq \lambda_j$, then $|\lambda_i| \neq |\lambda_j|$ for all $1 \leq i, j \leq k$.

For the convenience of the reader we will divide Case 1 into two subcases.

Subcase (a) Assume that all characteristic roots are distinct. So, by relabeling them, one may write the characteristic roots in descending order

$$|\lambda_1| > |\lambda_2| > \cdots > |\lambda_k|.$$

Then the general solution of (8.2.1) is given by

$$x(n) = c_1 \lambda_1^n + c_2 \lambda_2^n + \cdots + c_k \lambda_k^n. \tag{8.2.3}$$

Hence if $c_1 \neq 0$, we have

$$\begin{aligned}\lim_{n\to\infty}\frac{x(n+1)}{x(n)} &= \lim_{n\to\infty}\frac{c_1\lambda_1^{n+1}+c_2\lambda_2^{n+1}+\cdots+c_k\lambda_k^{n+1}}{c_1\lambda_1^n+c_2\lambda_2^n+\cdots+c_k\lambda_k^n}\\ &= \lim_{n\to\infty}\lambda_1\left[\frac{c_1+c_2\left(\frac{\lambda_2}{\lambda_1}\right)^{n+1}+\cdots+c_k\left(\frac{\lambda_k}{\lambda_1}\right)^{n+1}}{c_1+c_2\left(\frac{\lambda_2}{\lambda_1}\right)^{n}+\cdots+c_k\left(\frac{\lambda_k}{\lambda_1}\right)^{n}}\right]\\ &= \lambda_1, \qquad \text{since} \quad \left|\frac{\lambda_i}{\lambda_1}\right| < 1, \quad i = 2,\ldots,k.\end{aligned}$$

Similarly, if $c_1 = 0, c_2 \neq 0$, we obtain

$$\lim_{n\to\infty}\frac{x(n+1)}{x(n)} = \lambda_2.$$

And, in general, if $c_1 = c_2 = \cdots = c_{i-1} = 0$, $c_i \neq 0$, then

$$\lim_{n\to\infty}\frac{x(n+1)}{x(n)} = \lambda_i.$$

Subcase (b) Now suppose that there are some repeated characteristic roots. For simplicity assume that λ_1 is of multiplicity r, so $\lambda_1 = \lambda_2 = \cdots = \lambda_r, |\lambda_1| = |\lambda_2| = \cdots = |\lambda_r| > |\lambda_{r+1}| > \cdots > |\lambda_k|$. Then the general solution of (8.2.1) is given by

$$x(n) = (c_1 + c_2 n + \cdots + c_r n^{r-1})\lambda_1^n + c_{r+1}\lambda_{r+1}^n + \cdots + c_k\lambda_k^n.$$

Then one may show easily that this case is similar to Subcase (a) (Exercises 8.2, Problem 1).

Case 2. There exist two distinct characteristic roots λ_r, λ_j with $|\lambda_r| = |\lambda_j|$. This may occur if λ_r and λ_j are conjugates, i.e., $\lambda_r = \alpha + i\beta, \lambda_j = \alpha - i\beta$ for some real numbers α and β. For simplicity, let us assume that $r = 1, j = 2$, so $\lambda_r \equiv \lambda_1$ and $\lambda_j \equiv \lambda_2$. We write $\lambda_1 = \alpha + i\beta = re^{i\theta}, \lambda_2 = \alpha - i\beta = re^{-i\theta}$, where $r = (\alpha^2+\beta^2)^{1/2}, \theta = \tan^{-1}\left(\frac{\beta}{\alpha}\right)$. Then the general solution of (8.2.1) is given by

$$x(n) = c_1 r^n e^{in\theta} + c_2 r^n e^{-in\theta} + c_3\lambda_3^n + \cdots + c_k\lambda_k^n.$$

Hence

$$\begin{aligned}&\lim_{n\to\infty}\frac{x(n+1)}{x(n)}\\ &= \lim_{n\to\infty}\frac{r^{n+1}(c_1 e^{i(n+1)\theta} + c_2 e^{-i(n+1)\theta}) + c_3\lambda_3^{n+1} + \cdots + c_k\lambda_k^{n+1}}{r^n(c_1 e^{in\theta} + c_2 e^{-in\theta}) + c_3\lambda_3^n + \cdots + c_k\lambda_k^n}. \qquad (8.2.4)\end{aligned}$$

Since $e^{in\theta} = \cos n\theta + i\sin n\theta, e^{-in\theta} = \cos n\theta - i\sin n\theta$ do not tend to definite limits as $n \to \infty$, we conclude that the limit (8.2.4) does not exist. For particular solutions the limit may exist. For example, if $|\lambda_1| = |\lambda_2| > |\lambda_3| > \cdots > |\lambda_k|$, and

(a) $c_1 \neq 0, c_2 = 0$, then $\lim_{n\to\infty} \dfrac{x(n+1)}{x(n)} = re^{i\theta} = \lambda_1$,

(b) $c_1 = 0, c_2 \neq 0$, then $\lim_{n\to\infty} \dfrac{x(n+1)}{x(n)} = re^{-i\theta} = \lambda_2$.

Case 2 may also occur if $\lambda_i = -\lambda_j$. It is left to the reader as Exercises 8.2, Problem 2, to verify that in this case, too, $\lim_{n\to\infty} x(n+1)/x(n)$ does not exist.

We now summarize the above discussion in the following theorem.

Theorem 8.7. *Let $x(n)$ be any nonzero solution of* (8.2.1). *Then*

$$\boxed{\lim_{n\to\infty} \frac{x(n+1)}{x(n)} = \lambda_m} \tag{8.2.5}$$

for some characteristic root λ_m, provided that distinct characteristic roots have distinct moduli. Moreover, if there are two or more distinct roots λ_r, λ_j with the same modulus ($|\lambda_r| = |\lambda_j|$), the limit (8.2.5) *may not exist in general, but particular solutions can always be found for which the limit* (8.2.5) *exists and is equal to a given characteristic root λ_m.*

Example 8.8. Consider the difference equation

$$x(n+2) + \mu x(n) = 0.$$

(a) If $\mu = \beta^2$, then the characteristic equation is given by

$$\lambda^2 + \beta^2 = 0.$$

Hence the characteristic roots are $\lambda_1 = \beta i = \beta e^{i\pi/2}$ and $\lambda_2 = -\beta i = \beta e^{-i\pi/2}$. The general solution is given by

$$x(n) = c_1 \beta^n e^{in\pi/2} + c_2 \beta^n e^{-in\pi/2}.$$

So

$$\lim_{n\to\infty} \frac{x(n+1)}{x(n)} = \beta \left(\frac{c_1 e^{i(n+1)\pi/2} + c_2 e^{-i(n+1)\pi/2}}{c_1 e^{in\pi/2} + c_2 e^{-in\pi/2}} \right),$$

which does not exist. However, if we pick the particular solution

$$\overline{x}(n) = c_1 \beta^n e^{in\pi/2},$$

then

$$\lim_{n\to\infty} \frac{\overline{x}(n+1)}{\overline{x}(n)} = \beta e^{in\pi/2} = \beta i.$$

Similarly, for the solution $\hat{x}(n) = c_2 \beta^n e^{-in\pi/2}$,

$$\lim_{n\to\infty} \frac{\hat{x}(n+1)}{\hat{x}(n)} = -\beta i.$$

(b) If $\mu = -\beta^2$, then the characteristic roots are $\lambda_1 = \beta, \lambda_2 = -\beta$. The general solution is given by $x(n) = c_1\beta^n + c_2(-\beta)^n$.

Hence

$$\lim_{n\to\infty} \frac{x(n+1)}{x(n)} = \lim_{n\to\infty} \frac{c_1\beta^{n+1} + c_2(-\beta)^{n+1}}{c_1\beta^n + c_2(-\beta)^n}$$
$$= \beta \lim_{n\to\infty} \frac{c_1 + c_2(-1)^{n+1}}{c_1 + c_2(-1)^n}. \tag{8.2.6}$$

The limit (8.2.6) does not exist, since $x(n+1)/x(n)$ oscillates between $\beta(c_1+c_2)/(c_1-c_2)$ and $\beta(c_1-c_2)/(c_1+c_2)$. Notice that for the solution $\overline{x}(n) = c_1\beta^n$,

$$\lim_{n\to\infty} \frac{\overline{x}(n+1)}{\overline{x}(n)} = \beta,$$

and for the solution $\tilde{x}(n) = c_2(-\beta)^n$,

$$\lim_{n\to\infty} \frac{\tilde{x}(n+1)}{\tilde{x}(n)} = -\beta.$$

In 1885 the French mathematician Henri Poincaré [123] extended the above observations to equations with nonconstant coefficients of the form

$$x(n+k) + p_1(n)x(n+k-1) + \cdots + p_k(n)x(n) = 0 \tag{8.2.7}$$

such that there are real numbers $p_i, 1 \le i \le k$, with

$$\lim_{n\to\infty} p_i(n) = p_i, \qquad 1 \le i \le k. \tag{8.2.8}$$

We shall call an equation of the form (8.2.7), (8.2.8) a difference equation of *Poincaré type.* The characteristic equation associated with (8.2.7) is

$$\lambda^k + p_1\lambda^{k-1} + \cdots + p_k = 0. \tag{8.2.9}$$

The underlying idea behind Poincaré's theorem is that since the coefficients of a difference equation of Poincaré type are nearly constant for large n, one would expect solutions of (8.2.7) to exhibit some of the properties of the solutions of the corresponding constant coefficient difference equation (8.2.1) as stated in Theorem 8.7.

An important observation which carries over from autonomous to nonautonomous systems is the following. If $\lim_{n\to\infty} \frac{x(n+1)}{x(n)} = \alpha$, then α must be a characteristic root, i.e., a root of (8.2.9).

Theorem 8.9 (Poincaré's Theorem). *Suppose that condition* (8.2.8) *holds and the characteristic roots* $\lambda_1, \lambda_2, \ldots, \lambda_k$ *of* (8.2.9) *have distinct moduli. If* $x(n)$ *is a solution of* (8.2.7), *then either* $x(n) = 0$ *for all large* n *or*

$$\boxed{\lim_{n\to\infty} \frac{x(n+1)}{x(n)} = \lambda_i} \tag{8.2.10}$$

for some i, $1 \le i \le k$.

PROOF. The proof will be given in Section 8.8. □

Note that Poincaré's Theorem does not tell us whether or not each characteristic root λ_i can be written in the form (8.2.10). In 1921, Oscar Perron [117] gave an affirmative answer to this question.

Theorem 8.10 (Perron's First Theorem). *Assume that $p_k(n) \neq 0$ for all $n \in \mathbb{Z}^+$ and the assumptions of Theorem 8.9 hold. Then (8.2.7) has a fundamental set of solutions $\{x_1(n), x_2(n), \ldots, x_k(n)\}$ with the property*

$$\boxed{\lim_{n\to\infty} \frac{x_i(n+1)}{x_i(n)} = \lambda_i, \quad 1 \le i \le k.} \tag{8.2.11}$$

PROOF. A proof of this theorem may be found in Meschkowski [99, p. 10]. Theorem 8.10 is commonly referred to as the Poincaré–Perron Theorem. Perron [117] later formulated and proved a result of a different nature that does not suffer from the restriction on the characteristic roots. □

Theorem 8.11 (Perron's Second Theorem). *Suppose that $p_k(n) \neq 0$ for all $n \in \mathbb{Z}^+$. Then (8.2.7) has a fundamental set of solutions $\{x_1(n), x_2(n), \ldots, x_k(n)\}$ such that*

$$\boxed{\limsup_{n\to\infty} \sqrt[n]{|x_i(n)|} = |\lambda_i|.} \tag{8.2.12}$$

It is questionable whether Poincaré–Perron Theorem remains valid if (8.2.7) has characteristic roots with equal moduli. Perron himself addressed this question and gave the following example, which shows that Poincaré's theorem may fail in this case.

But in order to understand this example we need to make a detour to infinite products.

8.2.1 *Infinite Products and Perron's Example*

An expression of the form

$$\prod_{n=1}^{\infty} (1 + a(n)), \quad a(n) \neq -1 \quad \text{for all } n \in \mathbb{Z}^+, \tag{8.2.13}$$

is called an infinite product. The partial products are $\prod_{j=1}^{n} (1 + a(j))$. The infinite product (8.2.13) is said to converge if $\lim_{n\to\infty} \prod_{j=1}^{n} (1 + a(j))$ is finite and nonzero. Otherwise, it is said to be divergent.

Theorem 8.12. *Consider the infinite series*

$$\sum_{n=1}^{\infty} a(n), \tag{8.2.14}$$

$$\sum_{n=1}^{\infty} a^2(n). \tag{8.2.15}$$

Then the following statements hold:

(i) *The convergence of any two of* (8.2.13), (8.2.14), (8.2.15) *implies that of the third.*

(ii) *If* $\sum_{n=1}^{\infty} |a(n)|$ *converges, then both* (8.2.13) *and* (8.2.15) *converge.*

(iii) *If* (8.2.14) *converges conditionally, then:*

(a) (8.2.13) *converges if* (8.2.15) *converges,*

(b) (8.2.13) *diverges to zero if* (8.2.15) *diverges.*

PROOF. See [109]. □

Example 8.13. Consider the difference equation

$$x(n+2) - \left(1 + \frac{(-1)^n}{n+1}\right) x(n) = 0, \quad n \geq 0. \tag{8.2.16}$$

Then the associated characteristic equation of (8.2.16) is $\lambda^2 - 1 = 0$. Hence the characteristic roots are $\lambda_1 = 1$ and $\lambda_2 = -1$, with $|\lambda_1| = |\lambda_2| = 1$. We now have two cases to consider:

(i) For $n = 2k$, $x(2k+2) = \left(1 + \frac{1}{2k+1}\right) x(2k)$, and hence

$$x(2k) = \prod_{j=1}^{k} \left(1 + \frac{1}{2j-1}\right) x(0). \tag{8.2.17}$$

(ii) For $n = 2k-1$, $x(2k+1) = \left(1 + \frac{1}{2k}\right) x(2k-1)$, and hence

$$x(2k-1) = \prod_{j=1}^{k-1} \left(1 - \frac{1}{2j}\right) x(1). \tag{8.2.18}$$

Hence

$$\lim_{k\to\infty} \frac{x(2k)}{x(2k-1)} = \lim_{k\to\infty} \left[\prod_{j=1}^{k} \left(1 + \frac{1}{2j-1}\right) x(0) \Big/ \prod_{j=1}^{k-1} \left(1 - \frac{1}{2j}\right) x(1) \right]. \tag{8.2.19}$$

In the sequel we will show that this limit does not exist. To accomplish this task, we need to evaluate the infinite products

$$\prod_{j=1}^{\infty}\left(1+\frac{1}{2j-1}\right) \tag{8.2.20}$$

and

$$\prod_{j=1}^{\infty}\left(1-\frac{1}{2j}\right). \tag{8.2.21}$$

Let us now apply Theorem 8.12(i) to the infinite product (8.2.20). Since $\sum_{j=1}^{\infty}\frac{1}{(2j-1)^2}$ converges, it follows by Theorem 8.12(i) that if (8.2.20) converges, then so does $\sum_{j=1}^{\infty}\frac{1}{2j-1}$, which is false. Thus the infinite product (8.2.20) diverges to ∞, since each term $\left(1+\frac{1}{2j-1}\right)$ is greater than 1.

Next we consider the infinite product (8.2.21). By a similar argument, we show that it diverges to zero, since each term $\left(1-\frac{1}{2j}\right)$ is less than 1.

It follows that the limit (8.2.21) does not exist.

Example 8.14. Consider the difference equation

$$x(n+2)-\frac{n}{n+1}x(n+1)+\frac{1}{n}x(n)=0.$$

The associated characteristic equation is given by

$$\lambda^2-\lambda=0$$

with characteristic roots $\lambda_1=1$, $\lambda_2=0$. Hence by Perron's theorem there exist solutions $x_1(n)$, $x_2(n)$ such that

$$\lim_{n\to\infty}\frac{x_1(n+1)}{x_1(n)}=1 \quad \text{and} \quad \lim_{n\to\infty}\frac{x_2(n+1)}{x_2(n)}=0.$$

What can we conclude about the solutions $x_1(n)$ and $x_2(n)$? The solution $x_1(n)$ may be equal to a constant c, a polynomial in n such as

$$a_k n^k + a_{k-1}n^{k-1}+\cdots+a_0,$$

or a function such as $\frac{1}{n}$, $\log n$, among others. The solution $x_2(n)$ may be equal to 0, e^{-2^n}, e^{-n^2}, etc.

The reader may correctly conclude from the preceding examples that Poincaré's or Perron's theorem provides only partial results about the asymptotic behavior of solutions of linear difference equations. The question remains whether we can use Perron's theorem to write an asymptotic expression of solutions of equations of Poincaré type. Using null sequences, Wimp [145] devised an elegant and simple method to address the above question. Recall that $\nu(n)$ is called a *null sequence* if $\lim_{n\to\infty}\nu(n)=0$.

Lemma 8.15. *Suppose that* $\lim_{n\to\infty} \frac{x(n+1)}{x(n)} = \lambda$.

(a) *If* $\lambda \neq 0$, *then*

$$\boxed{x(n) = \pm\lambda^n e^{n\nu(n)}} \tag{8.2.22}$$

for some null sequence $\nu(n)$.

(b) *If* $\lambda = 0$, *then*

$$\boxed{|x(n)| = e^{-n/\mu(n)}} \tag{8.2.23}$$

for some positive null sequence $\mu(n)$.

PROOF.

(a) Let

$$y(n) = \left|\frac{x(n)}{\lambda^n}\right|.$$

Then

$$\lim_{n\to\infty} \frac{y(n+1)}{y(n)} = \lim_{n\to\infty} \left|\frac{1}{\lambda}\frac{x(n+1)}{x(n)}\right| = 1.$$

If we let $z(n) = \log y(n)$, then we have

$$\begin{aligned}\lim_{n\to\infty} z(n+1) - z(n) &= \lim_{n\to\infty} \log\left(\frac{y(n+1)}{y(n)}\right)\\ &= \log \lim_{n\to\infty} \frac{y(n+1)}{y(n)}\\ &= 0.\end{aligned}$$

Hence for a given $\varepsilon > 0$ there exists a positive integer N such that

$$|z(n+1) - z(n)| < \varepsilon/2 \quad \text{for all } n \geq N.$$

Moreover, for $n \geq N$, we obtain

$$|z(n) - z(N)| \leq \sum_{r=N+1}^{n} |z(r) - z(r-1)| < \frac{\varepsilon}{2}(n-N).$$

Hence

$$\begin{aligned}\left|\frac{z(n)}{n}\right| &< \frac{\varepsilon}{2}\left(1 - \frac{N}{n}\right) + \left|\frac{z(N)}{n}\right|\\ &< \frac{\varepsilon}{2} + \frac{\varepsilon}{2} = \varepsilon,\end{aligned}$$

for sufficiently large n. It follows that $\lim_{n\to\infty} \frac{z(n)}{n} = 0$ or $z(n) = n\nu(n)$ for some null sequence $\nu(n)$.

(b) This is left to the reader as Exercises 8.2, Problem 6. □

Example 8.16. Use Lemma 8.15 and the Poincaré–Perron Theorem to find asymptotic estimates of a fundamental set of solutions of the difference equation

$$y(n+2) + \frac{n+1}{n+2}y(n+1) - \frac{2n}{n+2}y(n) = 0.$$

Solution The associated characteristic equation is given by

$$\lambda^2 + \lambda - 2 = 0$$

with roots $\lambda_1 = 1, \lambda_2 = -2$. By Perron's Theorem, there is a fundamental set of solutions $y_1(n), y_2(n)$ with

$$\lim_{n\to\infty} \frac{y_1(n+1)}{y_1(n)} = 1, \qquad \lim_{n\to\infty} \frac{y_2(n+1)}{y_2(n)} = -2.$$

Thus by Lemma 8.15 we obtain

$$y_1(n) = e^{n\nu(n)}, \qquad y_2(n) = (-2)^n e^{n\mu(n)},$$

for some null sequences $\nu(n)$ and $\mu(n)$.

For the curious reader we note that an exact fundamental set of solutions is given by

$$y_1(n) = \frac{1}{n}, \qquad y_2(n) = \frac{(-2)^n}{2}.$$

Exercises 8.2

1. Prove that each nontrivial solution $x(n)$ of the second-order difference equation

$$x(n+2) + p_1x(n+1) + p_2x(n) = 0$$

with double characteristic roots $\lambda_1 = \lambda_2 = \lambda$ satisfies $\lim_{n\to\infty}(x(n+1))/x(n) = \lambda$.

2. Suppose that the characteristic roots λ_1, λ_2 of

$$x(n+2) + p_1x(n+1) + p_2x(n) = 0$$

are such that $\lambda_1 = -\lambda_2$. Prove that $\lim_{n\to\infty}(x(n+1))/x(n)$ does not exist for some solution $x(n)$.

3. Consider the difference equation

$$x(n+3) - (\alpha+\beta+\gamma)x(n+2) + (\alpha\beta+\beta\gamma+\gamma\alpha)x(n+1) - \alpha\beta\gamma u(x) = 0,$$

where α, β, γ are constants.

(a) Show that the characteristic roots are $\lambda_1 = \alpha, \lambda_2 = \beta$, and $\lambda_3 = \gamma$.

(b) If $|\alpha| > |\beta| > |\gamma|$, find a fundamental set of solutions $x_1(n)$, $x_2(n)$, and $x_3(n)$ with

$$\lim_{n\to\infty} \frac{x_1(n+1)}{x_1(n)} = \alpha, \quad \lim_{n\to\infty} \frac{x_2(n+1)}{x_2(n)} = \beta,$$

$$\lim_{n\to\infty} \frac{x_3(n+1)}{x_3(n)} = \gamma.$$

(c) If $|\alpha| = |\beta|, \alpha \neq \beta, |\alpha| > |\gamma|$, find a fundamental set of solutions $x_1(n), x_2(n)$, and $x_3(n)$ such that $\lim_{n\to\infty} x_1(n+1)/x_1(n) = \alpha, \lim_{n\to\infty} x_2(n+1)/x_2(n) = \beta, \lim_{n\to\infty} x_3(n+1)/x_3(n) = \gamma$.

4. Consider the difference equation

$$x(n+2) + \frac{1}{n+4}x(n+1) - \frac{n+1}{n+4}x(n) = 0.$$

Use iteration to show that $\lim_{n\to\infty} x(n+1)/x(n)$ does not exist for every solution $x(n)$.

5. Consider the equation

$$x(n+2) - ((n+2) + 2(-1)^n)/(n+2)^3(n+3)x(n) = 0.$$

Use iteration to show that $\lim_{n\to\infty}(x(n+1))/x(n)$ does not exist for any solution $x(n)$.

6. Prove part (b) of Lemma 8.15.

7. Show that the difference equation

$$x(n+1) - \frac{n+7}{n+5}x(n) - \frac{n^2+1}{n^2+4}x(n-1) = 0$$

has an oscillatory solution and a nonoscillatory solution.

8. Consider the difference equation

$$x(n+2) - \left(3 + \frac{2n-1}{n^2-2n-1}\right)x(n+1) + 2\left(1 + \frac{2n-1}{n^2-2n-1}\right)x(n) = 0.$$

(a) Use Lemma 8.15 and Perron's theorem to find asymptotic estimates of a fundamental set of solutions of the equation.

(b) Verify that $x_1(n) = 2^n$ and $x_2(n) = n^2$ constitute a fundamental set of solutions.

9. Let $x(n)$ be a nontrivial solution of (8.2.7) such that $\lim_{n\to\infty} x(n+1)/x(n) = \alpha$. Show that α is a characteristic root, i.e., a root of (8.2.9).

10. Let α be a number whose modulus is greater than all of the characteristic roots of a difference equation of Poincaré type (8.2.7). Prove

that

$$\lim_{n\to\infty} \frac{x(n)}{\alpha^n} = 0$$

for any solution $x(n)$ of the equation.

11. Suppose that $\lim_{n\to\infty} x(n+1)/x(n) = \lambda > 0$. Prove that for any $\delta \in (0, \lambda)$:

    (i) $|x(n)| = O(\lambda + \delta)^n$, and

    (ii) $(\lambda + \delta)^n = O(x(n))$.

12. Consider the equation $x(n+2) - (n+1)x(n+1) - 2n^2x(n) = 0$.

    (a) Transform the equation into an equation of Poincaré type by letting $x(n) = (n-1)!\, y(n)$.

    (b) Use part (a) to get an asymptotic estimate of a fundamental set of solutions.

13. Use the scheme of Problem 11 to find an asymptotic set of a fundamental set of solutions of the equation

$$x(n+2) + 4^n x(n+1) + 4n(n-1)x(n) = 0.$$

14. Prove Theorem 8.12.

15. Consider the equation

$$(n+2)x(n+2) - (n+3)x(n+1) + 2x(n) = 0. \tag{8.2.24}$$

    (a) Show that 1, 0 are the characteristic roots of the equation.

    (b) Put

$$\frac{x(n+1)}{x(n)} = 1 + \mu(n) \tag{8.2.25}$$

    in (8.2.23), where $\mu(n)$ is a null sequence, and show that the equation becomes

$$(n+2)\mu(n+1) = 1 - \frac{2}{1+\mu(n)}. \tag{8.2.26}$$

    (c) Show that

$$\frac{2}{1+\mu(n)} = 2 + O(\mu(n)).$$

    (d) Use part (c) to show that (8.2.27) is equivalent to

$$\mu(n+1) = -\frac{1}{n+1} + O\left(\frac{1}{n^2}\right). \tag{8.2.27}$$

(e) Show that

$$x(n+1) = \frac{n}{n+1}\left(1 + O\left(\frac{1}{n^2}\right)\right)x(n). \tag{8.2.28}$$

(f) Prove that $x(n) \sim \frac{c}{n}, n \to \infty$.

16. Show that (8.2.24) has another solution $\overline{x} \sim c\frac{2^n}{n!}, n \to \infty$.

17. Use the scheme of Problem 15 to find asymptotic estimates of a fundamental set of solutions of the equation

$$(n+1)x(n+2) - (n+4)x(n+1) + x(n) = 0.$$

18. Show that the equation $x(n+2) - (n+1)x(n+1) + (n+1)x(n) = 0$ has solutions $x_1(n), x_2(n)$ with asymptotic estimates

$$x_1(n) \sim c(n-2)!, \qquad x_2(n) = an, \qquad n \to \infty.$$

*19. (Hard). Consider the equation of Poincaré type

$$x(n+2) - (2 + p_1(n))x(n+1) + (1 + p_2(n))x(n) = 0,$$

where $p_1(n) \geq p_2(n)$ for all $n \in \mathbb{Z}^+$. Show that if $x(n)$ is a solution that is not constantly zero for large values of n, then $\lim_{n\to\infty}(x(n+1))/x(n) = 1$.

*20. (Hard). Consider the equation

$$x(n+2) + P_1(n)x(n+1) + P_2(n)x(n) = 0$$

with $\lim_{n\to\infty} P_1(n) = p_1, \lim_{n\to\infty} P_2(n) = p_2$. Let η be a positive constant such that $|x(n+1)/x(n)|^2 > |p_2| + \eta$ for sufficiently large n. Suppose that the characteristic roots λ_1, λ_2 of the associated equation are such that $|\lambda_1| \geq |\lambda_2|$.

Prove that $\lim_{n\to\infty} x(n+1)/x(n) = \lambda_1$.

8.3 Asymptotically Diagonal Systems

In this section we derive conditions under which solutions of a perturbed diagonal system are asymptotic to solutions of the unperturbed diagonal system. As a byproduct we obtain asymptotic results for nonautonomous kth-order scalar difference equations.

We begin our study by considering the perturbed diagonal system

$$y(n+1) = (D(n) + B(n))y(n) \tag{8.3.1}$$

and the unperturbed diagonal system

$$x(n+1) = D(n)x(n), \tag{8.3.2}$$

where $D(n) = \text{diag}(\lambda_1(n), \lambda_2(n), \ldots, \lambda_k(n)), \lambda_i(n) \neq 0$, for all $n \geq n_0 \geq 0, 1 \leq i \leq k$, and $B(n)$ is a $k \times k$ matrix defined for $n \geq n_0 \geq 0$. The fundamental matrix of system (8.3.2) is given by

$$\Phi(n) = \text{diag}\left(\prod_{r=n_0}^{n-1} \lambda_1(r), \prod_{r=n_0}^{n-1} \lambda_2(r), \ldots, \prod_{r=n_0}^{n-1} \lambda_k(r)\right). \tag{8.3.3}$$

Let S be a subset of the set $\{1, 2, 3, \ldots, k\}$. Define

$$\Phi_1(n) = \text{diag}(\mu_1(n), \mu_2(n), \ldots, \mu_k(n)) \tag{8.3.4}$$

by letting

$$\mu_i(n) = \begin{cases} \prod_{r=n_0}^{n-1} \lambda_i(r), & \text{if } i \in S, \\ 0, & \text{otherwise.} \end{cases}$$

Define $\Phi_2(n) = \Phi(n) - \Phi_1(n)$.

We are now ready for the definition of the important notion of *dichotomy*.

Definition 8.17. System (8.3.2) is said to possess an *ordinary dichotomy* if there exists a constant M such that:

(i) $\|\Phi_1(n)\Phi^{-1}(m)\| \leq M$, for $n \geq m \geq n_0$,

(ii) $\|\Phi_2(n)\Phi^{-1}(m)\| \leq M$, for $m \geq n \geq n_0$.

Notice that if $D(n)$ is constant, then system (8.3.2) always possesses an ordinary dichotomy.

After wading through the complicated notation above, here is an example.

Example 8.18. Consider the difference system $x(n+1) = D(n)x(n)$ with

$$D(n) = \begin{pmatrix} 1 + \dfrac{1}{n+1} & 0 & 0 & 0 \\ 0 & 0.5 & 0 & 0 \\ 0 & 0 & n+1 & 0 \\ 0 & 0 & 0 & \dfrac{1}{n+2} \end{pmatrix}.$$

Then a fundamental matrix of the system may be given by

$$\begin{aligned} \Phi(n) &= \text{diag}\left(\prod_{j=0}^{n-1}\left(1 + \frac{1}{j+1}\right), (0.5)^n, \prod_{j=0}^{n-1}(j+1), \prod_{j=0}^{n-1}\left(\frac{1}{j+2}\right)\right) \\ &= \text{diag}\left(n+1, (0.5)^n, n!, \frac{1}{(n+1)!}\right). \end{aligned}$$

From this we deduce that

$$\Phi_1(n) = \text{diag}\left(0, (0.5)^n, 0, \frac{1}{(n+1)!}\right)$$

and

$$\Phi_2(n) = \text{diag}(n+1, 0, n!, 0).$$

Finally,

$$\Phi_1(n)\Phi^{-1}(m) = \text{diag}\left(0, (0.5)^{n-m}, 0, \frac{1}{(n+1)(n)\cdots(m+2)}\right).$$

Hence

$$\|\Phi_1(n)\Phi^{-1}(m)\| \leq 1, \quad \text{for } n \geq m \geq 0.$$

Similarly,

$$\Phi_2(n)\Phi^{-1}(m) = \text{diag}\left(\frac{n+1}{m+1}, 0, \frac{n!}{m!}, 0\right), \qquad \text{for } m \geq n \geq n_0.$$

Hence

$$\|\Phi_2(n)\Phi^{-1}(m)\| \leq 1, \qquad \text{for } m \geq n \geq n_0.$$

We are now ready to establish a new variation of constants formula that is very useful in asymptotic theory.

Theorem 8.19 (Variation of Constants Formula). *Suppose that system* (8.3.2) *possesses an ordinary dichotomy and the following condition holds:*

$$\sum_{n=n_0}^{\infty} \|B(n)\| < \infty. \tag{8.3.5}$$

Then for each bounded solution $x(n)$ of (8.3.2) *there corresponds a bounded solution $y(n)$ of* (8.3.1) *given by*

$$\boxed{\begin{aligned} y(n) = x(n) &+ \sum_{j=n_0}^{n-1} \Phi_1(n)\Phi^{-1}(j+1)B(j)y(j) \\ &- \sum_{j=n}^{\infty} \Phi_2(n)\Phi^{-1}(j+1)B(j)y(j). \end{aligned}} \tag{8.3.6}$$

The converse also holds; for each bounded solution $y(n)$ of (8.3.1) *there corresponds a bounded solution $x(n)$ of* (8.3.2).

PROOF. Let $x(n)$ be a bounded solution of (8.3.2). By using the method of successive approximation, we will produce a corresponding bounded solution $y(n)$ of (8.3.1). We define a sequence $\{y_i(n)\}$ $(i = 1, 2, \ldots)$ by letting

$y_1(n) = x(n)$ and

$$y_{i+1}(n) = x(n) + \sum_{j=n_0}^{n-1} \Phi_1(n))\Phi^{-1}(j+1)B(j)y_i(j)$$
$$- \sum_{j=n}^{\infty} \Phi_2(n)\Phi^{-1}(j+1)B(j)y_i(j). \tag{8.3.7}$$

First we prove that $y_i(n)$ is bounded on the discrete interval $[n_0, \infty)$. This task will be accomplished by induction on i. From our assumption we have $|y_1(n)| = |x(n)| \leq c_1$, for some constant c_1. Now assume that $|y_i(n)| \leq c_i$, for some constant c_i. Then by Definition 8.17 we have

$$|y_{i+1}(n)| \leq c_1 + Mc_i \sum_{j=n_0}^{\infty} \|B(j)\| = c_{i+1}.$$

Hence $y_i(n)$ is bounded for each i.

In the next step we show that the sequence $\{y_i(n)\}$ converges uniformly on the discrete interval $[n_0, \infty)$.

Using (8.3.7) we have, for $i = 1, 2, \ldots,$

$$|y_{i+2}(n) - y_{i+1}(n)| \leq M \sum_{j=n_0}^{\infty} \|B(j)\| |y_{i+1}(j) - y_i(j)|.$$

Hence by induction on i (Exercises 8.3, Problem 8)

$$|y_{i+1}(n) - y_i(n)| \leq \left[M \sum_{j=n_0}^{\infty} \|B(j)\| \right]^i c_1. \tag{8.3.8}$$

We choose n_0 sufficiently large such that

$$M \sum_{j=n_0}^{\infty} \|B(j)\| = \eta < 1. \tag{8.3.9}$$

Thus $|y_{i+1}(n) - y_i(n)| \leq c_1\eta^i$ and, consequently, $\sum_{i=1}^{\infty}\{y_{i+1}(n) - y_i(n)\}$ converges uniformly on $n \geq n_0$ (by the Weierstrass M-test).[1]

We define

$$y(n) = y_1(n) + \sum_{i=1}^{\infty}\{y_{i+1}(n) - y_i(n)\} = \lim_{i\to\infty} y_i(n).$$

---

[1] *Weierstrass M-test*: Let $u_n(x), n = 1, 2, \ldots,$ be defined on a set A with range in $\mathbb{R}$. Suppose that $|u_n(x)| \leq M_n$ for all n and for all $x \in A$. If the series of constants $\sum_{n=1}^{\infty} M_n$ converges, then $\sum_{n=1}^{\infty} u_n(x)$ and $\sum_{n=1}^{\infty} |u_n(x)|$ converge uniformly on A.

Hence $|y(n)| \leq L$, for some constant L. Letting $i \to \infty$ in (8.3.7), we obtain (8.3.6). The second part of the proof of the theorem is left to the reader as Exercises 8.3, Problem 10. □

If the condition of ordinary dichotomy is strengthened, then we obtain the following important result in asymptotic theory.

Theorem 8.20. *Suppose that the following assumption holds:*

$$\textit{Condition (H)} \begin{cases} \text{(i)} & \textit{Systems (8.3.2) posses an ordinary dichotomy;} \\ \text{(ii)} & \lim_{n\to\infty} \Phi_1(n) = 0. \end{cases}$$

If, in addition, condition (8.3.5) holds, then for each bounded solution $x(n)$ of (8.3.2) there corresponds a bounded solution $y(n)$ of (8.3.1) such that

$$y(n) = x(n) + o(1). \tag{8.3.10}$$

PROOF. Let $x(n)$ be a bounded solution of (8.3.2). Then by using formula (8.3.6) we obtain, for a suitable choice of m (to be determined later),

$$y(n) = x(n) + \Phi_1(n) \sum_{j=n_0}^{m-1} \Phi^{-1}(j+1)B(j)y(j) + \Psi(n), \tag{8.3.11}$$

where

$$\Psi(n) = \Phi_1(n) \sum_{j=m}^{n-1} \Phi^{-1}(j+1)B(j)y(j) - \sum_{j=n}^{\infty} \Phi_2(n)\Phi^{-1}(j+1)B(j)y(j). \tag{8.3.12}$$

Now recall that from Theorem 8.20, $\|y\| \leq L$, for some $L > 0$. Hence from formula (8.3.12) it follows that

$$|\Psi(n)| \leq ML \sum_{j=m}^{\infty} \|B(j)\|.$$

Thus for $\varepsilon > 0$, there exists a sufficiently large m such that $|\Psi(n)| < \varepsilon/2$. Since $\Phi_1(n) \to 0$ as $n \to \infty$, it follows from formula (8.3.11) that $|y(n) - x(n)| < \varepsilon$, for sufficiently large n. Therefore, $y(n) = x(n) + o(1)$.

Our next objective is to apply the preceding theorem to produce a discrete analogue of Levinson's theorem [91], [36]. We start our analysis by making the change of variables

$$y(n) = \prod_{r=n_0}^{n-1} \lambda_i(r) z(n), \qquad \text{for a specific } i, \quad 1 \leq i \leq k. \tag{8.3.13}$$

Then (8.3.1) becomes

$$z(n+1) = (D_i(n) + B_i(n))z(n), \tag{8.3.14}$$

where

$$D_i(n) = \operatorname{diag}\left(\frac{\lambda_1(n)}{\lambda_i(n)}, \dots, 1, \dots, \frac{\lambda_k(n)}{\lambda_i(n)}\right),$$

$$B_i(n) = \frac{1}{\lambda_i(n)} B(n).$$

Associated with (8.3.14) is the unperturbed diagonal system

$$x(n+1) = D_i(n)x(n). \tag{8.3.15}$$

To make the proof of our main theorem more transparent we introduce the following lemma.

Lemma 8.21 [9]. *Assumption* (H) *hold for every equation* (8.3.15), $1 \le i \le k$, *if the following conditions hold.*

There exist constants $\mu > 0$ *and* $K > 0$ *such that for each pair* λ_i, λ_j, $i \neq j$, *either*

$$\text{Condition (L)} \begin{cases} \displaystyle\prod_{r=0}^{n} \left|\frac{\lambda_i(r)}{\lambda_j(r)}\right| \to +\infty, & \text{as } n \to \infty, \\ \text{and } \displaystyle\prod_{r=n_1}^{n_2} \left|\frac{\lambda_i(r)}{\lambda_j(r)}\right| \ge \mu > 0, & \text{for all } 0 \le n_1 \le n_2, \\ \text{or } \displaystyle\prod_{r=n_1}^{n_2} \left|\frac{\lambda_i(r)}{\lambda_j(r)}\right| \le K, & \text{for all } 0 \le n_1 \le n_2. \end{cases}$$

The proof is omitted and left to the reader to do as Exercises 8.3, Problem 7.

Example 8.22. Consider the diagonal matrix

$$D(n) = \operatorname{diag}(\lambda_1(n), \lambda_2(n), \lambda_3(n))$$

where

$$\lambda_1(n) = 2 + \sin\left(\frac{2n+1}{2}\right)\pi,$$

$$\lambda_2(n) = 2 - \sin\left(\frac{2n+1}{2}\right)\pi,$$

$$\lambda_3(n) = 2.$$

Notice that:

(i)

$$\prod_{r=n_1}^{n_2} \left|\frac{\lambda_1(r)}{\lambda_2(r)}\right| = \begin{cases} 3 & \text{if both } n_1 \text{ and } n_2 \text{ are even,} \\ 1 & \text{if } n_1 \text{ is odd and } n_2 \text{ is even or vice versa,} \\ \frac{1}{3} & \text{if both } n_1 \text{ and } n_2 \text{ are odd,} \end{cases}$$

and

$$\prod_{r=0}^{n}\left|\frac{\lambda_1(r)}{\lambda_3(r)}\right| \to \infty \quad \text{as } n \to \infty.$$

(ii)

$$\prod_{r=n_1}^{n_2}\left|\frac{\lambda_2(r)}{\lambda_1(r)}\right| = \begin{cases} \dfrac{1}{3} & \text{if both } n_1 \text{ and } n_2 \text{ are even,} \\ 1 & \text{if } n_1 \text{ is odd and } n_2 \text{ is even or vice versa,} \\ 3 & \text{if both } n_1 \text{ and } n_2 \text{ are odd,} \end{cases}$$

and

$$\prod_{r=0}^{n}\left|\frac{\lambda_2(r)}{\lambda_3(r)}\right| \to \infty \quad \text{as } n \to \infty.$$

(iii)

$$\prod_{r=0}^{n}\left|\frac{\lambda_3(r)}{\lambda_1(r)}\right| \to \infty \quad \text{as } n \to \infty,$$

but no subsequence of the product converges to zero, and

$$\prod_{r=0}^{n}\left|\frac{\lambda_3(r)}{\lambda_2(r)}\right| \to \infty \quad \text{as } n \to \infty,$$

but no subsequence of the product converges to zero.

Thus the system $x(n+1) = D(n)x(n)$ satisfies Condition (L) and, consequently, it satisfies Condition (H).

Next we give the fundamental theorem in the asymptotic theory of difference equations; the discrete analogue of Levinson's theorem [91].

Theorem 8.23. *Suppose that Condition* (L) *holds and for each* $i, 1 \le i \le k$,

$$\sum_{n=n_0}^{\infty} \frac{1}{|\lambda_i(n)|}\|B(n)\| < \infty. \tag{8.3.16}$$

Then system (8.3.1) *has a fundamental set of* k *solutions* $y_i(n)$ *such that*

$$\boxed{y_i(n) = (e_i + o(1)) \prod_{r=n_0}^{n-1} \lambda_i(r),} \tag{8.3.17}$$

where e_i *is the standard unit vector in* $\mathbb{R}^k$ *where its components are all zero, except that the* i*th component is* 1.

Proof. Notice that under Condition (L) it follows from Lemma 8.21 that (8.3.15) satisfies Condition (H). Moreover, from assumption (8.3.16), $B_i(n)$

satisfies condition (8.3.5). Thus we can apply Theorem 8.20 to (8.3.14) and (8.3.15). Observe that since the ith diagonal element in $D_i(n)$ is 1, it follows that $x(n) = e_i$ is a bounded solution of (8.3.15). By Theorem 8.20, there corresponds a solution $z(n)$ of (8.3.14) such that $z(n) = e_i + o(1)$. Now conclusion (8.3.17) follows immediately by substituting for $z(n)$ from formula (8.3.13). □

Theorem 8.23 will be referred to as the Benzaid–Lutz theorem.

Example 8.24. Consider the difference system $y(n+1) = A(n)y(n)$, where

$$A(n) = \begin{pmatrix} \frac{n^2+2}{2n^2} & 0 & \frac{1}{n^3} \\ 0 & 1 & 0 \\ \frac{1}{2^n} & 0 & n \end{pmatrix}.$$

To apply Theorem 8.23 we need to write $A(n)$ in the form $D(n) + B(n)$ with $D(n)$ a diagonal matrix and $B(n)$ satisfying condition (8.3.16). To achieve this we let

$$D(n) = \begin{pmatrix} \frac{1}{2} & 0 & 0 \\ 0 & 1 & 0 \\ 0 & 0 & n \end{pmatrix}, \qquad B(n) = \begin{pmatrix} \frac{1}{n^2} & 0 & \frac{1}{n^3} \\ 0 & 0 & 0 \\ \frac{1}{2^n} & 0 & 0 \end{pmatrix}.$$

Hence $\lambda_1 = \frac{1}{2}, \lambda_2 = 1$, and $\lambda_3 = n$. Thus for $n_0 = 2$, our system satisfies the hypotheses of Theorem 8.23. Consequently, there are three solutions:

$$y_1(n) \sim \left(\frac{1}{2}\right)^n \begin{pmatrix} 1 \\ 0 \\ 0 \end{pmatrix},$$

$$y_2(n) \sim \begin{pmatrix} 0 \\ 1 \\ 0 \end{pmatrix},$$

$$y_3(n) \sim \left(\prod_{j=1}^{n-1} j\right) \begin{pmatrix} 0 \\ 0 \\ 1 \end{pmatrix} = (n-1)! \begin{pmatrix} 0 \\ 0 \\ 1 \end{pmatrix}.$$

Remark: Before ending this section we make one further comment on the conditions in Theorem 8.23. This comment concerns the necessity for some condition on $B(n)$ such as (8.3.16). Certainly, condition (8.3.16) holds when $B(n) = O(n^{-\alpha}), n \to \infty$, for some $\alpha > 1$, i.e., $n^\alpha \|B(n)\| \leq L$ for all $n \geq n_0$. On the other hand, the condition $B(n) = O(n^{-1})$ is not sufficient for formula (8.3.17) to hold, and a simple example illustrates this point. Let us take $k = 1, D(n) = 1$, and $B(n) = \frac{1}{n}$. Then (8.3.1) takes the form $y(n+1) = \left(\frac{n+1}{n}\right) y(n) = \left(1 + \frac{1}{n}\right) y(n)$, which has the general solution $y(n) = cn$, for some constant c. Hence no solution satisfies formula (8.3.17).

Exercises 8.3

In Problems 1 through 5 find asymptotic estimates (using Theorem 8.23) for a fundamental set of solutions of the given system.

1. $y(n+1) = (D(n) + B(n))y(n)$, where

$$D(n) = \begin{pmatrix} \dfrac{3}{n+2} & 0 \\ 0 & n+1 \end{pmatrix}, \qquad B(n) = \begin{pmatrix} \dfrac{1}{n^2} & \dfrac{3}{n^3} \\ 0 & \dfrac{5}{n^{3/2}} \end{pmatrix}.$$

2. $y(n+1) = (D(n) + B(n))y(n)$, where

$$D(n) = \begin{pmatrix} \cos \pi n & 0 & 0 \\ 0 & \dfrac{n}{n+1} & 0 \\ 0 & 0 & 3 \end{pmatrix}, \qquad B(n) = \begin{pmatrix} \dfrac{\sin n}{n^3} & \dfrac{n}{e^n} & 0 \\ 0 & 0 & \dfrac{n}{3^n} \\ \dfrac{1}{2^n} & 0 & \dfrac{n}{n^3+5} \end{pmatrix}.$$

3. $y(n+1) = A(n)y(n)$, where

$$A(n) = \begin{pmatrix} 1 + \dfrac{1}{n} & 0 & \dfrac{1}{n(n+1)} \\ 0 & \dfrac{1}{n} & 0 \\ 0 & 0 & 1 + (-1)^n \cos n\pi \end{pmatrix}.$$

4. $y(n+1) = A(n)y(n)$, where

$$A(n) = \begin{pmatrix} n & e^{-n} & 0 \\ 0 & 3 - e^{-2n} & 0 \\ 2^{-n} & 0 & 1+n \end{pmatrix}.$$

5. Give an example of a two-dimensional difference system where Theorem 8.20 does not hold.

6. Define a diagonal matrix $P = \operatorname{diag}(a_1, a_2, \dots, a_k)$, where

$$a_i = \begin{cases} 0 & \text{if } i \notin S, \\ 1 & \text{if } i \in S, \end{cases} \qquad \text{where } S \text{ is a subset of the set } \{1, 2, \dots, k\}.$$

Prove the following statements:

(a) $P^2 = P$ (a *projection matrix*).

(b) $\Phi(n)P = \Phi_1(n)$ as defined in (8.3.4).

(c) $\Phi(n)(I - P) = \Phi_2(n)$, where $\Phi_2(n) = I - \Phi_1(n)$.

(d) $\Phi_1(n)P = \Phi_1(n), \Phi_2(n)(I - P) = \Phi_2(n)$.

7. Prove Lemma 8.21.

8. Prove formula (8.3.8) using mathematical induction on i.

9. Prove that the solution $y(n)$ of (8.3.1) defined by (8.3.6) is bounded for $n \geq n_0 \geq 0$.

10. Prove that under the assumption of Theorem 8.19, for each bounded solution $y(n)$ of (8.3.1), there exists a bounded solution $x(n)$ of (8.3.2).

*11. (Open Problem). Improve Theorem 8.19 by relaxing condition (8.3.5), requiring only conditional convergence of $\sum_{n=n_0}^{\infty} B(n)$.

*12. (Hard). Extend Theorem 8.19 to the case where $D(n)$ is a constant matrix in a one-block Jordan form, then extend it to the case when $D(n)$ is a constant matrix in the general Jordan form.

*13. (Hard). Extend Theorem 8.19 to the case where $D(n)$ has an eigenvalue equal to zero.

*14. (Open Problem). Suppose that there are r distinct eigenvalues $\lambda_1(n)$, $\lambda_2(n),\ldots,\lambda_r(n)$ with distinct moduli. Prove that with the conditions of Theorem 8.19 holding for $1 \leq i \leq r$, there are solutions $y_i(n)$, $1 \leq i \leq r$, of system equation (8.3.1) that satisfy formula (8.3.12).

8.4 High-Order Difference Equations

In this section we turn our attention to the kth-order scalar equations of the form

$$y(n+k) + (a_1 + p_1(n))y(n+k-1) + \cdots + (a_k + p_k(n))y(n) = 0, \quad (8.4.1)$$

where $a_i \in \mathbb{R}$ and $p_i(n), 1 \leq i \leq k$, are real sequences. As we have seen in Chapter 3, (8.4.1) may be put in the form of a k-dimensional system of first-order difference equations that is asymptotically constant. Thus we are led to the study of a special case of (8.4.1), namely, the asymptotically constant system

$$y(n+1) = [A + B(n)]y(n), \quad (8.4.2)$$

where A is a $k \times k$ constant matrix that is not necessarily diagonal. This system is, obviously, more general than the system induced by (8.4.1). The first asymptotic result concerning system equation (8.4.2) is a consequence of Theorem 8.23.

Theorem 8.25 [9]. *Suppose that the matrix A has k linearly independent eigenvectors $\xi_1, \xi_2, \ldots, \xi_k$ and k corresponding eigenvalues $\lambda_1, \lambda_2, \ldots, \lambda_k$.*

If condition (8.3.16) holds for $B(n)$, then system equation (8.4.2) has solutions $y_i(n), 1 \le i \le k$, such that

$$\boxed{y_i(n) = [\xi_i + o(1)]\lambda_i^n.} \tag{8.4.3}$$

PROOF. In order to be able to apply Theorem 8.23 we need to diagonalize the matrix A. This may be accomplished by letting

$$y = Tz \tag{8.4.4}$$

in (8.4.2), where

$$T = (\xi_1, \xi_2, \ldots, \xi_k), \tag{8.4.5}$$

that is, the ith column of T is ξ_i.

Then we obtain

$$Tz(n+1) = [A + B(n)]Tz(n),$$

or

$$z(n+1) = [D + \tilde{B}(n)]z(n), \tag{8.4.6}$$

where $D = T^{-1}AT = \text{diag}(\lambda_1, \lambda_2, \ldots, \lambda_k)$ and $\tilde{B}(n) = T^{-1}B(n)T$. It is easy to see that $\tilde{B}(n)$ satisfies condition (8.3.16). Now formula (8.4.3) follows by applying Theorem 8.23. □

Example 8.26. Find an asymptotic estimate of a fundamental set of solutions of

$$y(n+1) = [A + B(n)]y(n), \tag{8.4.7}$$

where

$$A = \begin{pmatrix} 2 & 2 & 1 \\ 1 & 3 & 1 \\ 1 & 2 & 1, \end{pmatrix},$$

$$B(n) = \begin{pmatrix} 1/n^2 + 1 & 0 & (0.5)^n \\ 0 & (0.2)^n & 0 \\ e^{-n} & 0 & \log n/n^2 \end{pmatrix}.$$

Solution The eigenvalues of A are $\lambda_1 = 5, \lambda_2 = 1$, and $\lambda_3 = 1$, and the corresponding eigenvectors are $\xi_i = \begin{pmatrix} 1 \\ 1 \\ 1 \end{pmatrix}, \xi_2 = \begin{pmatrix} 1 \\ 0 \\ -1 \end{pmatrix}$, and $\xi_3 = \begin{pmatrix} 1 \\ 0 \\ -2 \end{pmatrix}$. Furthermore, $B(n)$ satisfies condition (8.3.16). Thus by Theorem 8.25, equation (8.4.7) has the solutions

$$y_1(n) = (1+o(1))(5^n)\begin{pmatrix}1\\1\\1\end{pmatrix} \sim \begin{pmatrix}1\\1\\1\end{pmatrix}(5^n),$$

$$y_2(n) = (1+o(1))\begin{pmatrix}1\\0\\-1\end{pmatrix} \sim \begin{pmatrix}1\\0\\-1\end{pmatrix},$$

$$y_3(n) = (1+o(1))\begin{pmatrix}1\\0\\-2\end{pmatrix} \sim \begin{pmatrix}1\\0\\-2\end{pmatrix}.$$

Next, we apply Theorem 8.23 to establish the following asymptotic result for (8.4.1).

Corollary 8.27. *Suppose that the polynomial*

$$p(\lambda) = \lambda^k + a_1\lambda^{k-1} + \cdots + a_k \tag{8.4.8}$$

has distinct roots $\lambda_1, \lambda_2, \ldots, \lambda_k$ *and that*

$$\sum_{n=1}^{\infty} |p_i(n)| < \infty, \quad \text{for } 1 \le i \le k. \tag{8.4.9}$$

Then (8.4.1) *has* k *solutions* $y_1(n), y_2(n), \ldots, y_k(n)$ *with*

$$y_i(n) = [1+o(1)]\lambda_i^n. \tag{8.4.10}$$

PROOF. First we put (8.4.1) into the form of a k-dimensional system

$$z(n+1) = [A + B(n)]z(n), \tag{8.4.11}$$

where

$$A = \begin{pmatrix} 0 & 1 & \ldots & 0\\ 0 & 0 & 1 & 0\\ \vdots & & & \vdots\\ -a_k & -a_{k-1} & \ldots & -a_1 \end{pmatrix},$$

$$B(n) = \begin{pmatrix} 0 & 0 & \ldots & 0\\ 0 & 0 & \ldots & 0\\ -p_k(n) & -p_{k-1}(n) & \ldots & -p_1(n) \end{pmatrix},$$

$$z(n) = (y(n), y(n+1), \ldots, y(n+k-1))^T.$$

Notice that polynomial (8.4.8) is the characteristic polynomial of the matrix A. Furthermore, for each eigenvalue λ_i there corresponds the eigenvector $\xi_i = (1, \lambda_i, \lambda_i^2, \ldots, \lambda_i^{k-1})^T$. In addition, the matrix $B(n)$ satisfies condition

(8.3.16). Hence one may apply Theorem 8.25 to conclude that there are k solutions $z_1(n), z_2(n), \ldots, z_k(n)$ of (8.4.11) such that, for $1 \le i \le k$,

$$z_i(n) = \begin{pmatrix} y_i(n) \\ y_i(n+1) \\ y_i(n+2) \\ \vdots \\ y_i(n+k-1) \end{pmatrix} = (1+o(1))\lambda_i^n \begin{pmatrix} 1 \\ \lambda_i \\ \lambda_i^2 \\ \vdots \\ \lambda_i^{k-1} \end{pmatrix}.$$

Hence $y_i(n) = [1+o(1)]\lambda_i^n$. □

Example 8.28. Find asymptotic estimates of fundamental solutions to the difference equation

$$y(n+3) - \left(2+e^{-n-2}\right) y(n+2) - \left(1+\frac{1}{n^2+1}\right) y(n+1) + 2y(n) = 0.$$

Solution The characteristic equation is given by $\lambda^3 - 2\lambda^2 - \lambda + 2 = 0$ with roots $\lambda_1 = 2, \lambda_2 = 1, \lambda_3 = -1$. Notice that $p_1(n) = -e^{-n-2}, p_2(n) = -\frac{1}{n^2+1}$, and $p_3(n) = 0$ all satisfy condition (8.4.8). Hence Corollary 8.27 applies to produce solutions $y_1(n), y_2(n)$, and $y_3(n)$ defined as follows:

$$y_1(n) = [1+o(1)]2^n, \qquad y_2(n) = 1+o(1), \qquad y_3(n) = [1+o(1)](-1)^n.$$

Corollary 8.27 is due to Evgrafov. It says that for each characteristic root of polynomial (8.4.8), at least one solution behaves as in formula (8.4.10), provided that the rate of convergence of the coefficients is not too slow.

What happens if all the roots of the characteristic equation (8.4.8) are equal? This same question was addressed by Coffman [22], where he obtained the following result.

Theorem 8.29. *Suppose that the polynomial* (8.4.8) *has a k-fold root of* 1 *and that*

$$\boxed{\sum_{n=1}^{\infty} n^{k-1}|p_i(n)| < \infty, \quad \text{for } 1 \le i \le k.} \tag{8.4.12}$$

Then (8.4.1) *has k solutions $y_1(n), y_2(n), \ldots, y_k(n)$ with*

$$y_i(n) = n^{i-1}(1+o(1)), \quad n \to \infty. \tag{8.4.13}$$

We remark here that the actual result of Coffman is stronger than the statement of Theorem 8.29. Indeed, he proved that

$$\Delta^m y_i(n) = \begin{cases} \begin{pmatrix} n \\ i-m \end{pmatrix} + o\begin{pmatrix} i-m \\ n \end{pmatrix} & \text{for } 1 \le m \le i, \\ o\begin{pmatrix} i-m \\ n \end{pmatrix} & \text{for } i \le m \le k-1. \end{cases}$$

The curious reader might wonder whether Coffman's theorem (Theorem 8.29) applies if the polynomial (8.4.8) has a k-fold root not equal to 1. Luckily, by a very simple trick, one is able to do exactly that. Assume that the characteristic equation (8.4.8) has a k-fold root $\mu \neq 1$. Then polynomial (8.4.8) may be written as

$$(\lambda - \mu)^k = 0. \tag{8.4.14}$$

Letting $y(n) = \mu^n x(n)$ in (8.4.1), we obtain

$$\mu^{n+k}x(n+k)+\mu^{n+k-1}(a_1+p_1(n))x(n+k-1)+\cdots+\mu^n(a_k+p_k(n))x(n) = 0,$$

or

$$x(n+k)-\frac{1}{\mu}(a_1+p_1(n))x(n+k-1)+\cdots+\frac{1}{\mu^k}(a_k+p_k(n))x(n) = 0. \tag{8.4.15}$$

The characteristic equation (8.4.15) is given by

$$\lambda_k + \frac{a_1}{\mu}\lambda^{k-1} + \frac{a_2}{\mu^2}\lambda^{k-2} + \cdots + \frac{a_k}{\mu^k} = 0,$$

which has a k-fold root $\lambda = 1$. Moreover, if $p_i(n), 1 \leq i \leq k$, satisfies condition (8.4.1), then so does $(1/\mu^i)p_i(n)$. Hence Theorem 8.29 applies to (8.4.15) to yield solutions $x_1(n), x_2(n), \ldots, x_k(n)$ with

$$x_i(n) = n^{i-1}(1 + o(1)), \qquad n \to \infty.$$

Consequently, there are solutions $y_1(n), y_2(n), \ldots, y_k(n)$ of (8.4.1) such that

$$y_i(n) = n^{i-1}(1 + o(1))\mu^n.$$

We now summarize the above observations in the following corollary.

Corollary 8.30. *Suppose that the polynomial* (8.4.8) *has a k-fold root μ and that condition* (8.4.12) *holds. Then* (8.4.1) *has k solutions $y_1(n), y_2(n), \ldots, y_k(n)$ such that*

$$\boxed{y_i(n) = n^{i-1}(1 + o(1))\mu^n.} \tag{8.4.16}$$

Example 8.31. Investigate the asymptotic behavior of solutions of the difference equation

$$y(n+3) - (6 + e^{-n-2})y(n+2) + \left(12 - \frac{1}{(n+1)^4}\right)y(n+1) - 8y(n) = 0.$$

Solution The characteristic equation is given by $\lambda^3 - 6\lambda^2 + 12\lambda - 8 = 0$ with roots $\lambda_1 = \lambda_2 = \lambda_3 = 2$. Also, $p_1(n) = -e^{-n-2}, p_2(n) = -1/(n+1)^4$, and $p_3(n) = 0$ all satisfy condition (8.4.12). Hence, by Corollary 8.30 there are three solutions $y_1(n) = (1 + o(1))2^n, y_2(n) = n(1 + o(1))2^n$, and $y_3(n) = n^2(1 + o(1))2^n$.

Example 8.32. Consider the difference equation

$$x(n+2) + p_1(n)x(n+1) + p_2(n)x(n) = 0, \tag{8.4.17}$$

where

$$p_1(n) \neq 0, \quad n \geq n_0 \geq 0,$$

and where

$$\lim_{n\to\infty} \frac{4p_2(n)}{p_1(n)p_1(n-1)} = p \tag{8.4.18}$$

exists. Let $\alpha(n)$ be defined by

$$\alpha(n) = \frac{4p_2(n)}{p_1(n)p_1(n-1)} - p. \tag{8.4.19}$$

Assume that $p \neq 0$, $p < 1$, and

$$\sum_{j=n_0}^{\infty} |\alpha(j)| < \infty. \tag{8.4.20}$$

Show that (8.4.17) has two solutions,

$$x_\pm(n) \sim \left(-\frac{1}{2}\right)^n \prod_{j=n_0}^{n-1} p_1(j) \left(1 \pm \nu \mp \left(\frac{\alpha(j)}{2\nu}\right)\right), \tag{8.4.21}$$

where $\nu = \sqrt{1-p}$.

Solution Let

$$x(n) = \left(-\frac{1}{2}\right)^n \left(\prod_{j=n_0}^{n-2} p_1(j)\right) y(n). \tag{8.4.22}$$

Then (8.4.17) is transformed to

$$y(n+2) - 2y(n+1) + (p + \alpha(n))y(n) = 0. \tag{8.4.23}$$

Let $z(n) = (z_1(n), z_2(n))^T = (y(n), y(n+1))^T$. Then (8.4.23) may be put into a system of the form

$$\begin{pmatrix} z_1(n+1) \\ z_2(n+1) \end{pmatrix} = \begin{pmatrix} 0 & 1 \\ \nu^2 - 1 - \alpha(n) & 2 \end{pmatrix} \begin{pmatrix} z_1(n) \\ z_2(n) \end{pmatrix}. \tag{8.4.24}$$

Again we let

$$\begin{pmatrix} z_1(n) \\ z_2(n) \end{pmatrix} = \begin{pmatrix} 1 & 1 \\ -(\nu-1) & \nu+1 \end{pmatrix} \begin{pmatrix} u_1(n) \\ u_2(n) \end{pmatrix}.$$

Then (8.4.24) becomes

$$\begin{pmatrix} u_1(n+1) \\ u_2(n+1) \end{pmatrix} = \begin{pmatrix} (1 - \nu + (\alpha(n)/2\nu)) & \alpha(n)/2\nu \\ -\alpha(n)/2\nu & (1 + \nu - (\alpha(n)/2\nu)) \end{pmatrix} \begin{pmatrix} u_1(n) \\ u_2(n) \end{pmatrix}. \tag{8.4.25}$$

If we let $u(n) = (u_1(n), u_2(n))^T$, then we may write (8.4.25) in the form

$$u(n+1) = (D(n) + B(n))u(n), \tag{8.4.26}$$

where

$$D(n) = \begin{pmatrix} (1 - \nu + \alpha(n)/2\nu) & 0 \\ 0 & (1 + \nu - (\alpha(n)/2\nu)) \end{pmatrix},$$

$$B(n) = \begin{pmatrix} 0 & \alpha(n)/2\nu \\ -\alpha(n)/2\nu & 0 \end{pmatrix}.$$

By Theorem 8.23, there are two solutions of (8.4.26) given by

$$u_+(n) \sim \left[\prod_{j=n_0}^{n-1} (1 - \nu + (\alpha(j)/2\nu))\right] \begin{pmatrix} 1 \\ 0 \end{pmatrix},$$

$$u_-(n) \sim \left[\prod_{j=n_0}^{n-1} (1 + \nu - (\alpha(j)/2\nu))\right] \begin{pmatrix} 0 \\ 1 \end{pmatrix}.$$

These two solutions produce two solutions of (8.4.24),

$$\begin{aligned} z_+(n) &= \begin{pmatrix} y_+(n) \\ y_+(n+1) \end{pmatrix} \\ &= \begin{pmatrix} 1 & 1 \\ -(\nu - 1) & \nu + 1 \end{pmatrix} \left(\prod_{j=n_0}^{n-1} (1 - \nu + (\alpha(j)/2\nu))\right) \begin{pmatrix} 1 \\ 0 \end{pmatrix}, \\ z_-(n) &= \begin{pmatrix} y_-(n) \\ y_-(n+1) \end{pmatrix} \\ &= \begin{pmatrix} 1 & 1 \\ -(\nu - 1) & \nu + 1 \end{pmatrix} \left(\prod_{j=n_0}^{n-1} (1 + \nu - (\alpha(j)/2\nu))\right) \begin{pmatrix} 1 \\ 0 \end{pmatrix}. \end{aligned}$$

Hence

$$y_+(n) \sim \prod_{j=n_0}^{n-1} (1 - \nu + (\alpha(j)/2\nu)),$$

$$y_-(n) \sim \prod_{j=n_0}^{n-1} (1 + \nu - (\alpha(j)/2\nu)).$$

Using (8.4.22) we obtain

$$x_+(n) \sim \left(-\frac{1}{2}\right)^n \prod_{j=n_0}^{n-2} p_1(j) \prod_{j=n_0}^{n-1} (1 - \nu + (\alpha(j)/2\nu)). \tag{8.4.27}$$

Similarly, one may show that

$$x_-(n) \sim \left(-\frac{1}{2}\right)^n \prod_{j=n_0}^{n-2} p_1(j) \prod_{j=n_0}^{n-1} (1+\nu-(\alpha(j)/2\nu)). \qquad (8.4.28)$$

(See Exercises 8.4, Problem 11.)

Exercises 8.4

In Problems 1 through 4 find an asymptotic estimate of a fundamental set of solutions of the given equation $y(n+1) = [A+B(n)]y(n)$.

1. $A = \begin{pmatrix} 2 & 0 \\ 0 & 3 \end{pmatrix}, \quad B(n) = \begin{pmatrix} e^{-n} & 0 \\ \dfrac{1}{(n+1)^2} & (0.1)^n \end{pmatrix}.$

2. $A = \begin{pmatrix} 1 & 6 \\ 5 & 2 \end{pmatrix}, \quad B(n) = \begin{pmatrix} 0 & e^{-n-1} \\ 2^{-n} & \dfrac{n}{e^n} \end{pmatrix}.$

3. $A = \begin{pmatrix} -1 & 0 & 0 \\ 0 & 1 & 0 \\ 0 & 0 & 4 \end{pmatrix}, \quad B(n) = \begin{pmatrix} 3^{-n} & 0 & 2^{-n} \\ \dfrac{\sin n}{(n+1)^2} & 0 & 0 \\ 0 & e^{-n} & \dfrac{1}{(n+1)^3} \end{pmatrix}.$

4. $A = \begin{pmatrix} 5 & 4 & 2 \\ 4 & 5 & 2 \\ 2 & 2 & 2 \end{pmatrix}, \quad B(n) = \begin{pmatrix} 0 & (0.2)^n & 0 \\ (0.1)^n & 0 & e^{-n^2} \\ 0 & \dfrac{1}{n^2+1} & 0 \end{pmatrix}.$

In Problems 5 through 10 investigate the asymptotic behavior of solutions of the given equation.

5. $y(n+2) - (5+e^{-n})y(n+1) + \left(6 - \dfrac{1}{(n+1)^2}\right) y(n) = 0.$

6. $y(n+2) - (4+ne^{-n})y(n) = 0.$

7. $y(n+2) + (4+ne^{-n})y(n) = 0.$

8. $y(n+3) - 6y(n+2) + (11+(\sin n)e^{-n})y(n+1) - 6y(n) = 0.$

9. $y(n+3) - (3+2^{-n})y(n+2) + 3y(n+1) - y(n) = 0.$

10. $y(n+3) - 15y(n+2) + 75y(n+1) - (125+(0.1)^n)y(n) = 0.$

11. Complete the proof of Example 8.32 by verifying formula (8.4.28).

*12. Consider the second-order difference equation

$$x(n+2) + p_1(n)x(n+1) + p_2(n)x(n) = 0. \tag{8.4.29}$$

Assume that $p_1(n) \neq 0$ for $n \geq n_0$ and that:

(i) $\lim_{n\to\infty} 4p_2(n)/(p_1(n)p_1(n-1)) = p$,

(ii) $\sum_{n=n_0}^{\infty} |\alpha(n)| < \infty$, where

$$\alpha(n) = [4p_2(n)/(p_1(n)p_1(n-1)] - p). \tag{8.4.30}$$

If p is neither 0 nor 1, show that (8.4.29) has two solutions

$$x_{\pm}(n) \sim \left(-\frac{1}{2}\right)^n \left(\prod_{j=n_0}^{n-2} p_1(j)\right) (1 \pm \nu)^n, \quad (n \to \infty),$$

where $\nu = \sqrt{1-p}$.

13. In Problem 12, suppose that $p = 1$ and that all the assumptions there hold except that the condition $\sum_{n=n_0}^{\infty} |\alpha(n)| < \infty$ is replaced by $\sum_{n=n_0}^{\infty} n|\alpha(n)| < \infty$.

Show that there are two solutions $x_1(n) \sim \left(-\frac{1}{2}\right)^n \prod_{j=n_0}^{n-2} p_1(j)$ and $x_2(n) \sim n\left(-\frac{1}{2}\right)^n \prod_{j=n_0}^{n-2} p_1(j)$, $n \to \infty$.

14. Consider the difference equation (8.4.29) such that $p_1(n) \neq 0$ for $n \geq n_0$. Assume that $\lim_{n\to\infty}(p_2(n))/(p_1(n)p_1(n-1)) = 0$ and $\alpha(n) = (p_2(n))/(p_1(n)p_1(n-1))$.

(a) Use the transformation $x(n) = \left(-\frac{1}{2}\right)^n \prod_{j=n_0}^{n-2} p_1(j) z(n)$ to transform (8.4.29) to $z(n+2) - 2z(n+1) + \alpha(n)z(n) = 0$.

(b) Show that (8.4.29) has two solutions $x_1(n) \sim (-1)^n \prod_{j=n_0}^{n-2} p_1(j)$ and $x_2(n) = o(\nu^n |x_1(n)|)$ for any ν with $0 < \nu < 1$.

*15. Consider the difference (8.4.17) with conditions (8.4.17) and (8.4.18) satisfied. If p is real and $p > 1$, show that formula (8.4.21) remains valid if we assume that

$$\sum_{j=n_0}^{\infty} j|\alpha(j)| < \infty. \tag{8.4.31}$$

*16. Show that formula (8.4.21) remains valid if one replaces hypothesis (8.4.20) by

$$\sum_{j=n_0}^{\infty} |\alpha(j)|^{\sigma} < \infty \tag{8.4.32}$$

for some real number σ with $1 \leq \sigma \leq 2$.

*17. Show that the conclusions of Problem 16 remain valid if condition (8.4.32) is replaced by

$$\sum_{j=n_0}^{\infty} |\alpha(j)|^{\sigma} j^{\tau-1}$$

for some real numbers σ and τ such that $1 \leq \sigma \leq 2$ and $\tau > \sigma$.

8.5 Second-Order Difference Equations

The asymptotics of second-order difference equations play a central role in many branches of pure and applied mathematics such as continued fractions, special functions, orthogonal polynomials, and combinatorics. In this section we will utilize the special characteristics of second-order equations to obtain a deeper understanding of the asymptotics of their solutions. Consider the difference equation

$$x(n+2) + p_1(n)x(n+1) + p_2(n)x(n) = 0. \tag{8.5.1}$$

One of the most effective techniques to study (8.5.1) is to make the change of variables

$$x(n) = \left(-\frac{1}{2}\right)^{n-1} \left(\prod_{j=n_0}^{n-2} p_1(j)\right) y(n). \tag{8.5.2}$$

Then (8.5.1) is transformed to

$$y(n+2) - 2y(n+1) + \frac{4p_2(n)}{p_1(n)p_1(n-1)}\, y(n) = 0. \tag{8.5.3}$$

Put

$$q = \lim_{n\to\infty} \frac{4p_2(n)}{p_1(n)p_1(n-1)}, \quad \alpha(n) = \frac{4p_2(n)}{p_1(n)p_1(n-1)} - q. \tag{8.5.4}$$

Then the characteristic roots associated with (8.5.3) are $\lambda_1 = 1 - \sqrt{1-q}$ and $\lambda_2 = 1 + \sqrt{1-q}$.

Here there are several cases to consider:

Case I. If $-\infty < q < 1$, then λ_1 and λ_2 are real distinct roots with $|\lambda_1| \neq |\lambda_2|$. Case I may be divided into subcases.

(a) If $\alpha(n) \to 0$, then by invoking the Poincaré–Perron theorem we obtain two linearly independent solutions $y_1(n)$ and $y_2(n)$ such that

$$\lim_{n\to\infty} \frac{y_1(n+1)}{y_1(n)} = \lambda_1, \quad \lim_{n\to\infty} \frac{y_2(n+1)}{y_2(n)} = \lambda_2. \tag{8.5.5}$$

Although this does not provide us with explicit representations of the solutions $y_1(n)$ and $y_2(n)$ it does guarantee the existence of a special solution, called a minimal solution. As we will see later, minimal solutions play a central role in the convergence of continued fractions and the asymptotics of orthogonal polynomials.

Definition 8.33. A solution $\varphi(n)$ of (8.5.1) is said to be *minimal* (subdominant, recessive) if

$$\boxed{\lim_{n\to\infty} \frac{\varphi(n)}{x(n)} = 0}$$

for any solution $x(n)$ of (8.5.1) that is not a multiple of $\varphi(n)$. A nonminimal solution is called *dominant*. One may show that a minimal solution is unique up to multiplicity (Exercises 8.5, Problem 1).

Returning to (8.5.3), let us assume that $|\lambda_1| < |\lambda_2|$. Then there exist μ_1, μ_2 such that $|\lambda_1| < \mu_1 < \mu_2 < |\lambda_2|$. By (8.5.5) it follows that, for sufficiently large n,

$$\frac{|y_1(n+1)|}{|y_1(n)|} \le \mu_1 \quad \text{and} \quad \frac{|y_2(n+1)|}{|y_2(n)|} \ge \mu_2.$$

Hence

$$|y_1(n)| \le \mu_1^n |y_1(0)|, \quad |y_2(n)| \ge \mu_2^n |y_2(0)|,$$

which implies that

$$\lim_{n\to\infty} \frac{|y_1(n)|}{|y_2(n)|} = \lim_{n\to\infty} \left(\frac{\mu_1}{\mu_2}\right)^n \frac{|y_1(0)|}{|y_2(0)|} = 0. \tag{8.5.6}$$

Therefore, $y_1(n)$ is a minimal solution of (8.5.3) (Why?) (Exercises 8.5, Problem 1).

(b) If $\alpha(n) \in l^1(\mathbb{Z}^+)$, that is, $\sum_{n_0}^{\infty} |\alpha(n)| < \infty$, then by Corollary 8.27, (8.5.3) has a fundamental set of solutions $y_1(n)$ and $y_2(n)$ such that

$$y_1(n) = \lambda_1^n(1 + o(1)), \quad y_2(n) = \lambda_2^n(1 + o(1)).$$

Hence

$$x_1(n) = \left(-\frac{1}{2}\right)^{n-1} \left(\prod_{j=n_0}^{n-2} p_1(j)\right) \lambda_1^n(1 + o(1)),$$

$$x_2(n) = \left(-\frac{1}{2}\right)^{n-1} \left(\prod_{j=n_0}^{n-2} p_1(j)\right) \lambda_2^n(1 + o(1)).$$

Notice that we have treated this case thoroughly in Example 8.32, where we obtained formulas (8.4.27) and (8.4.28).

(c) Suppose that $\alpha(n) \in l^2(\mathbb{Z}^+)$, that is, $\sum_{n_0}^{\infty} \alpha^2(n) < \infty$. Then using the scheme of Example 8.32, Elaydi [38] showed that (8.5.1) has two linearly independent solutions $x_1(n)$, $x_2(n)$ obeying formulas (8.4.27) and (8.4.28). In other words,

$$x_1(n) = \left(-\frac{1}{2}\right)^n \left(\prod_{j=n_0}^{n-2} p_1(j)\right) \prod_{j=n_0}^{n-1} (1 - \nu + (\alpha(j)/2\nu))(1 + o(1)), \tag{8.5.7}$$

$$x_2(n) = \left(-\frac{1}{2}\right)^n \left(\prod_{j=n_0}^{n-2} p_1(j)\right) \prod_{j=n_0}^{n-1} (1 + \nu - (\alpha(j)/2\nu))(1 + o(1)). \tag{8.5.8}$$

Moreover, $x_1(n)$ is a minimal solution, and $x_2(n)$ is a dominant solution.

We remark here that the above results may be extended to systems as well as to higher-order difference equations [38].

Case II. If $q = 1$, then $\lambda_1 = \lambda_2 = 1$. In this case we use Coffman's result (Theorem 8.29) to produce two solutions of (8.5.3),

$$y_1(n) \sim 1 \quad \text{and} \quad y_2(n) \sim n,$$

provided that $\sum\limits_{n_0}^{\infty} n|\alpha(n)| < \infty$.

Case III. If $q > 1$, then λ_1 and λ_2 are complex conjugates $\bar{\lambda}_1 = \lambda_2$, $|\lambda_1| = |\lambda_2|$. In this case we use another result from [43].

Theorem 8.34 [43]. *Suppose that $q > 1$ and the following condition holds.*

$$\sum_{n=n_0}^{\infty} |\alpha(n+1) - \alpha(n)| < \infty.$$

Then (8.5.3) has two solutions $y_1(n), y_2(n)$ with

$$\boxed{y_i(n) = (1 + o(1)) \prod_{m=n_0}^{n-1} \beta_i(m), \quad i = 1, 2, \ldots,} \tag{8.5.9}$$

where

$$\beta_1(n) = 1 - \sqrt{1 - q + \alpha(n)}, \quad \beta_2(n) = 1 + \sqrt{1 - q + \alpha(n)},$$

provided that $Re\sqrt{1 - q + \alpha(n)}$ is semidefinite for a fixed branch of the square root $(0 \leq \arg \sqrt{z} < \pi)$.

8.5.1 A Generalization of the Poincaré–Perron Theorem

[57] In many applications related to (8.5.1) the coefficients $p_1(n)$ and $p_2(n)$ are of the form

$$p_1(n) \sim an^{\alpha}, \quad p_2(n) \sim bn^{\beta}, \quad ab \neq 0, \quad \alpha, \beta \text{ real}; \quad n \to \infty.$$

The asymptotics of the solutions of (8.5.1) can be determined by means of the Newton–Puiseux diagram formed with the points $P_0(0,0)$, $P_1(1,\alpha)$, $P_2(2,\beta)$ (Figure 8.1)

Theorem 8.35 [114], [82].

(a) *If the point P_1 is above the line P_0P_2 (i.e., $\alpha > \beta/2$), then* (8.5.1) *has a fundamental set of solutions $x_1(n)$ and $x_2(n)$ such that*

$$\boxed{\lim_{n\to\infty} \frac{x_1(n+1)}{x_1(n)} = -an^{\alpha},} \boxed{\lim_{n\to\infty} \frac{x_2(n+1)}{x_2(n)} = \frac{-b}{a} n^{(\beta-\alpha)}.} \tag{8.5.10}$$

Moreover, $x_2(n)$ is a minimal solution.

(b) *Suppose that the points P_0, P_1, P_2 are collinear (i.e., $\alpha = \beta/2$). Let λ_1, λ_2 be the roots of the equation $\lambda^2 + a\lambda + b = 0$, such that $|\lambda_1| \geq |\lambda_2|$. Then* (8.5.1) *has a fundamental set of solutions $x_1(n)$ and $x_2(n)$ with*

$$\boxed{\lim_{n\to\infty} \frac{x_1(n+1)}{x_1(n)} = \lambda_1 n^{\alpha},} \boxed{\lim_{n\to\infty} \frac{x_2(n+1)}{x_2(n)} = \lambda_2 n^{\alpha},} \tag{8.5.11}$$

provided that $|\lambda_1| \neq |\lambda_2|$. Moreover, $x_2(n)$ is a minimal solution.

If $|\lambda_1| = |\lambda_2|$, then

$$\boxed{\limsup_{n\to\infty} \left[\frac{|x(n)|}{(n!)^{\alpha}}\right]^{1/n} = |\lambda_1|} \tag{8.5.12}$$

for all nontrivial solutions $x(n)$ of (8.5.1).

(c) *If the point P_1 lies below the line segment $\overline{P_0P_2}$, then*

$$\boxed{\limsup_{n\to\infty} \left[\frac{|x(n)|}{(n!)^{\beta/2}}\right]^{1/n} = \sqrt{|b|}} \tag{8.5.13}$$

for all nontrivial solutions of (8.5.1).

PROOF. Let $p_1(n) = an^{\alpha} + \nu(n)$, $P_2(n) = bn^{\beta} + \mu(n)$, where $\nu(n)$ and $\mu(n)$ are null sequences. Then we may write (8.5.1) as

$$x(n+2) + (a\,n^{\alpha} + \nu(n))\,x(n+1) + (b\,n^{\beta} + \mu(n))\,x(n) = 0. \tag{8.5.14}$$

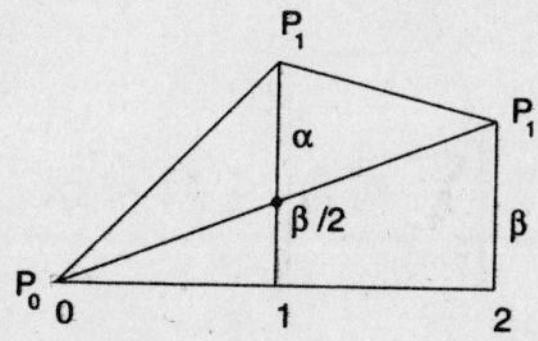

FIGURE 8.1. Newton–Puiseux diagram for (8.5.14).

Making the change of variable $x(n) = (n!)^{\alpha} y(n)$ in (8.5.14) yields

$$y(n+2) + \left[a\left(\frac{n}{n+2}\right)^{\alpha} + \frac{\nu(n)}{(n+2)^{\alpha}}\right] y(n+1) + \left[\frac{bn^{\beta}}{(n+1)^{\alpha}(n+2)^{\alpha}} + \frac{\mu(n)}{(n+1)^{\alpha}(n+2)^{\alpha}}\right] y(n) = 0. \quad (8.5.15)$$

(a) If $2\alpha > \beta$, the characteristic equation of (8.5.14) is $\lambda^2 + a\lambda = 0$. The first solution $x_1(n)$ in (8.5.1) corresponds to $\lambda_1 = -a$ in the Poincaré–Perron theorem. The second solution $x_2(n)$ may be obtained by using formula (2.2.18) and is left to the reader as Exercises 8.5, Problem 2.

The proofs of parts (b) and (c) are left to the reader as Problem 2. □

Remark: The above theorem is valid for kth-order scalar difference equations. The interested reader may consult [113], [82], [146].

Exercises 8.5

1. Consider a kth-order scalar difference equation of Poincaré type (8.2.7) such that its characteristic roots have distinct moduli.
   (a) Show that the equation has a minimal solution.
   (b) Show that the minimal solution is unique up to multiplicity.
2. Complete the proofs of parts (a), (b), (c) in Theorem 8.35.
3. Investigate the asymptotic behavior of the equation
$$y(n+2) - 2y(n+1) + \left(1 - \frac{1}{n^3}\right) y(n) = 0.$$
4. Investigate the asymptotic behavior of solutions of the equation
$$\Delta^2 y(n) = \frac{(-1)^n}{n^{\alpha+1}} y(n+1)$$
where $\alpha > 1$.
5. Investigate the asymptotic behavior of solutions of
$$\Delta^2 y(n) = \frac{p(n)}{n^{\alpha+1}} y(n+1),$$

where $\alpha > 1$, and

$$\left|\sum_{j=1}^{n} p(j)\right| \le M < \infty$$

for all $n > 1$.

6. Show that the difference equation

$$\Delta^2 x(n) = p(n)x(n+1)$$

has two linearly independent solutions $x_1(n)$ and $x_2(n)$ such that

$$\det \begin{bmatrix} x_1(n) & x_2(n) \\ \Delta x_1(n) & \Delta x_2(n) \end{bmatrix} = -1.$$

7. (*Multiple Summation*). Show that for any sequence $f(n)$, $n \in \mathbb{Z}^+$,

$$\boxed{\sum_{r=n_0}^{n-1} \sum_{j=n_0}^{r-1} f(j) = \sum_{j=n_0}^{n-1} (n-j)f(j).} \tag{8.5.16}$$

8. Consider the second-order difference equation [34], [35]

$$\Delta^2 y(n) + p(n)y(n) = 0 \tag{8.5.17}$$

such that

$$\sum_{j=1}^{\infty} j|p(j)| < \infty. \tag{8.5.18}$$

Show that (8.5.17) has two solutions $y_1(n) \sim 1$ and $y_2(n) \sim n$ as $n \to \infty$, *without using Coffman's theorem* (Theorem 8.29). You may use the following steps:

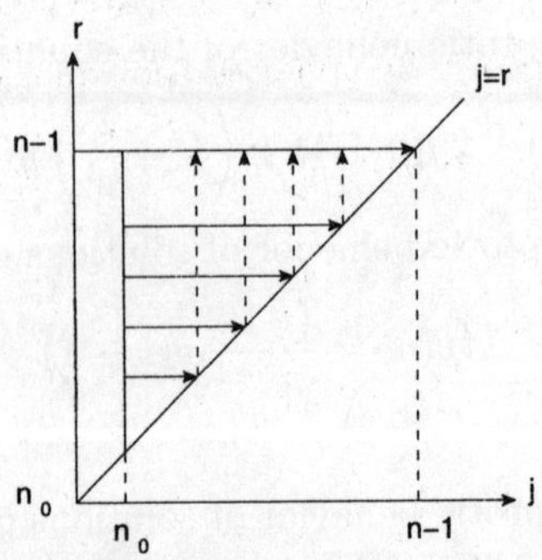

FIGURE 8.2.

(a) Use $\Delta^2 y(n) = -p(n)y(n)$ to show that

$$y(n) = c_1 + c_2(n) - \sum_{r=1}^{n-1}\sum_{j=1}^{r-1} p(j)y(j).$$

(b) Use formula (8.5.16) to show that

$$\frac{|y(n)|}{n} \le |c_1| + |c_2| + \sum_{j=1}^{n-1} j|p(j)|\frac{|y(j)|}{j}.$$

(c) Use the discrete Gronwall's inequality (Lemma 4.32) to show that

$$|y(n)| \le c_3 n.$$

(d) Substitute back into

$$\Delta y(n) = c_1 - \sum_{j=1}^{n-1} p(j)y(j)$$

to obtain

$$\lim_{n\to\infty} \Delta y(n) = c_2 - M.$$

*9. (*Generalized Gronwall's Inequality*). Suppose that

$$u(n) \le a + b\sum_{j=n_0}^{n-1} c(j)u^{\gamma}(j),$$

where $1 \ne \gamma > 0$, $a \ge 0$, $b > 0$, $c(j) > 0$, and $u(j) > 0$, for $j \ge n_0$.

Prove that

$$u(n) \le \left[a^{1-\gamma} + b(1-\gamma)\sum_{j=n_0}^{n-1} c(j)\right]^{1/(1-\gamma)}, \tag{8.5.19}$$

provided that, for $\gamma > 1$,

$$a^{1-\gamma} + b(1-\gamma)\sum_{j=n_0}^{n-1} c(j) > 0 \quad \text{for} \quad n \ge n_0.$$

*10. Generalize the result of Problem 8 to the so-called Emden–Fowler equation

$$\Delta^2 y(n) + p(n)|y(n)|^{\gamma}\operatorname{sgn} y(n) = 0, \tag{8.5.20}$$

where $\gamma \ne 1$ is a positive real number, and

$$\operatorname{sgn} y(n) = \begin{cases} 1 & \text{if } y(n) > 0, \\ -1 & \text{if } y(n) < 0. \end{cases}$$

Show that if

$$\sum_{j=n_0}^{\infty} j^{\gamma} p(j) = M < \infty,$$

then each solution $y(n)$ with the initial condition $y(n_0)$ with

$$\left[\Delta y(n_0) + \left|\frac{y(n_0)}{n_0} - \Delta y(n_0)\right|\right]^{1-\gamma} + 2(1-\gamma)M > 0$$

is such that $y(n) \sim n$ as $n \to \infty$. You may use the following steps.

(a) Let $A(n) = \Delta y(n), B(n) = y(n) - n\Delta y(n)$. Show that $y(n) = nA(n) + B(n)$.

(b) Show that

$$\Delta A(n) = p(n)[|nA(n) + B(n)|]^{\gamma} \text{ sgn } y(n),$$
$$\Delta B(n) = (n+1)p(n)[|nA(n) + B(n)|]^{\gamma} \text{ sgn } y(n).$$

(c) Use the antidifference operator Δ^{-1} to obtain $A(n)$ and $B(n)$ and then use the generalized Gronwall's inequality.

(d) Suppose that $x_1(n)$ and $x_2(n)$ are two linearly independent solutions of the equation

$$\Delta^2 x(n) = p(n)x(n+1).$$

In addition, assume that for a sequence $q(n)$ we have

$$\sum_{j=n_0}^{\infty} |q(j)|u(j) = M < \infty,$$

where for a specific $m \in \mathbb{Z}^+$,

$$u(n) = \max\left\{|x_1(n+1)||x_1(n)|^{2m+1},\ |x_1(n+1)||x_2(n)|^{2m+1},\right.$$
$$\left.|x_2(n+1)||x_1(n)|^{2m+1},\ |x_2(n+1)||x_2(n)|^{2m+1}\right\}.$$

Show that there exist solutions $y(n)$ of the equation

$$\Delta^2 y(n) = p(n)y(n+1) + q(n)y^{2m+1}(n)$$

such that

$$y(n) = \alpha(n)x_1(n) + \beta(n)x_2(n),$$

with

$$\lim_{n\to\infty} \alpha(n) = a, \quad \lim_{n\to\infty} \beta(n) = b,$$

for some constants a, b.

8.6 Birkhoff's Theorem

Consider again the second-order difference equation

$$x(n+2) + p_1(n)x(n+1) + p_2(n)x(n) = 0, \tag{8.6.1}$$

where $p_1(n)$ and $p_2(n)$ have asymptotic expansions of the form

$$p_1(n) \sim \sum_{j=0}^{\infty} \frac{a_j}{n^j}, \quad p_2(n) \sim \sum_{j=0}^{\infty} \frac{b_j}{n^j}, \tag{8.6.2}$$

with $b_0 \neq 0$.

The characteristic equation associated with (8.6.1) is $\lambda^2 + a_0\lambda + b_0 = 0$ with roots

$$\lambda_1, \lambda_2 = -\frac{1}{2}a_0 \pm \sqrt{\frac{1}{4}a_0^2 - b_0}. \tag{8.6.3}$$

Extensive work by Birkhoff [11], [12] Birkhoff and Trjitzinsky [13], and Adams [2] has been done concerning the asymptotics of equations of type (8.6.1) with expansions (8.6.2). Due to the limitations imposed by the introductory nature of this book, we will restrict our exposition to second-order difference equations. Our presentation here follows closely the excellent papers by Wong and Li [147], [148].

Theorem 8.36 (Birkhoff–Adams).

(a) *If $\lambda_1 \neq \lambda_2$, i.e., $a_0^2 \neq 4b_0$, then equation* (8.6.1) *has two linearly independent solutions $x_1(n), x_2(n)$, which will be called normal solutions, of the form*

$$\boxed{x_i(n) \sim \lambda_i^n n^{\alpha_i} \sum_{r=0}^{\infty} \frac{c_i(r)}{n^r}, \quad i = 1, 2,} \tag{8.6.4}$$

$$\alpha_i = \frac{a_1\lambda_i + b_1}{a_0\lambda_i + 2b_0}, \quad i = 1, 2, \tag{8.6.5}$$

$$\sum_{j=0}^{s-1} \left[\lambda_i^2 2^{s-j} \binom{\alpha_i - j}{s - j} + \lambda_i \sum_{r=j}^{s} \binom{\alpha_i - j}{r - j} a_{s-r} + b_{s-j} \right] \tag{8.6.6}$$

$c_i(j) = 0,\ c_i(0) = 1.$

In particular, we obtain

$$c_i(1) = \frac{-2\lambda_i^2\alpha_i(\alpha_i - 1) - \lambda_i(a_2 + \lambda_i a_1 + \alpha_i(\alpha_i - 1)a_0/2) - b_2}{2\lambda_i^2(\alpha_i - 1) + \lambda_i(a_1 + (\lambda_i - 1)a_0) + b_1}.$$

(b) *If $\lambda_1 = \lambda_2 = \lambda$ but $\lambda = -\frac{1}{2}a_0$ is not a root of the equation $a_1\lambda + b_1 = 0$ (i.e., $2b_1 \neq a_0a_1$), then equation (8.6.1) has two linearly independent solutions, $x_1(n)$, $x_2(n)$, which will be called subnormal solutions, of the form*

$$\boxed{x_i(n) \sim \lambda^n e^{\gamma_i \sqrt{n}} n^{\alpha} \sum_{j=0}^{\infty} \frac{c_i(j)}{n^{j/2}}, \quad i = 1, 2,} \tag{8.6.7}$$

where

$$\alpha = \frac{1}{4} + \frac{b_1}{2b_0}, \quad \gamma_1 = 2\sqrt{\frac{a_0a_1 - 2b_1}{2b_0}}, \quad \gamma_2 = -2\sqrt{\frac{a_0a_1 - 2b_1}{2b_0}}, \tag{8.6.8}$$

$$\begin{aligned} c_0 = 1, \quad c_i(1) = \frac{1}{24b_0^2\gamma_i}&(a_0^2a_1^2 - 24a_0a_1b_0 + 8a_0a_1b_1 \\ &- 24a_0a_2b_0 - 9b_0^2 - 32b_1^2 + 24b_0b_1 + 48b_0b_2). \end{aligned} \tag{8.6.9}$$

(The general recursive formula for $c_i(n)$ is too complicated to be included here. The interested reader is referred to Wong and Li [147].)

(c) *If $\lambda_1 = \lambda_2 = \lambda$ and $2b_1 = a_0a_1$, then we consider the equation*

$$\alpha(\alpha - 1)\lambda^2 + (a_1\alpha + a_2)\lambda + b_2 = 0.$$

Let α_1, α_2 ($\operatorname{Re}\alpha_2 \geq \operatorname{Re}\alpha_1$) be the roots of this equation. Then there are three subcases to contemplate.

(c_1) *If $\alpha_2 - \alpha_1 \neq 0, 1, 2, \ldots$, then equation (8.6.1) has two linearly independent solutions of the form*

$$\boxed{x_i(n) \sim \lambda^n n^{\alpha_i} \sum_{j=0}^{\infty} \frac{c_i(j)}{n^j}.} \tag{8.6.10}$$

(c_2) *If $\alpha_2 - \alpha_1 = 1, 2, \ldots$, then equation (8.6.1) has two solutions, $x_1(n)$ given by (8.6.10) and $x_2(n) = z(n) + c(\ln\, n)x_1(n)$, where c is a constant that may be zero, and*

$$\boxed{z(n) \sim \lambda^n n^{\alpha_2} \sum_{s=0}^{\infty} \frac{d_s}{n^s}.} \tag{8.6.11}$$

(c_3) *If $\alpha_2 = \alpha_1$, then equation (8.6.1) has two solutions: $x_1(n)$ given by (8.6.10), and $x_2(n) = z(n) + c(\ln n)x_1(n)$, $c \neq 0$,*

$$\boxed{z(n) \sim \lambda^n n^{\alpha_1 - r + 2} \sum_{s=0}^{\infty} \frac{d_s}{n^s},} \tag{8.6.12}$$

where r is an integer ≥ 3.

PROOF. The main idea of the proof is to substitute the formal solutions (8.6.4) and (8.6.7) back into (8.6.1) and then compare coefficients of powers of n in the resulting expression. Details of the proof will not be included here, and we refer the interested reader to the paper of Wong and Li [147]. □

Example 8.37. The Apéry Sequence

The sequence [141]

$$u(n) = \sum_{k=0}^{n} \binom{n}{k}^2 \binom{n+k}{k}^2$$

satisfies the second-order difference equation

$$(n+2)^3 u(n+2) - (34n^3 + 153n^2 + 231n + 117)u(n+1) + (n+1)^3 u(n) = 0. \tag{8.6.13}$$

Writing (8.6.13) in the form (8.6.1), we have

$$p_1(n) = \frac{-(34n^3 + 153n^2 + 231n + 117)}{(n+2)^3}$$
$$= a_0 + \frac{a_1}{n} + \frac{a_2}{n^2} + \cdots, \tag{8.6.14}$$

$$p_2(n) = \left(\frac{n+1}{n+2}\right)^3 = b_0 + \frac{b_1}{n} + \frac{b_2}{n^2} + \cdots. \tag{8.6.15}$$

To find a_0 we just take the limit of both sides of (8.6.14) to obtain $a_0 = -34$. Subtracting a_0 from both sides, multiplying by n, and then taking the limit as $n \to \infty$ yields $a_1 = 51$. Repeating this process, we obtain $a_2 = -129$. Similarly, one may obtain $b_0 = 1, b_1 = -3, b_2 = 9$. Hence by formula (8.6.3) the characteristic roots are $\lambda_1, \lambda_2 = 17 \pm 12\sqrt{2}$. From formula (8.6.5) we have

$$\alpha_1 = \frac{51(17+12\sqrt{2}) - 3}{(-34)(17+12\sqrt{2}) + 2} = \frac{-3}{2},$$
$$\alpha_2 = \frac{51(17-12\sqrt{2}) - 3}{(-34)(17-12\sqrt{2}) + 2} = \frac{-3}{2}.$$

Hence we have two solutions $u_1(n)$ and $u_2(n)$ such that

$$u_1(n) \sim (17+12\sqrt{2})^n n^{-3/2} \left[1 + \frac{c_1(1)}{n} + \frac{c_1(2)}{n^2} + \cdots\right],$$

with $c_1(1) \approx -15.43155325$, and

$$u_2(n) \sim (17-12\sqrt{2})^n n^{-3/2} \left[1 + \frac{c_2(1)}{n} + \frac{c_2(2)}{n^2} + \cdots\right],$$

with $c_2(1) \approx -1.068446129$. Since $u_2(n) \to 0$, it follows that $u(n) = cu_1(n)$ for some constant c.

Example 8.38. Laguerre Polynomials [147]

Laguerre polynomials $L_n^\beta(x)$ are defined for $\beta > -1, 0 < x < \infty$, by the following formula, called *Rodrigues' formula*:

$$L_n^\beta(x) = \frac{1}{n!} e^x x^{-\beta} \frac{d^n}{d\,x^n} \left(e^{-x}\, x^{n+\alpha}\right) = \sum_{m=0}^{n} (-1)^m \binom{n+\beta}{n-m} \frac{x^m}{m!}.$$

It can be shown (see Appendix F) that $L_n^\beta(x)$ satisfies a second-order difference equation of the form

$$ny(n) + (x - 2n - \beta + 1)y(n-1) + (n + \beta - 1)y(n-2) = 0.$$

Writing this equation in the form (8.6.1) yields

$$y(n+2) + \frac{x - 2n - \beta - 3}{n+2} y(n+1) + \frac{n+\beta+1}{n+2} y(n) = 0. \qquad (8.6.16)$$

Following the procedure in the preceding example, we obtain

$$a_n = -2 + \frac{x - \beta + 1}{n} - \frac{2(x - \beta + 1)}{n^2} + \cdots,$$

$$b_n = 1 + \frac{\beta - 1}{n} - \frac{2(\beta - 1)}{n^2} + \cdots.$$

The characteristic equation is $\lambda^2 - 2\lambda + 1 = 0$, which has a multiple root $\lambda_1 = \lambda_2 = 1$. This root does not satisfy $(x - \alpha + 1)\lambda + \alpha - 1 = 0$, and hence we have two subnormal solutions of the form (8.6.7). Using formula (8.6.8) we obtain $\alpha = \frac{1}{2}$, $\beta - \frac{1}{4}$, $\gamma_1 = 2\sqrt{xi}$, $\gamma_2 = -2\sqrt{xi}$. Hence it follows from formula (8.6.7) that we have two solutions

$$y_r(n) = e^{(-1)^{r+1} 2\sqrt{nxi}}\, n^{\beta/2 - 1/4} \sum_{j=0}^{\infty} \frac{c_r(j)}{n^{j/2}}, \quad r = 1, 2, \qquad (8.6.17)$$

with $c_1(0) = c_2(0) = 1$,

$$c_r(1) = \frac{(-1)^r i}{48\sqrt{x}} \left(4x^2 - 12\beta^2 - 24x\beta - 24x + 3\right), \quad r = 1, 2.$$

Thus $y_2(n) = \overline{y_1(n)}$. But we know from [98, p. 245] that

$$L_n^\beta(x) = \pi^{-1/2} e^{x/2} x^{-\beta/2 - 1/4} n^{\beta/2 - 1/4} \cos\left(2\sqrt{nx} - \frac{1}{2}\beta\pi - \frac{1}{4}\right) + O(n^{\beta/2 - 3/4}).$$

Thus

$$L_n^\beta(x) = \frac{1}{2}\pi^{-1/2}e^{x/2}x^{-\beta/2-1/4}\left\{\bar{e}^{(\beta\pi/2+\pi/4)i}y_1(n) + e^{(\beta\pi/2+\pi/4)i}y_2(n)\right\}$$

$$= \pi^{1/2}e^{x/2}x^{-\beta/2-1/4}n^{\beta/2-1/4}\left\{\cos\left(2\sqrt{nx} - \frac{1}{2}\beta\pi - \frac{1}{4}\pi\right)\sum_{s=0}^{\infty}\frac{A_s(x)}{n^{s/2}}\right.$$

$$\left. + \sin\left(2\sqrt{nx} - \frac{1}{2}\beta\pi - \frac{1}{4}\pi\right)\sum_{s=0}^{\infty}\frac{B_s(x)}{n^{s/2}}\right\},$$

where $A_0(x) = 1$, $A_1(x) = 0$, $B_0(x) = 0$, and

$$B_1(x) = \frac{1}{48\sqrt{x}}(4x^2 - 12\beta^2 - 24x\beta - 24x + 3).$$

Remark: In [148] the authors extended their analysis to equations of the form

$$x(n+2) + n^r p_1(n)x(n+1) + n^s p_2(n)x(n) = 0,$$

with r and s integers and $p_1(n)$, $p_2(n)$ of the form (8.6.2).

Exercises 8.6

1. (Binomial Sums)

   (a) Verify that the sequence $u(n) = \sum_{k=0}^{n}\binom{n}{k}^3$ satisfies the equation
   $$(n+2)^2u(n+2) - (7n^2+21n+16)u(n+1) - 8(n+1)^2u(n) = 0.$$

   (b) Find an asymptotic representation of $u(n)$.

2. (a) Verify that the sequence $u(n) = \sum_{k=0}^{n}\binom{n}{k}^4$ satisfies the equation
   $$(n+2)^3u(n+2) - 12\left(n+\frac{3}{2}\right)\left(n^2+3n+\frac{7}{3}\right)u(n+1)$$
   $$- 64\left(n+\frac{3}{4}\right)(n+1)\left(n+\frac{5}{4}\right)u(n) = 0.$$

   (b) Find an asymptotic representation of $u(n)$.

3. Find asymptotic representations for the solutions of the equation
   $$(n+2)^5u(n+2) - ((n+2)^5 + (n+1)^5)u(n+1) + (n+1)^5u(n) = 0.$$

4. Find asymptotic representations for the solutions of the difference equation
   $$u(n+2) - u(n+1) - (n+1)u(n) = 0.$$

5. Find asymptotic representations for the solutions of the difference equation

$$(n+1)(n+2)x(n+2) - (n+1)[(2n+b+c+1)+z] \\ \times x(n+1) + n + b)(n+c)x(n) = 0, \quad z \neq 0.$$

6. Find asymptotic representations for the solutions of the second-order difference equation

$$(n+1)(n+2)y(n+2) - (n+1)(2n+2b+1)y(n+1) + (n+b)^2 y(n) = 0.$$

8.7 Nonlinear Difference Equations

In this section we consider the nonlinearly perturbed system

$$y(n+1) = A(n)y(n) + f(n, y(n)) \tag{8.7.1}$$

along with the associated unperturbed system

$$x(n+1) = A(n)x(n), \tag{8.7.2}$$

where $A(n)$ is an invertible $k \times k$ matrix function on $\mathbb{Z}^+$ and $f(n,y)$ is a function from $\mathbb{Z}^+ \times \mathbb{R}^k \to \mathbb{R}^k$ that is continuous in y. Let $\Phi(n)$ be the fundamental matrix of system (8.7.2). The first step in our analysis is to extend the variation of constants formula (Theorem 8.19) to system (8.7.1). Since $A(n)$ is not assumed here to be a diagonal matrix, we need to replace Definition 8.17 by a more general definition of dichotomy.

Definition 8.39. System (8.7.2) is said to possess an ordinary dichotomy if there exists a projection matrix P and a positive constant M such that

$$\begin{aligned} |\Phi(n)P\Phi^{-1}(m)| &\leq M, \quad \text{for } n_0 \leq m \leq n, \\ |\Phi(n)(I-P)\Phi^{-1}(m)| &\leq M, \quad \text{for } n_0 \leq n \leq m. \end{aligned} \tag{8.7.3}$$

Notice that if $A(n) = \operatorname{diag}(\lambda_1(n), \ldots, \lambda_k(n))$, then this definition reduces to Definition 8.17 if we let $\Phi_1(n) = \Phi(n)P$ and $\Phi_2(n) = \Phi(n)(I-P)$.

Theorem 8.40 [44], [121], [131]. *Suppose that system* (8.7.2) *possesses an ordinary dichotomy. If, in addition,*

$$\sum_{j=n_0}^{\infty} |f(j,0)| < \infty \tag{8.7.4}$$

and

$$|f(n,x) - f(n,y)| \leq \gamma(n)|x-y|, \tag{8.7.5}$$

where $\gamma(n) \in l^1([n_0,\infty))$ i.e., $\sum_{j=n_0}^{\infty} \gamma(j) < \infty$, then for each bounded solution $x(n)$ of (8.7.2) *there corresponds a bounded solution $y(n)$ of* (8.7.1) *and vice versa. Furthermore, $y(n)$ is given by the formula*

$$y(n) = x(n) + \sum_{j=n_0}^{n-1} \Phi(n)P\Phi^{-1}(j+1)f(j,y(j)) - \sum_{j=n}^{\infty} \Phi(n)(I-P)\Phi^{-1}(j+1)f(j,y(j)). \tag{8.7.6}$$

PROOF. The proof mimics that of Theorem 8.19 with some obvious modifications. Let $x(n)$ be a bounded solution of (8.7.2). Define a sequence $\{y_i(n)\}$ $(i = 1, 2, \dots)$ successively by letting $y_1(n) = x(n)$ and

$$y_{i+1}(n) = x(n) + \sum_{j=n_0}^{n-1} \Phi(n)P\Phi^{-1}(j+1)f(j,y_i(j)) - \sum_{j=n}^{\infty} \Phi(n)(I-P)\Phi^{-1}(j+1)f(j,y_i(j)). \tag{8.7.7}$$

We use mathematical induction to show that $y_i(n)$ is bounded on $[n_0, \infty)$ for each i. First we notice that by assumption, $|y_1(n)| \leq c_1$. Now suppose that $|y_i(n)| \leq c_i$. Then it follows from (8.7.4), (8.7.5), and (8.7.7) that

$$\begin{aligned} |y_{i+1}(n)| &\leq c_1 + M \sum_{j=n_0}^{\infty} [\gamma(j)|y_i(j)| + |f(j,0)|] \\ &\leq c_1 + M \left[\sum_{j=n_0}^{\infty} c_i\gamma(j) + \tilde{M} \right] = c_{i+1}, \end{aligned}$$

where

$$\sum_{j=n_0}^{\infty} |f(j,0)| = \tilde{M}.$$

Hence $y_i(n)$ is bounded for each i.

As in the proof of Theorem 8.19, one may show that the sequence $\{y_i(n)\}$ converges uniformly on $[n_0, \infty)$ to a bounded solution $y(n)$ of (8.7.1). Conversely, let $y(n)$ be a bounded solution of (8.7.1). Then one may verify easily that

$$\tilde{y}(n) = \sum_{j=n_0}^{n-1} \Phi(n)P\Phi^{-1}(j+1)f(j,\tilde{y}(j)) - \sum_{j=n}^{\infty} \Phi(n)(I-P)\Phi^{-1}(j+1)f(j,\tilde{y}(j))$$

is another bounded solution of (8.7.1). Hence $x(n) = y(n) - \tilde{y}(n)$ is a bounded solution of (8.7.2). □

The preceding result does not provide enough information about the asymptotic behavior of solutions of system equation (8.7.1). In order to obtain such results we need one more assumption on (8.7.2).

Theorem 8.41 [44]. *Let all the assumptions of Theorem* 8.40 *hold. If* $\Phi(n)P \to 0$ *as* $n \to \infty$, *then for each bounded solution* $x(n)$ *of* (8.7.2) *there corresponds a bounded solution* $y(n)$ *of* (8.7.1) *such that*

$$y(n) = x(n) + o(1), \tag{8.7.8}$$

or

$$y(n) \sim x(n).$$

PROOF. The proof is similar to the proof of Theorem 8.20 and is left to the reader as Exercises 8.7, Problem 7. □

Example 8.42. Consider the equation

$$\begin{pmatrix} y_1(n+1) \\ y_2(n+1) \end{pmatrix} = \begin{pmatrix} 3 & 0 \\ 0 & 1/2 \end{pmatrix} \begin{pmatrix} y_1(n) \\ y_2(n) \end{pmatrix} + \begin{pmatrix} \sin y_1(n)/n^2 \\ (1-\cos y_2(n))/n^2 \end{pmatrix}. \tag{8.7.9}$$

Here

$$A(n) = \begin{pmatrix} 3 & 0 \\ 0 & 1/2 \end{pmatrix}, \quad f(n,y) = \begin{pmatrix} \sin y_1/n^2 \\ (1-\cos y_2)/n^2 \end{pmatrix}.$$

Using the Euclidean norm we obtain $\sum_{j=1}^{\infty} |f(j,0)| = 0$. Moreover, for

$$x = \begin{pmatrix} x_1 \\ x_2 \end{pmatrix}, \quad y = \begin{pmatrix} y_1 \\ y_2 \end{pmatrix},$$

we have

$$\begin{aligned} |f(n,x) - f(n,y)| &= \frac{1}{n^2} \left| \begin{matrix} \sin x_1 - \sin y_1 \\ \cos x_2 - \cos y_2 \end{matrix} \right| \\ &= \frac{1}{n^2} \sqrt{(\sin x_1 - \sin y_1)^2 + (\cos x_2 - \cos y_2)^2}. \end{aligned} \tag{8.7.10}$$

By the Mean Value Theorem,

$$\begin{aligned} \frac{|\sin x_1 - \sin y_1|}{|x_1 - y_1|} &= |\cos c|, \quad \text{for some } c \text{ between } x_1 \text{ and } y_1, \\ &\le 1, \end{aligned}$$

and

$$\frac{|\cos x_2 - \cos y_2|}{|x_2 - y_2|} \le 1.$$

Hence substituting into (8.7.10), we obtain

$$|f(n,x) - f(n,y)| \le \frac{1}{n^2}|x - y|.$$

The associated homogeneous equation

$$x(n+1) = A(n)x(n)$$

has a fundamental matrix $\Phi(n) = \begin{pmatrix} 3^n & 0 \\ 0 & (1/2)^n \end{pmatrix}$ and two solutions; one bounded, $x_1(n) = \binom{0}{1}\left(\frac{1}{2}\right)^n$, and one unbounded, $x_2(n) = \binom{1}{0}3^n$. If we let the projection matrix be

$$P = \begin{pmatrix} 0 & 0 \\ 0 & 1 \end{pmatrix},$$

then

$$\Phi(n)P = \begin{pmatrix} 0 & 0 \\ 0 & (1/2)^n \end{pmatrix} \to 0$$

as $n \to \infty$.

Hence all the conditions of Theorem 8.41 hold. Thus corresponding to the bounded solution $x_1(n) = \binom{0}{1}\left(\frac{1}{2}\right)^n$ there corresponds a solution $y(n)$ of (8.7.9) such that

$$y(n) \sim \begin{pmatrix} 0 \\ 1 \end{pmatrix} \left(\frac{1}{2}\right)^n.$$

Next we specialize Theorem 8.41 to the following kth-order nonlinear equation of Poincaré type:

$$y(n+k) + (a_1 + p_1(n))y(n+k-1) + \cdots + (a_k + p_k(n))y(n) = f(n, y(n)). \tag{8.7.11}$$

Corollary 8.43. *Suppose that the characteristic equation* $\lambda^k + a_1\lambda^{k-1} + \cdots + a_k = 0$ *has distinct roots* $\lambda_i, 1 \le i \le k$, *and* $\sum_{n=1}^{\infty} |p_j(n)| < \infty, 1 \le j \le k$. *Assume further that conditions* (8.7.4) *and* (8.7.5) *hold. Then for each* λ_j *with* $|\lambda_j| \le 1$ *there corresponds a solution* y_j *of* (8.7.11) *such that* $y_j(n) \sim \lambda_j^n$.

PROOF. By Corollary 8.27, the homogeneous part of (8.7.11),

$$x(n+k) + (a_1 + p_1(n))x(n+k-1) + \cdots + (a_k + p_k(n))x(n) = 0,$$

has solutions $x_1(n), x_2(n), \ldots, x_k(n)$ with $x_j(n) \sim \lambda_j^n$. If $|\lambda_j| \le 1$, then $x_j(n)$ is bounded. Corresponding to this bounded solution $x_j(n)$ there is a solution $y_j(n)$ of (8.7.11) with $y_j(n) \sim x_j(n)$. Thus $y_j(n) \sim \lambda_j^n$. □

Example 8.44. Investigate the asymptotic behavior of solutions of the equation

$$y(n+2) - \frac{3}{2}y(n+1) + \frac{1}{2}y(n) = \frac{e^{-n}}{1 + y^2(n)}. \tag{8.7.12}$$

Solution The characteristic equation is given by

$$\lambda^2 - \frac{3}{2}\lambda + \frac{1}{2} = 0$$

with distinct roots

$$\lambda_1 = 1, \quad \lambda_2 = \frac{1}{2}.$$

Now,

$$\sum_{n=0}^{\infty} f(n,0) = \sum_{n=0}^{\infty} e^{-n} < \infty.$$

Moreover,

$$\begin{aligned}|f(n,x) - f(n,y)| &= e^{-n}\left|\frac{1}{1+x^2} - \frac{1}{1+y^2}\right| \\ &= e^{-n}\frac{|x+y|}{(1+x^2+y^2+x^2y^2)} \cdot |x-y| \\ &\leq |x-y|.\end{aligned}$$

Hence all the assumptions of Corollary 8.43 are satisfied. Consequently, (8.7.12) has two solutions $y_1(n) \sim 1$ and $y_2(n) \sim (\frac{1}{2})^n$.

Exercises 8.7

In Problems 1 through 3 investigate the asymptotic behavior of solutions of the equation $y(n+1) = A(n)y(n) + f(n, y(n))$.

1. $A(n) = \begin{pmatrix} \frac{1}{2} & 0 \\ 0 & \frac{1}{n+2} \end{pmatrix}, \qquad f(n,y) = \begin{pmatrix} e^{-n}\cos y_1 \\ 0 \end{pmatrix}.$

2. $A(n) = \begin{pmatrix} 2 & 3 \\ 3 & -1 \end{pmatrix}, \qquad f(n,y) = \begin{pmatrix} \frac{1}{n^3} \\ 0 \end{pmatrix}.$

3. $A(n) = \begin{pmatrix} 3 & 2 \\ 2 & 1 \end{pmatrix}, \qquad f(n,y) = \begin{pmatrix} y_1 n e^{-n} \\ y_2 e^{-n} \end{pmatrix}.$

4. Study the asymptotic behavior of solutions of

$$y(n+2) + \left(\frac{3}{2} + \frac{1}{n^2}\right) y(n+1) - (1+e^{-n})y(n) = \frac{\sin y(n)}{n^2}.$$

5. Study the asymptotic behavior of solutions of

$$y(n+2) - 4y(n+1) + 3y(n) = \frac{1}{n^2+y^2}, \qquad n \geq 1.$$

6. Study the asymptotic behavior of solutions of $y(n+2) + (1+e^{-n})y(n) = e^{-n}$.

7. Prove Theorem 8.41.

In Problems 8 through 10 investigate the asymptotic behavior of the difference equation.

8. $\Delta^2 y(n) + e^{-n} y^3(n) = 0.$

9. $\Delta^2 y(n) - \dfrac{1}{n^2}(|y(n)|)^{1/2} = 0.$

10. $\Delta^2 y(n) + \dfrac{1}{n^2}(y(n))^{1/3} = 0.$

In Problems 11 and 12 we consider the nonlinear equation

$$x(n+1) = f(x(n)) \tag{8.7.13}$$

with $f(0) = 0,\ 0 < |f'(0)| < 1$, and f' continuous at 0.

11. Show that there exist $\delta > 0$ and $0 < \alpha < 1$ such that

$$|x(n)| \le \alpha^n |x(0)|, \quad n \in \mathbb{Z}^+,$$

for all solutions $x(n)$ of (8.7.13) with $|x(0)| < \delta$.

12. Suppose that f'' is bounded near 0, and $|f'(x_0)| < 1$ for $|x_0| < \delta$. Prove that for any solution $x(n)$ of (8.7.11) with $|x(0)| < \delta$ we have $x(n) \sim cx(0)(f'(0))^n$ as $n \to \infty$, where c depends on $x(0)$.

13. Use Problem 12 to find an asymptotic representation of solutions of the equation

$$x(n+1) = x(n)/(1 + x(n)), \quad x(0) = 0.1.$$

14. Find an asymptotic representation of solutions of the equation

$$u(n+1) = u(n) + \frac{1}{u(n)}.$$

8.8 Extensions of the Poincaré and Perron Theorems

8.8.1 An Extension of Perron's Second Theorem

Coffman [22] considers the nonlinear system

$$y(n+1) = Cy(n) + f(n, y(n)) \tag{8.8.1}$$

where C is a $k \times k$ matrix and $f : \mathbb{Z}^+ \times \mathbb{R}^k \to \mathbb{R}^k$ is continuous. The following result extends Perron's Second Theorem to nonlinear systems.

Theorem 8.45 [22]. *Suppose that* $\frac{\|f(n,y)\|}{\|y\|} \to 0$ *as* $(n, y) \to (\infty, 0)$. *If* $y(n)$ *is a solution of* (8.8.1) *such that* $y(n) \ne 0$ *for all large* n *and* $y(n) \to 0$ *as* $n \to \infty$, *then*

$$\boxed{\lim_{n\to\infty} \sqrt[n]{\|y(n)\|} = |\lambda_i|} \tag{8.8.2}$$

for some eigenvalue λ_i *of* A. *Moreover,* $|\lambda_i| \le 1$.

Using the above theorem, Pituk [122] was able to improve Perron's Second Theorem such that conclusion (8.2.12) is now valid for all nontrivial solutions. As a bonus we get a system version of this new result.

Consider again the perturbed linear system

$$y(n+1) = [A + B(n)]y(n) \tag{8.8.3}$$

such that A is a $k \times k$ constant matrix and

$$\lim_{n\to\infty} \|B(n)\| = 0. \tag{8.8.4}$$

Theorem 8.46 [122]. *Under condition* (8.8.4), *for any solution* $y(n)$ *of* (8.8.3), *either* $y(n) = 0$ *for all large* n *or*

$$\boxed{\lim_{n\to\infty} \sqrt[n]{\|y(n)\|} = |\lambda_i|} \tag{8.8.5}$$

for some eigenvalue λ_i *of* A.

PROOF. Let $y(n)$ be a solution of (8.8.3). Clearly if $y(N) = 0$ for some N, then $y(n) = 0$ for all $n \geq N$. Hence we assume without loss of generality that $y(n) \neq 0$ for $n \geq n_0$. Let $\lambda_1, \lambda_2, \ldots, \lambda_k$ be the eigenvalues of A. Let $\mu > \max\limits_{1\leq i\leq k} |\lambda_i|$ and let

$$z(n) = x(n)/\mu^n, \quad n \geq n_0. \tag{8.8.6}$$

Substituting in (8.8.3) yields $z(n+1) = \left[\frac{1}{\mu}A + \frac{1}{\mu}B(n)\right] z(n)$ or

$$z(n+1) = Cz(n) + E(n)z(n) \tag{8.8.7}$$

where $C = \frac{1}{\mu}A$, $E(n) = \frac{1}{\mu}B(n)$. Notice that the eigenvalues of C are $\frac{1}{\mu}\lambda_1, \frac{1}{\mu}\lambda_2, \ldots, \frac{1}{\mu}\lambda_k$, where λ_i, $1 \leq i \leq k$, are the eigenvalues of A. Moreover $\left|\frac{1}{\mu}\lambda_i\right| < 1$, for $1 \leq i \leq k$. By virtue of (8.8.4), $\frac{\|f(n,z)\|}{\|z\|} \leq \mu^{-1}\|B(n)\| \to 0$ as $n \to \infty$. Hence by Corollary 4.34, the zero solution of (8.8.7) is (globally) exponentially stable. Thus $z(n) \to 0$ as $n \to \infty$ for every solution $z(n)$ of (8.8.7). By Coffman's Theorem 8.45 we have $\lim_{n\to\infty} \sqrt[n]{\|z(n)\|} = \frac{1}{\mu}|\lambda_i|$, for some $1 \leq i \leq k$. This implies that $\lim_{n\to\infty} \sqrt[n]{\|y(n)\|} = |\lambda_i|$, for some $1 \leq i \leq k$. □

Now we specialize the preceding result to the scalar difference equation of Poincaré type

$$x(n+k) + p_1(n)x(n+k-1) + \cdots + p_k(n)x(n) = 0. \tag{8.8.8}$$

Using the l_∞-norm $\|y(n)\|_\infty = \max\{|x_i(n)| \mid 1 \leq i \leq k\}$, we obtain the following extension of Perron's Second Theorem.

Theorem 8.47. *Consider the difference equation of Poincaré type* (8.8.8). *If $x(n)$ is a solution, then either $x(n) = 0$ for all large n or*

$$\boxed{\limsup_{n\to\infty} \sqrt[n]{|x(n)|} = |\lambda_i|,} \tag{8.8.9}$$

for some characteristic root λ_i of (8.8.8). [2] [3]

PROOF. We first convert (8.8.8) to a system of the form (8.8.3), where $y(n) = (x(n), x(n+1), \ldots, x(n+k-1))^T$. Notice that $\|y(n)\|_\infty = \max\{|x(n)|, |x(n+1)|, \ldots, |x(n+k-1)|\}$. Conclusion (8.8.9) follows from Theorem 8.46. □

Using the l_1-norm $\|y(n)\|_1$ we obtain the following interesting result.

Theorem 8.48. *If $x(n)$ is a solution of the difference equation of Poincaré type* (8.8.8), *then either $x(n) = 0$ for all large n or*

$$\lim_{n\to\infty} \sqrt[n]{|x(n)| + |x(n+1)| + \cdots + |x(n+k-1)|}.$$

Remarks:

(i) The conclusion (8.8.9) cannot be improved, the lim sup cannot be replaced by lim as shown by considering the equation

$$x(n+2) - x(n) = 0.$$

This equation has the solution $x(n) = 1 + (-1)^n$, where $\limsup_{n\to\infty} \sqrt[n]{|x(n)|} = 1 = |\lambda_1| = |\lambda_2|$. However, $\lim_{n\to\infty} \sqrt[n]{|x(n)|}$ does not exist.

(ii) For a direct proof of Theorem 8.46 without the use of Coffman's Theorem, the reader may consult the paper by Pituk [122].

8.8.2 *Poincaré's Theorem Revisited*

The main objective of this subsection is to extend Poincaré's Theorem to systems of the form (8.8.3). So as a by-product we prove Poincaré's Theorem for scalar difference equations. The exposition here is based on a recent paper by Abu-Saris, Elaydi, and Jang [1]. The following definitions were developed in a seminar at Trinity University led by Ulrich Krause of the University of Bremen and the author.

---

[2] This was conjectured by U. Krause and S. Elaydi in a seminar at Trinity University.

[3] Mihály Pituk, a Professor of Mathematics at the University of Veszprém, received the best paper award (2002) from the International Society of Difference Equations for proving Theorem 8.47 and other related results.

Definition 8.49. Let $y(n)$ be a solution of (8.8.3). Then $y(n)$ is said to be of:

(1) Weak Poincaré type (WP) if

$$\boxed{\lim_{n\to\infty} \sqrt[n]{\|y(n)\|} = |\lambda|}$$

for some eigenvalue λ of A.

(2) Poincaré type (P) if

$$\boxed{\lim_{n\to\infty} \frac{\|y(n+1)\|}{\|y(n)\|} = |\lambda|}$$

for some eigenvalue λ of A.

(3) Strong Poincaré type (SP) if

$$\boxed{\lim_{n\to\infty} \frac{y(n)}{\lambda^n} = C}$$

for some eigenvalue λ of A and a nonzero vector C.

(4) Ergodic Poincaré type (EP) if

$$\boxed{\lim_{n\to\infty} \frac{y(n)}{\|y(n)\|} = v}$$

for some eigenvector v of A.

The following examples [1] illustrate the interrelationship among the above concepts.

Example 8.50. Consider the system

$$y(n+1) = \begin{pmatrix} 1 & 2 \\ 0 & 1 \end{pmatrix} y(n), \quad n \geq 0.$$

Then

$$y(n) = \alpha(-1)^n \begin{pmatrix} 1 \\ 0 \end{pmatrix} + \beta \begin{pmatrix} 1 \\ 1 \end{pmatrix} = \begin{pmatrix} \beta + \alpha(-1)^n \\ \beta \end{pmatrix}$$

is a solution. Notice that

$$\lim_{n\to\infty} \sqrt[n]{\|y(n)\|} = 1$$

but

$$\lim_{n\to\infty} \frac{\|y(n+1)\|}{\|y(n)\|} = \begin{cases} \beta/(\beta+\alpha) & \text{if } n \text{ is even,} \\ (\beta+\alpha)/\beta & \text{if } n \text{ is odd.} \end{cases}$$

Hence $y(n)$ is weak Poincaré but not Poincaré.

Example 8.51. Consider the system

$$y(n+1) = \begin{pmatrix} -\dfrac{n+1}{2n} & 0 \\ 0 & 1 \end{pmatrix} y(n), \quad n \geq 1, \quad y(1) = \begin{pmatrix} 1 \\ 0 \end{pmatrix}.$$

The solution is given by

$$y(n) = \frac{(-1)^{n-1} n}{2^{n-1}} \begin{pmatrix} 1 \\ 0 \end{pmatrix}, \quad n \geq 1.$$

Notice that $\lim_{n\to\infty} \frac{\|y(n+1)\|}{\|y(n)\|} = \frac{1}{2}$, where $-\frac{1}{2}$ is an eigenvalue. However, $\lim_{n\to\infty} \frac{y(n)}{\|y(n)\|} = \lim_{n\to\infty} (-1)^{n-1} \begin{pmatrix} 1 \\ 0 \end{pmatrix}$ does not exist. Thus $y(n)$ is Poincaré but not ergodic Poincaré.

Example 8.52. Contemplate the system

$$y(n+1) = \begin{pmatrix} \dfrac{n+1}{n} & 0 \\ 0 & 1 \end{pmatrix} y(n), \quad n \geq 1, \quad y(1) = \begin{pmatrix} 1 \\ 0 \end{pmatrix}.$$

The solution is given by

$$y(n) = n \begin{pmatrix} 1 \\ 0 \end{pmatrix}, \quad n \geq 1.$$

Notice that $\frac{y(n)}{\|y(n)\|} = \begin{pmatrix} 1 \\ 0 \end{pmatrix}$, where $\begin{pmatrix} 1 \\ 0 \end{pmatrix}$ is an eigenvector that corresponds to the eigenvalue 1. However $\lim_{n\to\infty} \frac{y(n)}{1^n}$ diverges. Hence $y(n)$ is ergodic Poincaré but not strong Poincaré.

To facilitate the proof of the main result we present a definition and two lemmas.

Definition 8.53. A solution $y(n) = (y_1(n), y_2(n), \ldots, y_i(n))^T$ of (8.8.3) is said to have the *index for maximum property* (IMP) if there exists an index $r \in \{1, 2, \ldots, k\}$ such that, for sufficiently large n,

$$\boxed{\|y(n)\| = \max_{1\leq i\leq k} |y_i(n)| = |y_r(n)|.}$$

Observe that solutions in Examples 8.51 and 8.52 possess the IMP, while the solution in Example 8.50 does not possess the IMP.

Lemma 8.54. *Suppose that* $A = \text{diag}(\lambda_1, \lambda_2, \ldots, \lambda_k)$ *such that* $0 < |\lambda_1| < |\lambda_2| < \cdots < |\lambda_k|$ *and* (8.8.4) *holds. Then every solution of* (8.8.3) *possesses the IMP.*

PROOF. Since $\lim_{n\to\infty}\|B(n)\| = 0$, for any $\varepsilon > 0$, there exists $N_1 > 0$ such that $\|B(n)\| = \max_{1\le i\le k}\sum_{j=1}^{k}|b_{ij}(n)| < \varepsilon$ for $n \ge N_1$. We choose $\varepsilon > 0$ such that

$$\frac{|\lambda_i| + \varepsilon}{|\lambda_j| - \varepsilon} < 1 \quad \text{for } 1 \le i < j \le k.$$

Let $y(n)$ be a nontrivial solution of (8.8.3) and let r_n be the first index such that

$$\|y(n)\| = |y_{r_n}(n)|.$$

We claim that r_n is nondecreasing. To see this suppose that $r_{n+1} < r_n$, then

$$|y_i(n+1)| \le |\lambda_i||y_i(n)| + \varepsilon|y_{r_n}(n)|,$$
$$|y_i(n+1)| \ge |\lambda_i||y_i(n)| - \varepsilon|y_{r_n}(n)|,$$

for all $n \ge N_1$. This implies that

$$\begin{aligned}\frac{|y_{r_{n+1}}(n+1)|}{|y_{r_n}(n+1)|} &\le \frac{|\lambda_{r_{n+1}}|\,|y_{r_{n+1}}(n)| + \varepsilon|y_{r_n}(n)|}{|\lambda_{r_n}|\,|y_{r_n}(n)| - \varepsilon|y_{r_n}(n)|} \\ &= \frac{|\lambda_{r_{n+1}}|\,|y_{r_{n+1}}(n)|/|y_{r_n}(n)| + \varepsilon}{|\lambda_{r_n}| - \varepsilon} \\ &\le \frac{|\lambda_{r_{n+1}}|| + \varepsilon}{|\lambda_{r_n}| - \varepsilon} < 1\end{aligned}$$

which contradicts the definition of r_{n+1}. Since r_n assumes only finitely many values, the result follows. □

Lemma 8.55. *Let* $\lim_{n\to\infty}\|B(n)\| = 0$. *Suppose that* $A = \text{diag}(\lambda_1, \ldots, \lambda_k)$ *such that* $0 < |\lambda_1| \le \cdots \le |\lambda_k|$. *Then every nonzero solution* $y(n)$ *of* (8.8.3) *that has the IMP with* $\|y(n)\| = |y_r(n)|$ *for all large n satisfies*

$$\lim_{n\to\infty}\frac{|y_j(n)|}{|y_r(n)|} = 0 \quad \textit{for } |\lambda_j| \ne |\lambda_r|.$$

PROOF. Let $y(n)$ be a nonzero solution of (8.8.3) that has the IMP. Since $\lim_{n\to\infty}\|B(n)\| = 0$, for any $\varepsilon > 0$ there exists $N > 0$ such that $\|B(n)\| < \varepsilon$ and $\|y(n)\| = |y_r(n)|$ for $n \ge N$. We choose $\varepsilon > 0$ so that $|\lambda_i|/(|\lambda_j| - \varepsilon) < 1$ for $1 \le i < j \le k$ and $|\lambda_i| \ne |\lambda_j|$. Observe that, for $n \ge N$,

$$|y_i(n+1)| \le |\lambda_i||y_i(n)| + \varepsilon|y_{r_n}(n)|,$$
$$|y_i(n+1)| \ge |\lambda_i||y_i(n)| - \varepsilon|y_{r_n}(n)|,$$

for $1 \le i \le k$. Suppose that $|\lambda_j| \ne |\lambda_r|$. We first consider the case when $j > r$. Let

$$s = \sup_n \frac{|y_j(n)|}{|y_r(n)|}.$$

Then there exists a subsequence n_i such that

$$\lim_{n_i\to\infty} \frac{|y_j(n_i)|}{|y_r(n_i)|} = s.$$

Observe that

$$\frac{|y_j(n_i+1)|}{|y_r(n_i+1)|} \geq \frac{|\lambda_j|\,|y_j(n_i)| - \varepsilon|y_r(n_i)|}{(|\lambda_r|+\varepsilon)|y_r(n_i)|} = \frac{|\lambda_j|\,|y_j(n_i)|/|y_r(n_i)| - \varepsilon}{|\lambda_r|+\varepsilon}$$

for $n_i > N$. Therefore,

$$s \geq \frac{|\lambda_j|s - \varepsilon}{|\lambda_j| + \varepsilon}$$

and, consequently,

$$s \leq \frac{\varepsilon}{|\lambda_j| - |\lambda_r| - \varepsilon}$$

for all sufficiently small ε. This implies that $s = 0$ and the assertion is shown.

On the other hand, if $j < r$, then

$$\frac{|y_j(n+1)|}{|y_r(n+1)|} \leq \frac{|\lambda_j|\,|y_j(n)| + \varepsilon|y_r(n)|}{(|\lambda_r|-\varepsilon)|y_r(n)|} = \left(\frac{|\lambda_j|}{|\lambda_r|-\varepsilon}\right)\frac{|y_j(n)|}{|y_r(n)|} + \frac{\varepsilon}{|\lambda_r|-\varepsilon}$$

for $n > N$. Thus

$$\frac{|y_j(n)|}{|y_r(n)|} \leq \left(\frac{|\lambda_j|}{|\lambda_r|-\varepsilon}\right)^{n-N}\frac{|y_j(N)|}{|y_r(N)|} + \left[\frac{1-\left(\frac{|\lambda_j|}{|\lambda_r|-\varepsilon}\right)^{n-N}}{1-\frac{|\lambda_j|}{|\lambda_r|-\varepsilon}}\right]\frac{\varepsilon}{|\lambda_r|-\varepsilon}$$

and as a result

$$\limsup_{n\to\infty} \frac{|y_j(n)|}{|y_r(n)|} \leq \frac{\varepsilon}{|\lambda_r| - |\lambda_j| - \varepsilon}$$

for all sufficiently small ε. This implies that

$$\limsup_{n\to\infty} \frac{|y_j(n)|}{|y_r(n)|} = 0$$

and complete the proof. □

By using Lemmas 8.54 and 8.55 we present a sufficient condition for which (8.8.3) has the Poincaré property.

Theorem 8.56. *Suppose that the eigenvalues of A have distinct moduli and* $\lim_{n\to\infty}\|B(n)\| = 0$. *Then* (8.8.3) *possesses the Poincaré property P.*

PROOF. We may assume, without loss of generality, that A is in diagonal form, i.e., $A = \mathrm{diag}(\lambda_1, \lambda_2, \ldots, \lambda_k)$, where $0 < |\lambda_1| < \cdots < |\lambda_k|$. Let $y(n)$ be a nontrivial solution of (8.8.3). It follows from Lemma 8.54 that

$$\|y(n)\| = |y_r(n)|$$

for all large n, for some $1 \le r \le k$. Moreover, Lemma 8.55 implies

$$\lim_{n\to\infty} \frac{|y_i(n)|}{|y_r(n)|} = 0$$

for $1 \le i \le k$ such that $i \ne r$. Therefore, if $i \ne r$, then

$$\lim_{n\to\infty} \frac{y_i(n+1)}{|y_r(n)|} = \lim_{n\to\infty} \left[\lambda_i \frac{y_i(n)}{|y_r(n)|} + \sum_{j=1}^{k} b_{ij}(n) \frac{y_i(n)}{|y_r(n)|} \right] = 0$$

and if $i = 1$, we have

$$\lim_{n\to\infty} \frac{y_r(n+1) - \lambda_r y_r(n)}{|y_r(n)|} = \lim_{n\to\infty} \sum_{j=1}^{k} b_{ij}(n) \frac{y_j(n)}{|y_r(n)|} = 0.$$

Consequently,

$$\lim_{n\to\infty} \frac{\|y(n+1) - \lambda_r y(n)\|}{\|y(n)\|} = 0.$$

Since

$$\frac{\|y(n+1)\|}{\|y(n)\|} - |\lambda_r| \le \frac{\|y(n+1) - \lambda_r\|}{\|y(n)\|},$$

it follows that

$$\lim_{n\to\infty} \frac{\|y(n+1)\|}{\|y(n)\|} = |\lambda_r|.$$

The proof is now complete. □

As an immediate consequence of the above result, we obtain the original Poincaré' Theorem.

Proof of Theorem 8.9. Write equation (8.2.7) as a system of the form (8.8.3). Then a solution $y(n)$ of (8.8.3) is of the form $y(n) = (x(n), x(n+1), \ldots, x(n+k-1))^T$. By Theorem 8.56 we have

$$\lim_{n\to\infty} \frac{\|y(n+1)\|}{\|y(n)\|} = |\lambda| \tag{8.8.10}$$

for some eigenvalue λ. By Lemma 8.54, there exists $r \in \{1, 2, \ldots, k\}$ such that

$$\|y(n)\| = |y_r(n)| = |x(n+r-1)|.$$

Substituting (8.8.10) yields

$$\lim_{n\to\infty} \frac{|x(n+r)|}{|x(n+r-1)|} = \lim_{n\to\infty} \frac{|x(n+1)|}{|x(n)|} = |\lambda|$$

where λ is a characteristic root of (8.2.7). Since $\frac{x(n+1)}{x(n)} \leq \left|\frac{x(n+1)}{x(n)}\right|$, $\lim_{n\to\infty} \frac{x(n+1)}{x(n)}$ exists. Hence $\lim_{n\to\infty} \frac{x(n+1)}{x(n)} = \lambda$. (See Exercises 8.2, Problem 9.) □

Term Projects 8.8

1. Find the relationships among the notions of WP, P, SP, and EP for both scalar equations and systems of Poincaré type.
2. Find sufficient conditions for EP and SP.
3. Extend Perron's First Theorem to systems.

9

Applications to Continued Fractions and Orthogonal Polynomials

9.1 Continued Fractions: Fundamental Recurrence Formula

Continued fractions are intimately connected with second-order difference equations. Every continued fraction may be associated with a second-order difference equation; and conversely, every homogeneous second-order difference equation may be derived from some continued fraction. The first point of view is useful for computing continued fractions, the second for computing the minimal solutions.

Let $\{a_n\}_{n=1}^{\infty}$ and $\{b_n\}_{n=0}^{\infty}$ be two sequences of real or complex numbers. A continued fraction is of the form

$$b_0 + \cfrac{a_1}{b_1 + \cfrac{a_2}{b_2 + \ddots + \cfrac{a_n}{b_n + \ddots}}}$$

or, in compact form,

$$b_0 + \frac{a_1}{b_1 +}\;\frac{a_2}{b_2 +}\;\frac{a_3}{b_3 +}\cdots \tag{9.1.1}$$

or

$$b_0 + K(a_n/b_n)$$

or

$$b_0 + K_{n=1}^{\infty}(a_n/b_n).$$

The nth *approximant* of a continued fraction is defined as

$$C(n) = \frac{A(n)}{B(n)} = b_0 + K_{j=1}^{n}(a_j/b_j)$$
$$= b_0 + \frac{a_1}{b_1 +} \frac{a_2}{b_2 +} \cdots \frac{a_n}{b_n}. \tag{9.1.2}$$

The sequences $A(n)$ and $B(n)$ are called the nth *partial numerator* and the nth *partial denominator* of the continued fraction, respectively. It is always assumed that $\frac{A(n)}{B(n)}$ is in reduced form, that is, $A(n)$ and $B(n)$ are coprime (i.e., they have no common factors). An alternative way of defining a continued fraction is through the use of a "Möbius" transformation, which is defined as

$$t_0(u) = b_0 + u, \quad t_n(u) = \frac{a_n}{b_n + u}, \quad n = 1,\ 2,\ 3,\ \ldots. \tag{9.1.3}$$

Then it is easy to see that the nth approximant is given by

$$C(n) = (t_0 \circ t_1 \circ \cdots \circ t_n)(0). \tag{9.1.4}$$

Definition 9.1. The continued fraction (9.1.1) is said to converge to a finite limit L if $\lim_{n\to\infty} C(n) = L$, and it is said to diverge otherwise.

Next we show that both the nth partial numerator $A(n)$ and the nth partial denominator $B(n)$ of the continued fraction (9.1.1) satisfy a second-order difference equation commonly known in the literature as the *fundamental recurrence formula* for continued fractions. The explicit statement of this important result now follows.

Theorem 9.2. *Consider the continued fraction* $b_0 + K_{n=1}^{\infty}(a_n/b_n)$ *with* n*th approximant* $C(n) = A(n)/B(n)$. *Then* $A(n)$ *and* $B(n)$ *satisfy, respectively, the difference equations*

$$\boxed{\begin{aligned} A(n) &= b_n A(n-1) + a_n A(n-2),\ A(-1) = 1,\ A(0) = b_0, \\ B(n) &= b_n B(n-1) + a_n B(n-2),\ B(-1) = 0,\ B(0) = 1. \end{aligned}} \tag{9.1.5}$$

PROOF. The proof of (9.1.5) will be established using mathematical induction on n.

Observe that from (9.1.5) we obtain

$$A(1) = b_1 A(0) + a_1 A(-1) = b_1 b_0 + a,$$
$$B(1) = b_1 B(0) + a_1 B(-1) = b_1.$$

Hence, (9.1.5) is valid for $n = 1$. Next, we assume that it is true for $n = m$, that is,

$$A(m) = b_m A(m-1) + a_m A(m-2), \quad A(-1) = 1,$$
$$B(m) = b_m B(m-1) + a_m B(m-2), \quad B(-1) = 0.$$

Now $A(m+1)/B(m+1)$ is obtained from $A(m)/B(m)$ by replacing b_m by $b_m + \frac{a_{m+1}}{b_{m+1}}$. Then we can write $A(m+1)/B(m+1) = A^*(m)/B^*(m)$, where

$$\begin{aligned} A^*(m) &= \left(b_m + \frac{a_{m+1}}{b_{m+1}}\right) A(m-1) + a_m A(m-2) \\ &= b_{m+1}^{-1} \left(b_{m+1}\ A(m) + a_{m+1}\ A(m-1)\right). \end{aligned}$$

Similarly,

$$B^*(m) = b_{m+1}^{-1} \left(b_{m+1}\ B(m) + a_{m+1}\ B(m-1)\right).$$

Hence

$$C(m+1) = \frac{A(m+1)}{B(m+1)} = \frac{b_{m+1}A(m) + a_{m+1}A(m-1)}{b_{m+1}B(m) + a_{m+1}B(m-1)},$$

which establishes (9.1.5), and the proof of the theorem is now complete. □

The converse of the preceding theorem is also true. In other words, every homogeneous second-order difference equation gives rise to an associated continued fraction.

Suppose now we are given the difference equation

$$x(n) - b_n x(n-1) - a_n x(n-2) = 0, \quad a_n \neq 0 \quad \text{for } n \in \mathbb{Z}^+. \tag{9.1.6}$$

Dividing (9.1.6) by $x(n-1)$ and then setting $y(n) = \frac{x(n)}{x(n-1)}$ yields $y(n) - b_n = \frac{a_n}{y(n-1)}$.

Hence,

$$y(n-1) = \frac{a_n}{-b_n + y(n)}.$$

Applying this formula repeatedly, with n successively increasing, we obtain

$$y(n-1) = \frac{a_n}{-b_n+} \frac{a_{n+1}}{-b_{n+1}+} \frac{a_{n+2}}{-b_{n+2}+\dots}.$$

In particular, when $n = 1$, we have

$$y(0) = \frac{x(0)}{x(-1)} = \frac{a_1}{-b_1+} \frac{a_2}{-b_2+} \frac{a_3}{-b_3+}. \tag{9.1.7}$$

Remark 9.3. We would like to make the following important observations concerning (9.1.7):

(a) Formula (9.1.7) is formal in the sense that it does not tell us whether the continued fraction $K(a_n/-b_n)$ converges or diverges.

(b) Even if the continued fraction $K(a_n/-b_n)$ converges, formula (9.1.7) does not show us how to pick the particular solution $x(n)$ for which $x(0)/x(-1)$ is the limit of the continued fraction.

(c) It is not difficult to show that if $K(a_n/-b_n)$ converges to $x(0)/x(-1)$, then $K(a_n/b_n)$ converges to $-x(0)/x(-1)$.

A Formula for $C(n)$

We end this section by providing a formula for computing the nth approximant $C(n) = \frac{A(n)}{B(n)}$ of the continued fraction $b_0 + K(a_n/b_n)$.

To find the formula, we multiply the first equation in (9.1.5) by $B(n-1)$ and the second by $A(n-1)$ and then subtract one from the other. This yields

$$A(n)B(n-1) - B(n)A(n-1) = -a_n[A(n-1)B(n-2) - B(n-1)A(n-2)],$$

which is equivalent to

$$u(n) = -a_n u(n-1), \quad u(0) = -1,$$

where $u(n) = A(n)B(n-1) - B(n)A(n-1)$. Hence,

$$u(n) = A(n)B(n-1) - B(n)A(n-1) = (-1)^{n+1} a_1 a_2 \cdots a_n, \quad n \geq 1. \tag{9.1.8}$$

Dividing both sides by $B(n)B(n-1)$ yields

$$\Delta \left(\frac{A(n-1)}{B(n-1)} \right) = \frac{(-1)^{n+1} a_1 a_2 \cdots a_n}{B(n-1)B(n)}. \tag{9.1.9}$$

Taking the antidifference Δ^{-1} of both sides of (9.1.9), we obtain (see formula (2.1.16))

$$C(n-1) = \frac{A(n-1)}{B(n-1)} = \frac{A(0)}{B(0)} + \sum_{k=1}^{n-1} \frac{(-1)^{k+1} a_1 a_2 \cdots a_k}{B(k-1)B(k)}.$$

This produces the desired formula

$$\boxed{C(n) = b_0 + \sum_{k=1}^{n} \frac{(-1)^{k+1} a_1 a_2 \cdots a_k}{B(k-1)B(k)}.} \tag{9.1.10}$$

9.2 Convergence of Continued Fractions

Two continued fractions $K(a_n/b_n)$ and $K(a_n^*/b_n^*)$ are said to be *equivalent*, denoted by the symbol $K(a_n/b_n) \approx K(a_n^*/b_n^*)$, if they have the same sequence of approximants.

Let $\{d_n\}_{n=1}^{\infty}$ be any sequence of nonzero complex numbers. Then the Möbius transformation $t_n(u) = \frac{a_n}{b_n+u}$ can be represented as $t_n(u) = \frac{d_n a_n}{d_n b_n + d_n u}$; which may be repeated as a composition of two transformations $t_n = s_n \circ r_n$, where

$$s_n(u) = \frac{d_n a_n}{d_n b_n + u} \quad \text{and} \quad r_n(u) = d_n u.$$

Hence we have

$$t_1 \circ t_2 \circ \cdots \circ t_n = s_1 \circ r_1 \circ s_2 \circ r_2 \circ \cdots \circ s_n \circ r_n$$
$$= s_1 \circ (r_1 \circ s_2) \circ (r_2 \circ s_3) \circ \cdots \circ (r_{n-1} \circ s_n) \circ r_n.$$

Define $t_n^*(u) := r_{n-1} \circ s_n(u) = \frac{d_{n-1}}{d_n}\frac{d_n a_n}{b_n+u}$. Then if $d_0 := 1$,

$$C(n) = (t_1 \circ t_2 \circ \cdots \circ t_n)(0) = (t_1^* \circ t_2^* \circ \cdots \circ t_n^*)(0).$$

This yields the important equivalence relation

$$K(a_n/b_n) \approx K\left(\frac{d_{n-1}d_n a_n}{d_n b_n}\right), \tag{9.2.1}$$

which holds for any arbitrary sequence of nonzero complex numbers $d_0 = 1$, $d_1, d_2, \ldots$.

Observe that if we choose the sequence $\{d_n\}$ such that $d_n b_n = 1$, then (9.2.1) becomes

$$K(a_n/b_n) \approx K\left(\frac{b_{n-1}b_n a_n}{1}\right). \tag{9.2.2}$$

Similarly, one can show that

$$K(a_n/b_n) \approx K(1/b_n d_n), \tag{9.2.3}$$

where $d_1 = \frac{1}{a_1}$, $d_n = \frac{1}{a_n d_{n-1}}$. Hence

$$d_{2n} = \frac{a_1 a_3 \cdots a_{2n-1}}{a_2 a_4 \cdots a_{2n}}, \quad d_{2n+1} = \frac{a_2 a_4 \cdots a_{2n}}{a_1 a_3 \cdots a_{2n+1}}, \tag{9.2.4}$$

(Exercises 9.1 and 9.2, Problem 8).

We are now ready to give the first convergence theorem.

Theorem 9.4. *Let $b_n > 0$, $n = 1, 2, 3, \ldots$. Then the continued fraction $K(1/b_n)$ is convergent if and only if the infinite series $\sum_{n=1}^{\infty} b_n$ is divergent.*

PROOF. From (9.1.10) we have

$$K(1/b_n) = \sum_{r=1}^{\infty} \frac{(-1)^{r+1}}{B(r-1)B(r)}. \tag{9.2.5}$$

Hence $K(1/b_n)$ converges if and only if the alternating series on the right-hand side of (9.2.5) converges. Now the fundamental recurrence formula for $K(1/b_n)$ is

$$B(n) = B(n-2) + b_n B(n-1), \quad B(0) = 1, \quad B(1) = b_1. \tag{9.2.6}$$

This implies that $B(n+1) > B(n-1)$, $n = 1, 2, 3, \ldots$, and, consequently, $B(n)B(n+1) > B(n-1)B(n)$, $n = 1, 2, 3, \ldots$. Thus $\left|\frac{(-1)^{n+1}}{B(n-1)B(n)}\right|$ is monotonically decreasing. Hence the series (9.2.5) converges if and only if

$$\lim_{n\to\infty} B(n-1)B(n) = \infty. \tag{9.2.7}$$

Again from (9.2.6) we have $B(n) \geq \gamma = \min(1, b_1)$, $n = 1, 2, 3, \ldots$, and, consequently,

$$\begin{aligned} B(n-1)\; B(n) &= B(n-2)B(n-1) + b_n B^2(n-1) \\ &\geq B(n-2)B(n-1) + b_n\gamma^2 \\ &\geq (b_1 + b_2 + \cdots + b_n)\gamma^2. \end{aligned}$$

Thus if $\sum_{i=1}^{\infty} b_i$ diverges, (9.2.7) holds and $K(1/b_n)$ converges. On the other hand, we have, from (9.2.6),

$$B(n-1)+B(n) = B(n-2)+(1+b_n)B(n-1) \leq (1+b_n)[B(n-1)+B(n-2)].$$

It follows by induction that

$$B(n-1) + B(n) \leq (1+b_1)(1+b_2)\cdots(1+b_n) < e^{b_1+b_2+\cdots+b_n}.$$

Thus if $\sum_{n=1}^{\infty} b_n$ converges to L, then $B(n-1) + B(n) \leq e^L$. Therefore,

$$B(n-1)B(n) \leq \frac{1}{4}(B(n-1) + B(n))^2 \leq \frac{1}{4}e^{2L}.$$

Consequently, (9.2.7) does not hold, and hence the continued fraction diverges. □

A more general criterion for convergence was given by Pincherle in his fundamental work [120] on continued fractions. Consider again the difference equation

$$x(n) - b_n x(n-1) - a_n x(n-2) = 0, \quad a_n \neq 0 \quad \text{for } n \in \mathbb{Z}^+. \tag{9.2.8}$$

Theorem 9.5 (Pincherle). *The continued fraction*

$$\frac{a_1}{b_1+}\frac{a_2}{b_2+}\frac{a_3}{b_3+}\cdots \tag{9.2.9}$$

converges if and only if (9.2.8) has a minimal solution $\varphi(n)$, with $\varphi(0) \neq 0$. In case of convergence, moreover, one has

$$-\frac{\varphi(n-1)}{\varphi(n-2)} = \frac{a_n}{b_n+}\frac{a_{n+1}}{b_{n+1}+}\frac{a_{n+2}}{b_{n+2}+}\cdots, \quad n = 1, 2, 3, \ldots. \tag{9.2.10}$$

PROOF.

(a) Assume that the continued fraction (9.2.9) converges. Hence if $A(n)$ and $B(n)$ are the the nth partial numerator and nth partial denominator of (9.2.9), respectively, then

$$\lim_{n\to\infty} \frac{A(n)}{B(n)} = L.$$

It follows from Theorem 9.2 that $A(n)$ and $B(n)$ are solutions of (9.2.8) with $A(-1) = 1$, $A(0) = 0$, and $B(-1) = 0$, $B(0) = 1$.

We claim that

$$\varphi(n) = A(n) - LB(n) \tag{9.2.11}$$

is a minimal solution of (9.2.8).

To prove the claim, let $y(n) = \alpha A(n) + \beta B(n)$, for some scalars α and β, be any other solution of (9.2.8). Then

$$\lim_{n\to\infty} \frac{\varphi(n)}{y(n)} = \lim_{n\to\infty} \frac{A(n) - LB(n)}{\alpha A(n) + \beta B(n)} = \lim_{n\to\infty} \frac{(A(n)/B(n)) - L}{\alpha\,(A(n)/B(n)) + \beta} = 0.$$

This establishes the claim. Furthermore, $\varphi(-1) = 1 \neq 0$.

(b) Conversely, assume that (9.2.8) possesses a minimal solution $\varphi(n)$, with $\varphi(-1) \neq 0$. From (9.1.7), the associated continued fraction to (9.2.8) is $K(a_n/-b_n)$ with the nth approximant $C^*(n) = \frac{A^*(n)}{B^*(n)}$. Since $A^*(n)$ and $B^*(n)$ are two linearly independent solutions of (9.2.8) with $A^*(-1) = 1$, $A^*(0)$, $B^*(-1) = 0$, $B^*(0) = 1$ (Theorem 9.2), it follows that

$$\varphi(n) = A^*(n) - LB^*(n), \quad n \geq 0.$$

Observe that

$$0 = \lim_{n\to\infty} \frac{\varphi(n)}{B^*(n)} = \lim_{n\to\infty} \frac{A^*(n)}{B^*(n)} - L.$$

Hence

$$\lim_{n\to\infty} \frac{A^*(n)}{B^*(n)} = \frac{\varphi(0)}{\varphi(-1)}. \tag{9.2.12}$$

From Remark 9.3(c), we conclude that

$$\lim_{n\to\infty} C(n) = \lim_{n\to\infty} \frac{A(n)}{B(n)} = -\frac{\varphi(0)}{\varphi(-1)}.$$

This proves the first part of the theorem as well as (9.2.10) for $n = 1$. The proof of (9.2.10) for $n > 1$ is left to the reader as Exercises 9.1 and 9.2 Problem 5. □

The following example illustrates Theorem 9.5.

Example 9.6. Contemplate the continued fraction

$$\frac{a}{1+}\frac{a}{1+}\frac{a}{1+}\cdots, \tag{9.2.13}$$

where a is any complex number. Find conditions on a under which the continued fraction converges.

Solution

Method 1: Let $A(n)$ and $B(n)$ be the nth partial numerator and denominator, respectively. Then from (9.1.5) we have

$$A(n) - A(n-1) - aA(n-2) = 0, \quad A(-1) = 1, \quad A(0) = 0,$$
$$B(n) - B(n-1) - aB(n-2) = 0, \quad B(-1) = 1, \quad B(0) = 1.$$

The characteristic equation of either equation is given by $\lambda^2 - \lambda - a = 0$, whose roots are

$$\lambda_{1,2} = \frac{1 \pm \sqrt{1+4a}}{2}.$$

Now, if $|\lambda_1| \neq |\lambda_2|$, then the difference equation

$$x(n) - x(n-1) - ax(n-2) = 0 \qquad (9.2.14)$$

has a minimal solution and, consequently by Pincherle's theorem the continued fraction (9.2.13) converges.

Suppose that $|\lambda_2| < |\lambda_1|$. Then $\varphi(n) = \lambda_2^n$ is a minimal solution of (9.2.14). Hence by (9.2.12) the continued fraction (9.2.13) converges to $-\lambda_2 = -\frac{1}{2} + \frac{1}{2}\sqrt{1+4a}$.

On the other hand, if $a = -\frac{1}{4}$, then $\lambda_1 = \lambda_2 = \frac{1}{2}$. Thus $A(n) = c_1\left(\frac{1}{2}\right)^n + c_2 n\left(\frac{1}{2}\right)^n$. Using the initial conditions $A(-1) = 1$, $A(0) = 0$, we get $c_1 = 0$, $c_2 = -\frac{1}{2}$. Hence $A(n) = -n\left(\frac{1}{2}\right)^{n+1}$. Similarly, we obtain $B(n) = (n+1)\left(\frac{1}{2}\right)^n$. Thus

$$K(a/1) = \lim_{n\to\infty} \frac{A(n)}{B(n)} = -\frac{1}{2} = -\lambda_2.$$

Conclusion. If a is complex, $K(a/1)$ converges to $-\frac{1}{2} + \sqrt{\frac{1}{4}+a}$ if and only if $a \notin \left\{x \in \mathbb{R}: x < -\frac{1}{4}\right\}$.

Method 2: Let

$$x = \frac{a}{1+}\frac{a}{1+}\frac{a}{1+}\cdots.$$

Then

$$x = \frac{a}{1+x}, \quad x^2 + x - a = 0.$$

Hence

$$x_1 = -\frac{1}{2} + \sqrt{\frac{1}{4}+a}, \quad x_2 = -\frac{1}{2} - \sqrt{\frac{1}{4}+a},$$

are two solutions. If a is real we require that $a \geq -\frac{1}{4}$ in order for x to be real. By inspection, we conclude that the continued function converges to x_1.

Exercises 9.1 and 9.2

1. Show that

$$1 - \frac{a}{1-}\frac{a}{1-}\cdots$$

converges to $(1+\sqrt{1-4a})/2$ if $0 < 4a < 1$.

2. Prove that the continued fraction

$$\frac{1}{b_1+}\frac{1}{b_2+}\frac{1}{b_3+}\cdots$$

converges if $b_i \geq 1$ for $i = 1, 2, 3, \ldots$.

3. Discuss the convergence of the continued fraction

$$\frac{a}{b+}\frac{a}{b+}\frac{a}{b+}\cdots,$$

where a, b are complex numbers, $a \neq 0$.

4. Show that the continued fraction

$$\frac{\lambda_1}{(x-c_1)-}\frac{\lambda_2}{(x-c_2)-}\frac{\lambda_3}{(x-c_3)-}\cdots$$

is equivalent to

$$\frac{\alpha_0(x)}{1-}\frac{\alpha_1(x)}{1-}\frac{\alpha_2(x)}{1-},$$

where $\alpha_0(x) = \frac{\lambda_1}{x-c_1}, \alpha_n(x) = \frac{\lambda_{n+1}}{(c_n-x)(c_{n+1}-x)}, n = 1, 2, 3, \ldots$.

5. Prove (9.2.10) for $n > 1$.

6. [19] Prove that if $b_0 + K_{n=1}^{\infty}(a_n/b_n) = L \neq 0$, then

$$K_{n=0}^{\infty}(a_n/b_n) = \frac{a_0}{L}.$$

7. Consider the continued fraction $b_0 + K(a_n/b_n)$ and let $A(n)$ and $B(n)$ be the nth partial numerator and denominator, respectively.

Show that

$$A(n) = \begin{vmatrix} b_0 & 1 & & & & & 0 \\ a_1 & b_1 & 1 & & & & \\ 0 & a_2 & b_2 & 1 & & & \\ & & \cdots & \cdots & \cdots & & \\ & & & a_{n-1} & b_{n-1} & 1 & \\ 0 & & & & a_n & b_n & \end{vmatrix},$$

$$B(n) = \begin{vmatrix} b_1 & 1 & & & & 0 \\ a_2 & b_2 & 1 & & & \\ & a_3 & b_3 & 1 & & \\ & \cdots & \cdots & \cdots & & \\ & & a_{n-1} & b_{n-1} & 1 & \\ 0 & & & a_n & b_n & \end{vmatrix}.$$

8. Prove (9.2.4).

9. Let $\{t_n\}$ be a sequence of Möbius transformations defined as

$$t_n(u) = \frac{a_n}{b_n + u}, \quad a_n \neq 0, \quad n = 0, 1, 2, \ldots.$$

Let $T_0(u) = t_0(u), T_n(u) = T_{n-1}(t_n(u))$. Show that

$$T_n(u) = \frac{A(n) + A(n-1)u}{B(n) + B(n-1)u}$$

and $A(n)B(n-1) - A(n-1)B(n) \neq 0$, $n = 0, 1, 2, \ldots$, where $A(n)$ and $B(n)$ are the nth partial numerator and denominator of $b_0 + K(a_n/b_n)$, respectively.

10. Let $\{A(n)\}$ and $\{B(n)\}$ be sequences of complex numbers such that

$$A(-1) = 1, \quad A(0) = b_0, \quad B(-1) = 0, \quad B(0) = 1,$$

and

$$A(n)B(n-1) - A(n-1)B(n) \neq 0, \quad n = 0, 1, 2, \ldots.$$

(a) Show that there exists a uniquely determined continued fraction $b_0 + K(a_n/b_n)$ with nth partial numerator $A(n)$ and nth partial denominator $B(n)$.

(b) Show that

$$b_0 = A(0), \quad b_1 = B(1), \quad a_1 = A(1) - A(0)B(1),$$

$$a_n = \frac{A(n-1)B(n) - A(n)B(n-1)}{A(n-1)B(n-2) - A(n-2)B(n-1)},$$

$$b_n = \frac{A(n)B(n-2) - A(n-2)B(n)}{A(n-1)B(n-2) - A(n-2)B(n-1)}.$$

11. Show that the nth partial denominator of the continued fraction

$$1 - \frac{1}{1-}\frac{a_1}{1-}\frac{(1-a_1)a_2}{1-}\cdots$$

is

$$B(n) = (1-a_1)(1-a_2)\cdots(1-a_{n-1}), \quad n \geq 1.$$

12. [19] Let $\alpha_n = (1-a_{n-1})a_n$, $a_0 = 0$, $0 < a_n < 1$, $n = 1, 2, 3, \ldots$. Prove that the continued fraction

$$1 - \frac{\alpha_1}{1-}\frac{\alpha_2}{1-}\frac{\alpha_3}{1-}\cdots$$

converges to $(1+L)^{-1}$, where

$$L = \sum_{n=1}^{\infty} \frac{a_1\, a_2 \cdots a_n}{(1-a_1)(1-a_2)\cdots(1-a_n)}.$$

13. Let $\beta_n = (1-b_{n-1})b_n$, $0 \leq b_0 < 1$, $0 < b_n < 1$, for $n \geq 1$. Prove that

$$1 - \frac{\beta_1}{1-}\frac{\beta_2}{1-}\frac{\beta_3}{1-}\cdots$$

converges to $(b_0 + \frac{1-b_0}{1+B})$, where

$$B = \sum_{n=1}^{\infty} \frac{b_1 b_2 \cdots b_n}{(1-b_1)(1-b_2)\cdots(1-b_n)}.$$

14. Let $a_n > 0$, $b_n > 0$, $n = 1, 2, 3, \ldots$, and let

$$\sum_{n=1}^{\infty} \sqrt{\frac{b_n b_{n+1}}{a_{n+1}}} = \infty.$$

Show that the continued fraction

$$\frac{a_1}{b_1+}\frac{a_2}{b_2+}\frac{a_3}{b_3+}\cdots$$

converges.

15. Show that the continued fraction

$$\frac{1^k}{b+}\frac{2^k}{b+}\frac{3^k}{b+}\cdots$$

converges for $k \leq 2$ and diverges for $k > 2$.

*16. (Term Project, [31]). Consider the continued fraction

$$\tau_2(a) = 2 - \frac{a}{2-}\frac{a}{2-}\cdots.$$

Let $t_1(a) = 2$, $t_2(a) = 2 - a/2$, $t_3(a) = 2 - a/(2 - a/2), \ldots$ be the approximant sequence.

(a) Show that $t_{n+1}(a) = 2 - \frac{a}{t_n(a)}$, $t_1 = 2$, $n = 1, 2, 3, \ldots$.

(b) Show that if $a \leq 1$, the continued fraction converges to $1+\sqrt{1-a}$.

(c) A number a is said to be periodic (of period n) if $t_{n+k}(a) = t_k$ for $k = 1, 2, 3, \ldots$.

Show that if a is of period n, then $t_{n-1}(a) = 0$.

(d) Let $P_n(x) = \sum_{k=0}^{\lfloor \frac{n-1}{2} \rfloor} (-1)^k \binom{n}{2k+1} x^k$, where $\lfloor \ \rfloor$ denotes the greatest integer function. Show that

$$P_{n+2}(x) = 2P_{n+1}(x) - (x+1)P_n(x), \quad P_1(x) = 1, \quad P_2(x) = 2.$$

(e) Prove that $t_n(a) = P_{n+1}(a-1)/P_n(a-1), n = 1, 2, 3, \ldots$.

(f) Prove that a number a is periodic if and only if $(a - 1)$ is a zero of the polynomial $P_n(x)$.

9.3 Continued Fractions and Infinite Series

Our main objective in this section is to show that every infinite series can be represented by a continued fraction and vice versa. Let $\{c_n\}$ be a sequence of complex numbers, with $c_n \neq 0$, $n = 1, 2, 3, \ldots$, and let

$$u_n = \sum_{k=0}^{n} c_k, \quad n = 0, 1, 2, \ldots.$$

Let $\rho_0 = c_0$, $\rho_1 = c_1$, $\rho_n = c_n/c_{n-1}$. Then $c_0 = \rho_0$, $c_n = \rho_1\rho_2\cdots\rho_n$. Moreover,

$$\rho_0 + \sum_{k=1}^{n} \rho_1\rho_2\cdots\rho_k = \rho_0 + \frac{\rho_1}{1-}\frac{\rho_2}{(1+\rho_2)-}\frac{\rho_3}{(1+\rho_3)-}\cdots\frac{\rho_n}{1+\rho_n}.$$

Hence

$$\sum_{k=0}^{\infty} c_k = b_0 + K(a_n/b_n), \tag{9.3.1}$$

where $b_0 = c_0$, $a_1 = c_1$, $b_1 = 1$, and

$$a_n = -\frac{c_n}{c_{n-1}}, \quad b_n = 1 + \frac{c_n}{c_{n-1}}, \quad n = 2, 3, 4, \ldots.$$

To illustrate the above method observe that

$$\sum_{k=0}^{\infty} c_k z^k = c_0 + \frac{c_1 z}{1-} \frac{c_2 z/c_1}{1 + \frac{c_2 z}{c_1} -} \frac{c_3 z/c_2}{1 + \frac{c_3 z}{c_2}} \cdots.$$

Here is a more interesting example.

Example 9.7. Consider the Riemann zeta function, defined by

$$\zeta(k) = \sum_{r=1}^{\infty} r^{-k} = 1 + 2^{-k} + 3^{-k} + \cdots, \quad k = 2, 3, 4, \ldots.$$

Then $b_0 = 0$, $a_1 = 1$, $b_1 = 1$,

$$a_n = -\left(\frac{n-1}{n}\right)^k, \quad \text{and} \quad b_n = 1 + \left(\frac{n-1}{n}\right)^k.$$

Thus

$$\zeta(k) = K(a_n/b_n).$$

If we let $u(n) = K_{j=1}^{n}(a_j/b_j)$, then it follows from (9.1.6) that

$$u(n+2) - \left(1 + \left(\frac{n+1}{n+2}\right)^k\right) u(n+1) + \left(\frac{n+1}{n+2}\right)^k u(n) = 0. \tag{9.3.2}$$

An equivalent representation of $\zeta(k)$ may be obtained using (9.3.2):

$$\zeta(k) = K(1/b_n d_n),$$

$$d_1 = 1, \quad d_{2n} = \frac{(2/3)^k (4/5)^k \cdots (2n-2/(2n-1))^k}{(1/2)^k (3/4)^k \cdots (2n-1/(2n))^k},$$

$$d_{2n+1} = \frac{(1/2)^k (3/4)^k \cdots (2n-1/(2n))^k}{(2/3)^k (4/5)^k \cdots (2n/(2n+1))^k}.$$

Example 9.8. (Regular continued fractions) [73].

A regular continued fraction $b_0 + K(1/b_n)$ of a positive real number x is defined by letting

$$b_n = \lfloor x_n \rfloor, \quad n = 0, 1, 2, \ldots,$$

where

$$x_0 = x, \quad x_n = \frac{1}{\text{Frac}(x_{n-1})}, \quad n = 1, 2, 3, \ldots,$$

where $\lfloor\ \rfloor$ denotes the greatest integer function, and $\text{Frac}(x_n)$ denotes the fractional part of x_n. If $\text{Frac}(x_{n-1}) = 0$, the regular continued fraction expansion terminates with b_{n-1}. Suppose now that $x = k/l$ is a rational number.

Set $x_0 = x$ and $r_1 = l$. Then by the Euclidean algorithm,

$$x_0 = b_0 + r_2/r_1 = b_0 + 1/x_1, \text{ with } x_1 = r_1/r_2, \ r_2 < r_1,$$
$$x_1 = b_1 + r_3/r_2 = b_1 + 1/x_2, \text{ with } x_2 = r_2/r_3, \ r_3 < r_2,$$
$$\vdots$$
$$x_m = b_m + r_{m+1}/r_m = b_m + 1/x_m, \text{ with } x_m = r_m/r_{m+1}, \ r_{m+1} < r_m.$$

Since $\{r_i\}$ is a decreasing sequence of positive integers, this algorithm necessarily terminates; that is, there exists n such that $r_{n+2} = 0$; the last relation would be $x_n = b_n$.

Let us define the Möbius transformation

$$t_0(u) = b_0 + u, \quad t_m(u) = \frac{1}{b_m + u}, \quad m = 1, 2, \ldots, n.$$

Then

$$1/x_n = t_n(0) \qquad \text{and} \quad 1/x_m = t_m\left(1/x_{m+1}\right),$$
$$m = 1, 2, \ldots, n-1, \qquad x_0 = t_0\left(1/x_1\right).$$

It follows that

$$x = (t_0 \circ t_1 \circ \cdots \circ t_n)(0),$$

and, consequently,

$$x = b_0 + \frac{1}{b_1+}\,\frac{1}{b_2+}\,\cdots\frac{1}{b_n}. \tag{9.3.3}$$

The continued fraction (9.3.3) obtained in the above procedure is called *regular*.

Conclusion. A real number is rational if and only if it is the value of a terminating regular continued fraction. One may easily show that

$$\frac{61}{48} = 1 + \frac{1}{3+}\,\frac{1}{1+}\,\frac{1}{2+}\,\frac{1}{4},$$
$$\frac{12}{55} = \frac{1}{4+}\,\frac{1}{1+}\,\frac{1}{1+}\,\frac{1}{2+}\,\frac{1}{2}.$$

Using the same procedure one may find a regular continued fraction representation of irrational numbers. It turns out that every irrational number can be represented by a nonterminating regular continued fraction and, conversely, the value of every nonterminating regular continued fraction is irrational [73]. For example,

$$\sqrt{7} = 2 + \frac{1}{1+}\,\frac{1}{1+}\,\frac{1}{1+}\,\frac{1}{4+}\,\frac{1}{1+}\,\frac{1}{1+}\,\frac{1}{1+}\,\frac{1}{4+}\cdots, \tag{9.3.4}$$

$$e = 2 + \frac{1}{1+}\,\frac{1}{2+}\,\frac{1}{1+}\,\frac{1}{1+}\,\frac{1}{4+}\,\frac{1}{1+}\,\frac{1}{1+}\,\frac{1}{6+}\,\frac{1}{1+}\,\frac{1}{1+}\,\frac{1}{8+}\cdots. \tag{9.3.5}$$

Exercises 9.3

1. Show that:

$$\frac{1+\sqrt{5}}{2} = 1 + \frac{1}{1+}\,\frac{1}{1+}\,\cdots \quad \text{(golden ratio)}.$$

2. Verify (9.3.4) and (9.3.5).

3. Show that:

   (a) $\pi = \frac{4}{1+}\,\frac{1^2}{3+}\,\frac{2^2}{5+}\,\frac{3^2}{7+}\cdots$.

   (b) $\sqrt{2} = 1 + \frac{1}{2+}\,\frac{1}{2+}\,\frac{1}{2+}\cdots$.

4. Show that, for $-1 \le x \le 1$,

$$\arctan x = \frac{x}{1+}\,\frac{1x^2}{3-x^2+}\,\frac{9x^2}{5-3x^2+}\,\frac{25x^2}{7-5x^2+}\cdots,$$

   and, consequently,

$$\frac{\pi}{4} = \frac{1}{1+}\,\frac{1}{2+}\,\frac{9}{2+}\,\frac{25}{2+}\,\frac{49}{2+}\cdots.$$

5. Prove for $z \neq 0$,

$$e^z = 1 + \frac{z}{1-}\,\frac{1z}{2+z-}\,\frac{2z}{3+z-}\,\frac{3z}{4+z-}\cdots.$$

6. Let $u_0, u_1, u_2, \ldots$ be numbers such that $u_i \neq 0$, $i = 1, 2, 3, \ldots$, and $U_n = u_0 + u_1 + \cdots + u_n$. Let

$$b_0 = u_0, \qquad b_1 = 1, \qquad a_1 = u_1/2,$$
$$b_{2n} = 2, \qquad b_{2n+1} = \frac{u_n + u_{n+1}}{u_n}, \qquad n = 1, 2, \ldots,$$
$$a_{2_n} = -1, \quad a_{2_{n+1}} = -\frac{u_{n+1}}{u_n}.$$

   Show that the $(2n)$th and $(2n+1)$th approximants of $b_0 + K\,(a_n/b_n)$ are U_n and $U_n + \frac{u_{n+1}}{2}$, respectively.

*7. (Open Problem). It is known [141] that $\zeta(2)$ and $\zeta(3)$ are irrational numbers. Show that $\zeta(5)$ is irrational.

In Problems 8 through 11 we will deal with Diophantine equations and their generalizations.

Diophantine Equations. Let k and l be two positive integers that are coprime. The problem that we are interested in is to find all pairs of integers (x, y) that solve the equation, called a *Diophantine equation*,

$$kx - ly = 1. \tag{9.3.6}$$

Let

$$\frac{k}{l} = b_0 + \frac{1}{b_1+}\,\frac{1}{b_2+}\,\cdots\,\frac{1}{+b_n}.$$

If $A(n)$ and $B(n)$ are the nth partial numerator and denominator, respectively, then from (9.1.8) we have

$$A(m)B(m-1) - A(m-1)B(m) = (-1)^{m-1}, \quad m = 1, 2, \ldots, n. \tag{9.3.7}$$

Moreover,

$$\frac{k}{l} = \frac{A(n)}{B(n)}. \tag{9.3.8}$$

Observe that if $m = n$, then (9.3.7) becomes

$$A(n)B(n-1) - A(n-1)B(n) = (-1)^{n-1}. \tag{9.3.9}$$

Now, if $A(n)$ and $B(n)$ have a common divisor d, then $\frac{A(n)}{B(n)} = \frac{d\cdot\tilde{A}(n)}{d\cdot\tilde{B}(n)} = \frac{\tilde{A}(n)}{\tilde{B}(n)}$. Hence

$$\tilde{A}(n)\tilde{B}(n-1) - \tilde{A}(n-1)\tilde{B}(n) = (-1)^{n-1}.$$

But this is impossible, since from (9.3.9) we get

$$d^2\left(\tilde{A}(n)\tilde{B}(n-1) - \tilde{A}(n-1)\tilde{B}(n)\right) = (-1)^{n-1}.$$

Hence it follows from (9.3.8) that $A(n) = k$ and $B(n) = l$.

Now, to find a solution of (9.3.6) we consider two cases.

Case (a). If n is odd, we let $x = B(n-1)$, $y = A(n-1)$. Then from (9.3.9) we have

$$kx - ly = A(n)x - B(n)y = (-1)^{n-1} = 1.$$

Case (b). If n is even, we let $x = l - B(n-1)$, and $y = k - A(n-1)$. Then

$$\begin{aligned} kx - ly &= A(n)(B(n) - B(n-1)) - B(n)(A(n) - A(n-1)) \\ &= -(-1)^{n-1} = 1. \end{aligned}$$

8. Show that the general solution of (9.3.6) is given by

$$(x, y) = (x_0, y_0) + m(l, k),$$

where m is an arbitrary integer and (x_0, y_0) is any special solution.

9. Solve the equation $61x - 48y = 1$ by finding all pairs of integers (x, y) that satisfy it.

*10. Solve Pell's equation

$$x^2 - ly^2 = 1,$$

where l is an integer, not a perfect square. You need to find all pairs of integers (x, y) that solve the equation.

*11. Solve $x^2 - 7y^2 = 1$.

9.4 Classical Orthogonal Polynomials

Let $w(x)$ be a positive function on a given finite or infinite interval (a, b) such that it is continuous, except possibly at a finite set of points. Moreover, we assume that the "*moments*"

$$\boxed{\mu_n = \int_a^b x^n w(x)\, dx} \qquad (9.4.1)$$

exist and are finite. Then a sequence $\{P_n(x)\}_{n=0}^{\infty}$ of polynomials, $P_n(x)$ of degree n, such that

$$\boxed{\int_a^b P_n(x)P_m(x)w(x)\, dx = \lambda_n \delta_{nm},} \qquad n, m = 0, 1, 2 \ldots, \qquad (9.4.2)$$

is said to be orthogonal over (a, b) with a weight function $w(x)$, where

$$\delta_{nm} = \begin{cases} 1, & n = m, \\ 0, & n \neq m, \end{cases}$$

is the Kronecker delta function. A polynomial $P_n(x) = \sum_{k=0}^{n} a_{nk}x^k$ is said to be *monic* if the coefficient a_{nn} of the leading term x^n is 1.

Example 9.9. (The Chebyshev Polynomials Revisited).

The Chebyshev polynomials of the first and second kind are defined, respectively, as follows (See Exercises 2.3, Problem 11):

$$T_n(x) = \cos n\theta, \quad U_n(x) = \sin(n+1)\theta/\sin\theta, \quad n = 0, 1, 2, \ldots ,$$

where $\theta = \cos^{-1} x$ and $|x| < 1$.

Observe that $\{T_n(x)\}$ is orthogonal on the interval $(-1, 1)$ with the weight function $w(x) = (1 - x^2)^{-\frac{1}{2}}$.

To show this, note that

$$\int_{-1}^{1} T_n(x)T_m(x)(1 - x^2)^{-\frac{1}{2}}\, dx = \int_0^{\pi} \cos n\theta \cos m\theta\, d\theta = 0 \quad \text{if } n \neq m.$$

Similarly, we may show that $\{U_n(x)\}$ is orthogonal on $(-1, 1)$ with the weight function $w(x) = (1 - x^2)^{\frac{1}{2}}$.

Next, we address the question of uniqueness of orthogonal polynomials.

Theorem 9.10. *If the coefficient a_{nn} of x^n in $P_n(x)$ is prescribed for each n, then the sequence of orthogonal polynomials relative to a weight function $w(x)$ exists and is unique. Moreover, each $P_n(x)$ is orthogonal to all polynomials of lower degree.*

PROOF. We will use mathematical induction on n to prove the first part. Suppose that a_{00} and a_{11} are known; $P_0(x) = a_{00}$ and $P_1(x) = a_{11}x + a_{10}$. Then from (9.4.2), $\int_a^b w(x)(a_{00}a_{11}x + a_{00}a_{10})\, dx = 0$, which gives a_{10}. Assume now that $P_0(x), P_1(x), \ldots, P_{n-1}(x)$ are determined such that they satisfy pairwise (9.4.2).

Then $P_n(x) = a_{nn}x^n + b_{n,n-1}P_{n-1}(x) + \cdots + b_{n,0}P_0(x)$, where $b_{n,s}$ are independent of x. From (9.4.2), we have for $r = 0, 1, \ldots, n-1$,

$$a_{nn}\int_a^b w(x)x^n P_r(x)\, dx + b_{nr}\int_a^b w(x)(P_r(x))^2\, dx = 0.$$

Since $\int_a^b w(x)(P_r(x))^2\, dx > 0$, it follows that b_{nr} exists and is uniquely determined for $r = 0, 1, \ldots, n-1$. This establishes the first part. The second part is left to the reader as Exercises 9.5, Problem 1. □

We are now ready to present some of the main classical orthogonal polynomials.

1. **Jacobi polynomials $P_n^{(\alpha,\beta)}(x)$**, $\alpha > -1$, $\beta > -1$.
   These polynomials are orthogonal on $(-1, 1)$ with the weight function $w(x) = (1-x)^\alpha(1+x)^\beta$, and $a_{nn} = 2^{-n}\begin{pmatrix} 2n+\alpha+\beta \\ n \end{pmatrix}$. An explicit expression for the Jacobi polynomials $P_n^{(\alpha,\beta)}(x)$ are given by Rodrigues' formula

$$\boxed{P_n^{(\alpha,\beta)}(x) = (-1)^n \frac{(1-x)^{-\alpha}(1+x)^{-\beta}}{2^n n!}\frac{d^n}{dx^n}\left\{(1-x)^{n+\alpha}(1+x)^{n+\beta}\right\}.} \tag{9.4.3}$$

   To write $P_n^{(\alpha,\beta)}(x)$ more explicitly, we need to utilize Leibniz's formula

$$\boxed{\frac{d^n}{dx^n}(uv) = \sum_{k=0}^{n}\binom{n}{k}\frac{d^{n-k}u}{dx^{n-k}}\frac{d^k v}{dx^k}.} \tag{9.4.4}$$

   (See Appendix G.)

Hence,

$$\begin{aligned}\frac{d^n}{dx^n}&(1-x)^{n+\alpha}(1+x)^{n+\beta}\\&=\sum_{k=0}^{n}\binom{n}{k}D^{n-k}(1-x)^{n+\alpha}D^k(1+x)^{n+\beta}\\&=(-1)^n(1-x)^{\alpha}(1+x)^{\beta}n!\\&\quad\times\sum_{k=0}^{n}\binom{n+\alpha}{n-k}\binom{n+\beta}{k}(x-1)^k(x+1)^{n-k}.\end{aligned}$$

Therefore,

$$P_n^{(\alpha,\beta)}(x)=2^{-n}\sum_{k=0}^{n}\binom{n+\alpha}{n-k}\binom{n+\beta}{k}(x-1)^k(x+1)^{n-k} \tag{9.4.5}$$

with the leading coefficient

$$a_{nn}=2^{-n}\sum_{k=0}^{n}\binom{n+\alpha}{n-k}\binom{n+\beta}{k}=2^{-n}\binom{2n+\alpha+\beta}{n}. \tag{9.4.6}$$

(See G.)

To verify (9.4.3), let $Q_n(x)$ denote the right-hand side of the equation and let $g(x)$ be another polynomial. Then successive application of integration by parts yields

$$\begin{aligned}\int_{-1}^{1}&(1-x)^{\alpha}(1-x)^{\beta}Q_n(x)g(x)\,dx\\&=\frac{1}{2^n n!}\int_{-1}^{1}(1-x)^{n+\alpha}(1+x)^{n+\beta}g^{(n)}(x)\,dx,\end{aligned} \tag{9.4.7}$$

where $g^{(n)}(x)$ denotes the nth derivative of $g(x)$.

Observe that if $g(x)$ is a polynomial of degree less than n, then $g^{(n)}(x)=0$. Hence, $Q_n(x)$ satisfies (9.4.2). Furthermore, the coefficient of x^n in $Q_n(x)$ is

$$2^{-n}\binom{2n+\alpha+\beta}{n}.$$

Hence, by uniqueness (Theorem 9.5), $Q_n(x)=P_n^{(\alpha,\beta)}(x)$.

2. **Legendre polynomials** $P_n(x)$: These are special Jacobi polynomials obtained by letting $\alpha=\beta=0$. Hence (9.4.3) is reduced to

$$\boxed{P_n(x)=P_n^{(0,0)}(x)=\frac{(-1)^n}{2^n n!}\frac{d^n}{dx^n}\{(1-x^2)^n\}.} \tag{9.4.8}$$

The Legendre polynomials are orthogonal on $(-1,1)$ with the weight function $w(x)=1$. Moreover, using Leibniz's formula yields

$$P_n(x) = 2^{-n}\sum_{k=0}^{n}\binom{n}{n-k}\binom{n}{k}(x-1)^k(x+1)^{n-k}, \tag{9.4.9}$$

with leading coefficient $a_{nn} = 2^{-n}(2n)!/(n!)^2$.

3. **Gegenbauer (or ultraspherical) polynomials** $P_n^{\nu}(x)$: These are special Jacobi polynomials obtained by setting $\alpha=\beta$ and $\alpha=\nu-\frac{1}{2}$. The Gegenbauer polynomials are orthogonal on $(-1,1)$ with the weight function $w(x) = (1-x^2)^{\nu-1/2}$, $\nu > -\frac{1}{2}$, and $a_{nn} = 2^n\binom{\nu+n-1}{\nu-1}$. By Rodrigues' formula we have

$$\begin{aligned} P_n^{\nu}(x) &= \frac{(-1)^n}{n!}(1-x^2)^{1/2-\nu}\frac{d^n}{dx^n}(1-x^2)^{\nu+n-1/2} \\ &= \binom{2\nu-1}{\nu}^{-1}\binom{n+2\nu-1}{\nu-1/2}P_n^{(\nu-1/2,\nu-1/2)}(x). \end{aligned} \tag{9.4.10}$$

Using Leibniz's formula yields

$$\begin{aligned} P_n^{\nu}(x) = 2^{-n}&\binom{2\nu-1}{\nu}^{-1}\binom{n+2\nu-1}{\nu-1/2} \\ &\times\sum_{k=0}^{n}\binom{n+\nu-1/2}{n-k}\binom{n+\nu-1/2}{k}(x-1)^k(x+1)^k. \end{aligned} \tag{9.4.11}$$

4. **Laguerre polynomials** $L_n^{\alpha}(x), \alpha > -1$, are orthogonal on $(0,\infty)$ with the weight function $w(x) = e^{-x}x^{\alpha}$, and $a_{nn} = (-1)^n/n!$. Moreover,

$$L_n^{\alpha}(x) = \frac{e^x x^{-\alpha}}{n!}\frac{d^n}{dx^n}(e^{-x}x^{n+\alpha}). \tag{9.4.12}$$

By Leibniz's formula we can show that

$$L_n^{\alpha}(x) = \sum_{k=0}^{n}\binom{n+\alpha}{n-k}\frac{(-x)^k}{k!}. \tag{9.4.13}$$

5. **Hermite polynomials** $H_n(x)$ are orthogonal on $(-\infty,\infty)$ with the weight function $w(x) = e^{-x^2}$, and $a_{nn} = 2^n$. These are given by Rodrigues' formula

$$H_n(x) = (-1)^n e^{x^2}\frac{d^n}{dx^n}(e^{-x^2}). \tag{9.4.14}$$

By Taylor's theorem,

$$e^{2xw-w^2} = \sum_{n=0}^{\infty} H_n(x)\frac{w^n}{n!}.$$

Expanding e^{2xw} and e^{-w^2} as power series in w and taking the Cauchy product[1] of the result gives

$$e^{2xw-w^2} = \sum_{n=0}^{\infty} \sum_{k=0}^{\lfloor n/2 \rfloor} \frac{(-1)^k (2x)^{n-2k} w^n}{(n-2k)!\, k!}.$$

Hence,

$$\boxed{H_n(x) = n! \sum_{k=0}^{\lfloor n/2 \rfloor} \frac{(-1)^k (2x)^{n-2k}}{(n-2k)!\, k!}.} \tag{9.4.15}$$

9.5 The Fundamental Recurrence Formula for Orthogonal Polynomials

We now show why difference equations, particularly those of second order, are of paramount importance in the study of orthogonal polynomials. The following is the main result.

Theorem 9.11. *Any sequence of monic orthogonal polynomials $\{P_n(x)\}$ with $P_n(x) = \sum_{k=0}^{n} a_k x^k$ must satisfy a second-order difference equation of the form*

$$\boxed{P_{n+1}(x) - (A_n x + B_n)P_n(x) + C_n P_{n-1}(x) = 0} \tag{9.5.1}$$

with

$$A_n = \frac{a_{n+1,n+1}}{a_{n,n}}. \tag{9.5.2}$$

Proof. Choose A_n such that $P_{n+1}(x) - A_n x P_n(x)$ does not possess any term in x^{n+1}. Hence, we may write

$$P_{n+1}(x) - A_n x P_n(x) = \sum_{r=0}^{n} d_{nr} P_r(x). \tag{9.5.3}$$

Multiplying both sides of (9.5.3) by $w(x)P_s(x)$ and integrating from a to b yields

$$d_{ns} \int_a^b w(x)\{P_s(x)\}^2\, dx = -A_n \int_a^b x w(x) P_s(x) P_n(x)\, dx. \tag{9.5.4}$$

---

[1]Given two series $\sum_{n=0}^{\infty} a_n$ and $\sum_{n=0}^{\infty} b_n$, we put $c_n = \sum_{k=0}^{n} a_k b_{n-k}$, $n = 0, 1, 2, \ldots$. Then $\sum_{n=0}^{\infty} c_n$ is called the Cauchy product of the two series.

Since $xP_s(x)$ is of degree $s+1$, and $P_n(x)$ is orthogonal to all polynomials of degree less than n, it follows that the right-hand side of (9.5.4) vanishes except possibly when $s = n-1$ and $s = n$. Hence, d_{nn} and $d_{n,n-1}$ are possibly not zero. Therefore, from (9.5.3) we have

$$P_{n+1}(x) - (A_n x + d_{nn})P_n(x) - d_{n,n-1}P_{n-1}(x) = 0,$$

which is (9.5.1) with $B_n = d_{nn}$, $C_n = -d_{n,n-1}$. □

Remark:

(i) A monic sequence of orthogonal polynomials $\{\hat{P}_n(x)\}$ satisfies the difference equation

$$\hat{P}_{n+1}(x) - (x - \beta_n)\hat{P}_n(x) + \gamma_n \hat{P}_{n-1}(x) = 0, \tag{9.5.5}$$

where

$$\beta_n = \frac{-B_n a_{nn}}{a_{n+1,n+1}}, \quad \gamma_n = \frac{C_n a_{n+1,n+1}}{a_{n-1,n-1}}. \tag{9.5.6}$$

This may be shown easily if one writes $\hat{P}(x) = a_{nn}^{-1}P_n(x)$.

(ii) If $P_n(-x) = (-1)^n P_n(x)$, then $\{P_n(x)\}$ is often called symmetric. In this case, one may show that $B_n = \beta_n = 0$. To show this, let $Q_n(x) = (-1)^n P_n(-x)$. Then

$$Q_{n+1}(x) + (B_n - A_n x)Q_n(x) + C_n Q_{n-1}(x) = 0. \tag{9.5.7}$$

If $Q_n(x) = P_n(x)$, then subtracting (9.5.1) from (9.5.7) yields $B_n = 0$.

(iii) The converse of Theorem 9.11 also holds and is commonly referred to as Favard's theorem. Basically, this theorem states that any polynomial sequence that satisfies a difference equation of the form of (9.5.1) must be an orthogonal polynomial sequence.

Let us illustrate the preceding theorem by an example.

Example 9.12. Find the difference equation that is satisfied by the Legendre polynomials $P_n(x)$.

Solution From (9.4.8) the coefficients of x^n, x^{n-1}, x^{n-2} are, respectively,

$$a_{nn} = \frac{(2n)!}{2^n (n!)^2}, \qquad a_{n,n-1} = 0, \quad a_{n,n-2} = \frac{(2n-2)!}{2^n (n-2)!\,(n-1)!}.$$

Furthermore, $\{P_n(x)\}$ is symmetric, since $P_n(x) = (-1)^n P_n(-x)$. Thus, from Remark (ii) above, we have $B_n = 0$. From (9.5.2), we have $A_n = \frac{2n+1}{n+1}$. It remains to find C_n. For this purpose, we compare the coefficients of x^{n-1} in (9.5.1). This yields

$$a_{n+1,n-1} - A_n a_{n,n-2} + C_n a_{n-1,n-1} = 0.$$

Thus,

$$C_n = \frac{A_n a_{n,n-2} - a_{n+1,n-1}}{a_{n-1,n-1}} = \frac{-n}{n+1}.$$

Hence the Legendre polynomials satisfy the difference equation

$$(n+1)P_{n+1}(x) - (2n+1)xP_n(x) + nP_{n-1}(x) = 0. \tag{9.5.8}$$

Example 9.13. Find the difference equation that is satisfied by the Jacobi polynomials $P_n^{(\alpha,\beta)}(x)$.

Solution This time we will use a special trick! Notice that from (9.4.5) we obtain

$$P_n^{(\alpha,\beta)}(-x) = (-1)^n P_n^{(\beta,\alpha)}(x), \tag{9.5.9}$$

$$P_n^{(\alpha,\beta)}(1) = \binom{n+\alpha}{n}, \tag{9.5.10}$$

$$P_n^{(\alpha,\beta)}(-1) = \binom{n+\beta}{n}. \tag{9.5.11}$$

From (9.4.6) and (9.5.2) we get

$$A_n = \frac{(2n+2+\alpha+\beta)(2n+1+\alpha+\beta)}{2(n+1)(n+1+\alpha+\beta)}. \tag{9.5.12}$$

Using (9.5.10) and setting $x=1$ in (9.5.1) yields

$$\binom{n+\alpha+1}{n+1} - (A_n+B_n)\binom{n+\alpha}{n} + \binom{n+\alpha-1}{n-1} = 0. \tag{9.5.13}$$

Similarly, setting $x=1$ in (9.5.1) and using (9.5.11) yields

$$(-1)^{n+1}\binom{n+\beta+1}{n+1} - (-A_n+B_n)(-1)^n\binom{n+\beta}{n} + C_n(-1)^{n-1}\binom{n+\beta-1}{n-1} = 0. \tag{9.5.14}$$

Multiplying (9.5.13) by $\binom{n+\beta}{n}$ and (9.5.14) by $\binom{n+\alpha}{n}$ and adding, produces an equation in A_n and C_n which, by substitution for A_n from (9.5.12), gives

$$C_n = \frac{(n+\alpha)(n+\beta)(2n+2+\alpha+\beta)}{(n+1)(n+\alpha+\beta+1)(2n+\alpha+\beta)}. \tag{9.5.15}$$

Substituting for C_n in (9.5.13) yields

$$B_n = \frac{(2n+1+\alpha+\beta)(\alpha^2-\beta^2)}{2(n+1)(n+\alpha+\beta+1)(2n+\alpha+\beta)}. \tag{9.5.16}$$

Exercises 9.4 and 9.5

1. Let $\{P_n(x)\}$ be a sequence of orthogonal polynomials on the interval (a,b) relative to the weight function $w(x)$. Prove that $P_n(x)$ is orthogonal to any polynomial of lower degree.

2. Verify formula (9.4.9).

3. Verify formula (9.4.12).

4. Verify formula (9.4.14).

5. Find the difference equation that represents the Gegenbauer polynomials $\{P_n^\nu(x)\}$.

6. Find the difference equation that represents the Laguerre polynomials $\{L_n^\alpha(x)\}$.

7. Find the difference equation that represents the Hermite polynomials $H_n(x)$.

8. (Charlier polynomials). Let $C_n^{(a)}(x)$ denote the monic Charlier polynomials defined by

$$\boxed{C_n^{(a)}(x) = n!\sum_{k=0}^{n}\binom{x}{k}\frac{(-a)^{n-k}}{(n-k)!}.} \tag{9.5.17}$$

Show that $\{C_n^{(a)}(x)\}$ satisfies the difference equation

$$C_{n+1}^{(a)}(x) = (x-n-a)C_n^{(a)}(x) - anC_{n-1}^{(a)}(x), \quad n \geq 0.$$

9. (The Bessel function). Let $n \in \mathbb{Z}$, $z \in \mathbb{C}$. The Bessel function $J_n(z)$ is defined by

$$\boxed{J_n(z) = (z/2)^n\sum_{j=0}^{\infty}\frac{(-1)^j(z^2/4)^j}{j!(n+j)!}, \qquad n = 0,1,2,\ldots.}$$

Find the corresponding difference equation.

*10. (Christoffel–Darboux identity). Let $\{P_n(x)\}$ satisfy

$$\begin{aligned} &P_n(x) = (x-c_n)P_{n-1}(x) - \lambda P_{n-2}(x), \quad n = 1,2,3,\ldots,\\ &P_{-1}(x) = 0, \quad P_0(x) = 1, \quad \lambda \neq 0. \end{aligned} \tag{9.5.18}$$

Prove that

$$\sum_{k=0}^{n}\frac{P_k(x)(P_k(u)}{\lambda_1\lambda_2\cdots\lambda_{k+1}} = (\lambda_1\lambda_2\cdots\lambda_{n+1})^{-1} \times \frac{P_{n+1}(x)P_n(u) - P_n(x)P_{n+1}(u)}{x-u}. \tag{9.5.19}$$

11. (Confluent form of (9.5.19)). Show that

$$\sum_{k=0}^{n} \frac{P_k^2(x)}{\lambda_1\lambda_2\cdots\lambda_{k+1}} = \frac{P_{n+1}^{'}(x)P_n^{'}(x) - P_n(x)P_{n+1}(x)}{\lambda_1\lambda_2\cdots\lambda_{n+1}}.$$

12. Consider the sequence $\{P_n(x)\}$ satisfying (9.5.18) and let $Q_n(x) = a^{-n}P_n(ax+b)$, $a \neq 0$.

    (a) Show that $Q_n(x) = (x - \frac{c_n - b}{a})Q_{n-1}(x) - \frac{\lambda_n}{a^2}Q_{n-2}(x)$.

    (b) If $\{P_n(x)\}$ is an orthogonal polynomial sequence with respect to the moments μ_n, show that $\{Q_n(x)\}$ is an orthogonal polynomial sequence with respect to the moments

$$\nu_n = a^{-n}\sum_{k=0}^{n}\binom{n}{k}(-b)^{n-k}\mu_k.$$

13. Suppose that $\{Q_n(x)\}$ satisfies (9.5.18), but with the initial conditions $Q_{-1}(x) = -1$ and $Q_0(x) = 0$.

    (a) Show that $Q_n(x)$ is a polynomial of degree $n-1$.

    (b) Put $P_n^{(1)}(x) = \lambda_1^{-1}Q_{n+1}(x)$ and write the difference equation corresponding to $\{P_n^{(1)}(x)\}$.

14. Let $\{P_n(x)\}$ be a sequence of orthogonal polynomials on the interval (a,b). Show that the zeros of $P_n(x)$ are real, distinct, and lie in (a,b).

15. In the following justify that $y(x)$ satisfies the given differential equations:

    (a) $y^{''} - 2xy' + 2ny = 0$; $y(x) = H_n(x)$.

    (b) $xy^{''} + (\alpha + 1 - x)y' + ny = 0$; $y(x) = L_n^{\alpha}(x)$.

    (c) $(1-x^2)y^{''} + \{(\beta - \alpha) - (\alpha + \beta + 2)x\}y' + n(n + \alpha + \beta + 1)y = 0$; $y(x) = P_n^{(\alpha,\beta)}(x)$.

9.6 Minimal Solutions, Continued Fractions, and Orthogonal Polynomials

The intimate connection between continued fractions and orthogonal polynomials is now apparent in light of the fundamental recurrence formulas for continued fractions and orthogonal polynomials. If $\{P_n(x)\}$ is a monic orthogonal polynomial sequence on the interval (a,b), with $P_{-1}(x) = 0$ and $P_0(x) = 1$, then it must satisfy the difference equation

$$P_{n+1} - (x - \beta_n)P_n(x) + \gamma_n P_{n-1}(x) = 0, \quad n \in \mathbb{Z}^+. \tag{9.6.1}$$

To find the continued fraction that corresponds to (9.6.1) we take $b_n = -(x-\beta_n), a_n = \gamma_n$ in (9.1.6). We then have the continued fraction

$$K = \frac{\gamma_0}{(x-\beta_0)-}\ \frac{\gamma_1}{(x-\beta_1)-}\ \frac{\gamma_2}{(x-\beta_2)-}\ \cdots. \tag{9.6.2}$$

Moreover, $P_n(x)$ is the nth partial denominator of the continued fraction (9.6.2).

Next we focus our attention on finding a minimal solution of (9.6.1). Recall from Pincherle's theorem that (9.6.1) has a minimal solution if and only if the continued fraction K converges. To accomplish our task we need to find another polynomial solution $Q_n(x)$, called the associated polynomials, that forms with $P_n(x)$ a fundamental set of solutions of (9.6.1).

Define

$$\boxed{Q_n(x) = \int_a^b \frac{(P_n(x) - P_n(t))}{x-t} w(t)\, dt.} \tag{9.6.3}$$

Lemma 9.14. *The set $\{P_n(x), Q_n(x)\}$ is a fundamental set of solutions of (9.6.1).*

PROOF. From (9.6.1), we have

$$\begin{aligned} P_{n+1}(x) - P_{n+1}(t) = (x-t)P_n(t) &+ (x-\beta_n)[P_n(x) - P_n(t)] \\ &- \gamma_n[P_{n-1}(x) - P_{n-1}(t)]. \end{aligned}$$

Dividing by $x - t$ and integrating yields

$$Q_{n+1}(x) = \int_a^b P_n(t)w(t)\, dt + (x-\beta_n)Q_n(x) - \gamma_n Q_{n-1}(x), \quad n = 0, 1, 2, \ldots,$$

with

$$Q_{-1}(x) = 0, \quad Q_0(x) = 0.$$

Notice that by the orthogonality of $\{P_n(x)\}$ we have

$$\int_a^b P_n(t)w(t)\, dt = \int_a^b P_n(t)P_0(t)w(t)\, dt = \begin{cases} 0 & \text{if } n > 0, \\ \mu_0 & \text{if } n = 0. \end{cases}$$

Hence, we obtain

$$\begin{aligned} Q_{n+1}(x) - (x-\beta_n)Q_n(x) + \gamma_n Q_{n-1}(x) &= 0, \\ Q_0(x) = 0, \quad Q_1(x) &= \mu_0. \end{aligned} \tag{9.6.4}$$

Since $P_0(x) = 1, P_1(x) = x - \beta_0$, the Casoratian $W(0)$ of P_n and Q_n at $n = 0$ is equal to $\mu_0 \neq 0$, which implies that $\{P_n(x)\}$ and $\{Q_n(x)\}$ are two linearly independent solutions of (9.6.1).

Observe that the polynomial sequence $Q_n(x)$ is the nth partial numerator of the continued fraction K. Hence if

$$\lim_{n\to\infty} \frac{Q_n(x)}{P_n(x)} = F(x)$$

exists, then by Pincherle's theorem, the minimal solution of (9.6.1) exists and is defined by

$$S_n(x) = F(x)P_n(x) - Q_n(x). \tag{9.6.5}$$

Furthermore, if we let $\gamma_0 = \mu_0$ and $Q_{-1}(x) = -1$, then $S_n(x)$ satisfies (9.6.1) not only for $n \geq 1$, but also for $n = 0$, with $S_{-1}(x) = -1$.

To find an explicit formula for the minimal solution $S_n(x)$ we need to utilize complex analysis. Henceforth, we replace the variable x by z in all the considered functions. From a result of [113], $F(z)$ has the integral representation

$$F(z) = \int_a^b \frac{w(t)}{z-t}\, dt, \quad z \notin [a,b]. \tag{9.6.6}$$

Combining this formula with (9.6.3) produces the following integral representation of $S_n(z)$:

$$S_n(z) = \int_a^b \frac{P_n(t)}{(z-t)} w(t)\, dt. \tag{9.6.7}$$

□

Remark:

(i) If (a, b) is a finite interval, then the existence of the minimal solution $S_n(z)$ is always guaranteed (see [113]).

(ii) If (a, b) is a half-infinite interval of the form (a, ∞) or $(-\infty, b)$, then a sufficient condition for the existence of the minimal solution is

$$\sum_{n=1}^{\infty} \mu_n^{-1/2n} = \infty$$

(see [66]).

(iii) If $(a, b) = (-\infty, \infty)$, then a sufficient condition for the existence of the minimal solution $S_n(z)$ is

$$\sum_{n=1}^{\infty} \mu_{2n}^{-1/2n} = \infty, \quad \text{or} \quad \sum_{n=1}^{\infty} \gamma_n^{-1/2} = \infty.$$

Proofs of these remarks are beyond the scope of this book and will be omitted. For details the reader may consult [66], [113].

Example 9.15 [145]. Consider the difference equation

$$P_{n+1}(z) - zP_n(z) + \gamma_n P_{n-1}(z) = 0, \quad n \geq 0, \tag{9.6.8}$$

with

$$\gamma_n = \frac{n(n+2\nu-1)}{4(n+\nu)(n+\nu-1)}, \quad \nu > 1.$$

Notice that $P_n(z)$ is related to the Gegenbauer polynomials $P_n^\nu(z)$ by the relation

$$P_n(z) = \frac{n!}{2^n(\nu)_n} P_n^\nu(z), \tag{9.6.9}$$

where

$$(\nu)_n = \nu(\nu+1)(\nu+2)\cdots(\nu+n-1) = \frac{\Gamma(\nu+n)}{\Gamma(\nu)} \tag{9.6.10}$$

denotes the Pochhammer symbol. Hence by formula (9.6.7), a minimal solution of (9.6.8) is given by

$$\boxed{S_n(z) = \frac{n!}{2^n(\nu)_n} \int_{-1}^{1} \frac{P_n^\nu(t)(1-t^2)^{\nu-1/2}}{z-t} dt, \quad z \notin [-1,1].}$$

Using the value of the integral found in Erdélyi et al. ([54, p. 281]) yields

$$S_n(z) = \frac{(n+\nu)n}{2^{n+\nu-3/2}} n!\, \sqrt{\pi}\, e^{i\pi(\nu-1/2)}(z^2-1)^{(2\nu-1)/4} Q_{n+\nu-1/2}^{1/2-\lambda}(z), \tag{9.6.11}$$

where $Q_\alpha^\beta(z)$ is a Legendre function defined as

$$Q_\alpha^\beta(z) = 2^\alpha \Gamma(\alpha+1) \frac{(z-1)^{(\beta/2)-\alpha-1}}{(z+1)^{\beta/2}} F\left(\alpha+1, \alpha-\beta+1; 2\alpha+2; \frac{2}{1-z}\right),$$

with

$$F(a,b;c;z) = \sum_{s=0}^{\infty} \frac{(a)_s(b)_s z^s}{\Gamma(c+s)s!} \quad (|z|<1). \tag{9.6.12}$$

The function $F(a,b;c;z)$ is called a *hypergeometric* function [110].

Quite often it is possible to find an asymptotic representation of the minimal solution using the methods of Chapter 8. The following example demonstrates this point.

Example 9.16. (Perturbations of Chebyshev Polynomials [71]).

Consider again a monic orthogonal polynomial sequence $\{P_n(x)\}$ represented by the second-order difference equation

$$Q_{n+1}(z) - (z-a_n)Q_n(z) + b_n Q_{n-1}(z) = 0 \tag{9.6.13}$$

with $Q_{-1}(z) = 0$, $Q_0(z) = 1$.

The (complex) Nevai class $M(a,b)$ [106] consists of all those orthogonal polynomial sequences such that $\lim_{n\to\infty} a_n = a$, $\lim_{n\to\infty} b_n = b$. Without loss of generality we take $a = 0$ and $b = \frac{1}{4}$. (Why?) Then the limiting equation associated with (9.6.1) is given by

$$P_{n+1}(z) - zP_n(z) + \frac{1}{4}P_{n-1}(z) = 0. \tag{9.6.14}$$

Observe that $\{P_n(z)\}$ are the monic Chebyshev polynomials, which can be obtained (Appendix F, No. 4) by setting $P_n(z) = 2^{1-n}\, T_n(z)$ (since 2^{n-1} is the leading coefficient in $T_n(z)$), where $\{T_n(x)\}$ are the Chebyshev polynomials of the first kind. Equation (9.6.14) has two solutions,

$$P_n^+(z) = \lambda_+^n(z), \quad \text{and} \quad P_n^-(z) = \lambda_-^n(z), \quad z \neq 1, -1, \tag{9.6.15}$$

where

$$\lambda_+(z) = (z + \sqrt{z^2-1})/2 \quad \text{and} \quad \lambda_-(z) = (z - \sqrt{z^2-1})/2.$$

Observe that if $z \in \mathbb{C}\backslash[-1,1]$, we can choose the square root branch such that

$$\left|\frac{\lambda_+(z)}{\lambda_-(z)}\right| < 1.$$

Hence $P_n^-(z)$ is a minimal solution on $z \in \mathbb{C}\backslash[-1,1]$, and $P_n^+(z)$ is a dominant solution on $z \in \mathbb{C}\backslash[-1,1]$.

Now, the boundary values of the minimal solution on the cut

$$P_n^-(x+i0) = \lim_{\varepsilon\to 0^+} P_n^-(x+i\varepsilon) = (\lambda_-(x))^n,$$
$$P_n^-(x-i0) = \lim_{\varepsilon\to 0^+} P_n^-(x-i\varepsilon) = (\lambda_+(x))^n,$$

with

$$\lambda_-(x) = (x - \sqrt{1-x^2})/2, \quad \lambda_+(x) = (x + i\sqrt{1-x^2})/2, \quad x \in (-1,1),$$

yields a ratio of solutions that oscillates as $n \to \infty$. Thus there is no minimal solution for $z = x \in (-1,1)$.

Next we turn our attention to (9.6.13). Since $|\lambda_+(z)| \neq |\lambda_-(z)|$, by virtue of the Poincaré–Perron theorem there are two linearly independent solutions $Q_n^+(z)$, $Q_n^-(z)$ of (9.6.13) such that

$$\lim_{n\to\infty} \frac{Q_{n+1}^+(z)}{Q_n^+(z)} = \lambda_+(z), \quad \lim_{n\to\infty} \frac{Q_{n+1}^-(z)}{Q_n^-(z)} = \lambda_-(z).$$

Furthermore, $Q_n^+(z)$ is a dominant solution and $Q_n^-(z)$ is a minimal solution of (9.6.1) for $z \in \mathbb{C}\backslash[-1,1]$.

Moreover, if

$$\sum_{n=0}^{\infty} |a_n| + \left|b_n - \frac{1}{4}\right| < \infty, \tag{9.6.16}$$

then by Corollary 8.30, we have

$$Q_n^-(z) = \lambda_-^n(z)(1+o(1)), \quad z \in \mathbb{C}\backslash[-1,1],$$
$$Q_n^+(z) = \lambda_+^n(z)(1+o(1)), \quad z \in \mathbb{C}\backslash[-1,1],$$

where Q_n^- and Q_n^+ are, respectively, minimal and dominant solutions of (9.6.13). Furthermore, for $z = x \in (-1,1)$, there are two linearly independent solutions

$$Q_n(x+i0) = (\lambda_-(x))^n(i+o(1)), \quad Q_n(x-i0) = (\lambda_+(x))^n(i-o(1)),$$

where

$$\lambda_-(x) = x - i\sqrt{1-x^2}, \quad \lambda_+(x) = x + i\sqrt{1-x^2}.$$

For relaxing condition (9.6.16) and more generalizations, the reader is referred to [38].

Exercises 9.6

1. Show that

$$H_{2n}(x) = (-1)^n 2^{2n} n!\, L_n^{(-1/2)}(x^2),$$
$$H_{2n+1}(x) = (-1)^n 2^{2n+1} n!\, x L_n^{(-1/2)}(x^2).$$

2. Show that

$$T_n(x) = \frac{2^{2n}(n!)^2}{(2n)!} P_n^{(-1/2,-1/2)}(x),$$
$$U_n(x) = \frac{2^{2n} n!(n+1)!}{(2n+1)!} P_n^{(1/2,1/2)}(x).$$

In Problems 3 through 6 investigate the existence of a minimal solution for the given polynomial. If a minimal solution exists, find an explicit representation for it.

3. Legendre polynomials $\{P_n(x)\}$ (See Appendix F, No. 3).
4. Hermite polynomials $\{H_n(x)\}$ (See Appendix F, No. 6).
5. Laguerre polynomials $\{L_n^\alpha(x)\}$ (See Appendix F, No. 7).
6. Charlier polynomials $\{C_n^{(a)}\}$ (See Appendix F, No. 8).

*7. Use Rodrigues' formula for the Legendre polynomial $P_n(x)$ and the Cauchy integral formula[2] for the nth derivative to derive the formula (*Schäfli's integral*)

$$\boxed{P_n(x) = \frac{1}{2^{n+1}\pi i} \oint_\gamma \frac{(t^2-1)^n}{(t-x)^{n+1}} dt,} \tag{9.6.17}$$

where γ is any positively directed simple closed curve enclosing the point x (x may be real or complex).

---

[2] $f^{(n)}(z_0) = \frac{n!}{2\pi i} \int_\gamma \frac{f(z)}{(z-z_0)^{n+1}} dz$, $n = 0, 1, 2, \ldots$, where γ is any positively directed closed curve enclosing z_0.

Generating functions [110]. Suppose that $G(x,u)$ is a polynomial with the Maclaurin expansion

$$G(x,u) = \sum_{n=0}^{\infty} Q_n(x)u^n.$$

Then $G(x,u)$ is called a *generating function* for $\{Q_n(x)\}$.

*8. (a) Show that

$$\frac{1}{2\pi i}\int_\gamma \left\{1 - \frac{(t^2-1)h}{2(t-x)}\right\}^{-1} \frac{dt}{t-x} = \sum_{n=0}^{\infty} P_n(x)h^n = G(x,h).$$

(b) Deduce from (a) that the generating function of $P_n(x)$ is given by

$$G(x,u) = (1 - 2xu + u^2)^{-1/2}.$$

*9. Consider the Chebyshev polynomials of the first kind $\{T_n(x)\}$.

(a) Show that $T_n(x) = \frac{(-1)^n n!}{1\cdot 3\cdots(2n-1)} \frac{\sqrt{1-x^2}}{2\pi i}\oint_\gamma \frac{(1-z^2)^{n-1/2}}{(z-x)^{n+1}}dz.$

(b) Then use (a) to verify that the generating function of $T_n(x)$ is given by

$$G(x,u) = \frac{1-u^2}{2(1-2xu+u^2)} - \frac{1}{2}.$$

*10. Consider the Gegenbauer polynomials $P_n^\nu(x)$.

(a) Show that

$$P_n^\nu(x) = \frac{(-1)^n(2\nu+n-1)!(\nu-\frac{1}{2})!}{(2\nu-1)!(\nu+n-\frac{1}{2})!(1-x^2)}\oint_\gamma \frac{(1-z^2)^{\nu+n-1/2}}{(z-x)^{n+1}}dz.$$

(b) Show that the generating function of $P_n^\nu(x)$ is $G(x,u) = (1 - 2xu + u^2)^{-\nu}$.

*11. Consider the Hermite polynomials $\{H_n(x)\}$.

(a) Show that

$$H_n(x) = \frac{(-1)^n n!}{n\pi i}\oint_\gamma \frac{\exp(x^2-z^2)}{(z-x)^{n+1}}dz.$$

(b) Show that

$$\exp(2ux - u^2) = \sum_{n=0}^{\infty} \frac{H_n(x)}{n!}u^n.$$

12. Consider the Laguerre polynomials $\{L_n^\alpha(x)\}$.

(a) Show that

$$L_n^\alpha(x) = \frac{x^{-\alpha}}{2\pi i}\oint_\gamma \frac{z^{n+\lambda}\exp(x-z)}{(z-x)^{n+1}}\,dz.$$

(b) Show that the generating function of $\{L_n^\alpha(x)\}$ is

$$G(x,u) = (1-u)^{-\alpha-1} \exp\left\{\frac{-xu}{1-x}\right\}.$$

10

Control Theory

10.1 Introduction

In the last three decades, control theory has gained importance as a discipline for engineers, mathematicians, scientists, and other researchers. Examples of control problems include landing a vehicle on the moon, controlling the economy of a nation, manufacturing robots, and controlling the spread of an epidemic. Though a plethora of other books discuss continuous control theory [6], [75], [96], we will present here an introduction to discrete control theory.

We may represent a physical system that we intend to control by the homogeneous difference system

$$x(n+1) = Ax(n), \tag{10.1.1}$$

where A is a $(k \times k)$ matrix. We extensively studied this equation in Chapters 3 and 4; here we will refer to it as an *uncontrolled system.*

To control this system, or to induce it to behave in a predetermined fashion, we introduce into the system a forcing term, or a *control,* $u(n)$. Thus, the *controlled system* is the nonhomogeneous system

$$x(n+1) = Ax(n) + u(n). \tag{10.1.2}$$

In realizing system (10.1.2), it is assumed that the control can be applied to affect directly each of the state variables $x_1(n), x_2(n), \ldots, x_k(n)$ of the system. In most applications, however, this assumption is unrealistic. For example, in controlling an epidemic, we cannot expect to be able to affect directly all of the state variables of the system.

We find another example in the realm of economics. Economists, and certain politicians even, would pay dearly to know how the rate of inflation can be controlled, especially by altering some or all of the following variables: taxes, the money supply, bank lending rates. There probably is no equation like (10.1.2) that accurately describes the rate of inflation. Thus, a more reasonable model for the controlled system may be developed: We denote it by

$$x(n+1) = Ax(n) + Bu(n), \tag{10.1.3}$$

where B is a $(k \times m)$ matrix sometimes called the input matrix, and $u(n)$ is an $m \times 1$ vector. In this system, we have m control variables, or components, $u_1(n), u_2(n), \ldots, u_m(n)$, where $m \leq k$.

In engineering design and implementation, the system is often represented by a block diagram, as in Figures 10.1, 10.2.

The delay is represented traditionally by z^{-1}, since $\frac{1}{z}Z[x(n+1)] = Z[x(n)]$. (See Figure 10.3.)

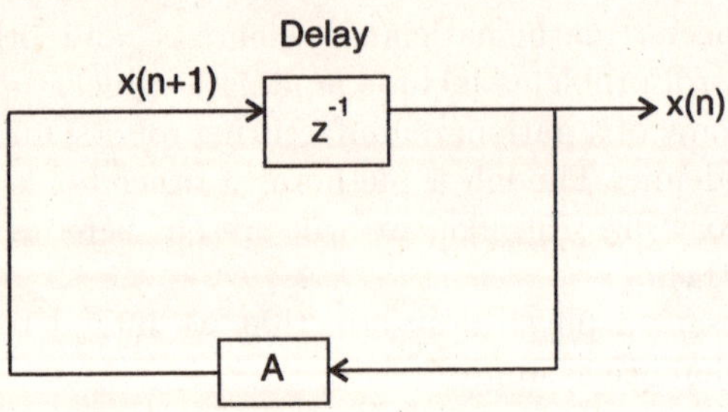

FIGURE 10.1. Uncontrolled system.

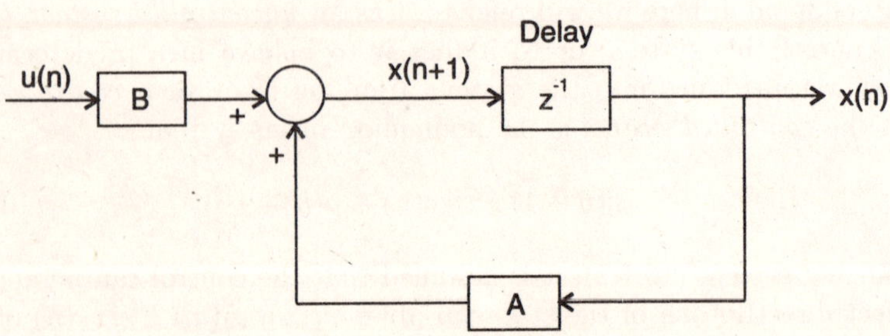

FIGURE 10.2. Controlled system.

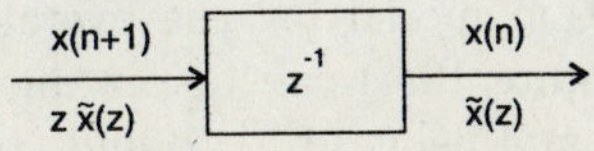

FIGURE 10.3. Representation of system delay.

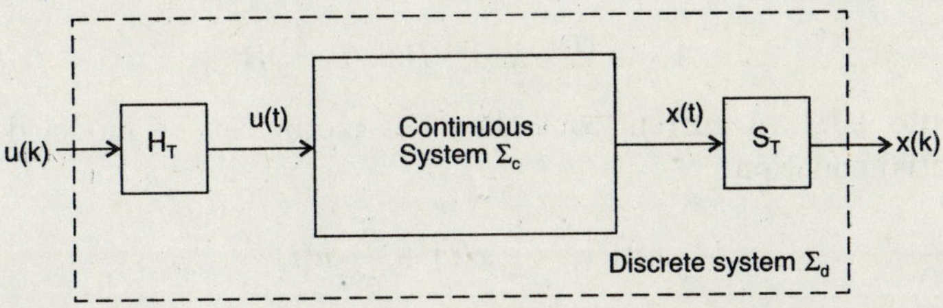

FIGURE 10.4. A continuous system with ideal sampler and zero-order hold.

10.1.1 *Discrete Equivalents for Continuous Systems*

One of the main areas of application for the discrete control methods developed in this chapter is the control of continuous systems, i.e., those modeled by differential and not difference equations. The reason for this is that while most physical systems are modeled by differential equations, control laws are often implemented on a digital computer, whose inputs and outputs are sequences. A common approach to control design in this case is to obtain an equivalent difference equation model for the continuous system to be controlled.

The block diagram of Figure 10.4 shows a common method of interfacing a continuous system to a computer for control. The system $\sum_c$ has state vector $x(t)$ and input $u(t)$ and is modeled by the differential equation

$$\dot{x}(t) = \hat{A}(t)x(t) + \hat{B}u(t). \tag{10.1.4}$$

The system S_T is an *ideal sampler* that produces, given a continuous signal $x(t)$, a sequence $x(k)$ defined by

$$x(k) = x(kT). \tag{10.1.5}$$

The system H_T is a *zero-order hold* that produces, given a sequence $u(k)$, a piecewise-constant continuous signal $u_c(t)$ defined by

$$u(t) = u(k), \quad t \in [kT, (k+1)T). \tag{10.1.6}$$

It is not hard to check that the solution to (10.1.4) for $t \in [kT, (k+1)T)$ is given by

$$\boxed{x(t) = e^{\hat{A}t}x(kT) + \int_{kT}^{t} e^{\hat{A}(t-\tau)}\hat{B}u(\tau)\,d\tau.} \tag{10.1.7}$$

Thus a difference equation model for the overall system $\sum_d$ (indicated by the dotted box in Figure 10.4) can be obtained by evaluating formula (10.1.7) at $t = (k+1)T$ and using (10.1.5) and (10.1.6):

$$x(k+1) = Ax(k) + Bu(k), \tag{10.1.8}$$

where

$$A = e^{\hat{A}T} \quad \text{and} \quad B = Te^{\hat{A}T}\hat{B}. \tag{10.1.9}$$

Example 10.1. A current-controlled DC motor can be modeled by the differential equation

$$\dot{x}(t) = -\frac{1}{\tau}\, x(t) + \frac{K}{\tau}\, u(t),$$

where x is the motor's angular velocity, u is the applied armature current, and K and τ are constants. A difference equation model suitable for the design of a discrete control system for this motor can be found using (10.1.8) and (10.1.9):

$$x(k+1) = Ax(k) + Bu(k),$$

where

$$A = e^{\hat{A}T} = e^{-T/\tau} \quad \text{and} \quad B = Te^{\hat{A}T}\hat{B} = \frac{KT}{\tau}\, e^{-T/\tau}.$$

10.2 Controllability

In this section we are mainly interested in the problem of whether it is possible to steer a system from a given initial state to any arbitrary state in a finite time period. In other words, we would like to determine whether a desired objective can be achieved by manipulating the chosen control variables. Until 1960, transform methods were the main tools in the analysis and design of controlled systems. Such methods are referred to now as classical control theory. In 1960, the Swiss mathematician/engineer R.E. Kalman [77] laid down the foundation of modern control theory by introducing state space methods. Consequently, matrices have gradually replaced transforms (e.g., Z-transform, Laplace transform), as the principal mathematical machinery in modern control theory [88], [108], [142].

Definition 10.2. System (10.1.3) is said to be *completely controllable* (or simply controllable) if for any $n_0 \in \mathbb{Z}^+$, any initial state $x(n_0) = x_0$, and any given final state (the desired state) x_f, there exists a finite time $N > n_0$ and a control $u(n)$, $n_0 < n \leq N$, such that $x(N) = x_f$.[1]

---

[1]In some books such a system may be referred to as completely reachable.

Remark: Since system (10.1.3) is completely determined by the matrices A and B, we may speak of the controllability of the pair $\{A, B\}$.

In other words, there exists a sequence of inputs $u(0), u(1), \ldots, u(N-1)$ such that this input sequence, applied to system (10.1.3), yields $x(N) = x_f$.

Example 10.3. Consider the system governed by the equations

$$\begin{aligned} x_1(n+1) &= a_{11}x_1(n) + a_{12}x_2(n) + bu(n), \\ x_2(n+1) &= a_{22}x_2(n). \end{aligned}$$

Here

$$A = \begin{pmatrix} a_{11} & a_{12} \\ 0 & a_{22} \end{pmatrix}, \quad B = \begin{pmatrix} b \\ 0 \end{pmatrix}.$$

It will not take much time before we realize that this system is not completely controllable, since $u(n)$ has no influence on $x_2(n)$. Moreover, $x_2(n)$ is entirely determined by the second equation and is given by $x_2(n) = a_{22}^n x_2(0)$.

The above example was easy enough that we were able to determine controllability by inspection. For more complicated systems, we are going to develop some simple criteria for controllability.

The *controllability matrix* W of system (10.1.3) is defined as the $k \times km$ matrix

$$\boxed{W = [B, AB, A^2B, \ldots, A^{k-1}B].} \tag{10.2.1}$$

The controllability matrix plays a major role in control theory, as may be seen in the following important basic result.

Theorem 10.4. *System* (10.1.3) *is* completely controllable *if and only if* rank $W = k$.

Before proving the theorem, we make a few observations about it and then prove a preliminary result.

First, consider the simple case where the system has only a single input, and thus the input matrix B reduces to an $m \times 1$ vector b. Hence the controllability matrix becomes the $k \times k$ matrix

$$\boxed{W = [b, Ab, \ldots, A^{k-1}b].}$$

The controllability condition that W has rank k means that the matrix W is nonsingular or its columns are linearly independent. For the general case, the controllability condition is that from among the km columns there are k linearly independent columns. Let us now illustrate the theorem by an example.

Example 10.5. Contemplate the system

$$y_1(n+1) = ay_1(n) + by_2(n),$$
$$y_2(n+1) = cy_1(n) + dy_2(n) + u(n),$$

where $ad - bc \neq 0$. Here

$$A = \begin{pmatrix} a & b \\ c & d \end{pmatrix}, \quad B = \begin{pmatrix} 0 \\ 1 \end{pmatrix},$$

and $u(n)$ is a scalar control. Now,

$$W = (B, AB) = \begin{pmatrix} 0 & b \\ 1 & d \end{pmatrix}$$

has rank 2 if $b \neq 0$. Thus the system is completely controllable by Theorem 10.4 if and only if $b \neq 0$.

Lemma 10.6. *For any $N \geq k$, the rank of the matrix*

$$[B, AB, A^2B, \ldots, A^{N-1}B]$$

is equal to the rank of the controllability matrix W.

PROOF. (I) Consider the matrix $W(n) = [B, AB, \ldots, A^{n-1}B], n = 1, 2, 3, \ldots$. As n increases by 1, either the rank of $W(n)$ remains constant or increases by at least 1. Suppose that for some $r > 1$, rank $W(r+1) =$ rank $W(r)$. Then every column in the matrix A^rB is linearly dependent on the columns of $W(r) = [B, AB, \ldots, A^{r-1}B]$. Hence

$$A^rB = BM_0 + ABM_1 + \cdots + A^{r-1}BM_{r-1}, \tag{10.2.2}$$

where each M_i is an $m \times m$ matrix. By premultiplying both sides of (10.2.2) by A, we obtain

$$A^{r+1}B = ABM_0 + A^2BM_1 + \cdots + A^rBM_{r-1}.$$

Thus the columns of $A^{r+1}B$ are linearly dependent on the columns of $W(r+1)$. This implies that rank $W(r+2) =$ rank $W(r+1) =$ rank $W(r)$. By repeating this process, one may conclude that

$$\text{rank } W(n) = \text{ rank } W(r) \quad \text{for all } n > r.$$

We conclude from the above argument that rank $W(n)$ increases by at least 1 as n increases by 1 until it attains its maximum k. Hence the rank maximum of $W(n)$ is attained in at most k steps. Therefore, the maximum rank is attained at $n \leq k$ and, consequently, rank W ($\equiv$ rank $W(k)$) $=$ rank $W(N)$ for all $N \geq k$. □

PROOF. (II) In the second proof we apply the Cayley–Hamilton theorem (Chapter 3). So if $p(\lambda) = \lambda^k + p_1\lambda^{k-1} + \cdots + p_k$ is the characteristic polynomial of A, then $p(A) = 0$, i.e.,

$$A^k + p_1A^{k-1} + \cdots + p_kI = 0,$$

or

$$A^k = \sum_{i=1}^{k} q_i A^{k-1}, \tag{10.2.3}$$

where $q_i = -p_i$. Multiplying expression (10.2.3) by B, we obtain

$$A^k B = \sum_{i=1}^{k} q_i A^{k-i} B. \tag{10.2.4}$$

Thus the columns of $A^k B$ are linearly dependent on the columns of $W(k) \equiv W$. Therefore, rank $W(k+1) =$ rank W. By multiplying expression (10.2.4) by A we have

$$A^{k+1}B = q_1 A^k + q_2 A^{k-1} + \cdots + q_k A.$$

Consequently, rank $W(k+2) =$ rank $W(k+1) =$ rank W. By repeating the process, one concludes that rank $W(N) =$ rank W for all $N \geq k$. □

We are now ready to prove the theorem.

PROOF OF THEOREM 10.4.

Sufficiency Suppose that rank $W = k$. Let x_0 and x_f be two arbitrary vectors in $\mathbb{R}^k$. Recall that by the variation of constants formula (3.2.14) we have

$$x(k) - A^k x(0) = \sum_{r=0}^{k-1} A^{k-r-1} Bu(r),$$

or

$$x(k) - A^k x(0) = W\bar{u}(k), \tag{10.2.5}$$

where

$$\bar{u}(k) = \begin{pmatrix} u(k-1) \\ u(k-2) \\ \vdots \\ u(0) \end{pmatrix}.$$

Since rank $W = k$, range $W = R^k$. Hence if we let $x(0) = x_0$ and $x(k) = x_f$, then $x_f - A^k x_o \in$ range W. Thus $x_f - A^k x_0 = W\bar{u}$ for some vector $\bar{u} \in \mathbb{R}^k$. Consequently, system (10.1.3) is completely controllable.

Necessity Assume that system (10.1.3) is completely controllable and rank $W < k$. From the proof of Lemma 10.6 (Proof I) we conclude that there exists $r \in \mathbb{Z}^+$ such that

$$\begin{aligned}\text{rank } W(1) < \text{ rank } W(2) < \cdots < \text{ rank } W(r)\\ = \text{ rank } W(r+1) = \cdots = \text{ rank } W.\end{aligned}$$

Moreover, rank $W(n) =$ rank W for all $n > k$. Furthermore, since $W(j+1) = (W(j), A^j B)$, it follows that

$$\begin{aligned}\text{range } W(1) \subset \text{range } W(2) \subset \cdots \subset \text{ range } W(r)\\ = \text{ range } W(r+1) = \cdots = \text{range } W = \cdots = \text{ range } W(n)\end{aligned}$$

for any $n > k$.

Since rank $W < k$, range $W \neq R^k$. Thus there exists $\xi \notin$ range W. This implies that $\xi \notin$ range $W(n)$ for all $n \in \mathbb{Z}^+$. If we let $x_0 = 0$ in formula (10.2.5) with k replaced by n, we have $x(n) = W(n)\bar{u}(n)$. Hence for ξ to be equal to $x(n)$ for some n, ξ must be in the range of $W(n)$. But $\xi \notin$ range $W(n)$ for all $n \in \mathbb{Z}^+$ implies that ξ may not be reached at any time from the origin, which is a contradiction. Therefore, rank $W = k$. □

Remark 10.7. There is another definition of complete controllability in the literature that I will call here "*controllability to the origin.*" A system is *controllable to the origin* if, for any $n_0 \in \mathbb{Z}^+$ and $x_0 \in \mathbb{R}^k$, there exists a finite time $N > n_0$ and a control $u(n), n_0 < n \leq N$, such that $x(N) = 0$.

Clearly, complete controllability is a stronger property than controllability to the origin. The two notions coincide in continuous-time systems. (See [75].) However, for the discrete-time system (10.1.3), controllability to the origin does not imply complete controllability unless A is nonsingular (Exercises 10.1 and 10.2, Problem 13). The following example illustrates our remark.

Example 10.8. Consider the control system $x(n+1) = Ax(n) + Bu(n)$ with

$$A = \begin{pmatrix} 0 & 1 \\ 0 & 0 \end{pmatrix}, \quad B = \begin{pmatrix} 1 \\ 0 \end{pmatrix}. \tag{10.2.6}$$

Now, for

$$x(0) = x_0 = \begin{pmatrix} x_{01} \\ x_{02} \end{pmatrix},$$

we have, from (10.2.6),

$$\begin{aligned} x(1) &= Ax_0 + Bu(0) \\ &= \begin{pmatrix} 0 & 1 \\ 0 & 0 \end{pmatrix}\begin{pmatrix} x_{01} \\ x_{02} \end{pmatrix} + \begin{pmatrix} 1 \\ 0 \end{pmatrix} u(0) \\ &= \begin{pmatrix} x_{02} \\ 0 \end{pmatrix} + \begin{pmatrix} u(0) \\ 0 \end{pmatrix}. \end{aligned}$$

So if we pick $u(0) = -x_{02}$, then we will have $x(1) = 0$. Therefore, system (10.2.6) is controllable to zero. Observe, however, that

$$\text{rank } (B, AB) = \text{ rank } \begin{pmatrix} 1 & 0 \\ 0 & 0 \end{pmatrix} = 1 < 2.$$

Thus by Theorem 10.4, system (10.2.6) is not completely controllable.

Example 10.9. Contemplate the system $y(n+1) = Ay(n) + Bu(n)$, where

$$A = \begin{pmatrix} 0 & 1 \\ 2 & -1 \end{pmatrix}, \quad B = \begin{pmatrix} 1 \\ 1 \end{pmatrix}.$$

Now, $W(1) = \begin{pmatrix} 1 \\ 1 \end{pmatrix}$ is of rank 1, and $W(2) = \begin{pmatrix} 1 & 1 \\ 1 & 1 \end{pmatrix}$ is also of rank 1, since it is now equivalent to $\begin{pmatrix} 1 & 1 \\ 0 & 0 \end{pmatrix}$. Hence according to Theorem 10.4 the system is not controllable. Notice, however, that the point $\begin{pmatrix} -4 \\ 0 \end{pmatrix}$ is reachable from $\begin{pmatrix} 1 \\ 2 \end{pmatrix}$ under the control $u(n) = -2$ in time $n = 2$.

Example 10.10. Figure 10.5 shows a cart of mass m attached to a wall via a flexible linkage. The equation of motion for this system is

$$m\ddot{x} + b\dot{x} + kx = u, \tag{10.2.7}$$

where k and b are the stiffness and damping, respectively, of the linkage, and u is an applied force. Equation (10.2.7) can be written in state variable

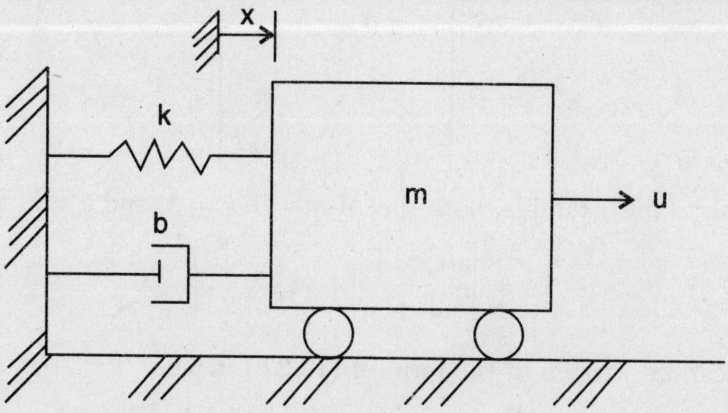

FIGURE 10.5. A cart attached to a wall via a flexible linkage.

form as

$$\begin{bmatrix} \dot{x} \\ \dot{v} \end{bmatrix} = \begin{bmatrix} 0 & 1 \\ -k/m & -b/m \end{bmatrix} \begin{bmatrix} x \\ v \end{bmatrix} + \begin{bmatrix} 0 \\ 1/m \end{bmatrix} u. \tag{10.2.8}$$

Thus

$$\hat{A} = \begin{bmatrix} 0 & 1 \\ -k/m & -b/m \end{bmatrix}, \quad \hat{B} = \begin{bmatrix} 0 \\ 1/m \end{bmatrix}.$$

Recall that, given a sample period T, the matrices A and B of the equivalent discrete system are given by

$$A = e^{\hat{A}T}, \quad B = Te^{\hat{A}T}\hat{B},$$

so that their computation requires finding the exponential of the matrix $\hat{A}$. This is not so difficult as it sounds, at least when $\hat{A}$ can be diagonalized, for then we can find a matrix P such that

$$\hat{A} = P\Lambda P^{-1}, \tag{10.2.9}$$

where Λ is a diagonal matrix

$$\Lambda = \begin{bmatrix} \lambda_1 & \cdots & 0 \\ \vdots & \ddots & \vdots \\ 0 & \cdots & \lambda_k \end{bmatrix}. \tag{10.2.10}$$

By definition,

$$e^{\hat{A}T} = I + \hat{A}T + \frac{1}{2!}\hat{A}^2T^2 + \frac{1}{3!}\hat{A}^3T^3 + \cdots,$$

so that substituting using (10.2.9) gives

$$e^{\hat{A}T} = Pe^{\Lambda T}P^{-1},$$

and the diagonal form (10.2.10) of Λ gives

$$e^{\hat{A}T} = P \begin{bmatrix} e^{\lambda_1 T} & \cdots & 0 \\ \vdots & \ddots & \vdots \\ 0 & \cdots & e^{\lambda_k T} \end{bmatrix} P^{-1}.$$

Returning to our example, note that if $m = 1, k = 2$, and $b = 3$, then

$$\hat{A} = \begin{bmatrix} 0 & 1 \\ -2 & -3 \end{bmatrix}, \quad \hat{B} = \begin{bmatrix} 0 \\ 1 \end{bmatrix}.$$

Thus $\hat{A}$ can be written in the form of (10.2.9), where

$$\Lambda = \begin{bmatrix} -1 & 0 \\ 0 & -2 \end{bmatrix}, \quad P = \begin{bmatrix} 1 & 1 \\ -1 & -2 \end{bmatrix}.$$

Hence

$$A = e^{\hat{A}T} = \begin{bmatrix} e^{-T} & 0 \\ 0 & e^{-2T} \end{bmatrix} P^{-1}$$

$$= \begin{bmatrix} 2e^{-T} - e^{-2T} & e^{-T} - e^{-2T} \\ -2e^{-T} + 2e^{-2T} & -e^{-T} + 2e^{-2T} \end{bmatrix},$$

$$B = Te^{\hat{A}T}\hat{B} = T\begin{bmatrix} e^{-T} - e^{-2T} \\ -e^{-T} + 2e^{-2T} \end{bmatrix}.$$

The controllability of the discrete equivalent system can then be checked by computing

$$W = [B \quad AB] = T\begin{bmatrix} e^{-T} - e^{-2T} & e^{-2T} - e^{-4T} \\ -e^{-T} + 2e^{-2T} & -e^{-2T} + 2e^{-4T} \end{bmatrix}.$$

Checking the determinant gives

$$\det\ W = -T^2e^{-4T}(1 - e^{-T} + e^{-2T}),$$

which is zero only if $T = 0$. Thus the cart is controllable for any nonzero sample period.

10.2.1 Controllability Canonical Forms

Consider the second-order difference equation

$$z(n+2) + p_1z(n+1) + p_2z(n) = u(n).$$

Recall from Section 3.2 that this equation is equivalent to the system

$$x(n+1) = Ax(u) + Bu(n),$$

where

$$A = \begin{pmatrix} 0 & 1 \\ -p_2 & -p_1 \end{pmatrix}, \quad B = \begin{pmatrix} 0 \\ 1 \end{pmatrix}, \quad x = \begin{pmatrix} z(n) \\ z(n+1) \end{pmatrix}.$$

Clearly,

$$W(2) = \begin{pmatrix} 0 & 1 \\ 1 & -p_1 \end{pmatrix}$$

has rank 2 for all values of p_1 and p_2. Consequently, this equation is always completely controllable.

The preceding example may be generalized to the kth-order equation

$$z(n+k) + p_1z(n+k+1) + \cdots + p_kz(n) = u(n), \tag{10.2.11}$$

which is equivalent to the system

$$x(n+1) = Ax(n) + bu(n),$$

where

$$A = \begin{pmatrix} 0 & 1 & 0 & \cdots & 0 \\ \vdots & 0 & 1 & & \vdots \\ \cdot & & & & 1 \\ -p_k & -p_{k-1} & \cdots & & -p_1 \end{pmatrix}, \quad B = e_k = \begin{pmatrix} 0 \\ \vdots \\ 1 \end{pmatrix},$$

$$x(n) = \begin{pmatrix} z(n) \\ z(n+1) \\ \vdots \\ z(n+k-1) \end{pmatrix}.$$

Notice that

$$AB = \begin{pmatrix} 0 \\ 0 \\ \vdots \\ 1 \\ -p_1 \end{pmatrix}, \quad A^2B = \begin{pmatrix} 0 \\ 0 \\ \vdots \\ 1 \\ * \\ * \end{pmatrix}, \quad \ldots, \quad A^{k-1}B = \begin{pmatrix} 1 \\ * \\ \vdots \\ * \end{pmatrix},$$

where the $*$'s are some combinations of the products of the p_i's. It follows that

$$W = \begin{pmatrix} 0 & 0 & \cdots & 1 \\ \vdots & \vdots & & \vdots \\ 0 & 1 & \cdots & * \\ 1 & * & \cdots & * \end{pmatrix}$$

is of rank k, and so the equation, and thus the system, is completely controllable. The converse of the above statement is also valid. That is to say, if system (10.1.3), with $k \times 1$ vector $B \equiv b$, is completely controllable, then it can be put in the form of a kth scalar equation (10.2.11) by a similarity transformation. To accomplish this task we start with the $k \times k$ controllability matrix $W = (b, Ab, \ldots, A^{k-1}b)$. Since system (10.1.3) is completely controllable, it follows from Theorem 10.4 that W is nonsingular. Let us write W^{-1} in terms of its rows as

$$W^{-1} = \begin{pmatrix} w_1 \\ w_2 \\ \vdots \\ w_k \end{pmatrix}.$$

We claim that the set $\{w_k, w_kA, \ldots, w_kA^{k-1}\}$ generated by the last row of W^{-1} is linearly independent. To show this, suppose that for some constants $a_1, a_2, \ldots, a_k$ we have

$$a_1 w_k + a_2 w_k A + \cdots + a_k w_k A^{k-1} = 0. \tag{10.2.12}$$

Multiplying (10.2.12) from the right by b yields

$$a_1 w_k b + a_2 w_k Ab + \cdots + a_k w_k A^{k-1} b = 0. \tag{10.2.13}$$

Since $W^{-1}W = I$, it follows that $w_k b = w_k Ab = \cdots = w_k A^{k-2} b = 0$ and $w_k A^{k-1} b = 1$. Hence it follows from (10.2.13) that $a_k = 0$. One may repeat this procedure by multiplying (10.2.12) by Ab (and letting $a_k = 0$) to conclude that $a_{k-1} = 0$. Continuing this procedure, one may show that $a_i = 0$ for $1 \le i \le k$. This proves our claim that the vectors $w_k, w_kA, \ldots, w_kA^{k-1}$ are linearly independent. Hence the $k \times k$ matrix

$$P = \begin{pmatrix} w_k \\ w_k A \\ \vdots \\ w_k A^{k-1} \end{pmatrix}$$

is nonsingular. Define a change of coordinates for system (10.1.3) by

$$z(n) = Px(n), \tag{10.2.14}$$

which gives

$$z(n+1) = PAP^{-1} z(n) + Pbu(n),$$

or

$$z(n+1) = \hat{A} z(n) + \hat{b} u(n), \tag{10.2.15}$$

where

$$\hat{A} = PAP^{-1}, \quad \hat{b} = Pb. \tag{10.2.16}$$

clearly,

$$\hat{b} = Pb = \begin{pmatrix} 0 \\ 0 \\ \vdots \\ 0 \\ 1 \end{pmatrix}.$$

Now,

$$\hat{A} = PAP^{-1} = \begin{pmatrix} w_k A \\ w_k A^2 \\ \vdots \\ w_k A^k \end{pmatrix} P^{-1}.$$

Since $w_k A$ is the second row in P, it follows that

$$w_k A P^{-1} = (0 \ \ 1 \ \ 0 \ \ \cdots \ \ 0).$$

Similarly,

$$w_k A^2 P^{-1} = (0 \ \ 0 \ \ 1 \ \ 0 \ \ \cdots \ \ 0),$$

$$\vdots$$

$$w_k A^{k-1} P^{-1} = (0 \ \ 0 \ \ \cdots \ \ 0 \ \ 1).$$

However,

$$w_k A^k P^{-1} = (-p_k \ \ -p_{k-1} \ \ \cdots \ \ -p_1),$$

where the p_i's are some constants. Thus

$$\hat{A} = \begin{pmatrix} 0 & 1 & 0 & \cdots & 0 \\ 0 & 0 & 1 & \cdots & 0 \\ \vdots & & & & \vdots \\ 0 & 0 & 0 & \cdots & 1 \\ -p_k & -p_{k-1} & -p_{k-2} & \cdots & -p_1 \end{pmatrix}$$

with characteristic equation

$$\lambda^k + p_1 \lambda^{k-1} + p_2 \lambda^{k-2} + \cdots + p_k = 0.$$

Observe that A and $\hat{A}$ have the same characteristic equation. The above discussion proves the following.

Theorem 10.11. *A system $x(n+1) = Ax(n) + bu(n)$ is completely controllable if and only if it is equivalent to a kth-order equation of the form* (10.2.11).

System (10.2.15) is said to be in a *controllable canonical form.*

Another controllable canonical form may be obtained by using the change of variables $x(n) = Wz(n)$, where W is the controllability matrix of the system. This is a more popular form among engineers due to its simple derivative. The reader is asked in Exercises 10.1 and 10.2, Problem 20, to show that the obtained controllable canonical pair $\{\tilde{A}, \tilde{b}\}$ are given by

$$\tilde{A} = \begin{pmatrix} 0 & 0 & 0 & \dots & -p_k \\ 1 & 0 & 0 & \dots & -p_{k-1} \\ 0 & 1 & 0 & \dots & -p_{k-2} \\ \vdots & \vdots & & & \\ 0 & 0 & & \dots & -p_1 \end{pmatrix}, \quad \tilde{b} = \begin{pmatrix} 1 \\ 0 \\ \vdots \\ 0 \end{pmatrix}. \tag{10.2.17}$$

Exercises 10.1 and 10.2

In Problems 1 through 6 determine whether or not the system $x(n+1) = Ax(n) + Bu(n)$ is completely controllable.

1. $A = \begin{pmatrix} -2 & 2 \\ 1 & -1 \end{pmatrix}, \quad B = \begin{pmatrix} 1 \\ 0 \end{pmatrix}.$

2. $A = \begin{pmatrix} -1 & 0 \\ 0 & -2 \end{pmatrix}, \quad B = \begin{pmatrix} 2 \\ 3 \end{pmatrix}.$

3. $A = \begin{pmatrix} -1 & 0 \\ 0 & -2 \end{pmatrix}, \quad B = \begin{pmatrix} 2 \\ 0 \end{pmatrix}.$

4. $A = \begin{pmatrix} -2 & 1 & 0 & 0 & 0 \\ 0 & -2 & 1 & 0 & 0 \\ 0 & 0 & -2 & 0 & 0 \\ 0 & 0 & 0 & -5 & 1 \\ 0 & 0 & 0 & 0 & -5 \end{pmatrix}, \quad B = \begin{pmatrix} 0 & 1 \\ 0 & 0 \\ 3 & 0 \\ 0 & 0 \\ 2 & 1 \end{pmatrix}.$

5. $A = \begin{pmatrix} -2 & 1 & 0 & 0 & 0 \\ 0 & -2 & 1 & 0 & 0 \\ 0 & 0 & -2 & 0 & 0 \\ 0 & 0 & 0 & -5 & 1 \\ 0 & 0 & 0 & 0 & -5 \end{pmatrix}, \quad B = \begin{pmatrix} 0 & 1 \\ 3 & 0 \\ 0 & 0 \\ 2 & 1 \\ 0 & 0 \end{pmatrix}.$

6. $A = \begin{pmatrix} A_{11} & A_{12} \\ 0 & A_{22} \end{pmatrix}, \quad B = \begin{pmatrix} B_1 \\ 0 \end{pmatrix},$

   A_{11} is an $r \times r$ matrix, A_{12} is an $r \times s$ matrix, A_{22} is an $s \times s$ matrix, B_1 is an $r \times m$ matrix, where $r + s = k$.

   We say that a state x_f is *reachable* from an initial state x_0 if there exists $N \in \mathbb{Z}^+$ and a control $u(n), n = 0, 1, \dots, N-1$, such that $x(N, x_0) = x_f$.

7. Prove that a state x_f is reachable from x_0 in time N if and only if $x_f - A^N x_0 \in$ range $W(N)$.

8. Consider the system $x(n+1) = Ax(n) + Bu(n)$, where

$$A = \begin{pmatrix} 1 & 2 & -1 \\ 0 & 1 & 0 \\ 1 & -4 & 3 \end{pmatrix}, \quad B = \begin{pmatrix} 0 \\ 0 \\ 1 \end{pmatrix}.$$

Find a basis for the set of vectors in $\mathbb{R}^3$ that are reachable from the origin.

9. Consider the system

$$x(n+1) = \begin{pmatrix} -1 & -1 \\ 2 & -4 \end{pmatrix} x(n) + Bu(n).$$

Find for what vectors B in $\mathbb{R}^2$ the system is not completely controllable.

10. Obtain a necessary and sufficient condition for

$$x_1(n+1) = a_{11}x_1(n) + a_{12}x_2(n) + u(n),$$
$$x_2(n+1) = a_{21}x_1(n) + a_{22}x_2(n) - u(n),$$

to be controllable.

11. Obtain a necessary and sufficient condition for $x(n+1) = Ax(n) + Bu(n)$ to be completely controllable, where

$$A = \begin{pmatrix} a_{11} & a_{12} \\ a_{21} & a_{22} \end{pmatrix}, \quad B = \begin{pmatrix} 1 \\ 1 \end{pmatrix}.$$

12. Consider the system

$$x_1(n+1) = x_2(n) + u_1(n) + u_2(n),$$
$$x_2(n+1) = x_3(n) + u_1(n) - u_2(n),$$
$$x_3(n+1) = u_1(n).$$

(a) Prove that the system is completely controllable in two steps.

(b) If $u_2(n) \equiv 0$, show that the system is completely controllable in three steps.

(c) If $u_1(n) \equiv 0$, show that the system is not completely controllable.

13. Show that if the matrix A in (10.1.3) is nonsingular, then complete controllability and controllability to the origin are equivalent.

14. Prove that if $U = WW^T$ is positive definite, then W has rank k, where W is the controllability matrix of (10.1.3). Prove also the converse.

15. Show that system (10.1.3) is completely controllable if $[B, AB, \ldots, A^{k-r}]$ has rank k, where $r = \text{rank } B$.

16. A polynomial $\varphi(\lambda) = \lambda^m + a_1\lambda^{m-1} + \cdots + a_m$ is said to be the *minimal polynomial* of a $k \times k$ matrix A if $\varphi(\lambda)$ is the lowest-degree polynomial for which $\varphi(A) = 0$. It follows that $m \leq k$. Prove that system $\{A, B\}$ is completely controllable if and only if $\text{rank}[B, AB, \ldots, A^{m-r}B] = k$, where rank $B = r$.

17. For a $k \times k$ matrix A and a $k \times m$ matrix B, prove that the following statements are true:

    (i) If $\{A, B\}$ is completely controllable, then so is $\{A + BC, B\}$ for any $m \times 1$ vector C.

    (ii) If $\{A + BC, B\}$ is completely controllable for some $m \times 1$ vector C_0, then $\{A + BC, B\}$ is completely controllable for any $m \times 1$ vector C.

18. Consider the system $x(n+1) = Ax(n) + Bu(n)$, where A is a $k \times k$ matrix, B is a $k \times m$ matrix, and such that A has k linearly independent eigenvectors. Prove that $\{A, B\}$ is completely controllable if and only if no row of $P^{-1}B$ has all zero elements, where $P = (\xi_1, \ldots, \xi_k)$, the ξ_i's being the eigenvectors of A.

19. Suppose that in Problem 18, A does not possess k linearly independent eigenvectors and that there exists a nonsingular matrix P where $P^{-1}AP = J$ is the Jordan canonical form of A. Prove that $\{A, B\}$ is completely controllable if and only if:

    (i) no two Jordan blocks in J are associated with the same eigenvalue,

    (ii) the elements of any row of $P^{-1}B$ that correspond to the last row of each Jordan block are not all zero, and

    (iii) the elements of each row of $P^{-1}B$ that correspond to distinct eigenvalues are not all zero.

20. (Another controllability canonical form). Consider the completely controllable system $x(n+1) = x(n) + bu(n)$, where b is a $k \times 1$ vector. Let $x(n) = Wz(n)$, where W is the controllability matrix. Then the system becomes $z(n+1) = \tilde{A}z(n) + \tilde{b}u(n)$. Show that $\tilde{A}$ and $\tilde{b}$ are given by (10.2.17).

21. Consider the system $x(n+1) = Ax(n) + Bu(n)$, where

$$A = \begin{pmatrix} 1 & 0.6 \\ 0 & 0.4 \end{pmatrix}, \quad B = \begin{pmatrix} 0.4 \\ 0.6 \end{pmatrix}.$$

The reachability set is defined by $R(n) = \{x(0)\colon x(0) \text{ is reached from the origin in } N \text{ steps with } |u(i)| \leq 1, 1 \leq i \leq N\}$.

    (a) Find $R(1)$ and plot it.

    (b) Find $R(2)$ and plot it.

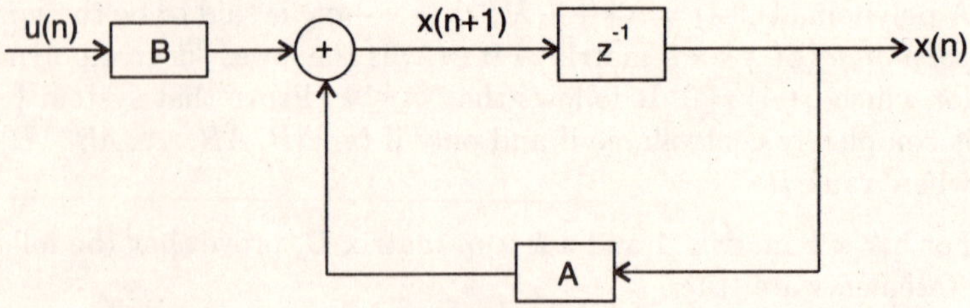

FIGURE 10.6. Output is the same as the state: $y(n) = x(n)$.

10.3 Observability

In the previous section it was assumed that (the observed) output of the control system is the same as that of the state of the system $x(n)$. In practice, however, one may not be able to observe the state of the system $x(n)$ but rather an output $y(n)$ that is related to $x(n)$ in a specific manner. The mathematical model of this type of system is given by

$$\begin{aligned} x(n+1) &= Ax(n) + Bu(n), \\ y(n) &= Cx(n), \end{aligned} \tag{10.3.1}$$

where $A(n)$ is a $k \times k$ matrix, B a $k \times m$ matrix, $u(n)$ an $m \times 1$ vector, and C an $r \times k$ matrix. The control $u(n)$ is the input of the system, and $y(n)$ is the output of the system, as shown in Figures 10.6, 10.7.

Roughly speaking, observability means that it is possible to determine the state of a system $x(n)$ by measuring only the output $y(n)$. Hence it is useful in solving the problem of reconstructing unmeasurable state variables from measurable ones. The input–output system (10.3.1) is *completely observable* if for any $n_0 \geq 0$, there exists $N > n_0$ such that the knowledge of $u(n)$ and $y(n)$ for $n_0 \leq n \leq N$ suffices to determine $x(n_0) = x_0$.

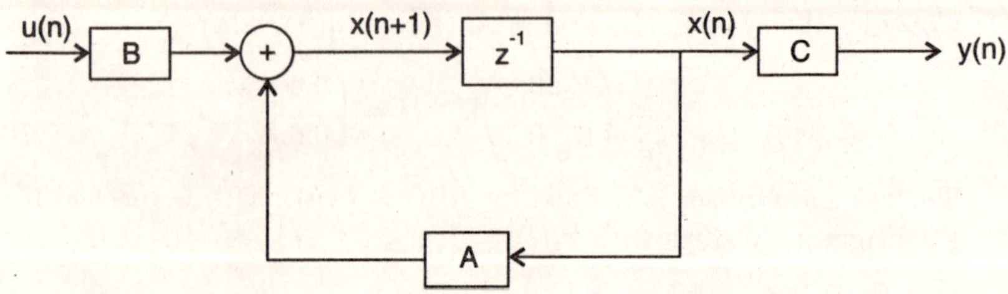

FIGURE 10.7. Input–output system: $y(n) = Cx(n)$.

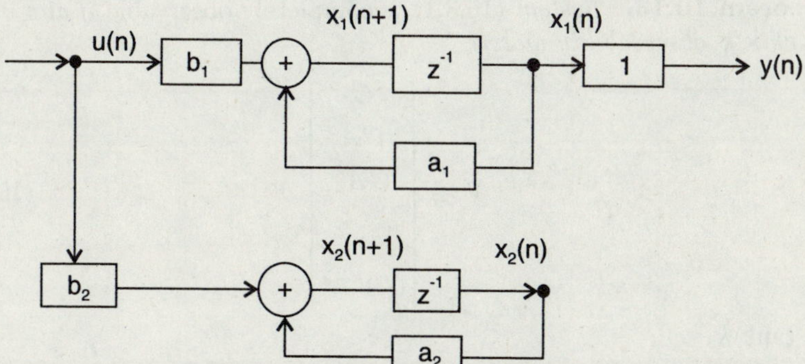

FIGURE 10.8. A nonobservable system.

Example 10.12. Consider the system (Figure 10.8)

$$\begin{aligned} x_1(n+1) &= a_1x_1(n) + b_1u(n), \\ x_2(n+1) &= a_2x_2(n) + b_2u(n), \\ y(n) &= x_1(n). \end{aligned}$$

This system is not observable, since the first equation shows that $x_1(n) = y(n)$ is completely determined by $u(n)$ and $x_1(0)$ and that there is no way to determine $x_2(0)$ from the output $y(n)$.

In discussing observability, one may assume that the control $u(n)$ is identically zero. This obviously simplifies our exposition. To explain why we can do this without loss of generality, we write $y(n)$ using the variation of constants formula (3.2.14) for $x(n)$:

$$y(n) = Cx(n),$$

or

$$\boxed{y(n) = CA^{n-n_0}x_0 + \sum_{j=n_0}^{n-1} CA^{n-j-1}Bu(j).}$$

Since C, A, B, and u are all known, the second term on the right-hand side of this last equation is known. Thus it may be subtracted from the observed value $y(n)$. Hence, for investigating a necessary and sufficient condition for complete observability it suffices to consider the case where $u(n) \equiv 0$.

We now present a criterion for complete observability that is analogous to that of complete controllability.

Theorem 10.13. *System* (10.3.1) *is* completely observable *if and only if the* $rk \times k$ *observability matrix*

$$V = \begin{bmatrix} C \\ CA \\ CA^2 \\ \vdots \\ CA^{k-1} \end{bmatrix} \tag{10.3.2}$$

has rank k.

PROOF. By applying the variation of constants formula (3.2.14) to (10.3.1) we obtain

$$y(n) = Cx(n) = C\left[A^n x_0 + \sum_{r=0}^{n-1} A^{n-r-1}Bu(r)\right]. \tag{10.3.3}$$

Let

$$\hat{y}(n) = y(n) - \sum_{r=0}^{n-1} CA^{n-r-1}Bu(r). \tag{10.3.4}$$

Using formula (10.3.3), equation (10.3.4) may now be written as

$$\hat{y}(n) = CA^n x_0. \tag{10.3.5}$$

Putting $n = 0, 1, 2, \ldots, k-1$ in (10.3.5) yields

$$\begin{bmatrix} \hat{y}(0) \\ \hat{y}(1) \\ \vdots \\ \hat{y}(k-1) \end{bmatrix} = \begin{bmatrix} C \\ CA \\ \vdots \\ CA^{k-1} \end{bmatrix} x_0. \tag{10.3.6}$$

Suppose that rank $V = k$. Then range $V = R^k$. Now, if $y(n), u(n)$ are given for $0 \le n \le k-1$, then it follows from (10.3.4) that $\hat{y}(n), 0 \le n \le k-1$, is also known. Hence there exists $x_0 \in \mathbb{R}^k$ such that (10.3.6) holds. Hence system (10.3.1) is completely observable. Conversely, suppose system (10.3.1) is completely observable. Let us write

$$V(N) = \begin{bmatrix} C \\ CA \\ CA^2 \\ \vdots \\ CA^{N-1} \end{bmatrix} = \left(C^T, A^T C^T, (A^T)^2 C^T, \ldots, (A^T)^{N-1} C^T\right)^T. \tag{10.3.7}$$

Then from Theorem 10.13, $V(N)$ is of rank k if and only if the observability matrix $V \equiv V(k)$ has rank k. Therefore, if x_0 can be uniquely determined from N observations $y(0), y(1), \ldots, y(N-1)$, it can be so determined from $y(0), y(1), \ldots, y(k-1)$. Thus rank $V = k$. □

Notice that the matrix B does not play any role in determining observability. This confirms our earlier remark that in studying observability, one may assume that $u(n) \equiv 0$. Henceforth, we may speak of the observability of the pair $\{A, C\}$.

Example 10.3 revisited. Consider again Example 10.3. The system may be written as

$$\begin{pmatrix} x_1(n+1) \\ x_2(n+1) \end{pmatrix} = \begin{pmatrix} a_1 & 0 \\ 0 & a_2 \end{pmatrix} \begin{pmatrix} x_1(n) \\ x_2(n) \end{pmatrix} + \begin{pmatrix} b_1 \\ b_2 \end{pmatrix} u(n),$$

$$y(n) = \begin{pmatrix} 1 & 0 \end{pmatrix} \begin{pmatrix} x_1(n) \\ x_2(n) \end{pmatrix}.$$

Thus $A = \begin{pmatrix} a_1 & 0 \\ 0 & a_2 \end{pmatrix}$ and $C = \begin{pmatrix} 1 & 0 \end{pmatrix}$. It follows that the observability matrix is given by $V = \begin{pmatrix} 1 & a_1 \end{pmatrix} \begin{pmatrix} 0 & 0 \end{pmatrix}$. Since rank $V = 1 < 2$, the system is not completely observable by virtue of Theorem 10.13.

Finally, we give an example to illustrate the above results.

Example 10.14. Consider the input–output system (Figure 10.9)

$$\begin{aligned} x_1(n+1) &= x_2(n), \\ x_2(n+1) &= -x_1(n) + 2x_2(n) + u(n), \\ y(n) &= c_1 x_1(n) + c_2 x_2(n). \end{aligned}$$

Then $A = \begin{pmatrix} 0 & 1 \\ -1 & 2 \end{pmatrix}$, $B = \begin{pmatrix} 0 \\ 1 \end{pmatrix}$, and $C = (c_1, c_2)$. The observability matrix is given by

$$V = \begin{pmatrix} C \\ CA \end{pmatrix} = \begin{pmatrix} c_1 & c_2 \\ -c_2 & c_1 + 2c_2 \end{pmatrix}.$$

By adding the first column to the second column in V we obtain the matrix

$$\hat{V} = \begin{pmatrix} c_1 & c_1 + c_2 \\ -c_2 & c_1 + c_2 \end{pmatrix}.$$

Observe that rank $\hat{V} = 2$ if and only if $c_1 + c_2 \neq 0$. Since rank $V =$ rank $\hat{V}$, it follows that the system is completely observable if and only if $c_1 + c_2 \neq 0$ (or $c_2 \neq -c_1$).

We may also note that the system is completely controllable.

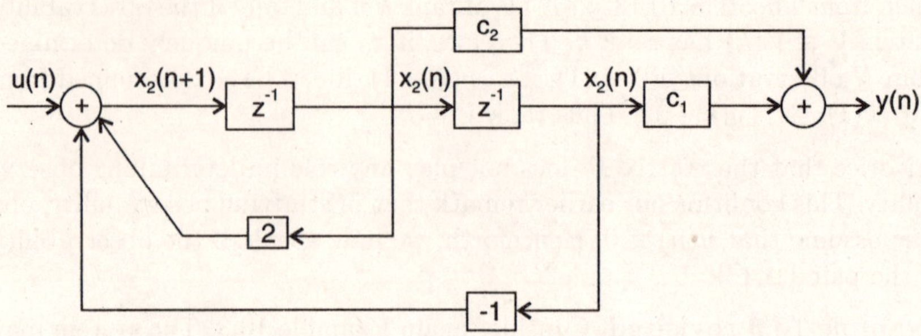

FIGURE 10.9. A completely observable and controllable system.

Example 10.15. Example 10.10 looked at the controllability of a cart attached to a fixed wall via a flexible linkage using an applied force u. A dual question can be posed: If the force on the cart is a constant, can its magnitude be observed by measuring the cart's position? In order to answer this question, the state equation (10.2.8) must be *augmented* with one additional equation, representing the assumption that the applied force is constant:

$$\begin{bmatrix} \dot{x} \\ \dot{v} \\ \dot{u} \end{bmatrix} = \begin{bmatrix} 0 & 1 & 0 \\ -k/m & -b/m & 1/m \\ 0 & 0 & 0 \end{bmatrix} \begin{bmatrix} x \\ v \\ u \end{bmatrix},$$

$$y = \begin{bmatrix} 1 & 0 & 0 \end{bmatrix} \begin{bmatrix} x \\ v \\ u \end{bmatrix}.$$

Using the same values $m = 1, k = 2$, and $b = 3$ as in Example 6.10,

$$\hat{A} = \begin{bmatrix} 0 & 1 & 0 \\ -2 & -3 & 1 \\ 0 & 0 & 0 \end{bmatrix}$$

can be written as

$$\hat{A} = P\Lambda P^{-1},$$

where

$$\Lambda = \begin{bmatrix} -1 & 0 & 0 \\ 0 & -2 & 0 \\ 0 & 0 & 0 \end{bmatrix}, \quad P = \begin{bmatrix} 1 & 1 & 1 \\ -1 & -2 & 0 \\ 0 & 0 & 2 \end{bmatrix}.$$

Thus

$$A = e^{\hat{A}T} = Pe^{\Lambda T}P^{-1}$$

$$= \begin{bmatrix} 2e^{-T} - e^{-2T} & e^{-T} - e^{-2T} & \frac{1}{2} + e^{-T} + \frac{1}{2}e^{-2T} \\ -2e^{-T} + 2e^{-2T} & -e^{-T} + 2e^{-2T} & -e^{-T} - e^{-2T} \\ 0 & 0 & 1 \end{bmatrix}$$

To check observability, we must compute

$$V = \begin{bmatrix} C \\ CA \\ CA^2 \end{bmatrix}$$

$$= \begin{bmatrix} 1 & 0 & 0 \\ 2e^{-T} - e^{-2T} & e^{-T} - e^{-2T} & \frac{1}{2} + e^{-T} + \frac{1}{2}e^{-2T} \\ 2e^{-2T} - e^{-4T} & e^{-2T} - e^{-4T} & \frac{1}{2} + 2e^{-T} + e^{-2T} + \frac{1}{2}e^{-4T} \end{bmatrix}$$

and its determinant

$$\begin{aligned} \det V &= e^{-T} + 2e^{-2T} - 4e^{-3T} - 2e^{-4T} + 3e^{-5T} \\ &= e^{-T}(1 + e^{-T})(1 - e^{-T})^2(1 + 3e^{-T}). \end{aligned}$$

The factored form above shows that since T is real, $\det V = 0$ only if $T = 0$. The system is therefore observable for all nonzero T.

Theorem 10.13 establishes a duality between the notions of controllability and observability. The following definition formalizes the notion of duality.

Definition 10.16. The dual system of (10.3.1) is given by

$$\begin{aligned} x(n+1) &= A^T x(n) + C^T u(n), \\ y(n) &= B^T x(n). \end{aligned} \tag{10.3.8}$$

Notice that the controllability matrix $\bar{W}$ of system (10.3.8) may be given by

$$\bar{W} = \left[C^T, A^T C^T, (A^T)^2 C^T, \ \ldots, (A^T)^{k-1} C^T\right].$$

Furthermore, the observability matrix V of system (10.3.1) is the transpose of $\bar{W}$, i.e.,

$$V = \bar{W}^T.$$

But since $\operatorname{rank} \bar{W} = \operatorname{rank} \bar{W}^T = \operatorname{rank} V$, we have the following conclusion.

Theorem 10.17 (Duality Principle). *System* (10.3.1) *is completely controllable if and only if its dual system* (10.3.8) *is completely observable.*

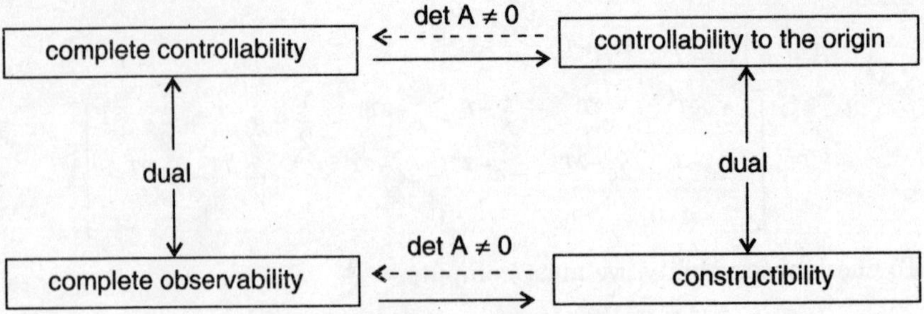

FIGURE 10.10.

Remark: In Remark 10.7 we introduced a weaker notion of controllability, namely, controllability to the origin. In this section we have established a duality between complete controllability and complete observability. To complete our analysis we need to find a dual notion for controllability to the origin. Fortunately, such a notion does exist, and it is called *constructibility*. System (10.3.1) is said to be *constructible* if there exists a positive integer N such that for given $u(0), u(1), \ldots, u(N-1)$ and $y(0), y(1), \ldots, y(N-1)$ it is possible to find the state vector $x(N)$ of the system. Since the knowledge of $x(0)$ yields $x(N)$ by the variation of constants formula, it follows that complete observability implies constructibility. The two notions are in fact equivalent if the matrix A is nonsingular. Figure 10.10 illustrates the relations among various notions of controllability and observability.

Finally, we give an example to demonstrate that constructibility does not imply complete observability.

Example 10.18. Contemplate a dual of the system in Example 10.8:

$$\begin{pmatrix} x_1(n+1) \\ x_2(n+1) \end{pmatrix} = \begin{pmatrix} 0 & 0 \\ 1 & 0 \end{pmatrix} \begin{pmatrix} x_1(n) \\ x_2(n) \end{pmatrix} + \begin{pmatrix} 1 \\ 0 \end{pmatrix} u(n),$$

$$y(n) = \begin{pmatrix} 1 & 0 \end{pmatrix} \begin{pmatrix} x_1(n) \\ x_2(n) \end{pmatrix}.$$

The observability matrix is given by

$$V = \begin{pmatrix} 1 & 0 \\ 0 & 0 \end{pmatrix},$$

whose rank is 1. It follows from Theorem 10.13 that the system is not completely observable. However, if we know $u(0), u(1)$ and $y(0), y(1)$, then from the second equation we find that $x_1(1) = y(1)$. The first equation yields $x_1(2) = u(1)$ and $x_2(2) = x_1(1)$. Thus

$$x(2) = \begin{pmatrix} x_1(2) \\ x_2(2) \end{pmatrix}$$

is now obtained and, consequently, the system is constructible.

10.3.1 Observability Canonical Forms

Consider again the completely observable system

$$\begin{aligned} x(n+1) &= Ax(n) + bu(n), \\ y(n) &= Cx(n), \end{aligned} \tag{10.3.9}$$

where b is a $k \times 1$ vector and C is a $1 \times k$ vector. Recall that in Section 10.2 we constructed two controllability canonical forms of system (10.3.1). By exactly parallel procedures we can obtain two *observability canonical forms* corresponding to system (10.3.9). Both procedures are based on the nonsingularity of the observability matrix

$$V = \begin{pmatrix} C \\ CA \\ \vdots \\ CA^{k-1} \end{pmatrix}.$$

If we let $z(n) = Vx(n)$ in (10.3.11), we obtain the first observability canonical form (Exercises 10.3, Problem 10)

$$\begin{aligned} z(n+1) &= \bar{A}\, z(n) + \bar{b}\, u(n), \\ y(n) &= \bar{c}\, z(n), \end{aligned} \tag{10.3.10}$$

where

$$\begin{aligned} \bar{A} &= \begin{pmatrix} 0 & 1 & 0 & \dots & 0 \\ 0 & 0 & 1 & \dots & 0 \\ \vdots & \vdots & \vdots & \dots & \vdots \\ 0 & 0 & 0 & \dots & 1 \\ -p_k & -p_{k-1} & -p_{k-2} & \dots & -p_1 \end{pmatrix}, \\ \bar{c} &= (1 \;\; 0 \;\; 0 \;\; \cdots \;\; 0), \\ \bar{b} &= Vb. \end{aligned} \tag{10.3.11}$$

In Exercises 10.3, Problem 10, the reader is asked to find a change of variable that yields the other observability canonical form $\{\tilde{A},\ \tilde{c}\}$, with

$$\tilde{A} = \begin{pmatrix} 0 & 0 & \dots & 0 & -p_k \\ 1 & 0 & \dots & 0 & -p_{k-1} \\ 0 & 1 & \dots & 0 & -p_{k-2} \\ \vdots & \vdots & & \vdots & \\ 0 & 0 & \dots & 1 & -p_1 \end{pmatrix}, \quad \tilde{c} = \begin{pmatrix} 0 & 0 & \cdots & 1 \end{pmatrix}. \tag{10.3.12}$$

Exercises 10.3

1. Consider the input–output system

$$\begin{aligned} x(n+1) &= Ax(n) + Bu(n), \\ y(n) &= Cx(n), \end{aligned}$$

where $A = \begin{pmatrix} 0 & 1 \\ 2 & -1 \end{pmatrix}$.

   (a) If $C = (0, 2)$, show that the pair$\{A, C\}$ is observable. Then find $x(0)$ if $y(0) = a$ and $y(1) = b$.

   (b) If $C = (2, 1)$, show that the pair $\{A, C\}$ is unobservable.

2. Determine the observability of the pair $\{A, C\}$, where

$$A = \begin{pmatrix} 0 & 1 & 0 \\ 0 & 0 & 1 \\ \frac{-1}{4} & \frac{1}{4} & 1 \end{pmatrix}, \quad C = \begin{pmatrix} 2 & -3 & -2 \\ 2 & 3 & 1 \end{pmatrix}.$$

3. Consider the system defined by

$$\begin{pmatrix} x_1(n+1) \\ x_2(n+1) \end{pmatrix} = \begin{pmatrix} a & b \\ c & d \end{pmatrix} \begin{pmatrix} x_1(n) \\ x_2(n) \end{pmatrix} + \begin{pmatrix} 1 \\ 1 \end{pmatrix} u(n),$$

$$y(n) = \begin{pmatrix} 1 & 0 \end{pmatrix} \begin{pmatrix} x_1(n) \\ x_2(n) \end{pmatrix}.$$

Determine the conditions on $a, b, c,$ and d for complete state controllability and complete observability. In Problems 4 and 5, determine the observability of the pair $\{A, C\}$.

4.

$$A=\begin{pmatrix}2&1&0&0&0\\0&2&1&0&0\\0&0&2&0&0\\0&0&0&-3&1\\0&0&0&0&-3\end{pmatrix},\quad C=\begin{pmatrix}1&1&1&0&1\\0&1&1&1&0\end{pmatrix}.$$

5.

$$A=\begin{pmatrix}2&1&0&0&0\\0&2&1&0&0\\0&0&2&0&0\\0&0&0&-3&1\\0&0&0&0&-3\end{pmatrix},\quad C=\begin{pmatrix}1&1&1&0&1\\0&1&1&0&0\end{pmatrix}.$$

6. Show that the pair $\{A, C\}$, where

$$\begin{pmatrix}A_{11}&0\\A_{21}&A_{22}\end{pmatrix},\quad C=(C_1\ \ 0),\quad \begin{matrix}A_{11}: r\times r; & A_{21}: m\times r;\\ A_{22}: m\times m; & C_1: p\times r;\end{matrix}$$

is not completely observable for any submatrices A_1, A_{21}, A_{22}, and C_1.

7. Prove that system (10.3.2) is completely observable if and only if $\operatorname{rank}[C, CA, \ldots, CA^{m-r}]^T = k$, where m is the degree of the minimal polynomial of A, and $r = \operatorname{rank} C$.

8. Prove that system (10.3.2) is completely observable if and only if the matrix V^TV is positive definite, where V is the observability matrix of $\{A, C\}$.

9. Show that the kth-order scalar equation

$$z(n+k)+p_1z(n+k-1)+\cdots+p_kz(n)=u(n),$$
$$y(n)=c\,z(n),$$

is completely observable if $c \neq 0$.

10. Verify that the change of variable $z(n) = Vx(n)$ produces the observability canonical pair $\{\bar{A}, \bar{c}\}$ defined in expression (6.3.13).

11. Consider system (10.3.2), where $P^{-1}AP$ is a diagonal matrix. Show that a necessary and sufficient condition for complete observability is that none of the columns of the $r \times k$ matrix CP consist of all zero elements.

12. Consider system (10.3.2), where $P^{-1}AP$ is in the Jordan form J. Show that necessary and sufficient conditions for complete observability of the system are:

(i) no two Jordan blocks in J correspond to the same eigenvalue of A,

(ii) none of the columns of CP that correspond to the first row of each Jordan block consists of all zero elements, and

(iii) no columns of CP that correspond to distinct eigenvalues consist of all zero elements.

13. Let P be a nonsingular matrix. Show that if the pair $\{A, C\}$ is completely observable, then so is the pair $\{P^{-1}AP, CP\}$.

14. Show that if the matrix A in system equation (10.3.2) is nonsingular, then complete observability and constructibility are equivalent.

15. Consider the completely observable system

$$x(n+1) = Ax(n) + Bu(n),$$
$$y(n) = Cx(n),$$

where a is a $k \times k$ matrix and C is a $1 \times k$ vector. Let $M = (C^T, A^T C^T, \ldots, (A^T)^{k-1} C^T)$.

(a) Show that $M^T A (M^T)^{-1} = \begin{pmatrix} 0 & 1 & 0 & \cdots & 0 \\ 0 & 0 & 1 & \cdots & 0 \\ \vdots & \vdots & \vdots & & \vdots \\ 0 & 0 & 0 & \cdots & 1 \\ -a_k & -a_{k-1} & -a_{k-2} & \cdots & -a_1 \end{pmatrix}$,
where $a_i, 1 \le i \le k$, are the coefficients of the characteristic polynomial

$$|\lambda I - A| = \lambda^k + a_1 \lambda^{k-1} + \cdots + a_k.$$

(b) Write down the corresponding *canonical controllable system* by letting $\tilde{x}(n) = M^T x(n)$. Then deduce a necessary condition on C for complete observability of the original system.

16. Consider the system

$$x(n+1) = Ax(n),$$
$$y(n) = Cx(n),$$

where A is a $k \times k$ matrix and C is a $1 \times k$ vector. Prove that the system is completely observable if and only if the matrix

$$G = (C, CA^{-1}, CA^{-2}, \ \ldots, \ CA^{-k+1}) \text{ is nonsingular.}$$

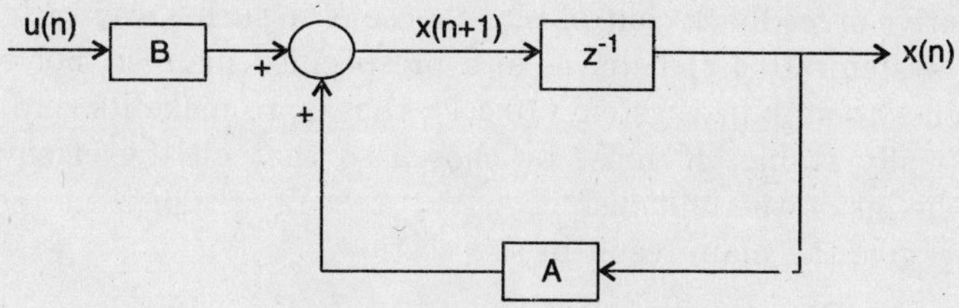

FIGURE 10.11. Open-loop system.

10.4 Stabilization by State Feedback (Design via Pole Placement)

Feedback controls are used in many aspects of our lives, from the braking system of a car to central air conditioning. The method has been used by engineers for many years. However, the systematic study of stabilization by state feedback control is of more recent origin (see [3], [4]) and dates to the 1960s. The idea of state feedback is simple: It is assumed that the state vector $x(n)$ can be directly measured, and the control $u(n)$ is adjusted based on this information. Consider the (open-loop) time-invariant control system shown in Figure 10.11, whose equation is

$$x(n+1) = Ax(n) + Bu(n), \tag{10.4.1}$$

where, as before, A is a $k \times k$ matrix and B a $k \times m$ matrix.

Suppose we apply linear feedback $u(n) = -Kx(n)$, where K is a real $m \times k$ matrix called the *state feedback* or *gain state* matrix. The resulting (closed-loop) system (Figure 10.12) obtained by substituting $u = -Kx$ into (10.4.1) is

$$x(n+1) = Ax(n) - BKx(n),$$

or

$$x(n+1) = (A - BK)x(n). \tag{10.4.2}$$

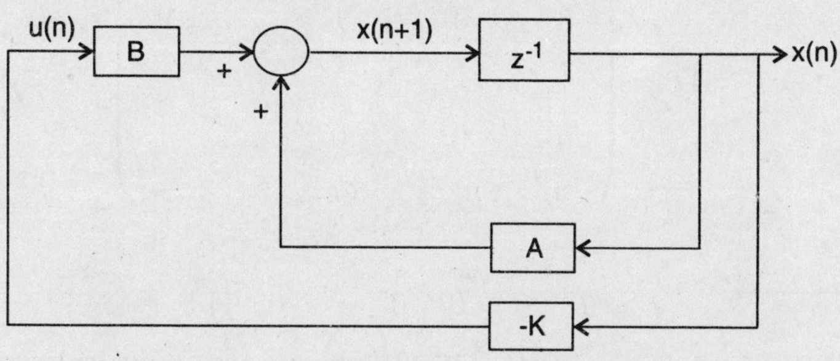

FIGURE 10.12. Closed-loop system.

The objective of feedback control is to choose K in such a way such that the resulting system (10.4.2) behaves in a prespecified manner. For example, if one wishes to stabilize system (10.4.1), that is, to make its zero solution asymptotically stable, K must be chosen so that all the eigenvalues of $A - BK$ lie inside the unit disk.

We now give the main result in this section.

Theorem 10.19. *Let $\Lambda = \{\mu_1, \mu_2, \ldots, \mu_k\}$ be an arbitrary set of k complex numbers such that $\bar{\Lambda} = \{\bar{\mu}_1, \bar{\mu}_2, \ldots, \bar{\mu}_k\} = \Lambda$. Then the pair $\{A, B\}$ is completely controllable if and only if there exists a matrix K such that the eigenvalues of $A - BK$ are the set Λ.*

Since the proof of the theorem is rather lengthy, we first present the proof for the case $m = 1$, i.e., when B is a $k \times 1$ vector and $u(n)$ is a scalar. We start the proof by writing the characteristic polynomial of $A, |A - \lambda I| = \lambda^k + a_1\lambda^{k-1} + a_2\lambda^{k-2} + \cdots + a_k$. Suppose also that

$$\prod_{i=1}^{k}(\lambda - \mu_i) = \lambda^k + b_1\lambda^{k-1} + b_2\lambda^{k-2} + \cdots + b_k.$$

Define $T = WM$, where W is the controllability matrix of rank k defined in (10.2.1) as

$$W = (B, AB, \ldots, A^{k-1}B)$$

and

$$M = \begin{pmatrix} a_{k-1} & a_{k-2} & \cdots & a_1 & 1 \\ a_{k-2} & a_{k-3} & \cdots & 1 & 0 \\ \vdots & \vdots & & \vdots & \vdots \\ a_1 & 1 & \cdots & 0 & 0 \\ 1 & 0 & & 0 & 0 \end{pmatrix}.$$

Then (Exercises 10.4, Problem 12)

$$\bar{A} = T^{-1}AT = \begin{pmatrix} 0 & 1 & 0 & \cdots & 0 \\ 0 & 0 & 1 & \cdots & 0 \\ \vdots & \vdots & \vdots & & \\ 0 & 0 & 0 & \cdots & 1 \\ -a_k & -a_{k-1} & -a_{k-2} & \cdots & -a_1 \end{pmatrix} \tag{10.4.3}$$

and

$$\bar{B} = T^{-1}B = \begin{pmatrix} 0 & 0 & \cdots & 0 & 1 \end{pmatrix}^T.$$

Letting $x(n) = T\bar{x}(n)$ in system (10.4.2), we get the equivalent system

$$\bar{x}(n+1) = (\bar{A} - \bar{B}\bar{K})\bar{x}(n), \tag{10.4.4}$$

where

$$\bar{K} = KT. \tag{10.4.5}$$

Choose

$$\bar{K} = (b_k - a_k, b_{k-1} - a_{k-1}, \ldots, b_1 - a_1). \tag{10.4.6}$$

Then

$$\bar{B}\bar{K} = \begin{pmatrix} 0 & 0 & \cdots & 0 \\ 0 & 0 & \cdots & 0 \\ \vdots & \vdots & & \vdots \\ 0 & 0 & \cdots & 0 \\ b_k - a_k & b_{k-1} - a_{k-1} & & b_1 - a_1 \end{pmatrix}.$$

Observe that $A - BK$ is similar to $\bar{A} - \bar{B}\bar{K}$, since $\bar{A} - \bar{B}\bar{K} = T^{-1}AT - T^{-1}BKT = T^{-1}(A - BK)T$. Thus

$$|\lambda I - A + BK| = |\lambda I - \bar{A} + \bar{B}\bar{K}| = \begin{vmatrix} \lambda & 1 & \cdots & 0 \\ 0 & \lambda & \cdots & 0 \\ \vdots & \vdots & & \vdots \\ 0 & 0 & \cdots & 1 \\ -b_k & -b_{k-1} & \cdots & \lambda - b_1 \end{vmatrix}$$

$$= \lambda^k + b_1 + \lambda^{k-1} + \cdots + b_k,$$

which has Λ as its set of eigenvalues. Hence the required feedback (gain) matrix is given by

$$K = \bar{K}T^{-1} = (b_k - a_k, b_{k-1} - a_{k-1}, \ \ldots, \ b_1 - a_1)T^{-1}.$$

Example 10.20. Consider the control system $x(n+1) = Ax(n) + Bu(n)$ with

$$A = \begin{pmatrix} 1 & -3 \\ 4 & 2 \end{pmatrix}, \quad B = \begin{pmatrix} 1 \\ 1 \end{pmatrix}.$$

Find a state feedback gain matrix K such that the eigenvalues of the closed loop system are $\frac{1}{2}$ and $\frac{1}{4}$.

Solution

Method 1

$$|A - \lambda I| = \begin{vmatrix} 1-\lambda & -3 \\ 4 & 2-\lambda \end{vmatrix} = \lambda^2 - 3\lambda + 14.$$

So

$$a_1 = -3, \quad a_2 = 14.$$

Also

$$\left(\lambda - \frac{1}{2}\right)\left(\lambda - \frac{1}{4}\right) = \lambda^2 - \frac{3}{4}\lambda + \frac{1}{8} \tag{10.4.7}$$

So

$$b_1 = -\frac{3}{4} \quad \text{and} \quad b_2 = \frac{1}{8}.$$

Now,

$$W = \begin{pmatrix} 1 & -2 \\ 1 & 6 \end{pmatrix}, \quad M = \begin{pmatrix} -3 & 1 \\ 1 & 0 \end{pmatrix}.$$

Hence

$$T = WM = \begin{pmatrix} 1 & -2 \\ 1 & 6 \end{pmatrix}\begin{pmatrix} -3 & 1 \\ 1 & 0 \end{pmatrix} = \begin{pmatrix} -5 & 1 \\ 3 & 1 \end{pmatrix}$$

and

$$T^{-1} = -\frac{1}{8}\begin{pmatrix} 1 & -1 \\ -3 & -5 \end{pmatrix}.$$

Therefore,

$$K = (b_2 - a_2, b_1 - a_1)T^{-1},$$

or

$$K = \begin{pmatrix} -13\frac{7}{8} & 2\frac{1}{4} \end{pmatrix} \cdot \left(-\frac{1}{8}\right)\begin{pmatrix} 1 & -1 \\ -3 & -5 \end{pmatrix} = \begin{pmatrix} \frac{165}{64} & \frac{-21}{64} \end{pmatrix}.$$

Method 2 In this method we substitute $K = (k_1 k_2)$ into the characteristic polynomial $|A - BK - \lambda I|$ and then match the coefficients of powers in λ with the desired characteristic polynomial (10.4.7).

$$\begin{vmatrix} 1 - k_1 - \lambda & -3 - k_2 \\ 4 - k_1 & 2 - k_2 - \lambda \end{vmatrix} = \lambda^2 - \lambda(3 - k_1 - k_2) + 14 - 5k_1 + 3k_2. \tag{10.4.8}$$

Comparing the coefficients of powers in λ in (10.4.7) and (10.4.8), we obtain

$$3 - k_1 - k_2 = \frac{3}{4},$$

$$14 - 5k_1 + 3k_2 = \frac{1}{8}.$$

This gives us $k_1 = \frac{165}{64}$ and $k_2 = \frac{-21}{64}$.

Hence

$$K = \begin{pmatrix} \frac{165}{64} & \frac{-21}{64} \end{pmatrix}.$$

To prove the general case $m > 1$ of Theorem 10.19 we need the following preliminary result.

Lemma 10.21. *If the pair $\{A, B\}$ is completely controllable and the columns of B, assumed nonzero, are $b_1, b_2, \ldots, b_m$, then there exist matrices $K_i, 1 \leq i \leq m$, such that the pairs $\{A - BK_i, b_i\}$ are completely controllable.*

PROOF. Let us consider the case $i = 1$. Since the controllability matrix W has rank k (full rank), one may select a basis of $\mathbb{R}^k$ consisting of k columns of W. One such selection would be the $k \times k$ matrix

$$M = \left(b_1, Ab_1, \ldots, A^{r_1-1}b_1, b_2, Ab_2, \ldots, a^{r_2-1}b_2, \ldots\right),$$

where r_i is the smallest integer such that $A^{r_i}b_i$ is linearly dependent on all the preceding vectors. Define an $m \times k$ matrix L having its r_1th column equal to $e_2 = (0, 1, \ldots, 0)^T$, its $(r_1 + r_2)$th column equal to $e_3 = (0, 0, 1, \ldots, 0)^T$, and so on, all its other columns being zero. We claim that the desired matrix K_1 is given by $K_1 = LM^{-1}$. To verify the claim, we compare the corresponding columns on both sides of $K_1M = L$. It follows immediately that

$$\begin{aligned}
&K_1b_1 = 0, \quad K_1Ab_1 = 0, \ldots, K_1A^{r_1-1}b_1 = e_2, \\
&K_1b_2 = 0, \quad K_1Ab_2 = 0, \ldots, K_1A^{r_2-1}b_2 = e_3, \\
&K_1b_3 = 0, \quad \text{etc.}
\end{aligned}$$

Thus we have

$$\left(b_1, (A - BK_1)b_1, (A - BK_2)^2b_1, \ldots, (A - BK_2)^{k-1}b_1\right) = W(k),$$

which has rank k by assumption. This proves our claim. We are now ready to give the proof of the general case $m > 1$ in Theorem 10.19. □

PROOF OF THEOREM 10.19. Let K_1 be the matrix in Lemma 10.21. Then by Lemma 10.21, it follows that the pair $\{A - BK_1, b_1\}$ is completely controllable. And by the proof of Lemma 10.21 for $m = 1$, there exists a $1 \times k$ vector ξ such that the eigenvalues of $A + BK_1 + b_1\xi$ are the set Λ. Let $\bar{K}$ be the $m \times k$ matrix whose first row is ξ and all other rows are zero. Then the desired feedback (gain) matrix is given by $K = K_1 + \bar{K}$. Since $u = -Kx$, this gives

$$x(n+1) = (A - BK)x(n) = (A - BK_1 - b_1\xi)x(n).$$

To prove the converse, select K_0 such that $(A - BK_0)^n \to 0$ as $n \to \infty$, that is, the spectral radius $\rho(A - BK_o)$ is less than 1, and select K_1 such that $\rho(A - BK_1) = \left\{\exp\left(\frac{2\pi n}{k}\right) : n = 0, 1, \ldots, k-1\right\}$, the kth roots of unity. Then clearly, $(A - BK_1)^k = 1$. Suppose that for some vector

$\xi \in R^k, \xi^T A^n B = 0$ for all $n \in \mathbb{Z}^+$. Then for any matrix K,

$$\begin{aligned}
\xi^T(A-BK)^n &= \xi^T(A-BK)(A-BK)^{n-1} \\
&= (\xi^T A - \xi^T BK)(A-BK)^{n-1} \\
&= \xi^T A(A-BK)^{n-1} \quad (\text{since } \xi^T B = 0) \\
&= \xi^T A(A-BK)(A-BK)^{n-2} \\
&= \xi^T A^2(A-BK)^{n-2} \quad (\text{since } \xi^T AB = 0).
\end{aligned}$$

Continuing this procedure we obtain

$$\xi^T(A-BK)^n = \xi^T A^n, \quad \text{for all } n \in \mathbb{Z}^+.$$

This implies that

$$\xi^T[(A-BK_o)^n - (A-BK_1)^n] = 0, \quad \text{for all } n \in \mathbb{Z}^+,$$

or

$$\xi^T[(A-BK_o)^{kr} - 1] = 0, \quad \text{for all } r \in \mathbb{Z}^+.$$

Letting $r \to \infty$, we have $(A-BK_0)^{kr} \to 0$ and, consequently, $\xi^T = 0$. This implies that the pair $\{A, B\}$ is completely controllable. □

An immediate consequence of Theorem 10.19 is a simple sufficient condition for *stabilizability.* A system $x(n+1) = Ax(n) + Bu(n)$ is *stabilizable* if one can find a feedback control $u(n) = -Kx(n)$ such that the zero solution of the resulting closed-loop system $x(n+1) = (A-BK)x(n)$ is asymptotically stable. In other words, the pair $\{A, B\}$ is stabilizable if for some matrix $K, A-BK$ is a stable matrix (i.e., all its eigenvalues lie inside the unit disk).

Corollary 10.22. *System* (10.4.1) *is stabilizable if it is completely controllable.*

The question still remains whether or not we can stabilize an uncontrollable system. The answer is yes and no, as may be seen by the following example.

Example 10.23. Consider the control system

$$x(n+1) = Ax(n) + Bu(n),$$

where

$$A = \begin{pmatrix} 0 & a & b \\ 1 & d & e \\ 0 & 0 & h \end{pmatrix}, \quad B = \begin{pmatrix} 1 & \beta_1 \\ 0 & \beta_2 \\ 0 & 0 \end{pmatrix}.$$

Let us write

$$A = \begin{pmatrix} A_{11} & A_{12} \\ 0 & A_{22} \end{pmatrix}, \quad B = \begin{pmatrix} B_1 \\ 0 \end{pmatrix},$$

where

$$A_{11} = \begin{pmatrix} 0 & a \\ 1 & d \end{pmatrix}, \quad A_{12} = \begin{pmatrix} b \\ e \end{pmatrix}, \quad A_{22} = (h), \quad B_1 = \begin{pmatrix} 1 & \beta_1 \\ 0 & \beta_2 \end{pmatrix}.$$

If $x = \begin{pmatrix} y \\ z \end{pmatrix}$, then our system may be written as

$$\begin{aligned} y(n+1) &= A_{11}y(n) + A_{12}z(n) + B_1u(n), \\ z(n+1) &= A_{22}z(n). \end{aligned}$$

It is easy to verify that the pair $\{A_{11}, B_1\}$ is completely controllable. Hence by Theorem 10.19, there is a 2×2 gain matrix $\bar{K}$ such that $A_{11} + B_1\bar{K}$ is a stable matrix. Letting $K = (\bar{K})(0)$, then

$$A - BK = \begin{pmatrix} A_{11} - B_1\bar{K} & * \\ 0 & h \end{pmatrix}.$$

Hence the matrix $A - BK$ is stable if and only if $|h| < 1$.

In the general case, a system is stabilizable if and only if the uncontrollable part is asymptotically stable (Exercises 10.4, Problem 8). In this instance, from the columns of the controllability matrix W we select a basis for the controllable part of the system and extend it to a basis S for $\mathbb{R}^k$. The change of variables $x = Py$, where P is the matrix whose columns are the elements of S, transforms our system to

$$y(n+1) = \bar{A}y(n) + \bar{B}u,$$

where

$$\bar{A} = \begin{pmatrix} A_{11} & A_{12} \\ 0 & A_{22} \end{pmatrix}, \quad \bar{B} = \begin{pmatrix} B_1 \\ 0 \end{pmatrix}.$$

Here the pair $\{A_{11}, B_1\}$ is controllable. Hence the system is stabilizable if and only if the matrix A_{22} is stable.

10.4.1 Stabilization of Nonlinear Systems by Feedback

Before ending this section, let us turn our attention to the problem of stabilizing a nonlinear system

$$x(n+1) = f(x(n), u(n)), \tag{10.4.9}$$

where $f : R^k \times R^m \to R^k$. The objective is to find a feedback control

$$u(n) = h(x(n)) \tag{10.4.10}$$

in such a way that the equilibrium point $x^* = 0$ of the closed-loop system

$$x(n+1) = f(x(n)), \quad h(x(n)), \tag{10.4.11}$$

is asymptotically stable (locally!). We make the following assumptions:

(i) $f(0,0) = 0$, and

(ii) f is continuously differentiable, $A = \frac{\partial f}{\partial x}(0,0)$ is a $k \times k$ matrix, $B = \frac{\partial f}{\partial u}(0,0)$ is a $k \times m$ matrix.

Under the above conditions, we have the following surprising result.

Theorem 10.24. *If the pair $\{A, B\}$ is controllable, then the nonlinear system* (10.4.9) *is stabilizable. Moreover, if K is the gain matrix for the pair $\{A, B\}$, then the control $u(n) = -Kx(n)$ may be used to stabilize system* (10.4.9).

PROOF. Since the pair $\{A, B\}$ is controllable, there exists a feedback control $u(n) = -Kx(n)$ that stabilizes the linear part of the system, namely,

$$y(n+1) = Ay(n) + Bv(n).$$

We are going to use the same control on the nonlinear system. So let $g: \mathbb{R}^k \to \mathbb{R}^k$ be a function defined by $g(x) = f(x, -Kx)$. Then system equation (10.4.9) becomes

$$x(n+1) = g(x(n)) \tag{10.4.12}$$

with

$$\left.\frac{\partial g}{\partial x}\right|_{x=0} = A - BK.$$

Since by assumption the zero solution of the linearized system

$$y(n+1) = (A - BK)y(n) \tag{10.4.13}$$

is asymptotically stable, it follows by Theorem 4.20 that the zero solution of system (10.4.12) is also asymptotically stable. This completes the proof of the theorem. □

Example 10.25. Consider the nonlinear difference system

$$x_1(n+1) = 2\sin(x_1(n)) + x_2 + u_1(n),$$
$$x_2(n+1) = x_1^2(n) - x_2(n) - u_2(n).$$

Find a control that stabilizes the system.

Solution One may check easily the controllability of the linearized system $\{A, B\}$, where

$$A = \begin{pmatrix} 2 & 1 \\ 0 & -1 \end{pmatrix}, \quad B = \begin{pmatrix} 1 \\ -1 \end{pmatrix},$$

after some computation. A gain matrix for the linearized system is $K = (2.015, 0.975)$, where the eigenvalues of $A - BK$ are $\frac{1}{2}$ and 0.1. As implied by

Theorem 10.24, the control $u(n) = -Kx(n)$ would stabilize the nonlinear system, where $K = (2.015, 0.975)$.

Exercises 10.4

In Problems 1 through 3 determine the gain matrix K that stabilizes the system $\{A, B\}$.

1. $A = \begin{pmatrix} 0 & 1 \\ -0.16 & -1 \end{pmatrix}, \quad B = \begin{pmatrix} 0 \\ 1 \end{pmatrix}.$

2. $A = \begin{pmatrix} 2 & 1 & 1 \\ -2 & 1 & 0 \\ -2 & -1 & 0 \end{pmatrix}, \quad B = \begin{pmatrix} 0 & 1 \\ 1 & 0 \\ 0 & -2 \end{pmatrix}.$

3. $A = \begin{pmatrix} 0 & 1 & 0 \\ 0 & 0 & 1 \\ -2 & 1 & 3 \end{pmatrix}, \quad B = \begin{pmatrix} 0 \\ 0 \\ 1 \end{pmatrix}.$

4. Determine the matrices B for which the system $\{A, B\}, A = \begin{pmatrix} 1 & -1 & 2 \\ 0 & \frac{1}{2} & 1 \\ \frac{1}{2} & \frac{-1}{2} & 1 \end{pmatrix}$, is (a) controllable and (b) stabilizable.

5. Consider the second-order equation
$$x(n+2) + a_1 x(n+1) + a_2 x(n) = u(n).$$
Determine a gain control $u(n) = c_1 x(n) + c_2 x(n+1)$ that stabilizes the equation.

6. Describe an algorithm for decomposing the system $x(n+1) = Ax(n) + Bu(n)$ into its controllable and uncontrollable parts when A is a 3×3 matrix and B is a 3×2 matrix.

7. Generalize the result of Problem 6 to the case where A is a $k \times k$ matrix and B is a $k \times r$ matrix.

*8. Show that the pair $\{A, B\}$ is stabilizable if and only if the uncontrollable part of the system is asymptotically stable.

9. *Deadbeat Response.* If the eigenvalues of the matrix $A - BK$ are all zero, then the solutions of the system $x(n+1) = (A - BK)x(n)$ will read 0 in finite time. It is then said that the gain matrix K produces a *deadbeat response.* Suppose that A is a 3×3 matrix and B a 3×1 vector.

(a) Show that the desired feedback matrix K for the deadbeat response is given by

$$K = \begin{bmatrix} 1 & 0 & 0 \end{bmatrix} \begin{bmatrix} \xi_1 & \xi_2 & \xi_3 \end{bmatrix}^{-1},$$

where

$$\xi_1 = A^{-1}B, \quad \xi_2 = (A^{-1})^2 B, \quad \xi_3 = (A^{-1})^3 B.$$

(b) Show that the vectors ξ_1, ξ_2, and ξ_3 are generalized eigenvectors of the matrix $A - BK$ [i.e., $(A - BK)\xi_1 = 0, (A - BK)\xi_2 = \xi_1, (A - BK)\xi_3 = \xi_2$].

10. Ackermann's Formula:
Let $\Lambda = \{\mu_1, \mu_2, \ldots, \mu_k\}$ be the desired eigenvalues for the completely controllable pair $\{A, B\}$, with $\Lambda = \bar{\Lambda}$. Show that the feedback (gain) matrix K can be given by

$$K = \begin{pmatrix} 0 & 0 & \cdots & 0 \end{pmatrix} \begin{pmatrix} B & AB & \cdots & A^{k-1}B \end{pmatrix}^{-1} p(A),$$

where

$$p(\lambda) = \prod_{i=1}^{k} (\lambda - \mu_i) = \lambda^k + \alpha_1 \lambda^{k-1} + \cdots + \alpha_k.$$

11. Let $\Lambda = \{\mu_1, \mu_2, \ldots, \mu_k\}$ be a set of complex numbers with $\Lambda = \bar{\Lambda}$. Show that if the pair $\{A, C\}$ is completely observable, then there exists a matrix L such that the eigenvalues of $A - LC$ are the set Λ.

12. Verify formula (10.4.3).

13. Find a stabilizing control for the system

$$\begin{aligned} x_1(n+1) &= 3x_1(n) + x_2^2(n) - \text{sat}(2x_2(n) + u(n)), \\ x_2(n+1) &= \sin\ x_1(n) - x_2(n) + u(n), \end{aligned}$$

where

$$\text{sat } y = \begin{cases} y & \text{if } |y| \le 1, \\ \text{sign } y & \text{if } |y| > 1. \end{cases}$$

14. Find a stabilizing control for the system

$$\begin{aligned} x_1(n+1) &= 2x_1(n) + x_2(n) + x_3^3(n) + u_1(n) + 2u_2(n), \\ x_2(n+1) &= x_1^2(n) + \sin\ x_2(n) + x_2^2(n) + u_1^2(n) + u_2(n), \\ x_3(n+1) &= x_1^4(n) + x_2^3(n) + \frac{1}{2}x_3(n) + u_1(n). \end{aligned}$$

15. (Research problem). Find sufficient conditions for the stabilizability of a time-variant system

$$x(n+1) = A(n)x(n) + B(n)u(n).$$

16. (Research problem). Extend the result of Problem 15 to nonlinear time-variant systems.

10.5 Observers

Theorem 10.19 provides a method of finding a control $u(n) = -Kx(n)$ that stabilizes a given system. This method clearly requires the knowledge of all state variables $x(n)$. Unfortunately, in many systems of practical importance, the entire state vector is not available for measurement. Faced with this difficulty, we are led to construct an estimate of the full state vector based on the available measurements. Let us consider again the system

$$\begin{aligned} x(n+1) &= Ax(n) + Bu(n), \\ y(n) &= Cx(n). \end{aligned} \tag{10.5.1}$$

To estimate the state vector $x(n)$ we construct the k-dimensional *observer* (Figure 10.13)

$$z(n+1) = Az(n) + E[y(n) - Cz(n)] + Bu(n), \tag{10.5.2}$$

where E is a $k \times r$ matrix to be determined later. Notice that unlike $x(n)$, the state observer $z(n)$ can be obtained from available data. To see this, let us write the observer (10.5.2) in the form

$$z(n+1) = (A - EC)z(n) + Ey(n) + Bu(n). \tag{10.5.3}$$

We observe here that the inputs to the observer involve $y(n)$ and $u(n)$, which are available to us.

The question remains whether the observer state $z(n)$ is a good estimate of the original state $x(n)$. One way to check the goodness of this estimator is to ensure that the error $e(n) = z(n) - x(n)$ goes to zero as $n \to \infty$. To achieve this objective we write the error equation in $e(n)$ by subtracting (10.5.2) from (10.5.1) and using $y(n) = Cx(n)$. Hence

$$z(n+1) - x(n+1) = [A - EC][z(n) - x(n)],$$

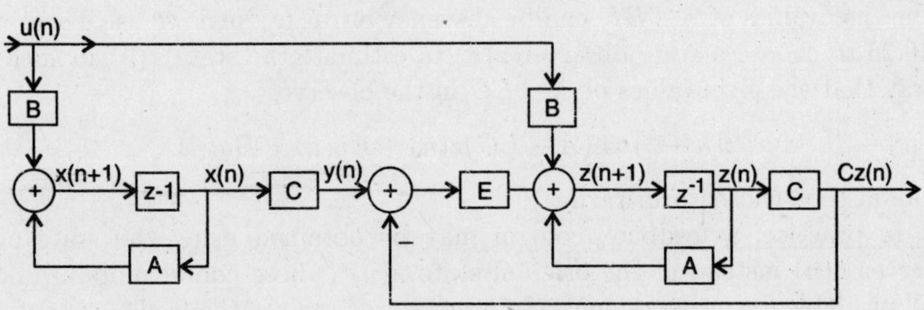

FIGURE 10.13. Observer.

or

$$e(n+1) = [A - EC]e(n). \tag{10.5.4}$$

Clearly, if the zero solution of system (10.5.4) is asymptotically stable (i.e., the matrix $A - EC$ is stable), then the error vector $e(n)$ tends to zero. Thus the problem reduces to finding a matrix E such that the matrix $A - EC$ has all its eigenvalues inside the unit disk. The following result gives a condition under which this can be done.

Theorem 10.26. *If system* (10.5.1) *is completely observable, then an observer* (10.5.2) *can be constructed such that the eigenvalues of the matrix $A - EC$ are arbitrarily chosen. In particular, one can choose a matrix E such that the error $e(n) = z(n) - x(n)$ in the estimate of the state $x(n)$ by the state observer $z(n)$ tends to zero.*

PROOF. Since the pair $\{A, C\}$ is completely observable, it follows from Section 4.3 that the pair $\{A^T, C^T\}$ is completely controllable. Hence by Theorem 10.19 the matrix E can be chosen such that $A^T - C^T E^T$ has an arbitrary set of eigenvalues, which is the same as the set of eigenvalues of the matrix $A - EC$.

Moreover, if we choose the matrix E such that all the eigenvalues of the matrix $A - EC$ are inside the unit disk, then $e(n) \to 0$ (see Corollary 3.24). □

10.5.1 Eigenvalue Separation Theorem

Suppose that the system

$$x(n+1) = Ax(n) + Bu(n),$$
$$y(n) = Cx(n),$$

is both completely observable and completely controllable. Assuming that the state vector $x(n)$ is available, we can use Theorem 10.24 to find a feedback control $u(n) = -Kx(n)$ such that in the closed-loop system

$$x(n+1) = (A - BK)x(n)$$

the eigenvalues of $A - BK$ can be chosen arbitrarily. Next we use Theorem 10.26 to choose a state observer $z(n)$ to estimate the state $x(n)$ in such a way that the eigenvalues of $A - EC$ in the observer

$$z(n+1) = (A - EC)z(n) + Ey(n) + Bu(n)$$

can also be chosen arbitrarily.

In practice, a feedback control may be obtained using the state observer $z(n)$ instead of the original state $x(n)$ (whose components are not all available for measurement). In other words, we use the feedback control

$$u(n) = -Kz(n). \tag{10.5.5}$$

The resulting composite system is given by

$$\begin{aligned} x(n+1) &= Ax(n) - BKz(n), \\ z(n+1) &= (A-EC)z(n) + ECx(n) - BKz(n). \end{aligned}$$

It follows that

$$e(n+1) = z(n+1) - x(n+1) = (A-EC)e(n).$$

Hence we have the following composite system:

$$\begin{aligned} x(n+1) &= (A-BK)x(n) + BKe(n), \\ e(n+1) &= (A-EC)e(n). \end{aligned}$$

The system matrix is given by

$$A = \begin{pmatrix} A-BK & BK \\ 0 & A-EC \end{pmatrix},$$

whose characteristic polynomial is the product of the characteristic polynomials of $(A-BK)$ and $(A-EC)$. Hence the eigenvalues of A are either eigenvalues of $A-BK$ or eigenvalues of $A-EC$, which we can choose arbitrarily. Thus we have proved the following result.

Theorem 10.27 (Eigenvalue Separation Theorem). *Consider the system*

$$\begin{aligned} x(n+1) &= Ax(n) + Bu(n), \\ y(n) &= Cx(n), \end{aligned}$$

with the observer

$$z(n+1) = (A-EC)z(n) + Ey(n) + Bu(n)$$

and the feedback control

$$u(n) = -Kz(n).$$

Then the characteristic polynomial of this composite system is the product of the characteristic polynomials of $A-BK$ and $A-EC$. Furthermore, the eigenvalues of the composite system can be chosen arbitrarily.

Example 10.28. Consider the system

$$\begin{aligned} x(n+1) &= Ax(n) + Bu(n), \\ y(n) &= Cx(n), \end{aligned}$$

where

$$A = \begin{pmatrix} 0 & -\frac{1}{4} \\ 1 & -1 \end{pmatrix}, \quad B = \begin{pmatrix} 0 \\ 1 \end{pmatrix}, \quad C = \begin{pmatrix} 0 & 1 \end{pmatrix}.$$

Design a state observer so that the eigenvalues of the observer matrix $A - EC$ are $\frac{1}{2} + \frac{1}{2}i$ and $\frac{1}{2} - \frac{1}{2}i$.

Solution The observability matrix is given by

$$\begin{pmatrix} C \\ CA \end{pmatrix} = \begin{pmatrix} 0 & 1 \\ 1 & -1 \end{pmatrix},$$

which has full rank 2. Thus the system is completely observable, and the desired observer feedback gain matrix E may be now determined. The characteristic equation of the observer is given by $\det(A - EC - \lambda I) = 0$.

If

$$E = \begin{pmatrix} E_1 \\ E_2 \end{pmatrix},$$

then we have

$$\left| \begin{pmatrix} 0 & -\frac{1}{4} \\ 1 & -1 \end{pmatrix} - \begin{pmatrix} E_1 \\ E_2 \end{pmatrix} (0 \ \ 1) - \begin{pmatrix} \lambda & 0 \\ 0 & \lambda \end{pmatrix} \right| = 0,$$

which reduces to

$$\lambda^2 + (1 + E_2)\lambda + E_1 + \frac{1}{4} = 0. \tag{10.5.6}$$

By assumption the desired characteristic equation is given by

$$\left(\lambda - \frac{1}{2} - \frac{1}{2}i\right)\left(\lambda - \frac{1}{2} + \frac{1}{2}i\right) = 0,$$

or

$$\lambda^2 - \lambda + \frac{1}{2} = 0. \tag{10.5.7}$$

Comparing (10.5.6) and (10.5.7) yields

$$E_1 = \frac{1}{4}, \quad E_2 = -2.$$

Thus $E = \begin{pmatrix} \frac{1}{4} \\ -2 \end{pmatrix}$.

Example 10.29. Figure 10.14 shows a metallic sphere of mass m suspended in a magnetic field generated by an electromagnet. The equation of motion for this system is

$$m\ddot{x}_t = mg - k\frac{u_t^2}{x_t}, \tag{10.5.8}$$

where x_t is the distance of the sphere from the magnet, u_t is the current driving the electromagnet, g is the acceleration of gravity, and k is a constant determined by the properties of the magnet.

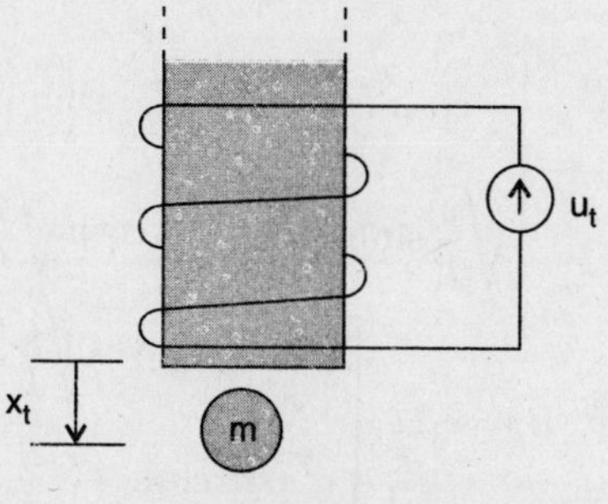

FIGURE 10.14. A metallic sphere suspended in a magnetic field.

It is easy to check that (10.5.8) has an equilibrium at

$$x_t = x_0 = 1,$$
$$u_t = u_0 = \sqrt{mg/k}.$$

Linearizing (10.5.8) about this equilibrium gives the following approximate model in terms of the deviations $x = x_t - x_0$ and $u = u_t - u_0$:

$$\ddot{x} - \frac{g}{k}x = -2\sqrt{kg/m}u,$$

or, in state variable form,

$$\begin{bmatrix} \dot{x} \\ \dot{v} \end{bmatrix} = \begin{bmatrix} 0 & 1 \\ \frac{g}{k} & 0 \end{bmatrix} \begin{bmatrix} x \\ v \end{bmatrix} + \begin{bmatrix} 0 \\ -2\sqrt{kg/m} \end{bmatrix} u.$$

Thus

$$\hat{A} = \begin{bmatrix} 0 & 1 \\ \frac{g}{k} & 0 \end{bmatrix}, \quad \hat{B} = \begin{bmatrix} 0 \\ -2\sqrt{kg/m} \end{bmatrix}.$$

The matrix $\hat{A}$ can be written in the form

$$\hat{A} = P\Lambda P^{-1},$$

where

$$\Lambda = \begin{bmatrix} \sqrt{\frac{g}{k}} & 0 \\ 0 & -\sqrt{\frac{g}{k}} \end{bmatrix}, \quad P = \frac{1}{\sqrt{2}} \begin{bmatrix} 1 & 1 \\ \sqrt{\frac{g}{k}} & -\sqrt{\frac{g}{k}} \end{bmatrix}.$$

So

$$A = e^{\hat{A}T} = \begin{bmatrix} \cosh\sqrt{\frac{g}{k}}T & \sqrt{\frac{k}{g}}\sinh\sqrt{\frac{g}{k}}T \\ \sqrt{\frac{g}{k}}\sinh\sqrt{\frac{g}{k}}T & \cosh\sqrt{\frac{g}{k}}T \end{bmatrix},$$

$$B = Te^{\hat{A}T}\hat{B} = -2T\begin{bmatrix} \sqrt{k/m}\sinh & \sqrt{\frac{g}{k}}T \\ \sqrt{g/m}\cosh & \sqrt{\frac{g}{k}}T \end{bmatrix}$$

The discrete equivalent system is thus controllable, since

$$\begin{aligned} \det W &= \left|[B \quad AB]\right| \\ &= \left| -2T\sqrt{k/m}\begin{bmatrix} \sinh\sqrt{\frac{g}{k}}T & 2\sinh\sqrt{\frac{g}{k}}T\cosh\sqrt{\frac{g}{k}}T \\ \sqrt{\frac{g}{k}}\sinh\sqrt{\frac{g}{k}}T & 2\sqrt{\frac{g}{k}}\sinh\sqrt{\frac{g}{k}}T\cosh\sqrt{\frac{g}{k}}T \end{bmatrix}\right| \\ &= ce\sqrt{\frac{g}{k}}T\sinh\sqrt{\frac{g}{k}}T, \end{aligned}$$

where $c = 0$ only if $T = 0$.

If the position deviation x of the ball from equilibrium can be measured, then the system is also observable, since then we have the measurement equation

$$y = [1 \quad 0]\begin{bmatrix} x \\ v \end{bmatrix}$$

and hence

$$C = [1 \quad 0].$$

Observability is easily verified by computing

$$\begin{aligned} \det V &= \left|\begin{bmatrix} C \\ CA \end{bmatrix}\right| = \left|\begin{bmatrix} 1 & 0 \\ \cosh\sqrt{\frac{g}{k}}T & \sqrt{\frac{k}{g}}\sinh\sqrt{\frac{g}{k}}T \end{bmatrix}\right| \\ &= \sqrt{\frac{k}{g}}\sinh\sqrt{\frac{g}{k}}T, \end{aligned}$$

which is zero only if $T = 0$. Before continuing, fix $m = k = 0.1, g = 10$, and $T = 0.01$. Thus

$$A = \begin{bmatrix} 1.0050 & 0.0100 \\ 1.0017 & 1.0050 \end{bmatrix}, \quad B = \begin{bmatrix} -0.0020 \\ -0.2010 \end{bmatrix}.$$

Note that A is unstable, with eigenvalues $\lambda_1 = 1.1052$ and $\lambda_2 = 0.9048$.

Controllability of $\{A, B\}$ implies that a stabilizing state feedback gain $K = [k_1, k_2]$ can be found. Moreover, the eigenvalues of the resulting system matrix $A - BK$ can be assigned arbitrarily. In our example,

$$A - BK = \begin{bmatrix} 1.0050 + 0.0020k_1 & 0.0100 + 0.0020k_2 \\ 1.00017 + 0.2010k_1 & 1.0050 + 0.2010k_2 \end{bmatrix},$$

so that

$$|\lambda I - A + BK| = \lambda^2 - (2.0100 + 0.002k_1 + 0.201k_2)\lambda + 0.2000k_2 + 1,$$

and eigenvalues $\lambda_1 = \frac{1}{2}$ and $\lambda_2 = -\frac{1}{2}$ (both inside the unit circle) can be obtained by choosing

$$K = [k_1, k_2] = [-376.2492 - 6.2500].$$

Observability of $\{A, C\}$ implies that an asymptotic observer can be constructed to produce an estimate of the system state vector from measurements of x. The observer gain $L = [l_1, l_2]^T$ can be chosen not only to ensure that the state estimate converges, but to place the observer eigenvalues arbitrarily. In our example,

$$A - LC = \begin{bmatrix} 1.0050 - l_1 & 0.0100 \\ 1.0017 - l_2 & 1.0050 \end{bmatrix},$$

so that

$$|\lambda I - A + LC| = \lambda^2 + (l_1 - 2.0100)\lambda - 1.0050l_1 + 0.0100l_2 + 1,$$

and eigenvalues $\lambda_1 = \frac{1}{4}$ and $\lambda_2 = -\frac{1}{4}$ can be obtained by choosing

$$L = \begin{bmatrix} l_1 \\ l_2 \end{bmatrix} \begin{bmatrix} 2.0100 \\ 95.5973 \end{bmatrix}.$$

The eigenvalue separation theorem ensures that combining this observer with the state feedback controller designed above will produce a stable closed-loop system with eigenvalues $\pm\frac{1}{2}$ and $\pm\frac{1}{4}$.

Exercises 10.5

In Problems 1 through 4 design an observer so that the eigenvalues of the matrix $A - EC$ are as given.

1. $A = \begin{pmatrix} 1 & 1 \\ 0 & -1 \end{pmatrix}, \quad C = \begin{pmatrix} 1 & 1 \end{pmatrix},$
   $\lambda_1 = \frac{1}{2}, \quad \lambda_2 = -\frac{1}{4}.$
2. $A = \begin{pmatrix} 0 & 1 \\ 1 & 0 \end{pmatrix}, \quad C = \begin{pmatrix} 0 & 1 \end{pmatrix},$
   $\lambda_1 = \frac{1}{2} - \frac{1}{4}i, \quad \lambda_2 = \frac{1}{2} + \frac{1}{4}i.$

3. $A = \begin{pmatrix} 1 & 0 & 0 \\ 0 & 0 & 1 \\ 0 & 1 & 0 \end{pmatrix}, \quad C = (0 \ \ 1 \ \ 0),$
$\lambda_1 = \frac{1}{2}, \quad \lambda_2 = \frac{1}{4} - \frac{1}{4}\,i, \quad \lambda_3 = \frac{1}{4} + \frac{1}{4}\,i.$

4. $A = \begin{pmatrix} 0 & 1 & 0 \\ 0 & -1 & 1 \\ 0 & 0 & -1 \end{pmatrix}, \quad C = \begin{pmatrix} 0 & 1 & 0 \\ 1 & 0 & 1 \end{pmatrix},$
$\lambda_1 = \frac{1}{2}, \quad \lambda_2 = -\frac{1}{4}, \quad \lambda_3 = -\frac{1}{2}.$

5. (Reduced-Order Observers):
Consider the completely observable system

$$\begin{aligned} x(n+1) &= Ax(n) + Bu(n), \\ y(n) &= Cx(n), \end{aligned} \tag{10.5.9}$$

where it is assumed that the $r \times k$ matrix C has rank r (i.e., the r measurements are linearly independent). Let H be a $(k-r) \times k$ matrix such that the matrix

$$P = \begin{pmatrix} H \\ C \end{pmatrix}$$

is nonsingular. Let

$$\bar{x}(n) = Px(n). \tag{10.5.10}$$

Then $\bar{x}$ may be written as

$$\bar{x} = \begin{pmatrix} w(n) \\ y(n) \end{pmatrix},$$

where $w(n)$ is $(k-r)$-dimensional and $y(n)$ is the r-dimensional vector of outputs.

(a) Use (10.5.10) to show that system equation (10.5.9) may be put in the form

$$\begin{pmatrix} w(n+1) \\ y(n+1) \end{pmatrix} = \begin{pmatrix} A_{11} & A_{12} \\ A_{21} & A_{22} \end{pmatrix} \begin{pmatrix} w(n) \\ y(n) \end{pmatrix} + \begin{pmatrix} B_1 \\ B_2 \end{pmatrix} u(n). \tag{10.5.11}$$

(b) Multiply the bottom part of (10.5.11) by any $(k-r) \times r$ matrix E to show that

$$\begin{aligned} W(n+1) - Ey(n+1) = {} & (A_{11} - EA_{21})[W(n) - Ey(n)] \\ & + [A_{11}E - EA_{21}E + A_{12} - EA_{22}]y(n) \\ & + (B_1 - EB_2)u(n). \end{aligned}$$

(c) If $v(n) = w(n) - Fy(n)$, show that

$$\begin{aligned} v(n+1) = {} & (A_{11} - EA_{21})v(n) \\ & + [A_{11}E - EA_{21}E + A_{12} - EA_{22}]y(n) \\ & + (B_1 - FB_2)u(n). \end{aligned}$$

(d) Explain why we can take an observer of system equation (10.5.9) as the $(k-r)$-dimensional system

$$\begin{aligned} z(n+1) = {} & (A_{11} - EA_{21})z(n) \\ & + [A_{11}E - EA_{21}E + A_{12} - EA_{22}]y(n) \\ & + (B_1 - FB_2)u(n). \end{aligned} \tag{10.5.12}$$

(e) Let $e(n) = z(n) - v(n)$. Show that

$$e(n+1) = (A_{11} - EA_{21})e(n). \tag{10.5.13}$$

6. Prove that if the system equation (10.5.9) is completely observable, then the pair $\{A_{11}, A_{21}\}$ in (10.5.11) is completely observable.

7. Prove the eigenvalue separation theorem, Theorem 10.27, for reduced-order observers.

8. Consider the system

$$\begin{aligned} x_1(n+1) &= x_2(n), \\ x_2(n+1) &= -x_1(n) + 2x_2(n) + u(n), \\ y(n) &= x_1(n). \end{aligned}$$

Construct a one-dimensional observer with a zero eigenvalue.

Appendix A
Stability of Nonhyperbolic Fixed Points of Maps on the Real Line

A.1 Local Stability of Nonoscillatory Nonhyperbolic Maps

Our aim in this appendix is to extend Theorems 1.15 and 1.16 to cover all the remaining unresolved cases. The exposition is based on the recent paper by Dannan, Elaydi and Ponomarenko [30]. The main tools used here are the Intermediate Value Theorem and Taylor's Theorem which we are going to state.

Theorem A.1 (The Intermediate Value Theorem). *Let f be a continuous function on an interval $I = [a, b]$ such that $f(a) \neq f(b)$. If c is between $f(a)$ and $f(b)$, then there exists $x_0 \in (a, b)$ such that $f(x_0) = c$. In particular, if $f(a)$ and $f(b)$ are of opposite sign, then since 0 is between $f(a)$ and $f(b)$, there exists x_0 between a and b such that $f(x_0) = 0$.*

Theorem A.2 (Taylor's Theorem). *Suppose that the $(n+1)$th derivative of the function f exists on an interval containing the points a and b. Then*

$$f(b) = f(a) + f'(a)(b-a) + \frac{f''(a)}{2!}(b-a)^2 + \frac{f^{(3)}(a)}{3!}(b-a)^3 + \cdots + \frac{f^{(n)}(a)}{n!}(b-a)^n + \frac{f^{(n+1)}(z)}{(n+1)!}(b-a)^{n+1} \qquad \text{(A.1.1)}$$

for some number z between a and b.

Notation: The notation of $f \in C^r$ means that the derivatives $f', f'', \ldots, f^{(r)}$ exist and are continuous.

Theorem A.3. *Let x^* be a fixed point of f, then the following statements hold true:*

(i) *Suppose that $f \in C^{2k}$. If $f'(x^*) = 1$, and $f''(x^*) = \cdots = f^{(2k-1)}(x^*) = 0$ but $f^{(2k)}(x^*) \neq 0$, then x^* is semi-asymptotically stable:*

 (a) *from the left if $f^{(2k)}(x^*) > 0$, and*

 (b) *from the right if $f^{(2k)}(x^*) < 0$.*

(ii) *Suppose that $f \in C^{(2k+1)}$. If $f'(x^*) = 1$, and $f''(x^*) = \cdots = f^{(2k)}(x^*) = 0$ but $f^{(2k+1)}(x^*) \neq 0$, then:*

 (a) *x^* is asymptotically stable if $f^{(2k+1)}(x^*) < 0$, and*

 (b) *x^* is unstable if $f^{(2k+1)}(x^*) > 0$.*

PROOF.

(i) Assume that $f'(x^*) = 1$, $f''(x^*) = \cdots = f^{(2k-1)}(x^*) = 0$ but $f^{(2k)}(x^*) \neq 0$.

 (a) If $f^{(2k)}(x^*) > 0$, then by Taylor's Theorem, for a sufficiently small number $\delta > 0$, we have

$$f(x^* + \delta) = f(x^*) + f'(x^*)\delta + \ldots + \frac{f^{(2k-1)}(x^*)\delta^{(2k-1)}}{(2k-1)!} + \frac{f^{(2k)}(\xi)\delta^{2k}}{(2k)!} \tag{A.1.2}$$

 for some $\xi \in (x^*, x^*+\delta)$. If δ is sufficiently small, we may conclude that $f^{(2k)}(\xi) > 0$. Substituting in (A.1.2) yields

$$f(x^* + \delta) = x^* + \delta + \frac{f^{(2k)}(\xi)\delta^{2k}}{(2k)!}. \tag{A.1.3}$$

 Similarly one may show that

$$f(x^* - \delta) = x^* - \delta + \frac{f^{(2k)}(\xi)\delta^{2k}}{(2k)!}. \tag{A.1.4}$$

 Hence from (A.1.3), it follows that $f(x^* + \delta) > x^* + \delta$. And from (A.1.4) we have $x^* - \delta < f(x^* - \delta) < x^*$. This proves semi-asymptotic stability from the left.

 (b) The proof of part (b) is analogous and will be left to the reader.

(ii) By the assumptions in (ii) we have, for some $\delta > 0$,

$$f(x^* + \delta) = x^* + \delta + \frac{f^{(2k+1)}(\xi)\delta^{2k+1}}{2k+1)!} \tag{A.1.5}$$

for some $\xi \in (x^*, x^* + \delta)$.
Furthermore,

$$f(x^* - \delta) = x^* - \delta + \frac{f^{(2k+1)}(\xi)\delta^{2k+1}}{2k+1)!}. \tag{A.1.6}$$

(a) If $f^{(2k+1)}(x^*) < 0$, then, from (A.1.5), $f(x^*+\delta) < x^*+\delta$ and, from (A.1.6), $f(x^* - \delta) > x^* - \delta$. Hence x^* is asymptotically stable.

(b) The proof of part (b) is analagous and will be left to the reader.

□

Consider the map $f(x) = x + (x-1)^4$, where $x^* = 1$ is a fixed point of f with $f'(x^*) = 1$, $f''(x^*) = f'''(x^*) = 0$, and $f^{(4)}(x^*) = 24 > 0$. Then by Theorem A.3, x^* is semi-asymptotically stable from the left.

A.2 Local Stability of Oscillatory Nonhyperbolic Maps

We now consider the case when $f'(x^*) = -1$. A nice trick here is to look at the map $g(x) = f(f(x)) = f^2(x)$.

A.2.1 Results with $g(x)$

Since x^* is a fixed point of f, it must be a fixed point of g and $g'(x^*) = 1$. Moreover, $g''(x^*) = 0$ and $g'''(x^*) = 2Sf(x^*)$. Notice that x^* is asymptotically stable {unstable} under g if, and only if, it is asymptotically stable {unstable} under f. This is due to the fact that $|f^n(x^*)| < 1$ if and only if $|g^n(x^*)| < 1$.

We can then apply the second half of Theorem A.3 to get the following result.

Theorem A.4. *Suppose that $f \in C^{(2k+1)}$ and x^* is a fixed point of f such that $f'(x^*) = -1$. If $g''(x^*) = \ldots = g^{(2k)}(x^*) = 0$ and $g^{(2k+1)}(x^*) \neq 0$, then:*

(1) *x^* is asymptotically stable if $g^{(2k+1)}(x^*) < 0$, and*

(2) *x^* is unstable if $g^{(2k+1)}(x^*) > 0$.*

Observe that this strategy does not use the other part of Theorem A.3– where x^* is semi-asymptotically stable under g. That is, the case where $f'(x^*) = -1$, $g''(x^*) = \ldots = g^{(2k-1)}(x^*) = 0$, and $g^{(2k)}(x^*) \neq 0$.

We now argue that this situation will never occur for analytic f.

Theorem A.5. *Let f be analytic with $f'(x^*) = -1$. Then, for some $k > 1$,*

(1) *If $g''(x^*) = \ldots = g^{(2k-1)}(x^*) = 0$, then $g^{(2k)}(x^*) = 0$.*

(2) *x^* cannot be semi-asymptotically stable under g.*

PROOF. By Taylor's Theorem, we have some small δ with

$$f(x^* + \delta) = f(x^*) + \delta f'(x^*) + \frac{\delta^2 f''(x^*)}{2!} + \dots$$
$$= x^* - \delta + 0(\delta^2).$$

Hence, for $x_0 = x^*+\delta > x^*$, we have $f(x_0) < x^*$ and for $x_0 = x^*-\delta < x^*$, we have $f(x_0) > x^*$. In other words, for $x_0 \in (x^* - \delta,\ x^* + \delta)$ either $f^{2k}(x^*) \in (x^*,\ x^* + \delta)$ and $f^{2k+1}(x^*) \in (x^* - \delta,\ x^*)$ for all $k \in \mathbb{Z}^+$ or $f^{2k}(x^*) \in (x^* - \delta,\ x^*)$ and $f^{2k+1}(x^*) \in (x^*,\ x^* + \delta)$ for all $k \in \mathbb{Z}^+$.

Now, if $f^{2k}(x_0) \to x^*$ as $k \to \infty$, then $f^{2k+1}(x_0) \to x^*$ as $k \to \infty$. Hence either x^* is asymptotically stable or x^* is unstable and, more importantly, it cannot be semi-asymptotically stable. □

These results using $g(x)$ are conclusive but not entirely satisfactory. For example, we return to $f(x) = -x + 2x^2 - 4x^3$. To determine the stability of $f(x)$ at 0, we need to find derivatives of $g(x) = -x + 4x^2 - 8x^3 + 64x^5 - 192x^6 + 384x^7 - 384x^8 + 256x^9$. It turns out that $g^5(0) = 7680$; hence, by Theorem A.4, 0 is an unstable fixed point. However, this was computationally difficult, and we would like an analogue of Theorem A.4 using only the derivatives of $f(x)$.

Remark: If $f'(x^*) = 1$, $f^{(k)}(x^*) = 0$ for all $k > 1$, and f is analytic, then $f(x) = x$. Consequently, every point in the vicinity of x^* is a fixed point and x^* is thus stable but not asymptotically stable. If $f'(x^*) = -1$, $g^{(k)}(x^*) = 0$ for all $k > 1$, and if f is analytic, then $g(x) = x$. Hence every point in the vicinity of x^* is periodic of period 2, and x^* is again stable but not asymptotically stable.

Example A.6. Consider the maps $f_1(x) = x+e^{-x^{-2}}$, $f_2(x) = x+xe^{-x^{-2}}$, $f_3(x) = x-xe^{-x^{-2}}$, with $f_i(0) = 0$. Each of these maps has $f_i'(0) = -1$, and $f_i^{(k)}(0) = 0$ for all $k > 1$. However, the fixed point 0 is semi-asymptotically stable from the left, unstable, and asymptotically stable, respectively.

Example A.7. Contemplate May's genotype selection model

$$x(n+1) = \frac{x(n)e^{\alpha(1-2x(n))}}{1 - x(n) + x(n)e^{\alpha(1-2x(n))}}, \quad \alpha > 0, \quad x \in (0,1). \qquad \text{(A.2.1)}$$

At the fixed point $x^* = \frac{1}{2}$, $f'(x^*) = 1-\frac{\alpha}{2}$. The fixed point is thus asymptotically stable for $1 < \alpha < 4$ by Theorem 1.13. At $\alpha = 4$, we have $f'(x^*) = -1$, $g''(x^*) = g'''(^*) = 0$, but $g'''(x^*) = -32 < 0$. Hence by Theorem A.4, the fixed point $x^* = \frac{1}{2}$ is asymptotically stable.

Appendix B
The Vandermonde Matrix

The generalized Vandermonde matrix is given by

$$V = \begin{pmatrix} 1 & 0 & \dots & 1 & 0 & \dots \\ \lambda_1 & 1 & \dots & \lambda_r & 1 & \dots \\ \lambda_1^2 & 2\lambda_1 & \dots & \lambda_r^2 & 2\lambda_r & \dots \\ \vdots & \vdots & & \vdots & \vdots & \\ \lambda_1^{k-1} & (k-1)\lambda_1^{k-2} & \dots & \lambda_r^{k-1} & (k-1)\lambda_r^{k-2} & \end{pmatrix} \tag{B.1}$$

and consists of $k \times m_i$ submatrices corresponding to the eigenvalues λ_i, $1 \le i \le r$, $\sum_{i=1}^{r} m_i = k$. The first column in the $k \times m_1$ submatrix is $c_1 = (1, \lambda_1, \lambda_1^2, \dots, \lambda_1^k)^T$, the second column is $c_2 = \frac{1}{1!} c_1'(\lambda_1) = (0, 1, 2\lambda_1, 3\lambda_1^2, \dots, k\lambda_1^{k-1})^T$, ..., the sth column is $c_s = \frac{1}{(s-1)!} c_1^{(s-1)}(\lambda_1)$, where $c_1^{(m)}(\lambda_1)$ is the mth derivative of column c_1. The extension of this definition to other $k \times m_i$ submatrices is done in the natural way. We are now going to prove the following result.

Lemma B.1 [76].

$$\boxed{W(0) = \det V = \prod_{1 \le i < j \le k} (\lambda_j - \lambda_i)^{m_i m_j}.} \tag{B.2}$$

Moreover, V is invertible.

PROOF. The proof proceeds by induction on the sum of multiplicities m_i of the eigenvalues λ_i, with $m_i > 1$. If all the m_i's are equal to 1, we have the regular Vandermonde matrix (2.3.3) and thus (B.2) holds. For the example $m_1 = 3$, $m_2 = 2$ ($\lambda_1 = \lambda_2 = \lambda_3, \lambda_4 = \lambda_5$), the sum of multiplicities which exceed 1 is $3 + 2 = 5$. To illustrate the induction step, let

$$\widetilde{W} = \begin{vmatrix} 1 & 0 & 1 & 1 & 0 \\ \lambda_1 & 1 & t & \lambda_2 & 1 \\ \lambda_1^2 & 2\lambda_1 & t^2 & \lambda_2^2 & 2\lambda_2^2 \\ \lambda_1^3 & 3\lambda_1^2 & t^3 & \lambda_2^3 & 3\lambda_2^2 \\ \lambda_1^4 & 4\lambda_1^3 & t^4 & \lambda_2^4 & 4\lambda_2^3 \end{vmatrix}$$

So that the sum of multiplicities greater than 1 is 4.

Assuming (B.2) for $\widetilde{W}$ yields

$$\widetilde{W} = (t - \lambda_1)^2(\lambda_2 - t)^2(\lambda_2 - \lambda_1)^4.$$

Note that

$$\begin{aligned} W(0) &= \frac{1}{2}\left(\frac{d^2}{dt^2}\right)\widetilde{W}\Big|_{t=\lambda_1} \\ &= \frac{1}{2}(\lambda_2 - \lambda_1)^4 \frac{d}{dt}\left[2(t - \lambda_1)(\lambda_2 - t)^2 - 2(t - \lambda_1)^2(\lambda_2 - t)\right]\Big|_{t=\lambda_1} \\ &= \frac{1}{2}(\lambda_2 - \lambda_1)^4\big[2(\lambda_2 - t)^2 - 4(t - \lambda_1)(\lambda_2 - t) \\ &\quad - 4(t - \lambda_1)(\lambda_2 - t) + 2(t - \lambda_1)^2\big]\Big|_{t=\lambda_1} \\ &= (\lambda_2 - \lambda_1)^6. \end{aligned}$$

In general, $W(0)$ is formed from $\widetilde{W}$ as long as there is one multiplicity $m_i > 1$. The general case may be proved in an analogous manner. □

Appendix C
Stability of Nondifferentiable Maps

The main objective of this appendix is to prove Theorem 4.8. In fact, we will prove a more general result which appeared in Elaydi and Sacker [50]. In the sequel we will assume that the $f : I \to I$ is continuous on the closed and bounded interval $I = [a, b]$. Clearly if $I = [a, \infty)$, and f is bounded and continuous, $f(I) \subset J \subset I$, where J is a closed and bounded interval, then $f : J \to J$. The following lemma and its corollary are immediate sequences of the Intermediate Value Theorem.

Lemma C.1. *Let $J = [c, d] \subset [a, b]$ such that either:*

(i) *$f(c) > c$ and $f(d) < d$, or*

(ii) *$f(c) < c$ and $f(d) > d$.*

Then f has a fixed point (c, d).

PROOF.

(i) Assume that $f(c) > c$ and $f(d) < d$. Then for the map $g(x) = f(x) - x$, $g(c) > 0$ and $g(d) < 0$. Hence by the Intermediate Value Theorem, there exists x^* between c and d such that $g(x^*) = 0$. Hence $f(x^*) = x^*$, and thus x^* is a fixed point of f.

The proof of (ii) is similar and will be left to the reader. □

Corollary C.2. *Suppose that $J = [c, d] \subset I$. If $f(d) > d$ and (c, d) is fixed point-free, then $f(x) > x$ for all $x \in (c, d)$.*

We are now ready to present the main result.

Theorem C.3. *Let $f : I \to I$ be continuous. Then the following statements are equivalent:*

(i) *f has no points of minimal period 2 in (a, b).*

(ii) *For all $x_0 \in (a, b)$, $\{f^n(x_0)\}$ converges in I.*

PROOF. (ii) ⇒ (i).
If $\{\bar{x}_1, \bar{x}_2\}$ is a periodic orbit of period 2 in (a, b), then $\{f^n(\bar{x})\}$ does not converge as it oscillates between $\bar{x}_1$ and $\bar{x}_2$.

(i) ⇒ (ii) .
Assume there exists $x_0 \in (a, b)$ such that $\{f^n(x_0)\}$ does not converge. Thus x_0 is not a fixed point or an eventually fixed point. Hence its orbit $O(x_0) = \{x_0, x(1), x(2), \dots\}$ can be partitioned into two sequences $A = \{x(k) | f(x(k)) > x(k)\}$ and $B = \{x(k) | f(x(k)) < x(k)\}$. Then $A \neq \emptyset$ and $B \neq \emptyset$. We claim that A is strictly monotonically increasing, i.e., $i < j$ implies $x(i) < x(j)$. Assume the contrary, that there exists $x(i), x(j) \in A$ such that $i < j$ but $x(i) > x(j)$. This means that $f^i(x_0) > f^j(x_0)$. Let $j = i + r$. Then $f^r(x(i)) < x(i)$. Since $x(i)$ is not a fixed point of f, there exists a small $\delta > 0$ such that the interval $(x(i) - \delta, x(i))$ is free of fixed points. Thus we may conclude that there exists a largest fixed point z of f in $[a, x(i)]$ (z may equal a). Hence the interval $(z, x(i))$ is fixed point-free. And since $f(x(i)) > x(i)$, it follows by Corollary C.2 that $f(x) > x$ for all $x \in (z, x(i))$.

Let z_n be a sequence in $(z, x(i))$ that converges to z. Then $f(z_n) > z_n$ and $\lim_{n\to\infty} f(z_n) = f(z) = z$. There exists $N_1 \in \mathbb{Z}^+$ such that $n > N_1$, $f(z_n) \in (z, x(i))$. For $n > N_1$, $f^2(z_n) > f(z_n) > z$ and $\lim_{n\to\infty} f^2(z_n) = f^2(z) = z$. There exists $N_2 \in \mathbb{Z}^+$ such that for $n > N_2$, $f^2(z_n) \in (z, x(i))$. Repeating this process, there are $N_3, N_4, \dots, N_r$ such that for $n > N_t$, $1 \leq t \leq r$, $f^t(z_n) \in (z, x(i))$. For $N = \max\{N_t | 1 \leq t \leq r\}$, $f^t(z_n) \in (z, x(i))$, $1 \leq t \leq r$. We conclude that there exists $y \in (z, x(i))$ such that $y, f(y), \dots, f^r(y) \in (z, x(i))$. Hence $f^r(y) > f^{r-1}(y) > \cdots > f(y) > y$. But $f^r(x(i)) < x(i)$ implies, by Lemma C.1, the existence of a fixed point in $(y, x(i))$, a contradiction which establishes our claim that A is strictly monotonically increasing.

Similarly, we may show that B is strictly monotonically decreasing. Define $\bar{x}_1 = \sup A$, $\bar{x}_2 = \inf B$. Then $\bar{x}_1 \leq \bar{x}_2$ and hence neither is an end point. Since $A \cup B = O(x_0)$, it follows that $\{\bar{x}_1, \bar{x}_2\} = \Omega(x_0)$, the set of all limit points of $O(x_0)$. Since $\Omega(x_0)$ is invariant, either:

(a) $f(\bar{x}_1) = \bar{x}_2$, $f(\bar{x}_2) = \bar{x}_1$, or

(b) $f(\bar{x}_1) = \bar{x}_1$, $f(\bar{x}_2) = \bar{x}_2$, or

(c) $\bar{x}_1 = \bar{x}_2$,

(d) $f(\bar{x}_1) = \bar{x}_1$ and $f(\bar{x}_2) = \bar{x}_1$ or $f(\bar{x}_1) = \bar{x}_2$ and $f(\bar{x}_2) = \bar{x}_2$.

Case (a) is excluded since there are no 2-cycles; cases (b) and (d) are also excluded since neither A nor B is invariant. Hence the only case left is case (c) which confirms the convergence of the sequence $\{f^n(x_0)\}$. □

As an immediate consequence of the preceding theorem, we have the following important result on global asymptotic stability.

Corollary C.4. *Let x^* be a fixed point of a continuous map on the closed and bounded interval $I = [a, b]$. Then x^* is globally asymptotically stable relative to (a, b) if and only if $f^2(x) > x$ for $x < x^*$ and $f^2(x) < x$ for $x > x^*$ for all $x \in (a, b)\backslash\{x^*\}$, and a, b are not periodic points.*

PROOF. The necessity is clear. To prove the sufficiency, notice that the given assumptions imply that there are no periodic points of minimal period 2. Hence by Theorem C.3, $\{f^n(x_0\}$ converges for every $x_0 \in I$. Now if $x_0 \in (a, x^*)$, $f(x_0) > x_0$. For, otherwise, we would have $f(x_0) < x_0 < f^2(x_0)$, which implies, by the Intermediate Value Theorem, the presence of a fixed point of the map f in the interval (a, x^*), a contradiction. Similarly, one may show that for all $x_0 \in (x^*, b)$, $f(x_0) < x_0$. Thus $\lim_{n\to\infty} f^n(x_0) = c$, where c is an interior point in the interval (a, b). Furthermore, since c is a fixed point of the map f, it follows that $c = x^*$. Hence x^* is globally attracting.

It remains to show that x^* is stable. We have two cases to consider.

Case (i): The map f is monotonically increasing in a small neighborhood $(x^* - \delta, x^*)$. Since $f(x) > x$ for all $x \in (x^* - \delta, x^*)$, it follows that for $x_0 \in (x^* - \delta, x^*)$, we have $x_0 < f(x_0) < f^2(x_0) < \cdots < x^*$.

Case (ii): The map of f is decreasing monotonically in $(x^* - \delta, x*)$. Given $\varepsilon > 0$, there exists $\delta > 0$, $\delta < \varepsilon$ such that $f(x^* - \delta) - x^* < \varepsilon$. Furthermore, $f(x_0) - x^* < \varepsilon$ for all $x_0 \in (x^* - \delta, x^*)$. Since $f(x_0) > x_0$, $f^2(x_0) < f(x_0)$ and since $x_0 < x^*$, $f^2(x_0) > x_0$. Thus $x_0 < f^2(x_0) < f(x_0)$ and, consequently, $f^2(x_0) - x^* < \delta < \varepsilon$. The same scenario occurs for $x_0 > x^*$. Hence x^* is stable. □

Appendix D
Stable Manifold and the Hartman–Grobman–Cushing Theorems

D.1 The Stable Manifold Theorem

Consider the nonlinear difference system

$$x(n+1) = f(x(n)) \tag{D.1.1}$$

such that f has a fixed point $x^* \in \mathbb{R}^k$ and $f \in C^2$ in an open neighborhood of x^*. Let $A = Df(x^*)$ be the Jacobian of f at x^*. Then (D.1.1) may be written in the form

$$x(n+1) = Ax(n) + g(x(n)). \tag{D.1.2}$$

The associated linear system is given by

$$z(n+1) = Az(n). \tag{D.1.3}$$

The fixed point x^* is assumed to be hyperbolic, where none of the eigenvalues of A lie on the unit circle. Arrange the eigenvalues of A into two sets: $S = \{\lambda_1, \lambda_2, \dots, \lambda_r\}$, $G = \{\lambda_{r+1}, \dots, \lambda_k\}$ with $|\lambda_i| < 1$ for $\lambda_i \in S$ and $|\lambda_j| > 1$ for $\lambda_j \in G$. Let E^s be the eigenspace spanned (generated) by the generalized eigenvectors corresponding to S and let E^u be the eigenspace spanned by the generalized eigenvectors corresponding to G. Then $\mathbb{R}^k = E^s \bigoplus E^u$. The sets E^s and E^u are called the stable and unstable subspaces of x^*, respectively.

The local stable manifold of x^* in an open neighborhood G defined as

$$W^s(x^*, G) \equiv W^s(x^*) = \{x_0 \in G \mid O(x_0) \subset G \text{ and } \lim_{n\to\infty} f^n(x_0) = x^*\}.$$

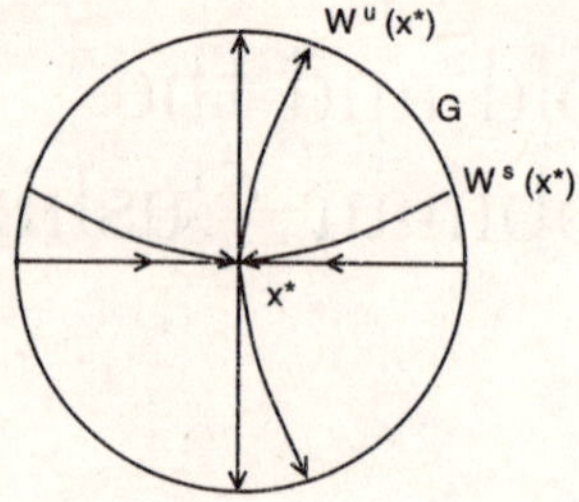

FIGURE D.1. E^s is tangent to $W^s(x^*)$ and E^u is tangent to $W^u(x^*)$.

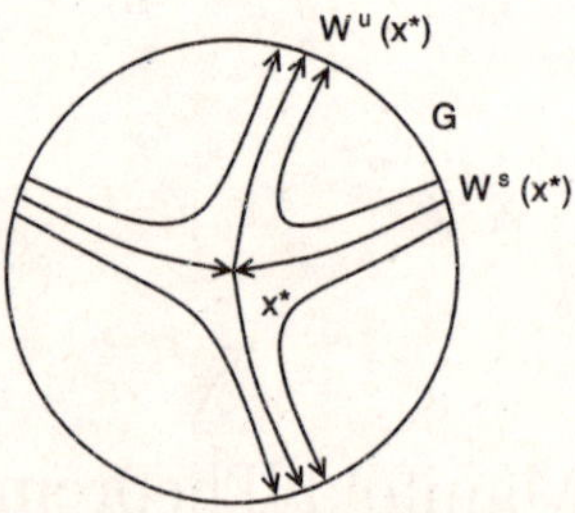

FIGURE D.2. Stable and unstable manifolds $W^s(x^*)$ and $W^u(x^*)$ in a neighborhood G of x^*.

To define the unstable manifold, we need to look at negative orbits. Since f is not assumed to be invertible, we have to define a principal negative orbit $O^-(x_0) = \{x(-n)\}$ of a point x_0 as follows. We let $x(0) = x_0$, and $f(x(-n-1)) = x(-n)$, $n \in \mathbb{Z}^+$. The local unstable manifold for x^* in G is defined to be the set

$$W^u(x^*, G) \equiv W^u(x^*) = \{x_0 \in G \mid \text{there exists a negative orbit,}\\ O^-(x_0) \subset G \text{ and } \lim_{n\to\infty} x(-n) = x^*\}.$$

The following theorem states that E^s is tangent to $W^s(x^*)$ and E^u is tangent to $W^u(x^*)$ at the fixed point x^* (see Figures D.1 and D.2).

Theorem D.1 (The Stable Manifold Theorem). *Let x^* be a hyperbolic fixed point of a C^2-map $f : \mathbb{R}^k \to \mathbb{R}^k$. Then in an open neighborhood G of x^* there exist two manifolds $W^s(x^*)$ of dimension E^s and $W^u(x^*)$ of dimension E^u such that:*

(i) *E^s is tangent to $W^s(x^*)$ at x^* and for any solution $x(n)$ of* (D.1.1) *with $x(0) \in W^s$, $\lim_{n\to\infty} x(n) = x^*$.*

(ii) *E^u is tangent to $W^u(x^*)$ at x^* and if $x(0) \in W^u(x^*)$, then there exists a principal negative solution $x(-n)$ with $\lim_{n\to\infty} x(-n) = x^*$.*

PROOF. See Cushing [24] and Robinson [128]. □

D.2 The Hartman–Grobman–Cushing Theorem

The Stable Manifold Theorem tells us what happens to solutions that lie on either the stable manifold W^s or the unstable manifold W^u in a neighborhood of a hyperbolic fixed point. The question that we are going to address here is: What happens to solutions whose initial points do not lie on either W^s or W^u?

The answer to this question is given by the classical Hartman–Grobman Theorem in differential equations and its analogue in difference equations. However, this theorem requires that the map is a diffeomorphism, that is differentiable and a homeomorphism. Two maps $f : X \to X$ and $g : Y \to Y$ are said to be topologically conjugate if there is a homeomorphism $h : Y \to X$ such that $f(h(y)) = h(g(y))$ for all $y \in Y$.

Theorem D.2 (Hartman–Grobman). *Let $f : \mathbb{R}^k \to \mathbb{R}^k$ be a C^r-diffeomorphism with hyperbolic fixed point x^*. Then there exists neighborhoods V of x^* and W of 0 and a homeomorphism $h : W \to V$ such that $f(h(x)) = h(Ax)$, where $A = Df(x^*)$.*

In other words, f is topologically conjugate in a neighborhood of the fixed point x^* to the linear map induced by the derivative at the fixed point (see Figure D.3).

PROOF. See Robinson [128]. □

As pointed out in Cushing [24], this classical theorem does not hold for noninvertible maps, as may be seen from the following example.

Example D.3. Consider the one-dimensional difference equation $x(n+1) = x^2(n)$. The Jacobian at the fixed point $x^* = 0$ is $A = 0$. If h is the conjugacy homeomorphism, then $f(h(x)) = h(Ax)$. Then $f(h(x)) = h(0) = 0$. Thus $[h(x)]^2 = 0$ and $h(x) = 0$ for all $x \in \mathbb{R}$, a contradiction, since h is one to one.

Cushing [24] extended the Hartman–Grobman Theorem to noninvertible maps and the new result will henceforth be called HGC (Hartman–

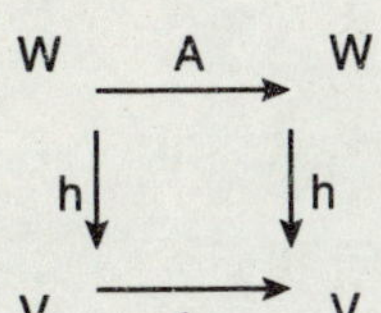

FIGURE D.3. f is conjugate to $A = Df(x^*)$, $f(h(x)) = h(Ax)$.

Grobman–Cushing). But before stating the theorem, we need to introduce a few definitions.

For a sequence $x(n) \in \mathbb{R}^k$, let $\|x\|_+ = \sup_{n \in \mathbb{Z}^+} |x(n)|$, where $|x(n)|$ is a norm on $\mathbb{R}^k$.

The sets

$$BS^+ = \{x(n) \mid \|x\|_+ < +\infty\},$$
$$BS_0^+ = \{x(n) \in BS^+ \mid \lim_{n \to +\infty} |x(n)| = 0\},$$

are (Banach)[1] spaces under the norm $\| \cdot \|_+$.

Similarly, we define

$$\|x\|_- = \sup_{n \in \mathbb{Z}^-} |x(n)|$$

and

$$BS^- = \{x(n) \mid \|x\|_- < +\infty\},$$
$$BS_0^- = \{x(n) \in BS^- \mid \lim_{n \to -\infty} |x(n)| = 0\}.$$

Define

$$\sum^{\pm}(\delta) = \{x(n) \in BS^{\pm} \mid \|x\|_{\pm} \le \delta\},$$
$$\sum^{\pm}(\delta) = \{x(n) \in BS_0^{\pm} \mid \|x\|_{\pm} \le \delta\}.$$

Theorem D.4 (Hartman–Grobman–Cushing). *Suppose that x^* is a hyperbolic fixed point of a map $f \in C^r$ and let $A = Df(x^*)$ be its Jacobian. There exists constants c and δ such that the following hold:*

(a) *There is a one to one bicontinuous map between a (forward) solution of* (D.1.1) *lying in $\sum^+(\delta)$ and a (forward) solution of its linearization* (D.1.3) *lying in $\sum^+(c\delta)$.*

(b) *There is a one to one bicontinuous map between a (forward) solution of* (D.1.1) *lying in $\sum_0^+(\delta)$ and a (forward) solution of* (D.1.3) *lying in $\sum_0^+(c\delta)$.*

Similar statements hold for $\sum^-(\delta)$ and $\sum_0^-(\delta)$.

PROOF. See Cushing [24]. □

---

[1]A Banach space is a complete space with a norm, where every Cauchy sequence converges in the space.

Appendix E

The Levin–May Theorem

To prove Theorem 5.2, we need the following result from Linear Algebra [68]:

P: "The k zeros of a polynomial of degree $k \geq 1$ with complex coefficients depend continuously upon the coefficients."

To make this more precise, let $x \in \mathbb{C}^k$, and $f(x) = (f_1(x), f_2(x), \ldots, f_k(x))^T$ in which $f_i : \mathbb{C}^k \to \mathbb{C}$, $1 \leq i \leq k$. The function f is continuous at x if each f_i is continuous at x, i.e., for each $\varepsilon > 0$ there exists $\delta > 0$ such that if $||y - x|| < \delta$, then $|f_i(x) - f_i(y)| < \varepsilon$ where $||\cdot||$ is a vector norm on $\mathbb{C}^k$. Now P may be stated intuitively by saying that the function $f : \mathbb{C}^k \to \mathbb{C}^k$ which takes the k coefficients (all but the leading one) of a monic polynomial of degree k to the k zeros of the polynomial, is continuous. Precisely, we have the following result.

Lemma E.1 [68]. *Let $k \geq 1$ and let*

$$p(x) = x^k + a_1 x^{k-1} + \cdots + a_{k-1} x + a_k$$

be a polynomial with complex coefficients. Then for every $\varepsilon > 0$, there is a $\delta > 0$ such that, for any polynomial,

$$q(x) = x^k + b_1 x^{k-1} + \cdots + b_{k-1} x + b_k$$

satisfying

$$\max_{1 \leq i \leq k} |a_i - b_i| < \delta$$

we have

$$\min_{\tau} \max_{1\le j\le k} |\lambda_j - \mu_{\tau(i)}| < \varepsilon,$$

where $\lambda_1, \lambda_2, \ldots, \lambda_k$ *are the zeros of* $p(x)$ *and* $\mu_1, \mu_2, \ldots, \mu_k$ *are the zeros of* $q(x)$ *in some order, counting multiplicities, and the minimum is taken over all permutations* τ *of* $1, 2, \ldots, k$.

The characteristic equation associated with equation (5.1.18) is given by

$$\lambda^{k+1} - \lambda^k + q = 0.$$

Since the characteristic roots may, in general, be complex, we may put $\lambda = re^{i\theta}$. This yields the equation

$$r^{k+1}e^{i\theta(k+1)} - r^k e^{i\theta k} + q = 0. \tag{E.1}$$

The general stability may be mapped out as a function of q and k as follows: if, for the dominant eigenvalues, $\theta = 0$ and $r < 1$ (λ is real and $|\lambda| < 1$), then there is monotonic damping; if, for the dominant eigenvalues, $\theta \neq 0$, $r < 1$ (λ is complex, $|\lambda| < 1$), then there is oscillatory damping; and if $r > 1$ ($|\lambda| > 1$) for any eigenvalue, the zero solution is unstable.

The next lemma shows that for

$$q_1 = \frac{k^k}{(k+1)^{k+1}} \tag{E.2}$$

there is monotonic damping if $0 < q \le q_1$.

Lemma E.2. *All solutions of* (5.1.20) *converge monotonically to the zero solution if*

$$0 < q \le q_1.$$

PROOF. To find the region of values of q where solutions of (5.1.20) converge monotonically to the zero solution we let $\theta = 0$ in (E.1). This yields

$$r^{k+1} - r^k + q = 0 \quad \text{or}$$
$$q = r^k - r^{k+1}. \tag{E.3}$$

Consider the function $q = h(r) = r^k - r^{k+1}$. Clearly, $h(0) = h(1) = 0$ and $h(r) > 0$ if and only if $0 < r < 1$. Moreover, if $q = 0$, then $r = 0$ is of multiplicity k. Since $h'(r) = r^{k-1}(k - (k+1)^r)$, we conclude that:

(i) If $0 < r < \frac{k}{k+1}$, then $h'(r) > 0$ and thus q is increasing.

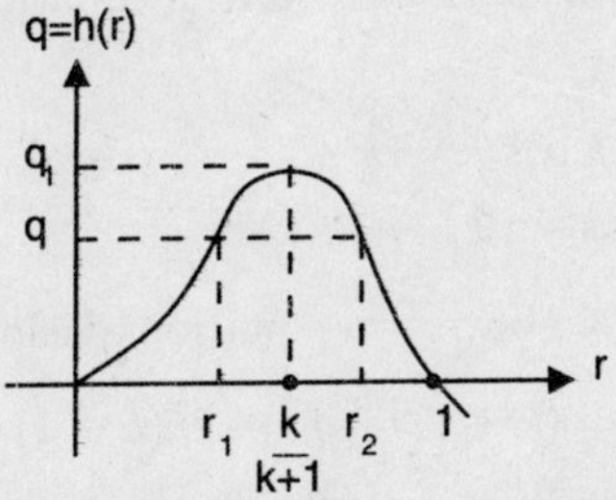

FIGURE E.1. For $q < q_1$, there are two positive solutions r_1, r_2 of (E.3).

(ii) If $r = \frac{k}{k+1}$, then $h'(r) = 0$ and q attains its maximal value

$$\begin{aligned} q_1 = h(r) = h\left(\frac{k}{k+1}\right) &= \left(\frac{k}{k+1}\right)^k - \left(\frac{k}{k+1}\right)^{k+1} \\ &= \left(\frac{k}{k+1}\right)^k \left[1 - \frac{k}{k+1}\right] \\ &= \frac{k^k}{(k+1)^{k+1}}. \end{aligned}$$

(iii) If $r > \frac{k}{k+1}$, then $h'(r) < 0$ and thus q is decreasing and intersects the r-axis at $r = 1$. –

Hence, for every $q > 0$, there are two positive real solutions r_1 and r_2 of (E.3) if and only if $q < q_1$ (see Figure E.1). As q tends to zero, $r_1 \to 0$ and $r_2 \to 1$; and as q tends to q_1, both roots coalesce to $\frac{k}{(k+1)}$.

It remains to show that the larger root r_2 of these two real roots is the dominant root for all $0 < q < q_1$. In fact, we will show that as q increases in the range $0 < q < q_1$, there will be one real root r with magnitude greater than $\frac{k}{k+1}$ but less than 1, and the remaining k roots have magnitude less than $\frac{k}{k+1}$. To accomplish this task, it suffices to show that as q increases in the interval $(0, q_1)$, no characteristic root crosses the circle centered at the origin and with radius $\frac{k}{k+1}$. To prove this statement we put $\mu = \lambda(k+1)/k$ and $p = q/q_1$ in (E.3). This yields

$$k\mu^{k+1} - (k+1)\mu^k + p = 0. \tag{E.4}$$

Notice that a characteristic root λ crosses the circle with radius $\frac{k}{k+1}$ if and only if the corresponding characteristic root μ crosses the unit circle. If such a crossing occurs, then, by Lemma E.1, there exists a characteristic root $\mu = e^{i\theta}$ of (E.4). Substituting in (E.4) gives

$$ke^{i\theta} = (k+1) - pe^{-ik\theta}.$$

Hence

$$k(\cos\theta + i\sin\theta) = k + 1 - p(\cos k\theta - i\sin k\theta).$$

Equating the real parts in both sides of the equation and similarly for the imaginary parts yield

$$k\cos\theta = (k+1) - p\cos k\theta,$$
$$k\sin\theta = p\sin k\theta.$$

Squaring and adding the above two equations yield

$$k^2 = (k+1)^2 + p^2 - 2p(k+1)\cos k\theta,$$
$$\cos k\theta = \frac{2k+1+p^2}{2p(k+1)}.$$

Since $|\cos k\theta| \leq 1$, it follows that

$$2k+1+p^2 \leq 2p(k+1)$$

or

$$(1-p)(2k+1-p) \leq 0. \tag{E.5}$$

Since $p = \frac{q}{q_1} < 1$, both terms in inequality (E.5) are positive and thus we have a contradiction. Hence no characteristic root μ crosses the unit circle and, consequently, no characteristic root λ crosses the circle with radius $\frac{k}{k+1}$.

This shows that for $0 < k < k_1$ there is a dominant positive real characteristic root of magnitude between $\frac{k}{k+1}$ and 1, whereas all the remaining k characteristic roots have modules less than $\frac{k}{k+1}$. □

Lemma E.3. *If* $0 < q < q_2$, *where*

$$q_2 = 2\cos\left(\frac{k\pi}{2k+1}\right), \tag{E.6}$$

then all solutions converge to the zero solution.

Proof. By virtue of Lemma E.2, it suffices to consider the case where $q_1 < q < q_2$. Consider again the function $q = h(r) = r^k - r^{k+1}$ whose derivative is given by $h'(r) = r^{k-1}(k - (k+1)r)$. For $r \leq 0$, the sign of $h'(r)$ depends very much on whether k is even or odd. If k is odd, then $h'(r) > 0$ and, consequently, $q = h(r)$ is increasing in the interval $(-\infty, 0]$ and equation (E.3) has all complex roots. On the other hand, if k is even, then $h'(r) < 0$ and thus h is decreasing on $(-\infty, 0]$. It follows that equation (E.3) has one real root and k complex roots. This real root $\lambda = re^{i\theta}$ must be negative and hence $\theta = \pi$. Substituting in (E.1) yields

$$-r^{k+1} - r^k + q = 0.$$

Hence

$$r^{k+1} + r^k = q. \tag{E.7}$$

Since $q < q_2 \leq 2$, it follows that $r < 1$. This implies that the real root is between -1 and 0. As q increases, the zero solution becomes unstable the first time r exceeds 1.

Putting $r = 1$ in (E.1) yields

$$e^{i\theta} - 1 + qe^{-ik\theta} = 0. \tag{E.8}$$

Equivalently, we have

$$(\cos\theta + i\sin\theta) = 1 - q(\cos k\theta - i\sin k\theta).$$

Equating the real part in the left side with the real part in the right side and similarly for the imaginary part yields

$$\cos\theta = 1 - q\cos k\theta,$$
$$\sin\theta = q\sin k\theta.$$

Squaring and adding yields

$$q = 2\cos k\theta. \tag{E.9}$$

Substituting (E.9) into (E.8) we obtain

$$\begin{aligned} e^{i\theta} &= 1 - 2(\cos^2 k\theta - i\cos k\theta\sin k\theta) \\ &= -\cos 2k\theta + i\sin 2k\theta \\ &= -e^{-2ik\theta}. \end{aligned}$$

Hence $\theta = (2n+1)\pi - 2k\theta$, when n is an integer. Solving for θ yields

$$\theta = \frac{(2n+1)\pi}{2k+1}. \tag{E.10}$$

By (E.9) we obtain

$$q = 2\cos\left(\frac{(2n+1)k\pi}{2k+1}\right). \tag{E.11}$$

Note that there may be several distinct values of q given by different values. The smallest of these values of q occurs at $n = 0$:

$$q = q_2 = 2\cos\left(\frac{k\pi}{(2k+1)}\right),$$

and this defines the upper boundary of the stability region. This is clear since $\frac{2n+1}{2} - \frac{(2n+1)k}{2k+1} = \frac{2n+1}{2(2k+1)}$ is increasing as n increases. This completes the proof of the Lemma. □

PROOF OF THEOREM 5.2. Lemma E.3 shows that, for $0 < q < q_2$, the zero solution of (5.1.18) is asymptotically stable. By examining the derivative of r with respect to q, we will show that at $r = 1$, $dr/dq > 0$. This would imply that as q increases, r can only cross the boundary $r = 1$

from below. Consequently, the complex pair of dominant eigenvalues cannot reenter the stable region once it leaves it, and so the zero solution of (5.1.18) is unstable for all $q > q_2$.

Equating the real part with the real part and likewise with the imaginary part in (E.1) yields

$$r = \frac{\sin k\theta}{\sin(k+1)\theta} \tag{E.12}$$

and

$$\begin{aligned} q &= r^k \cos k\theta - r^{k+1}\cos(k+1)\theta \\ &= (\sin k\theta)^k \left[\frac{\cos k\theta}{[\sin(k+1)\theta]^k} - \frac{\cos(k+1)\theta \sin k\theta}{[\sin(k+1)\theta]^{k+1}}\right] \\ &= \frac{(\sin k\theta)^k}{[\sin(k+1)\theta]^{k+1}}[\sin(k+1)\theta\cos k\theta - \cos(k+1)\theta \sin k\theta] \\ &= \frac{(\sin k\theta)^k \sin\theta}{[\sin(k+1)\theta]^{k+1}}. \end{aligned} \tag{E.13}$$

If $r = 1$, we obtain, for (E.12),

$$\sin k\theta = \sin(k+1)\theta \quad \text{and} \quad \cos k\theta = -\cos(k+1)\theta$$

and hence

$$\begin{aligned} \sin k\theta &= \sin k\theta \cos k + \cos k\theta \sin k, \\ \cos k\theta &= -\cos k\theta \cos\theta + \sin k\theta \sin\theta. \end{aligned}$$

Multiplying the first equation by $\cos k\theta$ and the second by $\sin k\theta$ and then adding yields

$$\sin\theta = \sin 2k\theta.$$

Multiplying the first equation by $\sin k\theta$ and the second by $\cos k\theta$ and then subtracting yields

$$\cos\theta = -\cos 2k\theta.$$

From (E.12) we have, for $r = 1$,

$$\begin{aligned} \frac{dr}{d\theta} &= k\cot(k\theta) - (k+1)\cot[(k+1)^\theta] \\ &= (2k+1)\cot k\theta \end{aligned}$$

and from (E.13) we have, for $r = 1$,

$$\begin{aligned}
\frac{dq}{d\theta} &= \frac{[\sin(k+1)\theta]^{k+1}[k^2(\sin k\theta)^{k-1}\cos k\theta \sin\theta + (\sin k\theta)^k \cos\theta]}{[\sin(k+1)\theta]^{2k+2}} \\
&\quad - \frac{(\sin k\theta)^k \sin\theta \cdot (k+1)^2[\sin(k+1)\theta]^k \cos(k+1)\theta}{[\sin(k+1)\theta]^{2k+2}} \\
&= \frac{k^2\cos k\theta \sin\theta}{[\sin(k+1)\theta]^2} + \frac{\cos\theta}{\sin(k+1)\theta} - \frac{(k+1)^2 \sin\theta \cos(k+1)\theta}{[\sin(k+1)\theta]^2} \\
&= (2k^2+2k+1)\cot k\theta - \cot 2k\theta \\
&= (2k^2+2k+1)\cot k\theta + \frac{1}{2}(\tan k\theta - \cot k\theta) \\
&= \frac{1}{2}\left(k+\frac{1}{2}\right)^2 \cot k\theta + \frac{1}{2}\tan k\theta.
\end{aligned}$$

Clearly, $\frac{dq}{d\theta}$ and $\frac{dr}{d\theta}$ both have the same sign as $\cot k\theta$, and hence $\frac{dr}{dq} = \frac{dr}{d\theta} / \frac{dq}{d\theta}$ is positive. This completes the proof. □

Appendix F
Classical Orthogonal Polynomials

This is a list of some classical orthogonal polynomials $Q_n(x)$, their definitions, the corresponding intervals of orthogonality (a, b), and difference equations $Q_{n+1}(x) - (A_n x + B_n)Q_n(x) + C_n Q_{n-1}(x) = 0$.

Name	Definition	(a, b)	Difference Equation
1. Jacobi:	$P_n^{\alpha,\beta}(x)$	$(-1, 1)$	see (9.5.12), (9.5.15), (9.5.16)
2. Gegenbauer: (ultraspherical)	$P_n^{\nu}(x)$ (see (9.4.10))	$(-1, 1)$	$A_n = 2\frac{\nu+n}{n+1}$, $B_n = 0$ $C_n = \frac{2\nu+n-1}{n+1}$
3. Legendre:	$P_n(x) = P_n^{(0,0)}$ (see (9.4.9))	$(-1, 1)$	$A_n = \frac{2n+1}{n+1}$, $B_n = 0$ $C_n = \frac{n}{n+1}$
4. Chebyshev: (First kind)	$T_n(x) = \cos n\theta$, $\theta = \cos^{-1}(x)$	$(-1, 1)$	$A_n = 2$, $B_n = 0$ $C_n = 1$
5. Chebyshev: (Second kind)	$U_n(x) = \frac{\sin(n+1)\theta}{\sin\theta}$, $\theta = \cos^{-1}(x)$	$(-1, 1)$	$A_n = 2$, $B_n = 0$ $C_n = 1$
6. Hermite:	$H_n(x)$ (see (9.4.15))	$(-\infty, \infty)$	$A_n = 2$, $B_n = 0$ $C_n = 2n$
7. Laguerre:	$L_n^{\alpha}(x)$ (see (9.4.13))	$(0, \infty)$	$A_n = \frac{2n+\alpha+1-x}{n+1}$ $B_n = 0$ $C_n = \frac{n+\alpha}{n+1}$
8. Charlier:	$C_n^{(\alpha)}(x)$ (see (9.5.17))	$(0, \infty)$	$A_n = 1$ $B_n = -n - \alpha$ $C_n = an$

Appendix G
Identities and Formulas

$$\binom{n+1}{r} = \binom{n}{r} + \binom{n}{r-1},$$

$$\sum_{k=0}^{n} \binom{n+\alpha}{k}\binom{n+\beta}{n-k} = \binom{2n+\alpha+\beta}{n}.$$

Leibniz's Formula

$$\frac{d^n}{dx^n}(uv) = \sum_{k=0}^{n} \binom{n}{k} \frac{d^{n-k}}{dx^{n-k}}\frac{d^k v}{dx^k},$$

$$\begin{aligned}\frac{d^n}{dx^n}(1-x)^{n+\alpha}(1+x)^{n+\beta} &= \sum_{k=0}^{n}\binom{n}{k} D^{n-k}(1-x)^{n+\alpha}D^k(1+x)^{n+\beta} \\ &= (-1)^n(1-x)^{\alpha}(1+x)^{\beta}n! \\ &\quad \sum_{k=0}^{n}\binom{n+\alpha}{n-k}\binom{n+\beta}{k}(x-1)^k(x+1)^{n-k}.\end{aligned}$$

Answers and Hints to Selected Problems

Exercises 1.1 and 1.2

1. (a) $cn!$
   (b) $c3^{\frac{n(n-1)}{2}}$
   (c) $ce^{n(n-1)}$
   (d) $\dfrac{c}{n}$
3. (a) $n!(2^n + c - 1)$
   (b) $c + \dfrac{e^n - 1}{e - 1}$
9. 38 payments + final payment \$52.29
11. (a) $A(n+1) = (1+r)A(n) + T$
    (b) \$25,000 $[(1.008)^n - 1]$
13. \$136,283.50
15. (a) $r = 1 - \left(\dfrac{1}{2}\right)^{\frac{1}{5700}}$
    (b) 2,933 years

Exercises 1.3

3. (a) $\dfrac{\alpha - 1}{\beta}$
   (b) $\dfrac{2x(n)}{1 + x(n)}$
5. (b) $\mu = 3.3$

7. (i) $D(n) = -p(n) + 15$
   $S(n+1) = 2p(n) + 3$
   (iii) $p_* = 4$, unstable
11. (a) $p(n+1) = -\frac{1}{2}p^2(n) + 1$
    (b) $p_* = -1 + \sqrt{3}$
    (c) asymptotically stable

Exercises 1.4

1. (a) $y(n+1) = y(n) - ky^2(n),\ y(0) = 1$
3. (a) $y(n+1) = y(n) + 0.25(1 - y(n)),\ y(0) = 2$
5. (a) $y(n+1) = y(n) + \frac{1}{4}y^2(n) + \frac{1}{2}$
7. $y(n+1) = \dfrac{5y(n)}{4 - y(n)}$
9. Nonstandard: $y(n+1) = \frac{5y(n)+n}{5-y(n)}$
   Euler: $y(n+1) = y(n) + hy^2(n) + hn$

Exercises 1.5

1. $\begin{cases} 0: & \text{asymptotically stable} \\ \pm 1: & \text{unstable} \end{cases}$
3. 0: asymptotically stable
5. 0: unstable
7. 0: unstable
9. Hint: Use L'Hôpital's rule
11. Hint: Consider monotonic and nonmonotonic functions
16. (a) from the left
    (b) from the right

Exercises 1.6

5. $\{0, 1\}$: asymptotically stable
7. $|b^2 - 3ab| < 1$
9. $\left\{\frac{1}{3}, \frac{2}{3}\right\}$
11. (b) unstable
13. $f(x) = -x$
17. $c = -\frac{7}{4}$
19. $\bar{x}_1 = \dfrac{\mu + 1 - \sqrt{((\mu+1)(\mu-3)}}{2\mu},\ \bar{x}_2 = \dfrac{\mu + 1 + \sqrt{((\mu+1)(\mu-3)}}{2\mu},$

Exercises 1.7

2. Hint: Let $x(n) = \sin^2\theta(n)$

5. Hint: Show that $0 < f'(x) < 1$ for $x^* < x < \dfrac{1}{2}$
7. Hint: Show that $Sf_\mu^2(x(0)) < 0$
9. Hint: Use a calculator or a computer
11. Hint: Let $y = -\mu x + \frac{1}{2}\mu$
13. $c = -\dfrac{3}{4}$
17. x_1^* is unstable
    x_2^* is asymptotically stable
    x_3^* is asymptotically stable
19. $x^* = 0$ is unstable

Exercises 1.8

1. There are fixed points $x_1^* = 0$, $x_2^* = 1$, $x_3^* = 9$
   $W^s(x_1^*) = \mathcal{B}(x_1^*) = (-1, 1)$
   x_2^* is unstable, -1 is eventually fixed
   $W^s(x_3^*) = [-3, -1) \cup (1, 9]$, $\mathcal{B}(x_3^*) = (1, 9]$
9. Hint: Consider three cases:
   (a) $0 < p < 1$
   (b) $p = 1$
   (c) $1 < p < 2$

Exercises 2.1

9. Hint: Write $f(n) = a_0 + a_1 n^{(1)} + a_2 n^{(2)} + \cdots + a_k n^{(k)}$
11. Hint: Use mathematical induction on m
13. $\begin{cases} x^3 = x^{(1)} + 3x^{(2)} + x^{(3)} \\ x^4 = x^{(1)} + 7x^{(2)} + 6x^{(3)} + x^{(4)} \\ x^5 = x^{(1)} + 15x^{(2)} + 25x^{(3)} + 10x^{(4)} + x^{(5)} \end{cases}$
15. $\dfrac{1}{2}n(n-1) + n(n-1)(n-2) + \dfrac{1}{4}n(n-1)(n-2)(n-3) + c$

Exercises 2.2

1. (a) 0, linearly dependent
   (b) $2(5^{3n+3})$, linearly independent
   (c) $(-1)^{n+1}(27)4^n$, linearly independent
   (d) 0, linearly dependent
3. (a) (i) linearly independent
   (ii) $c_1 + c_2 n + c_3 n^2$
   (b) (i) linearly independent
   (ii) $c_1 \cos\left(\dfrac{n\pi}{2}\right) + c_2 \sin\left(\dfrac{n\pi}{2}\right)$
   (c) (i) linearly dependent
   (ii) no general solutions

(d) (i) linearly independent

(ii) need one more solution

14. Hint: Consider $\Delta\left(\frac{u_2(n)}{u_1(n)}\right) = \frac{u_1(n)\Delta u_2(n) - u_2(n)\Delta u_1(n)}{u_1(n)u_1(n+1)}$
15. Hint: You may take $W(n_0) = 1$ or any of your favorite constants, use formula (2.2.10) to evaluate $W(n)$

(b) $\frac{2^n}{n!}\sum_{r=0}^{n-1}\frac{r!}{2^{r+1}}$

Exercises 2.3

1. (a) $x(n+2) - 7x(n+1) + 10x(n) = 0$

   (b) $x(n+2) + x(n) = 0$

   (c) $x(n+4) - 10\sqrt{2}x(n+3) + 160x(n+2) - 250\sqrt{2}x(n+1) + 625x(n) = 0$

   (d) $x(n+1) - 21x(n+2) + 49x(n+1) - 343x(n) = 0$

   (e) $x(n+4) - 4x(n+3) + 6x(n+2) - 4x(n+1) + x(n) = 0$
3. $c_1 4^n + c_2(-4)^n$
5. $c_1 3^n + c_2 n 3^n + c_3 2^n \cos\frac{n\pi}{2} + c_4 2^n \sin\frac{n\pi}{2}$
7. $c_1 2^{n/2}\cos\frac{n\pi}{2} + c_2 2^{n/2}\sin\frac{n\pi}{2} + c_3 2^{n/2} n\cos\frac{n\pi}{2} + c_4 2^{n/2} n\sin\frac{n\pi}{2}$
11. (d) $T_n : 1, x, 2x^2 - 1$

    $U_n : 1, 2x, 4x^2 - 1$
18. (a) Hint: $a^{p-1} = 1 \bmod p$ if a and p are relatively prime
21. Hint: $D(n) = bD(n-1) - a^2 D(n-2)$

    $a^n\frac{\sin(n+1)\theta}{\sin\theta}, \quad \theta = \cos^{-1}\left(\frac{b}{2a}\right)$

Exercises 2.4

1. $\frac{1}{2}n + \frac{5}{4}$
3. $\frac{1}{12}n4^n + \frac{7}{54}n - \frac{1}{18}n^2 + \frac{1}{9}n^3$
5. $\frac{1}{2}\cos\left(\frac{n\pi}{2}\right) - \frac{1}{2}n\cos\left(\frac{n\pi}{2}\right)$
7. $2 - 7n + 8n^2$
9. $\frac{6}{25}(3^n) - \frac{6}{25}\cos\left(\frac{n\pi}{2}\right) + \frac{9}{50}\sin\left(\frac{n\pi}{2}\right) + \frac{1}{30}n3^n$
11. $c_1(-7)^n + c_2(-1)^n + \frac{1}{27}n2^n - \frac{8}{243}2^n$
15. $y(2) = 1$

    $y(3+4n) = y(4+4n) = 0, \quad n = 0, 1, 2, 3, \ldots$

    $y(5+4n) = y(6+4n) = -1 \quad n = 0, 1, 2, 3, \ldots$
17. $y(n) = \frac{-n^2}{10} + \frac{3n}{50} - \frac{1}{125} + a + b(6^n)$

Exercises 2.5

1. repelling, oscillatory
3. attracting, oscillatory
13. (i) oscillatory
(ii) oscillate to ∞
(iii) y^* is asymptotically stable

Exercises 2.6

1. $c_1(3)^n + c_2(-1)^n$
3. $1/(c-n)$
5. $\dfrac{3(1+c(2/3)^{n+1})}{1+c(2/3)^n}$
7. $x_0e^{2^n}$
9. $\sin^2(c2^n)$
11. $e^{c(2^{n-1})}y(0)$
13. $\cot(c2^n)$
15. $\sin(c2^n)$

Exercises 2.7

1. (a) $s_1(n+2) - \sigma\gamma\alpha s_1(n+1) - \sigma^2\gamma\beta(1-\alpha)s_1(n) = 0$
(b) $\gamma > 50$
3. (a) $F(n+2) = F(n+1) + 2F(n)$
(b) 3, 5, 11
4. (ii) Hint: Let \$10 equal 1 unit, $n = 5$, $N = 10$
7. $\dfrac{13.9298}{9.66 \times 10^{235}}$
11. (a) $Y(n+3) - (a_1+1)Y(n+2) - (a_2 - a_1)Y(n+1) + a_2Y(n) = h$
(b) $Y(n) = c_1 + c_2\left(\dfrac{1+\sqrt{5}}{4}\right)^n + c_3\left(\dfrac{1-\sqrt{5}}{4}\right)^n + \alpha_1 + \beta n$
15. (a) $M(n) = M(n_0)2^{n-n_0}$
(b) $c = 1$

Exercises 3.1

1. $\begin{pmatrix} 2^{n+1} - 3^n & 3^n - 2^n \\ 2^{n+1} - 2(3^n) & 2(3^n) - 2^n \end{pmatrix}$
3. $\begin{pmatrix} 2^{n+1} - 3^n & -2 + 2^{n+1} & \frac{1}{2} - \frac{1}{2}3^n \\ (-2)^n + 3^n & 2 - 2^n & -\frac{1}{2} + \frac{1}{2}3^n \\ -2^{n+2} + 4(3^n) & 4 - 2^{n+2} & -1 + 2(3^n) \end{pmatrix}$

5. $\begin{pmatrix} \frac{1}{3}(2^{n+1} + (-1)^n) \\ 2^{n+1} \end{pmatrix}$

7. $\begin{pmatrix} 3 - 2^{n+1} \\ 2(1 - 2^n) \\ 2(-1 + 2^n) \end{pmatrix}$

10. (a) Hint: Use (3.1.18)

12. Hint: If $\lambda_1 = \lambda_2 = \lambda$ and $\lambda^n = a_0 + a_1\lambda + a_2\lambda^2 + \cdots + a_{k-1}r^{k-1}$, differentiate to get another equation $n\lambda^{n-1} = a_1 + 2a_2\lambda + \cdots + (k-1)a_{k-1}\lambda^{k-2}$

13. (i) $\begin{pmatrix} 2^{n+1} - 3^n & 3^n - 2^n \\ 2^{n+1} - 2(3^n) & 2(3^n) - 2^n \end{pmatrix}$

(ii) Same as Problem 3

15. (a) $\begin{pmatrix} 0 & 1 & 0 \\ 0 & 0 & 1 \\ \frac{1}{2} & \frac{1}{2} & 0 \end{pmatrix}$

(b) (2/5, 1/5)

Exercises 3.2

9. $\begin{pmatrix} \frac{11}{16} + \frac{3}{4}n - \frac{11}{16}5^n \\ \frac{-5}{16} - \frac{1}{4}n - \frac{11}{16}5^n \end{pmatrix}$

15. $a_1(-2)^n + a_2(-6)^n$

17. $a_1 + a_2 4^n + \frac{1}{3}n4^n$

19. $a_1\left(\frac{1-\sqrt{5}}{2}\right)^n + a_2\left(\frac{1+\sqrt{5}}{2}\right)^n$

Exercises 3.3

1. $\begin{pmatrix} 2^{n+1} - 4^n \\ 2(4^n) \end{pmatrix}$

3. $\begin{pmatrix} \frac{3}{7}[(-1)^{n+1} + 6^n] \\ \frac{3}{7}(-1)^n + \frac{4}{7}6^n \\ 0 \end{pmatrix}$

5. $c_1 2^n \begin{pmatrix} 0 \\ 1 \\ 0 \end{pmatrix} + c_2 \begin{pmatrix} 1 \\ -1 \\ 0 \end{pmatrix} + c_3 3^n \begin{pmatrix} 1 \\ 7 \\ 2 \end{pmatrix}$

9. $\begin{pmatrix} 2^{n/2}\left[-c_2 \sin\frac{n\pi}{4} + c_3 \cos\frac{n\pi}{4}\right] \\ 2^{n/2}\left[-c_2 \cos\frac{n\pi}{4} - c_3 \sin\frac{n\pi}{4}\right] \\ c_1 + 2^{n/2}\left[c_2 \cos\frac{n\pi}{4} + c_3 \sin\frac{n\pi}{4}\right] \end{pmatrix}$

11. (a) $\begin{pmatrix} 3^n & n3^{n-1} \\ 0 & 3^n \end{pmatrix}$

(b) $\begin{pmatrix} 2^n & n2^{n-1} & n(n-1)2^{n-3} \\ 0 & 2^n & n2^{n-1} \\ 0 & 0 & 2^n \end{pmatrix}$

(c) $\begin{pmatrix} 2 & 1 & -2 \\ -1 & 0 & 1 \\ 0 & 0 & \frac{2}{3} \end{pmatrix} \begin{pmatrix} 3^n & n3^{n-1} & \frac{n(n-1)}{2}3^{n-2} \\ 0 & 3^n & n3^{n-1} \\ 0 & 0 & 3^n \end{pmatrix}$

$\begin{pmatrix} 0 & -1 & \frac{3}{2} \\ 1 & 2 & 0 \\ 0 & 0 & \frac{3}{2} \end{pmatrix}$

(d) $\begin{pmatrix} 2^n & 0 & 0 & 0 \\ 0 & 2^n & n2^n & n(n-1)2^{n-2} \\ 0 & 0 & 2^n & n2^n \\ 0 & 0 & 0 & 2^n \end{pmatrix}$

13. $\begin{pmatrix} c_1(2^n - n2^{n-1}) + c_2 n2^n + c_3(3n^2 2^{n-1} + 3n(n-1)2^{n-3}) \\ -c_1 n2^{n-2} + c_2 2^n(1-n) - 3c_3 n(n-1)2^{n-4} \\ c_3 2^n \end{pmatrix}$

19. Hint: Use the similarity matrix $P = \text{diag}(1, \alpha, \alpha^2, \ldots, \alpha^{k-1})$

Exercises 3.4

5. Hint: First change the equation to a system and then show that the monodromy matrix is equal to $\Phi(N)$

Exercises 3.5

3. Hint: Consider $A^T\xi = \xi$ with $\xi = (1, 1, \dots, 1)^T$
5. (i) Hint: Consider $(I - A)x = 0$
   (ii) Hint: Use $(I - A)(I + A + A^2 + \cdots + A^{n-1}) = I - A^n$
7. $\begin{pmatrix} 5/9 \\ 2/9 \\ 2/9 \end{pmatrix}$
9. 0.25
11. 177.78; 272.22
13. 0

Exercises 4.1

1. (a) 3, 3, 3
   (b) 6, 4, $3\frac{3}{4}$
   (c) 6, 7, 5.21
3. Hint: Use $D = \text{diag}(1, \varepsilon, \varepsilon^2)$

Exercises 4.3

1. (a) unstable
   (b) asymptotically stable
   (c) asymptotically stable
   (d) stable
3. $\begin{pmatrix} \frac{5}{12} & 0 & \frac{1}{2} \\ -1 & -1 & \frac{5}{4} \\ \frac{1}{3} & 0 & 0 \end{pmatrix}$
5. (a) uniformly stable
   (b) no conclusion
   (c) asymptotically stable
   (d) no conclusion

Exercises 4.4

1. (a) asymptotically stable
   (b) unstable
   (c) unstable
   (d) asymptotically stable
3. unstable
5. stable, but not asymptotically stable

Exercises 4.5

1. Hint: Let $V(x) = x_1^2 + x_2^2$
   The fixed point $(0,0)$ is globally asymptotically stable.
13. Hint: Let $V = xy$ and then use Problem 11

Exercises 4.6

1. exponentially stable
3. Hint: $V = a^2\left(x_1 - \frac{1}{\sqrt{2}}\right)^2 + b^2\left(x_2 - \frac{1}{\sqrt{2}}\right)^2$
   The equilibrium point $\left(\frac{1}{\sqrt{2}}, \frac{1}{\sqrt{2}}\right)$ is unstable. The equilibrium point $\left(-\frac{1}{\sqrt{2}}, -\frac{1}{\sqrt{2}}\right)$ is unstable.
4. (a) $\begin{pmatrix}0\\0\\0\end{pmatrix}, \begin{pmatrix}1\\1\\0\end{pmatrix}$
   (b) undetermined, unstable
6. unstable
8. if $|a| < 1$ and $|b| < 1$, then the zero solution is asymptotically stable
10. (a) $(N, 0)$, $\left(\frac{N(\gamma+\beta)}{\alpha}, \beta N\left[\frac{\alpha - (\gamma+\beta)}{\alpha(\gamma+\beta)}\right]\right)$
    (b) The first point is asymptotically stable if $\frac{\alpha}{\gamma+\beta} < 1$ and unstable if $\frac{\alpha}{\gamma+\beta} > 1$. The second point is asymptotically stable.
12. Hint: Use the variation of constant formula (3.2.12) and then use Theorem 8.12

Exercises 5.1 and 5.2

3. $1 < \alpha < 2.62$
11. Hint: Let $g(z) = p_1 z^{k-1} + p_2 z^{k-2} + \cdots + p_k$, and $f(z) = z^k$ on the unit disk
12. Hint: Let $f(z) = p_1 z^{k-1}$, $g(z) = z^k - p_2 z^{k-2} + \cdots + p_k$, on the circle of radius $1 + \varepsilon$, for some appropriate $\varepsilon > 0$

Exercises 5.3

1. $-\frac{1}{2} < b < 0.78$
6. Hint: Make the change of variable $N(n) = N^* e^{x(n)}$
7. $x^* = \dfrac{\left(\sum_{i=0}^{k} a_i\right) - b}{2\sum_{i=0}^{k} b_i} + \dfrac{1}{2\sum_{i=0}^{k} b_i}\sqrt{\left(b - \sum_{i=0}^{k} a_i\right)^2 + 4a\sum_{i=0}^{k} b_i}$

Exercises 6.1

1. (a) $\dfrac{z(z-\cos\omega)}{z^2-2z\cos\omega+1}$, $|z|>1$

   (b) $\dfrac{z(z^2-1)\sin 2}{(z^2-2z\cos 2+1)^2}$, $|z|>1$

   (c) $\dfrac{z}{(z-1)^2}$, $|z|>1$

3. $\dfrac{-z+a^2+a}{z(z-a)}$, $|z|>|a|$

5. $\dfrac{(z+1)^2 z^{n-3}}{z^n-1}$

7. Hint: Use mathematical induction on k

9. $\dfrac{1}{(z-a)^3}$

15. Hint: $y(n)-y(n-1)=nx(n)$

17. (a) $\dfrac{z^2\sin\omega}{(z-a)(z^2-2z\cos\omega+1)}$

    (b) $\dfrac{z^2(z-\cos\omega)}{(z-1)(z^2-2z\cos\omega+1)}$

Exercises 6.2

1. (a) $2/3[2^{-n}-1]$

   (b) $-1/7(-2)^n+1/7n(-2)^n+6/7$

3. (a) $(-2)^{n-3}(3n^2-n)$

   (b) $2^{-n+1}+2\sin\left(\dfrac{(n-1)}{2}\pi\right)$

5. $\dfrac{1}{\sqrt{5}}\left[\left(\dfrac{1+\sqrt{5}}{2}\right)^n-\left(\dfrac{1-\sqrt{5}}{2}\right)^n\right]$

7. $\dfrac{1}{2}(n+1)$

11. Hint: Replace n by $n+1$
    $\dfrac{1-e}{2-e}+\left(\dfrac{1}{2-e}\right)(e-1)^n$

Exercises 6.3

1. $x(n)=\dfrac{1}{3}x(0)[1+2(4^n)]$
   unstable

3. Problem 1. unstable
   Problem 2. uniformly stable
5. unstable

Exercises 6.4

1. asymptotically stable
3. not asymptotically stable
4. Hint: $\sum_{n=0}^{\infty} na^n = \frac{a}{(1-a)^2}$ for $a < 1$, $\sum_{n=0}^{\infty} n^2a^n = \frac{a^2+a}{(1-a)^3}$

Exercises 6.5

3. asymptotically stable
5. uniformly stable

Exercises 6.6

4. Hint: See Theorem 4.9
5. (a) $x(n) = -\frac{1}{7}(-3)^n + \frac{1}{7}(4^n)$
7. (a) $\begin{pmatrix} (1+\sqrt{2})2^{n-1} + \dfrac{(1-\sqrt{2})}{2}(-1)^n & 0 \\ 0 & \left(\dfrac{3-\sqrt{6}}{5}\right)3^n + \left(\dfrac{2+\sqrt{6}}{5}\right)(-2)^n \end{pmatrix}$

   (b) $\begin{pmatrix} \dfrac{1+\sqrt{2}}{2}(2^n - n - 1) + \dfrac{1-\sqrt{2}}{8}[(-1)^n + 2n - 1] \\ 0 \end{pmatrix}$

11. $\begin{pmatrix} -2 + \dfrac{1}{12}(2^n) + \dfrac{3}{2}(3^n) \\ -1 + \dfrac{1}{12}(2^n) + \dfrac{1}{2}(3^n) \end{pmatrix}$

Exercises 7.1

5. Hint: Use Theorem 7.3
7. Hint: Consider the function $f(\lambda) = \lambda^{k+1} - \lambda^k + p$ and show that it attains its minimum when $\lambda = (k-1)/k$

Exercises 7.2

8. Hint: Use Problem 7
9. Hint: Use Problem 7
12. Hint: Use Theorem 7.16
14. Hint: Use Problem 13

Exercises 7.3

3. Hint: Let $x(n) = \alpha e^{y(n)}$ and then use Theorem 7.18
4. Hint: Let $x(n) = x^* e^{y(n)}$ and then apply Theorem 7.19
6. Hint: Let $z(n) = x(n)/x(n+1)$ and then mimic the proof of Theorem 7.18
7. Hint: Follow the hint in Problem 6

Exercises 8.1

12. Hint: Use $f(t) = (1 + o(1))g(t)$
14. (c) Hint: Show that $\int_1^\infty \frac{e^{-xt}}{t^n - 1} dt \leq \frac{e^{-x}}{n-2}$
15. Hint: Use integration by parts
16. Hint: Write $\sum_{k=1}^n k^k = n^n[1 + \frac{(n-1)^{n-1}}{n^n} + \cdots + \frac{1}{n^n}]$

Exercises 8.2

14. Hint: Notice first that $\log \prod_{i=n_0}^{n-1} (1 + u(i)) = \sum_{i=n_0}^{n-1} \log(1 + u(i))$
15. (c) Hint: Use the mean value theorem
    (e) Hint: Substitute (8.2.19) into (8.2.17)
    (f) Hint: Solve (8.2.28) and then use Problem 14
17. $x_1(n) \sim n^2, x_2(n) \sim \frac{1}{(n+2)!}, n \to \infty$
20. Hint: Let $y(n) = x(n+1)/x(n)$

Exercises 8.4

12. Hint: Let $x(n) = \left(-\frac{1}{2}\right)^n \left(\prod_{j=n_0}^{n-1} p_1(j)\right) z(n)$

Exercises 8.5

7. Hint: Reverse the order of summation on the left-hand side as in Figure 8.2
10. Hint: Use Problem 8 and then let $A(n) = x_2(n)\Delta y(n) - \Delta x_2(n)y(n)$ and $B(n) = \Delta x_1(n)y(n) - x_1(n)\Delta y(n)$, then mimic the proof of Problem 8

Exercises 9.3

9. $(x, y) = (37, 47) + m(48, 61)$
10. Hint: Consider the continued fraction representation of $\sqrt{\ell}$ Then $\sqrt{\ell} = \frac{A(m-1)+A(m)\xi_{m+1}}{B(m-1)+B(m)\xi_{m+1}}$,

$$\xi_{m+1} = 2b_0 + \frac{1}{b_1+}\frac{1}{b_2+}\cdots = \sqrt{\ell} + b_0;$$

show that

$$A(m)(A(m) - B(m)b_0) - B(m)(B(m)\ell - A(m)b_0) = (-1)^{m-1}$$

or

$$A^2(m) - \ell B^2(m) = (-1)^{m-1}$$

Conclude that $x = A(m), y = B(m)$ is a solution of Pell's equation if m is odd, and if m is even, $x = A(2m+1),\ y = B(2m+)$ is a solution

11. $x = 8, y = 3$

Exercises 9.4 and 9.5

5. $(n+1)P_{n+1}^{\nu}(x) = 2(\nu+n)xP_n^{\nu}(x) - (2\nu+n-1)P_{n-1}^{\nu}(x)$
6. $(n+1)L_{n+1}^{\alpha}(x) = (2n+\alpha+1-x)L_n^{\alpha}(x) - (n+\alpha)L_{n-1}^{\alpha}(x)$
7. $H_{n+1}(x) = 2xH_n(x) - 2nH_{n-1}(x)$
9. $J_{n+1}(z) = (\frac{2n}{z})J_n(z) - J_{n-1}(z)$
12. Hint: Use (9.5.18) and let $u \to x$

Exercises 9.6

7. Hint: Use the Cauchy integral formula: $\frac{d^n}{dx^n}(1-x^2)^n = \frac{n!}{2\pi i}\oint_c \frac{(1-t^2)^n}{(t-x)^{n+1}}dt$

Exercises 10.1 and 10.2

1. $W = \begin{bmatrix} 1 & -2 \\ 0 & 1 \end{bmatrix}, |W| = 1 \neq 0$, the system is completely controllable
3. Since A is diagonal and B has a row of zeros then, by inspection, the system is not completely controllable
5. $\text{rank}(W) = 4 < 5$, the system is not completely controllable
11. $W = \begin{bmatrix} 1 & a_{11}+a_{12} \\ 1 & a_{21}+a_{22} \end{bmatrix}, |W| = a_{21}+a_{22}-a_{11}-a_{12} \neq 0$ thus $a_{22}+a_{21} \neq a_{11}+a_{12}$

Exercises 10.3

1. (a) $V = \begin{bmatrix} 0 & 2 \\ 4 & -2 \end{bmatrix}, |V| = 2, x_0 = V^{-1}\begin{bmatrix} y(0) \\ y(1) \end{bmatrix} = \begin{bmatrix} 1/4(a+b) \\ 1/2b \end{bmatrix}$

   (b) $V = \begin{bmatrix} 2 & 1 \\ 2 & 1 \end{bmatrix}$, $\text{rank}(V) = 1 < 2$, the system is not observable

3. $W = \begin{bmatrix} 1 & a+b \\ 1 & c+d \end{bmatrix}, |W| = -(a+b)+c+d \neq 0$. Thus, for a system to be completely controllable $a+b \neq c+d$, and for a system to be completely observable, $V = \begin{bmatrix} 1 & 0 \\ a & b \end{bmatrix}, |V| = b \neq 0$
5. $\text{rank}(V) = 4$, the system is not completely observable

Exercises 10.4

1. $K = \begin{bmatrix} -0.1166 & -0.6982 \end{bmatrix}$
3. $K = \begin{bmatrix} -1.8599 & 0.5293 & 2.8599 \end{bmatrix}$
5. Hint: Put the equation into a system form

Exercises 10.5

1. $E = \begin{bmatrix} 0.875 \\ -1.125 \end{bmatrix}$
3. unsolvable

Maple Programs

(I) Solution of scalar difference equations and systems of difference equations using rsolve

```
> rsolve({x(n+1)-x(n)/(n+1)=1/(n+1)!,x(0)=1},x);
```

$$\frac{n+1}{\Gamma(n+1)}$$

```
> rsolve({x(n+1)=2*x(n)+y(n),y(n+1)=2*y(n),x(0)=a,y(0)=b},
{x,y});
```

$$\{y(n) = b2^n, x(n) = \frac{1}{2}b2^n + 2^n a\}$$

```
> rsolve(x(n+2)-5*x(n+1)+6*x(n)=0,x);
```

$$-(2x(0) - x(1))3^n - (-3x(0) + x(1))2^n$$

Cobweb Program

```
>#Cobweb Program
>#Call as: cobweb(F, n, s, l, u)
>#Where:  F: The one parameter function
>#    n: The number of iterations to be performed
>#    s: The initial value of x with which to start
>#    l: The lower bound value for x and y
>#    u: The upper bound value for x and y
>cobweb:=proc(function, iterations, initial, lowerbound,
         upperbound)
>local F, n, s, u, i, y, G, l;
>F:=function;
>n:=iterations;
>s:=initial;
>l:=lowerbound;
>u:=upperbound;
>with(plottools)
>y:=eval(subs(x=s,F));
>G:=[line([l,l], [u,u]), line([s,0], [s,y]),
     plot(F,x=l..u,color=black)];
>for i from 1 to n do
>    G:=[op(G), line([s,y], [y,y]))];
>    s:=y;
>    y:=evalf(subs(x=s,F));
>    G:=[op(G), line([s,s], [s,y])];
>od
>plots[display](G,tickmarks=[0,0]);
>end

> # Example: draw the cobweb diagram of the function
> # F(x)=3.7*x*(1-x) with initial point 0.1.
>cobweb(3.7*x*(1-x),10,0.1,0.1);
```

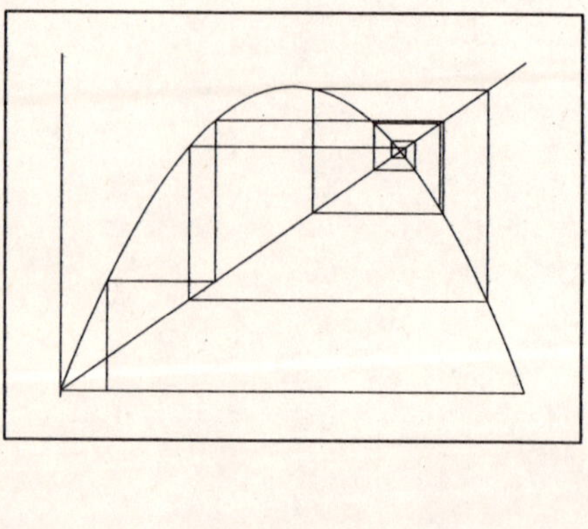

Bifurcation Diagram Program

```
>bifur(c*x*(1-x), 3.57, 4, 100, 50, .01, .5, 0, 1);
furcation Diagram Program
> # Bifurcation Diagram Program
> # Call as: bifur(F, l, r, N, T, d, s, b, t)
> #
> # Where:
> #       F: The one parameter function in terms of c and x
> #       l: The left bound on the graph
> #       r: The right bound on the graph
> #       N: The number of iterations to perform
> #       T: The number of iterations to discard
> #       d: The step size of the parameter (c)
> #       s: The value of x
> #       b: The bottom bound on the graph
> #       t: The top bound on the graph
> #
> bifur:=proc(function, left, right, iterations, discard,
step, start, bottom, top)
> local F, l, r, N, T, d, s, t, i, p, b, j, k, G;
> F:=function;
> l:=left;
> r:=right;
> N:=iterations;
> T:=discard;
> d:=step;
> s:=start;
> t:=top;
> b:=bottom; G:=[];
> with(plottools):
> for i from l by d*(r-l) to r do
>     p:=s;
>     for j to T do
>         p:=evalf(subs(x=p, c=i, F));
>     od;
>     for k to N do
>         p:=evalf(subs(x=p, c=i, F));
>         G:=[op(G), point([i,p])];
>     od;
> od;
> plots[display](G, axes=boxed, symbol=POINT,
view=[l..r, b..t]);
> end:
```

```
> # Example: Draw the graph of F(x)=c*x*(1-x), where c is
> # between 3.5 and 4 and initial point is 0.5.
> bifur(c*x*(1-x),3.5,4,200,50,.01,.5,0,1);
```

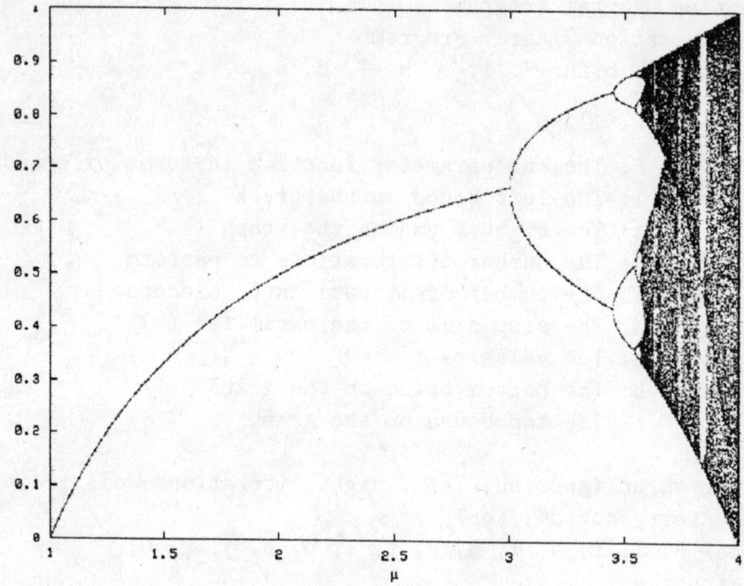

Phase Space Diagram with Four Initial Points

```
> # Phase Space Diagram Program (with 4 initial points)
> # Call as: phase4(A, x, y, z, v, n)
> #
> # Where:
> #        A: The matrix entries where f(x)=Ax
> #        v: The initial point (v1,v2)
> #        x: The initial point (x1,x2)
> #        y: The initial point (y1,y2)
> #        z: The initial point (z1,z2)
> #        n: The number of iterations to perform
> #
> phase4:=proc(matrix11, matrix12, matrix21, matrix22,
initial1, initial2, initial3, initial4, initial5, initial6,
initial7, initial8, iterations)
> local A, x, n, G, F, H, J, x1, x2, i, x3, x4, w1, w2, y3,
y4, z1, z2, z3, z4, y, z, v1, v2, v3, v4, v, K;
> A:=array(1..2,1..2,[[matrix11,matrix12],
[matrix21,matrix22]]);
> x:=array(1..2,1..1,[[initial1],[initial2]]);
y:=array(1..2, 1..1, [[initial3],[initial4]]);
z:=array(1..2, 1..1, [[initial5],[initial6]]);
v:=array(1..2, 1..1, [[initial7],[initial8]]);
> n:=iterations;
> x1:=x[1,1]; x2:=x[2,1]; w1:=y[1,1]; w2:=y[2,1]; z1:=z[1,1];
z2:=z[2,1]; v1:=v[1,1]; v2:=v[2,1];
> G:=[]; H:=[]; J:=[]; K:=[];
> with(plottools):
> for i from 1 to n do
>     F:=array(1..2, 1..1, [[(A[1,1]*x1)+(A[1,2]*x2)],
[(A[2,1]*x1)+(A[2,2]*x2)]]);
>     x3:=F[1,1]; x4:=F[2,1];
>     G:=[op(G), line([x1,x2],[x3,x4])];
>     x1:=x3; x2:=x4;
>     F:=array(1..2, 1..1, [[(A[1,1]*w1)+(A[1,2]*w2)],
[(A[2,1]*w1)+(A[2,2]*w2)]]);
>     y3:=F[1,1]; y4:=F[2,1];
>     H:=[op(H), line([w1,w2],[y3,y4])];
>     w1:=y3; w2:=y4;
>     F:=array(1..2, 1..1, [[(A[1,1]*z1)+(A[1,2]*z2)],
[(A[2,1]*z1)+(A[2,2]*z2)]]);
>     z3:=F[1,1]; z4:=F[2,1];
>     J:=[op(J), line([z1,z2],[z3,z4])];
>     z1:=z3; z2:=z4;
```

```
>     F:=array(1..2, 1..1, [[(A[1,1]*v1)+(A[1,2]*v2)],
[(A[2,1]*v1)+(A[2,2]*v2)]]);
>     v3:=F[1,1]; v4:=F[2,1];
>     K:=[op(K), line([v1,v2],[v3,v4])];
>     v1:=v3; v2:=v4;
> od;
> plots[display](G,H,J,K,tickmarks=[0,0],color=black);
> end:
```

```
> # Example: Draw the phase space diagram of the system
> # x(n+1)=A*x(n) where A=array(2,0,0,0.5).
> phase4(2,0,0,.5,1,3,-1,3,-1,-3,1,-3,6);
```

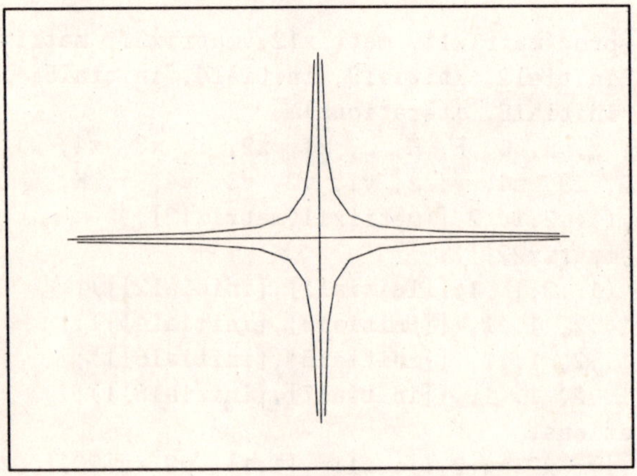

References

[1] Abu-Saris, R., S. Elaydi, and S. Jang, Poincaré type solutions of systems of difference equations, *J. Math. Anal. Appl.* **275** (2002), 69–83.

[2] Adams, C.R., On the irregular cases of linear ordinary difference equations, *Trans. Amer. Math. Soc.*, **30** (1928), 507–541.

[3] Agarwal, R.P., *Difference Equations and Inequalities*, Marcel Dekker, New York, 1992.

[4] Ahlbrandt, C., and A. Peterson, *Discrete Hamiltonian Systems*, Kluwer Academic, Dordrecht, 1996.

[5] Asplund, E., and L. Bungart, *A First Course in Integration*, Holt, Rinehart, and Winston, New York, 1966.

[6] Barnett, S., *Introduction to Mathematical Control Theory*, Clarendon Press, Oxford, 1975.

[7] Beddington, J.R., C.A. Free, and J.H. Lawton, Dynamic complexity in predator–prey models framed in difference equations, *Nature* **255** (1975), 58–60.

[8] Bender, E., and S.A. Orszag, *Advanced Mathematical Methods for Scientists and Engineers*, McGraw-Hill, New York, 1978.

[9] Benzaid, Z., and D.A. Lutz, Asymptotic representation of solutions of perturbed systems of linear difference equations, *Stud. Appl. Math.* **77** (1987), 195–221.

[10] Beverton, R.J., and S.J. Holt, *The theory of fishing, In sea fisheries; Their Investigation in the United Kingdom*, M. Graham, ed., pp. 372–441, Edward Arnold, London, 1956.

[11] Birkhoff, G.D., General theory of linear difference equations, *Trans. Amer. Math. Soc.* **12** (1911), 243–284.

[12] Birkhoff, G.D., Formal theory of irregular linear difference equations, *Acta Math.* **54** (1930), 205–246.

[13] Birkhoff, G.D., and W.J. Trjitzinsky, Analytic theory of singular difference equations, *Acta Math.* **60** (1932), 1–89.

[14] Brauer, A., Limits for the characteristic roots of a matrix, II, *Duke Math. J.* **14** (1947), 21–26.

[15] Brauer, F., and C. Castillo-Chávez, *Mathematical Models in Population Biology and Eqidemiology*, Texts in Applied Mathematics 40, Springer-Verlag, New York, 2001.

[16] Cadzow, J.A., *Discrete Time Systems*, Prentice-Hall, Englewood Cliffs, NJ, 1973.

[17] Carlson, D.C., *The Stability of Finite Difference Equations*, Master's thesis, University of Colorado, Colorado Springs, 1989.

[18] Carvalho, L.A.V., On a method to investigate bifurcation of periodic solutions in retarded differential equations, *J. Difference Equ. Appl.* **4**(1) (1998), 17–27.

[19] Chihara, T.S., *An Introduction to Orthogonal Polynomials*, Gordon and Breach, New York, 1978.

[20] Churchill, R.V., and J.W. Brown, *Complex Variables and Applications*, McGraw-Hill, New York, 2004.

[21] Clark, C.W., A delay recruitement model of population dynamics with an application to baleen whale population, *J. Math. Biol.* **3** (1976), 381–391.

[22] Coffman, C.V., Asymptotic behavior of solutions of ordinary difference equations, *Trans. Amer. Math. Soc.* **110** (1964), 22–51.

[23] Cooke, K.L., and L.A. Ladéira, Applying Carvalho's method to find periodic solutions of difference equations, *J. Difference Equ. Appl.* **2** (1996), 105–115.

[24] Cushing, J., *An Introduction to Structural Population Dynamics*, SIAM, Philadelphia, 1999.

[25] Cushing, J., Cycle chains and the LPA model, *J. Difference Equ. Appl.* **9** (2003), 655–670.

[26] Cushing, J., and S. Henson, Global dynamics of some periodically forced, monotone difference equations, *J. Difference Equ. Appl.* **7**(6) (2001), 859–872.

[27] Cushing, J.M., R.F. Costantino, B. Dennis, R.A. Desharnais, and S.M. Henson, *Chaos in Ecology: Experimental Nonlinear Dynamics*, Academic Press, New York, 2003.

[28] Dannan, F., The asymptotic stability of $x(n+k) + ax(n) + bx(n-l) = 0$, *J. Difference Equ. Appl.* **10**(6) (2004), 1–11.

[29] Dannan, F., and S. Elaydi, Asymptotic stability of linear difference equations of advanced type, *J. Comput. Anal. Appl.* **6**(2) (2004).

[30] Dannan, F., S. Elaydi, and V. Ponomarenko, Stability of hyperbolic and nonhyperbolic fixed points of one-dimensional maps, *J. Difference Equ. Appl.* **9** (2003), 449–457.

[31] Derrick, W., and J. Eidswick, Continued fractions, Chebyshev polynomials, and chaos, *Amer. Math. Soc. Monthly* **102**(4), 1995, 337–344.

[32] Devaney, R., *A First Course in Chaotic Dynamical Systems: Theory and Experiments*, Addison-Wesley, Reading, MA, 1992.

[33] Drozdowicz, A., On the asymptotic behavior of solutions of the second-order difference equations, *Glas. Mat.* **22**(42), 1987, pp. 327–333.

[34] Drozdowicz, A., and J. Popenda, Asymptotic behavior of the solutions of the second-order difference equations, *Proc. Amer. Math. Soc.* **99** (1987), 135–140.

[35] Drozdowicz, A., and J. Popenda, Asymptotic behavior of the solutions of the second-order difference equations, *Fasc. Math.* **17** (1987), 87–96.

[36] Eastham, M.S.P., *The Asymptotic Solutions of Linear Differential Systems*, Clarendon Press, Oxford, 1989.

[37] Edelstein-Keshet, L., *Mathematical Models in Biology*, Random House, New York, 1988.

[38] Elaydi, S., Asymptotics for linear difference equations I: Basic theory, *J. Difference Equ. Appl.* **5** (1999), 563–589.

[39] Elaydi, S., Stability of Volterra difference equations of convolution type, *Proceedings of the Special Program at Nankai Institute of Mathematics* (ed. Liao Shan-Tao et al.), World Scientific, Singapore, 1993, pp. 66–73.

[40] Elaydi, S., On a converse of Sharkovsky's theorem 423, *Amer. Math. Monthly* **103** (1996), 386–392.

[41] Elaydi, S., Periodicity and stability of linear Volterra difference systems, *J. Math. Anal. Appl.* **181** (1994), 483–492.

[42] Elaydi, S., An extension of Levinson's theorem to asymptotically Jordan difference equations, *J. Difference Equ. Appl.* **1** (1995), 369–390.

[43] Elaydi, S., Asymptotics for linear difference equations II: Applications, in *New Trends in Difference Equations*, Taylor & Francis, London, 2002, pp. 111–133.

[44] Elaydi, S., and I. Gyori, Asymptotic theory for delay difference equations, *J. Difference Equ. Appl.* **1** (1995), 99–116.

[45] Elaydi, S., and O. Hajek, Exponential dichotomy and trichotomy of nonlinear differential equations, *Differential Integral Equations* **3** (1990), 1201–1224.

[46] Elaydi, S., and W. Harris, On the computation of A^n, *SIAM Rev.* **40**(4) (1998), 965–971.

[47] Elaydi, S., and S. Jang, Difference equations from discretization of continuous epidemic models with disease in the prey, *Canad. Appl. Math. Quart.* (to appear).

[48] Elaydi, S., and K. Janglajew, Dichotomy and trichotomy of difference equations, *J. Difference Equ. Appl.* **3** (1998), 417–448.

[49] Elaydi, S., and S. Murakami, Asymptotic stability versus exponential stability in linear Volterra difference equations of convolution type, *J. Difference Equ. Appl.* **2** (1996), 401–410.

[50] Elaydi, S., and R. Sacker, Basin of attraction of periodic orbits and population biology (to appear).

[51] Elaydi, S., and S. Zhang, Stability and periodicity of difference equations with finite delay, *Funkcial. Ekvac.* **37** (1994), 401–413.

[52] Epstein, J.M., *Calculus of Conventional War*, Brookings Institute, Washington, DC, 1985.

[53] Erbe, L.H., and B.G. Zhang, Oscillation of discrete analogues of delay equations, *Differential Integral Equations* **2** (1989), 300–309.

[54] Erdélyi, A., W. Magnus, F. Oberhettinger, and F.G. Tricomi, *Tables of Integral Transforms*, Vol. 2, McGraw-Hill, New York, 1954.

[55] Evgrafov, M., The asymptotic behavior of solutions of difference equations, *Dokl. Akad. Nauk SSSR* **121** (1958), 26–29 (Russian).

[56] Feigenbaum, M., Quantitative universality for a class of nonlinear transformations, *J. Statist. Phys.* **19** (1978), 25–52.

[57] Gautschi, W., Computational aspects of three-term recurrence relations, *SIAM Rev.* **9** (1967), 24–82.

[58] Gautschi, W., Minimal solutions of three-term recurrence relations and orthogonal polynomials, *Math. Comp.* **36** (1981), 547–554.

[59] Goldberg, S., *Introduction to Difference Equations*, Dover, New York, 1986.

[60] Grove, E.A., and G. Ladas, *Periodicities in Nonlinear Difference Equations*, Taylor & Francis, London, to appear.

[61] Grove, E.A., C.M. Kent, G. Ladas, R. Levins, and S. Valicenti, Global stability in some population models, *Communications in Difference Equations* (Poznan, 1998), pp. 149–176, Gordon and Breach, Amsterdam, 2000.

[62] Gulick, D., *Encounters with Chaos*, McGraw-Hill, New York, 1992.

[63] Gyori, I., and G. Ladas, *Oscillation Theory of Delay Differential Equations with Applications*, Clarendon Press, Oxford, 1991.

[64] Hartman, P., Difference equations: Disconjugacy, principal solutions, Green's functions, complete monotonicity, *Trans. Amer. Math. Soc.* **246** (1978), 1–30.

[65] Hautus, M.L.J., and T.S. Bolis, Solution to problem E2721, *Amer. Math. Monthly* **86** (1979), 865–866.

[66] Henrici, P., *Applied and Computational Complex Analysis*, Vol. 2, Wiley-Interscience, New York, 1977.

[67] Hooker, J.W., and W.T. Patula, Riccati-type transformation for second-order linear difference equations, *J. Math. Anal. Appl.* **82** (1981), 451–462.

[68] Horn, R.A., and C.R. Johnson, *Matrix Analysis*, Cambridge University Press, Cambridge, 1999.

[69] Hurt, J., Some stability theorems for ordinary difference equations, *SIAM J. Numer. Anal.* **4** (1967), 582–596.

[70] Iggidr, A., and M. Bensoubaya, New results on the stability of discrete-time systems and applications to control problems, *J. Math. Anal. Appl.* **219** (1998), 392–414.

[71] Ismail, M.E.H., D.R. Masson, and E.B. Saff, A minimal solution approach to polynomial asymptotics, in *Orthogonal Polynomials and Their Applications*, (ed. C. Brezinski et al.), J.C. Bultzer AG, Scientific Publ. Co., IMACS, 1991, pp. 299–303.

[72] Johnson, W.P., The curious history of Faá di Bruno's formula, *Amer. Math. Monthly*, **109**(3) (2002), 217–234.

[73] Jones, W.B., and W.J. Thorn, *Continued Fractions, Analytic Theory and Applications*, Addison-Wesley, Reading, MA, 1980.

[74] Jury, E., *Theory and Applications of the Z-transform*, Wiley, New York, 1964.

[75] Kailath, T., *Linear Systems*, Prentice-Hall, Engelwood Cliffs, NJ, 1980.

[76] Kalman, D., The generalized Vandermonde matrix, *Math. Mag.* **57**(1) (1984), 15–21.

[77] Kalman, R.E., and J.E. Bertram, Control system analysis and design via the second method of Liapunov: I. Continuous-time systems; II. Discrete-time systems, *Trans. ASME Ser. D. J. Basic Engrg.* **82** (1960), 371–393, 394–400.

[78] Karakostas, G., H.G. Philos, and G.Y. Sficas, The dynamics of some discrete population models, *Nonlinear Anal.* **17** (1991), 1069–1084.

[79] Kelley, W.G., and A.C. Peterson, *Difference Equations, An Introduction with Applications*, Academic Press, New York, 1991.

[80] Kocic, V.L., and G. Ladas, *Global Behavior of Nonlinear Difference Equations of Higher Order with Applications*, Kluwer Academic, Dordrecht, 1993.

[81] Kot, M., *Elements of Mathematical Ecology*, Cambridge University Press, Cambridge, 2001.

[82] Kreuser, P., *Über das Verhalten der Integrale homogener linearer differenzengleichungen im Unendlichen*, Dissertation, University of Tübingen, Leipzig, 1914.

[83] Kuang, Y., and J. Cushing, Global stability in a nonlinear difference-delay equation model of flour beetle population growth, *J. Difference Equ. Appl.* **2**(1) (1996), 31–37.

[84] Kulenovic, M.R.S., and G. Ladas, *Dynamics of Second-Order Rational Difference Equations*, Chapman & Hall/CRC Press, Boca Raton, FL, 2002.

[85] Kulenovic, M.R.S., and G. Ladas, *Dynamics of Second-Order Rational Difference Equations with Open Problems and Conjectures*, Chapman & Hall/CRC Press, Boca Raton, FL, 2003.

[86] Kuruklis, S., The asymptotic stability of $x_{n+1} - ax_n + bx_{n-k} = 0$, *J. Math. Anal. Appl.* **188** (1994), 719–731.

[87] Lakshmikantham, V., and D. Trigiante, *Theory of Difference Equations: Numerical Methods and Applications*, Academic Press, New York, 1988.

[88] LaSalle, J.P., *The Stability and Control of Discrete Processes*, Applied Mathematical Sciences, Vol. 82, Springer-Verlag, New York, 1986.

[89] Lauwerier, H., *Mathematical Models of Epidemics*, Math. Centrum, Amsterdam, 1981.

[90] Levin, A., and R.M. May, A note on difference delay equations, *Theoret. Population Biol.* **9** (1976), 178–187.

[91] Levinson, N., The asymptotic behavior of a system of linear differential equations, *Amer. J. Math.* **68** (1946), 1–6.

[92] Li, T.Y., and J.A. Yorke, Period three implies chaos, *Amer. Math. Monthly* **82** (1975), 985–992.

[93] Liapunov, A., Problème général de la stabilité du movement, *Ann. of Math.*, Study #17 (1947).

[94] Ludyk, G., *Time-Variant Discrete-Time Systems*, Braunschweig, Wiesbaden: Vieweg, 1981.

[95] Ludyk, G., *Stability of Time-Variant Discrete-Time Systems*, Braunschweig, Wiesbaden: Vieweg, 1985.

[96] Luenberger, D.G., *Introduction to Dynamic Systems, Theory, Models and Applications*, Wiley, New York, 1979.

[97] Mackey, M.C., and L. Glass, Oscillation and chaos in physiological control systems, *Science* **197** (1997), 287–289.

[98] Magnus, W., Oberhettinger, F., and Soni, R.P., *Formulas and Theorems for Special Functions of Mathematical Physics*, Springer-Verlag, Berlin, 1996.

[99] Meschkowski, H., *Differenzengleichungen*, Vandehoeck and Ruprecht, Göttingen, 1959.

[100] Mickens, R., *Difference Equations*, Van Nostrand Reinhold, New York, 1990.

[101] Miller, K.S., *Linear Difference Equations*, Benjamin, New York, 1968.

[102] Miller, R.K., and A.N. Michael, *Ordinary Differential Equations*, Academic Press, New York, 1982.

[103] Milne-Thomson, L.M., *The Calculus of Finite Differences*, Macmillan, London, 1960.

[104] Murray, J.D., *Asymptotic Analysis*, Clarendon Press, Oxford, 1974.

[105] Neidinger, R.D., and R.J. Annen, *The road to chaos is filled with polynomial curves*, preprint.

[106] Nevai, P.G., Orthogonal polynomials, *Mem. Amer. Math. Soc.* **213** (1979).

[107] Nicholson, A.J., An outline of the dynamics of animal populations, *Austral. J. Zool.* **2** (1954), 9–65.

[108] Ogata, K., *Discrete-Time Control Systems*, Prentice-Hall, Englewood Cliffs, NJ, 1987.

[109] Olmsted, J.M.H., *Advanced Calculus*, Prentice-Hall, Englewood Cliffs, NJ, 1961.

[110] Olver, F.W.J., *Asymptotics and Special Functions*, Academic Press, New York, 1974.

[111] Ortega, J.M., *Matrix Theory, A Second Course*, Plenum, New York, 1987.

[112] Patula, W.T., Growth and oscillation properties of second-order linear difference equations, *SIAM J. Math. Anal.* **19** (1979), 55–61.

[113] Perron, O., *Die Lehre von den Kettenbruchen*, Vol. II, Teubner Verlag, Stuttgart, 1957.

[114] Perron, O., Über einen Satz des Herrn Poincaré, *J. Reine Angew. Math.* **136** (1909), 17–37.

[115] Perron, O., Über lineare Differenzengleichungen, *Acta Math.* **34** (1911), 109–137.

[116] Perron, O., Über Stabilität und Asymptotisches Verhalten der Integrale von Differential-Gleichungssystemen, *Math. Z.* **29** (1929), 129–160.

[117] Perron, O., Über Summengleichungen und Poincarésche Differenzengleichungen, *Math. Ann.* **84** (1921), 1.

[118] Petkovšek, M., H.S. Wilf, and D. Zeilberger, $A = B$, A.K. Peters, Wellesley, NY, 1996.

[119] Pielou, E.C., *An Introduction to Mathematical Ecology*, Wiley Interscience, New York, 1969.

[120] Pincherle, S. Delle, Funzioni ipergeometriche e de varie questioni ad esse attinenti, *Giorn. mat. Battaglini* **32** (1894), 209–291.

[121] Pinto, M., Discrete dichotomies, *Comput. Math. Appl.* **28** (1994), 259–270.

[122] Pituk, M., More on Poincaré's and Perron's Theorems for difference equations, *J. Difference Equ. Appl.* **8**(3) (2002), 201–216.

[123] Poincaré, H., Sur les equations linéaires aux differentielles ordinaires et aux différences finies, *Amer. J. Math.* **7** (1885), 203–258.

[124] Puu, T., *Nonlinear Economic Dynamics*, Lecture Notes in Economics and Mathematical Systems, Springer-Verlag, Berlin, 1989.

[125] Puu, T., Chaos in business cycles, *Chaos Solitons Fractals* **1** (1991), 457–473.

[126] Puu, T., and I. Sushko, A business cycle model with cubic nonlinearity, CERUM Working Paper 47, 2002.

[127] Ribenboim, R., *The Book of Prime Number Records*, Springer-Verlag, New York, 1988.

[128] Robinson, C., *Dynamical Systems*, 2nd ed., CRC Press, Boca Raton, FL, 1999.

[129] Samuelson, P.A., Interactions between the multiplier analysis and the principle of acceleration, *Rev. Econom. Statist.* **21** (1939), 75–78; reprinted in *Readings in Business Cycle Theory*, Blakiston, Philadelphia, 1944.

[130] Sandefur, J.T., *Discrete Dynamical Systems*, Clarendon Press, Oxford, 1990.

[131] Schinas, J., Stability and conditional stability of time-difference equations in Banach spaces, *J. Inst. Math. Appl.* **14** (1974), 335–346.

[132] Sedaghat, H., The impossibility of unstable globally attracting fixed points for continuous mappings of the line, *Amer. Math. Monthly* **104** (1997), 356–358.

[133] Sedaghat, H., Nonlinear difference equations, in *Theory with Applications to Social Science Models*, Kluwer Academic, Dordrecht, 2003.

[134] Shannon, C.E., and W. Weaver, *The Mathematical Theory of Communication*, University of Illinois, Urbana, 1949, pp. 7–8.

[135] Sharkovsky, A.N., Yu.L. Maistrenko, and E.Yu. Romanenko, *Difference Equations and Their Applications*, Kluwer Academic, Dordrecht, 1993.

[136] Smith, J.M., *Mathematical Ideas in Biology*, Cambridge University Press, Cambridge, 1968.

[137] Smith, G.D., *Numerical Solution of Partial Differential Equations: Finite Difference Methods*, 3rd ed., Clarendon Press, Oxford, 1985.

[138] Smith, H., Planar competitive and cooperative difference equations, *J. Difference Equ. Appl.* **3** (1998), 335–357.

[139] Szegö, G., *Orthogonal Polynomials*, Amer. Math. Soc. Colloq. Publ., Vol. 23, Providence, RI, 1959.

[140] Van Assche, W., *Asymptotics for Orthogonal Polynomials*, Lecture Notes in Mathematics, Vol. 1265, Springer-Verlag, Berlin, 1987.

[141] Van Der Poorten, A., A proof that Euler missed, Apéry's proof of the irrationality of $\zeta(3)$, *Math. Intelligencer* **1** (1979), 195–203.

[142] Weiss, L., Controllability, realization and stability of discrete-time systems, *SIAM J. Control* **10** (1972), 230–251.

[143] Williams, J.L., *Stability Theory of Dynamical Systems*, Nelson, London, 1970.

[144] Wimp, J., *Sequence Transformations*, Academic Press, New York, 1981.

[145] Wimp, J., *Computation with Recurrence Relations*, Pitman Advanced Publ., Boston, MA, 1984.

[146] Wimp, J., and D. Zeilberger, Resurrecting the asymptotics of linear recurrences, *J. Math. Anal. Appl.* **111** (1985), 162–176.

[147] Wong, R., and H. Li, Asymptotic expansions for second-order linear difference equations, *J. Comput. Appl. Math.* **41** (1992), 65–94.

[148] Wong, R., and H. Li, Asymptotic expansions for second-order difference equations, II, *Stud. Appl. Math.* **87** (1992), 289–324.

Index

(continued from page ii)

Frazier: An Introduction to Wavelets Through Linear Algebra
Gamelin: Complex Analysis.
Gordon: Discrete Probability.
Hairer/Wanner: Analysis by Its History. *Readings in Mathematics.*
Halmos: Finite-Dimensional Vector Spaces. Second edition.
Halmos: Naive Set Theory.
Hämmerlin/Hoffmann: Numerical Mathematics. *Readings in Mathematics.*
Harris/Hirst/Mossinghoff: Combinatorics and Graph Theory.
Hartshorne: Geometry: Euclid and Beyond.
Hijab: Introduction to Calculus and Classical Analysis.
Hilton/Holton/Pedersen: Mathematical Reflections: In a Room with Many Mirrors.
Hilton/Holton/Pedersen: Mathematical Vistas: From a Room with Many Windows.
Iooss/Joseph: Elementary Stability and Bifurcation Theory. Second edition.
Irving: Integers, Polynomials, and Rings: A Course in Algebra
Isaac: The Pleasures of Probability. *Readings in Mathematics.*
James: Topological and Uniform Spaces.
Jänich: Linear Algebra.
Jänich: Topology.
Jänich: Vector Analysis.
Kemeny/Snell: Finite Markov Chains.
Kinsey: Topology of Surfaces.
Klambauer: Aspects of Calculus.
Lang: A First Course in Calculus. Fifth edition.
Lang: Calculus of Several Variables. Third edition.
Lang: Introduction to Linear Algebra. Second edition.
Lang: Linear Algebra. Third edition.
Lang: Short Calculus: The Original Edition of "A First Course in Calculus."
Lang: Undergraduate Algebra. Third edition
Lang: Undergraduate Analysis.
Laubenbacher/Pengelley: Mathematical Expeditions.
Lax/Burstein/Lax: Calculus with Applications and Computing. Volume 1.
LeCuyer: College Mathematics with APL.
Lidl/Pilz: Applied Abstract Algebra. Second edition.
Logan: Applied Partial Differential Equations, Second edition.
Lovász/Pelikán/Vesztergombi: Discrete Mathematics.
Macki-Strauss: Introduction to Optimal Control Theory.
Malitz: Introduction to Mathematical Logic.
Marsden/Weinstein: Calculus I, II, III. Second edition.
Martin: Counting: The Art of Enumerative Combinatorics.
Martin: The Foundations of Geometry and the Non-Euclidean Plane.
Martin: Geometric Constructions.
Martin: Transformation Geometry: An Introduction to Symmetry.
Millman/Parker: Geometry: A Metric Approach with Models. Second edition.
Moschovakis: Notes on Set Theory.
Owen: A First Course in the Mathematical Foundations of Thermodynamics.
Palka: An Introduction to Complex Function Theory.
Pedrick: A First Course in Analysis.
Peressini/Sullivan/Uhl: The Mathematics of Nonlinear Programming.

Prenowitz/Jantosciak: Join Geometries.
Priestley: Calculus: A Liberal Art. Second edition.
Protter/Morrey: A First Course in Real Analysis. Second edition.
Protter/Morrey: Intermediate Calculus. Second edition.
Pugh: Real Mathematical Analysis.
Roman: An Introduction to Coding and Information Theory.
Roman: Introduction to the Mathematics of Finance: From Risk Management to Options Pricing.
Ross: Differential Equations: An Introduction with Mathematica®. Second edition.
Ross: Elementary Analysis: The Theory of Calculus.
Samuel: Projective Geometry. *Readings in Mathematics.*
Saxe: Beginning Functional Analysis
Scharlau/Opolka: From Fermat to Minkowski.
Schiff: The Laplace Transform: Theory and Applications.
Sethuraman: Rings, Fields, and Vector Spaces: An Approach to Geometric Constructability.
Sigler: Algebra.
Silverman/Tate: Rational Points on Elliptic Curves.
Simmonds: A Brief on Tensor Analysis. Second edition.
Singer: Geometry: Plane and Fancy.
Singer/Thorpe: Lecture Notes on Elementary Topology and Geometry.
Smith: Linear Algebra. Third edition.
Smith: Primer of Modern Analysis. Second edition.
Stanton/White: Constructive Combinatorics.
Stillwell: Elements of Algebra: Geometry, Numbers, Equations.
Stillwell: Elements of Number Theory.
Stillwell: Mathematics and Its History. Second edition.
Stillwell: Numbers and Geometry. *Readings in Mathematics.*
Strayer: Linear Programming and Its Applications.
Toth: Glimpses of Algebra and Geometry. Second Edition. *Readings in Mathematics.*
Troutman: Variational Calculus and Optimal Control. Second edition.
Valenza: Linear Algebra: An Introduction to Abstract Mathematics.
Whyburn/Duda: Dynamic Topology.
Wilson: Much Ado About Calculus.